Spider Ecophysiology

Wolfgang Nentwig
Editor

Spider Ecophysiology

Springer

Editor
Wolfgang Nentwig
Inst. Ecology and Evolution
University of Bern
Bern, Switzerland

ISBN 978-3-642-33988-2 ISBN 978-3-642-33989-9 (eBook)
DOI 10.1007/978-3-642-33989-9
Springer Heidelberg New York Dordrecht London

Library of Congress Control Number: 2013930769

Printed on acid-free paper

Springer is part of Springer Science+Business Media (www.springer.com)

Preface

Spiders are among the most successful groups of terrestrial organisms. With more than 42,000 species, spiders are the most numerous predacious arthropod group, only seconded by some insect families such as carabids' beetles or ants. This gives spiders, omnipresent in all terrestrial habitats, a key position in ecological networks and ecosystem functioning. During their evolution of more than 300 million years, spiders developed and improved unique features, the combination of which is regarded as entry for their unrivalled success story. Among the key achievements of spiders at least four have to be mentioned.

First, spiders possess up to six different silk gland types that allow them to use silk for a variety of web types not only to catch their prey but also to wrap their victims until they are defenceless. Spiders build silken retreats, sperm webs, cocoons and draglines, thus demonstrating the remarkable material properties of one of the most resistant and elastic biomaterials. Second, spiders are venomous animals and inject defined venom quantities into a prey item to paralyse or kill it. Spider venom is a complex mixture of hundreds of components, consisting of low molecular compounds, peptides and proteins, which target the extracellular matrix, membranes and a variety of receptors, quite often located in the nervous or muscular system. Third, the locomotion of spiders is driven by a combination of muscles and a hydraulic pressure system, since some leg segments only possess flexor muscles. Instead of extensor muscles, the hydraulic pressure of their haemolymph is fine-tuned by a well-balanced system of valves, which provides the necessary back-pressure. This reduces in major parts of the long leg tubes of spiders the muscle system and allows at the same time larger flexors, so that spiders in general are more powerful than comparable insects. Fourth, the distal end of the male pedipalp developed into a complex structure composed of fixed and movable sclerites that are used to transfer sperm to the female seminal receptacles during mating. This key-lock mechanism guarantees safe sperm transfer within the species, largely preventing mating outside the own species, and probably represents a major driver for the fast species radiation we observe in spiders.

Ecophysiology is a bridge from functional and evolutionary aspects of morphology, physiology, biochemistry and molecular biology to ecology. Currently, cutting-edge science in spiders focuses on the circulatory and respiratory system, locomotion and dispersal abilities, the immune system, endosymbionts and pathogens, chemical communication, gland secretions, venom components, silk structure, structure and perception of colours and colouration and nutritional requirements, to name only a few. Spiders became valuable indicator species in agroecosystems and for conservation biology. Modern transfer and application technologies consider spiders and their products with respect to biomimetics, material sciences and agrochemical and pharmaceutical industries.

It is now 26 years ago that I edited a first comprehensive book on ecophysiology of spiders, published also with Springer [Nentwig W (ed) (1987) Ecophysiology of spiders. Springer, Heidelberg]. Scientific progress since then was remarkable and an evaluation of the topics from that time and relevant publications over the last two decades showed the appearance of many new fascinating subjects. A new book on the old subject, therefore, is definitely not just a revised version but became something completely new. Seven subjects from the old book (on colouration, respiration system, reproductive glands, pheromones, venom, silk and dispersal) can also be found in this new book, most of them now represented by several much more detailed chapters and with a completely new content. Moreover, many additional and intriguing aspects are included. The innovative character of this book and of spider ecophysiology is also underlined by the fact that only two author teams from the old book contributed to this new book (Mark Townley with Ed Tillinghast and me, obviously dinosaurs in arachnology).

Special thanks go to Springer Publishers with Annika König and Jutta Lindenborn. I also want to acknowledge the support from Christian Kropf and Rita Schneider.

Bern, Switzerland
September 2012

Wolfgang Nentwig

Contents

Part I Respiration and Circulatory System

1 **Evolution and Adaptation of Hemocyanin Within Spiders** 3
Thorsten Burmester

2 **The Circulatory System of Spiders** 15
Christian S. Wirkner and Katarina Huckstorf

3 **Tracheae in Spiders: Respiratory Organs for Special Functions** 29
Anke Schmitz

Part II Locomotion and Dispersal

4 **Hydraulic System of Locomotion** 43
Christian Kropf

5 **Functional Aspects of Spider Scopulae** 57
Senta Niederegger

6 **Cost–Benefit Balance of Dispersal and the Evolution of Conditional Dispersal Strategies in Spiders** 67
Dries Bonte

Part III Immune System and Pathogens

7 **The Immune System of Spiders** 81
Lucia Kuhn-Nentwig and Wolfgang Nentwig

8 **Endosymbiont Infections in Spiders** 93
Sara L. Goodacre and Oliver Y. Martin

9 **Fungal Pathogens of Spiders** 107
Harry C. Evans

Part IV Chemical Communication and Reproduction

10 Chemical Communication and Contact Cuticular Compounds in Spiders 125
Marie Trabalon

11 Spider Olfaction: Attracting, Detecting, Luring and Avoiding 141
Gabriele Uhl

12 Karyotypes, Sex Chromosomes, and Meiotic Division in Spiders 159
Tereza Kořínková and Jiří Král

13 Male Reproductive System of Spiders 173
Peter Michalik and Elisabeth Lipke

Part V Venom

14 Main Components of Spider Venoms 191
Wolfgang Nentwig and Lucia Kuhn-Nentwig

15 The Neurotoxic Mode of Action of Venoms from the Spider Family Theraphosidae 203
Volker Herzig and Glenn F. King

16 The Cytotoxic Mode of Action of the Venom of *Cupiennius salei* (Ctenidae) 217
Lucia Kuhn-Nentwig and Wolfgang Nentwig

17 The Evolution of a Toxic Enzyme in Sicariid Spiders 229
Greta Binford

18 Predation by Spitting Spiders: Elaborate Venom Gland, Intricate Delivery System 241
Robert B. Suter and Gail E. Stratton

19 Spider Venoms Potentially Lethal to Humans 253
Wolfgang Nentwig and Lucia Kuhn-Nentwig

Part VI Silk

20 Spider Silk: Molecular Structure and Function in Webs 267
Todd A. Blackledge

21 Aggregate Silk Gland Secretions of Araneoid Spiders 283
Mark A. Townley and Edward K. Tillinghast

22 Cribellar Thread 303
Brent D. Opell

Part VII Colouration

23 Insect View of Orb Spider Body Colorations 319
I-Min Tso

24 Structural Colors in Spiders 333
Rainer F. Foelix, Bruno Erb, and David E. Hill

25 UV and Camouflage in Crab Spiders (Thomisidae) 349
Marie E. Herberstein and Felipe M. Gawryszewski

26 Rapid Colour Change in Spiders 361
Judith Wunderlin and Christian Kropf

Part VIII Nutrition

27 Nutritional Aspects of Spider Feeding 373
Søren Toft

28 Herbivory in Spiders 385
Dirk Sanders

29 Artificial Diets for Spiders 393
Divina M. Amalin

Part IX Ecotoxicology

30 Effects of Heavy Metal Contamination 403
Paweł Migula, Grażyna Wilczek, and Agnieszka Babczyńska

31 Side Effect of Synthetic Pesticides on Spiders 415
Stano Pekár

32 Side Effects of *Bacillus thuringiensis* Toxins on Spiders 429
Michael Meissle

33 Effects of Neem on Spiders 441
John D. Stark

Part X Applications

34 Modelling and Application of the Hydraulic Spider Leg Mechanism 451
Lena Zentner

35 Adhesion to Flat Surfaces: From Spiders to Stickers 463
Tobias Seidl and Renato Vidoni

36 Technical and Biomedical Uses of Nature's Strongest Fiber: Spider Silk 475
Christina Allmeling, Christine Radtke, and Peter M. Vogt

37 Spider Venom Components as Drug Candidates 491
Vera Oldrati, Estelle Bianchi, and Reto Stöcklin

Appendix. Spider Phylogeny 505

Subject Index 511

Part I
Respiration and Circulatory System

Different from insects, spiders do not transport aerial oxygen directly to the cells, but use their hemolymph as intermediate transport system, thus they more resemble crustacean, molluscs, or even vertebrates. Such a respiratory system demands structures to saturate carrier proteins with oxygen (hemocyanin) and a circulatory system for the hemolymph. Starting with a not very efficient ancestral type, spiders tuned their systems considerably: they reorganized their lung system, modified their circulatory system, and experimented with their tracheal system. Spiders did not really reach the high performance of many insects, but considering the evolutionary constraints, they did the best they could.

Chapter 1
Evolution and Adaptation of Hemocyanin Within Spiders

Thorsten Burmester

1.1 Introduction

The transport and storage of oxygen in animals are mediated by three distinct types of respiratory proteins (hemoglobins, hemerythrins or hemocyanins). In many arthropod and mollusc species, oxygen is bound to large copper proteins referred to as hemocyanins, which occur freely dissolved in the hemolymph (Markl and Decker 1992; van Holde and Miller 1995). However, arthropod and mollusc hemocyanins belong to different protein families and evolved from distinct types of copper-containing enzymes already in the Precambrian (Burmester 2001). Arthropod hemocyanins originated from the phenoloxidases, which are Cu^+-containing enzymes involved in immune response and cuticle sclerotisation (Burmester 2002). Some arthropod taxa have lost hemocyanin and rely on hemoglobin for oxygen supply (Weber and Vinogradov 2001) or lack any respiratory protein. In some decapod crustaceans and in the hexapods, hemocyanin-related proteins that do not bind O_2 occur. These proteins, referred to as pseudo-hemocyanins (cryptocyanins) or hexamerins, respectively, are considered mainly as storage proteins but may also have other, non-respiratory functions (Burmester 1999a, b; Terwilliger et al. 1999).

The general structure of an arthropod hemocyanin is highly conserved. The protein is built by hexamers or oligo-hexamers of subunits in the range of 70–85 kDa (Markl and Decker 1992; van Holde and Miller 1995). Each subunit can bind one O_2 molecule by the means of two Cu^+ ions, each of which being coordinated by three conserved histidines of the polypeptide chain (copper A and copper B binding sites). The arthropod hemocyanins may assemble to large structures (Fig. 1.1), which consist of up to 48 subunits (8×6-mer structure) and have a mass of up to 3,600 kDa (Markl and Decker 1992). The subunits that build an

T. Burmester (✉)
Institute of Zoology and Zoological Museum, University of Hamburg, Hamburg, Germany
e-mail: thorsten.burmester@uni-hamburg.de

W. Nentwig (ed.), *Spider Ecophysiology*,
DOI 10.1007/978-3-642-33989-9_1, © Springer-Verlag Berlin Heidelberg 2013

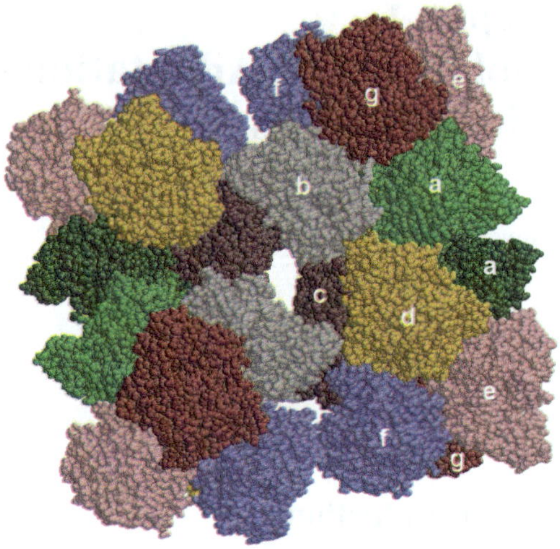

Fig. 1.1 Structure of *Aphonopelma hentzi* 4 × 6-mer hemocyanin. The seven subunit types are labeled on the two right-hand side hexamers and displayed in different colors: (a) *green*; (b) *gray*; (c) *brown*; (d) *yellow*; (e) *pink*; (f) *blue*; (g) *red*

oligomer may either be identical, deriving from the same gene, or structurally related and thus coded by distinct genes. Subunit composition and assembly in each hemocyanin is essentially taxon specific (Markl 1986; Markl et al. 1986). While in some decapod crustaceans, subunit composition and structure may change during development or in response to environmental challenges (Durstewitz and Terwilliger 1997a, b), the hemocyanins in the other arthropod taxa appear to be more conservative, without any known variations throughout the life cycle. Moreover, subunit compositions of the oligomeric hemocyanins in many chelicerate taxa have remained unchanged for several 100 million years (Markl 1986; Markl et al. 1986; Rehm et al. 2012).

1.2 Evolution of Chelicerate Hemocyanins

1.2.1 Occurrence and Evolution of Hemocyanins in Chelicerates

A typical chelicerate hemocyanin subunit comprises 620–630 amino acids (70–75 kDa). The evolutionary rate is comparatively low and in the range of $0.5–0.6 \times 10^{-9}$ amino acid replacements per site per year, which is about two- to threefold lower than in the crustacean hemocyanins (Burmester 2002). Biochemical and structural studies showed that a chelicerate hemocyanin may consist of up to eight distinct subunit types, which assemble to oligomers of either 1 × 6, 2 × 6, 4 × 6 or 8 × 6mers (Markl 1986; Markl et al. 1986). Each of the single subunits occupies a defined position in the native protein, which mirrors up to 520 million years of independent evolutionary history (Rehm et al. 2012). Hemocyanin proteins or cDNA sequences have been identified in the Pycnogonida (sea spiders or

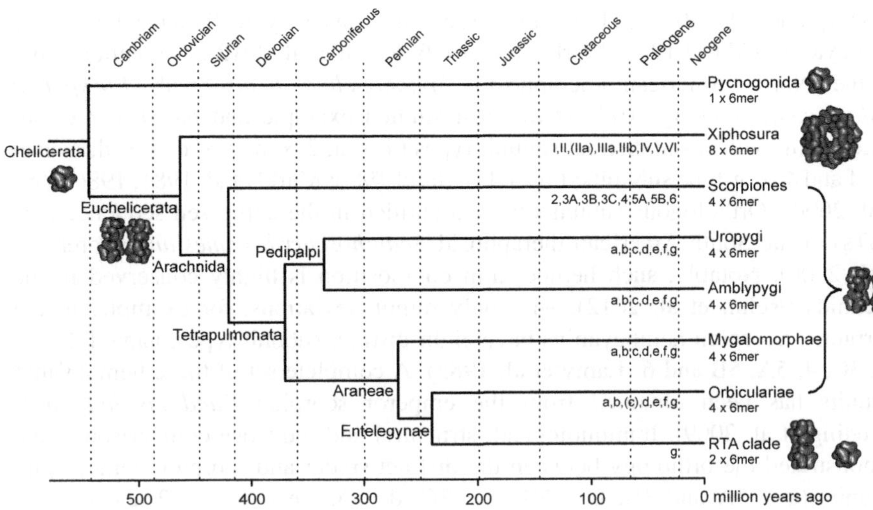

Fig. 1.2 Timescale of the evolution of chelicerates and evolution of hemocyanin structures. Adapted from Rehm et al. (2012). Note that only taxa for which hemocyanin sequences are available are displayed

Pantopoda), Xiphosura (horseshoe crabs), Scorpiones (scorpions), Amblypygi (whip spiders), Uropygi (whip scorpions) and Araneae (true spiders). By contrast, no hemocyanin has been found in the Opiliones (harvestmen), Pseudoscorpiones (false or book scorpions), Solifugae (camel or sun spiders) and Acari (mites and ticks) (Markl 1986; Markl et al. 1986; Rehm et al. 2012). There is at present no information on the hemocyanin status of the Palpigradi (microwhip scorpions) and Ricinulei (tickspiders).

Most studies agree that the Pycnogonida are the closest relatives of the Euchelicerata. A single hemocyanin subunit cDNA was found in the sea spider *Endeis spinosa*, indicating that the hemocyanin protein of this species is a homo-hexamer (Rehm et al. 2012) (Fig. 1.2). However, there are no biochemical studies to confirm this notion. Phylogenetic analyses identified this sequence in a sister group position to all other chelicerate hemocyanin subunits, indicating that the hemocyanins of the stem group chelicerates had a similar simple, possibly homo-hexameric structure. The horseshoe crabs (Xiphosura) have the largest hemocyanin molecules. These 8 × 6mers consist of seven or eight distinct subunits types (Markl 1986; Markl et al. 1986). The *Limulus polyphemus* hemocyanin, which is the best investigated hemocyanin of the Xiphosura, is composed of 6 × subunit type I, 8 × II, 2 × IIA, 8 × IIIA, 8 × IIIB, 8 × IV, 4 × V, 4 × VI. Orthologous subunits have been identified in other xiphosurans (Sugita and Shishikura 1995), which have only seven subunit types and no subunit IIa. The complete subunit sequences are available from *Carcinoscorpius rotundicauda* (GenBank acc. nos. DQ090469–DQ090484), which have, however, not been formally published.

Scorpiones, Amblypygi, Uropygi and several Araneae typically have 4 × 6-mer hemocyanins (Markl 1986; Markl et al. 1986; Rehm et al. 2012). The hemocyanin of the North American theraphosid *Aphonopelma hentzi* (sub *Eurypelma californicum*; Nentwig 2012) is the best studied example and has a 4 × 6-mer hemocyanin with seven distinct subunit types (4 × a, 2 × b, 2 × c, 4 × d, 4 × e, 4 × f and 4 × g-type subunits; Fig. 1.1) (Markl 1986; Markl et al. 1981, 1986; Voit et al. 2000). Orthologous subunits were identified in the expressed sequence tags (ESTs) of the South American theraphosid *Acanthoscurria gomesiana* (Lorenzini et al. 2006). Notably, such hemocyanin composition is highly conserved in the arachnids (Rehm et al. 2012), with only minor deviations, for example, in the scorpions. Scorpion hemocyanins have eight distinct subunit types, named 2, 3A, 3B, 3C, 4, 5A, 5B and 6 (Lamy et al. 1985). A complete set of these hemocyanin subunits has been obtained from the emperor scorpion *Pandinus imperator* (Roeding et al. 2009). Immunological, structural and sequence comparisons have demonstrated the orthology between the distinct spider and scorpion hemocyanin subunits (a = 3A and 3B, b = 5B, c = 3C, d = 5A, e = 6, f = 2 and g = 4), indicating a shared common ancestry (Lamy et al. 1981; Markl et al. 1984, 1986; Rehm et al. 2012).

Chelicerate hemocyanin subunits experienced a long independent evolutionary history (Markl 1986; Markl et al. 1986). In a recent phylogenetic analysis (Rehm et al. 2012), it has been demonstrated that the formation of distinct subunits commenced early in the evolution. Four distinct subunit types already emerged before Xiphosura and Arachnida separated ~470 million years ago (Fig. 1.3), indicating that an oligomeric hemocyanin (probably a 4 × 6mer) already existed at that time. The formation of an n × hexamer might be associated with an increase in the size of the animal and the demand for a better oxygen supply. Further independent gene duplications resulted in additional subunit types in the Xiphosura and Arachnida. A 4 × 6-mer hemocyanin consisting of seven subunit types (orthologs of the *A. hentzi* hemocyanin subunits a through g) is conserved in most arachnids since more than 450 million years (Rehm et al. 2012).

1.2.2 Spider Hemocyanins

True spiders (Araneae) either have the typical arachnid 4 × 6 hemocyanin with seven subunit types (a through g) or have distinct hemocyanins with more variable structures and subunit compositions (see below). There is no information on the hemocyanins in Mesothelae, which are considered as the most "primitive" spiders. The Mygalomorphae (Orthognatha), such as *A. hentzi*, possess the typical 4 × 6/seven subunit hemocyanin (Markl et al. 1981, 1986; Markl 1986; Voit et al. 2000). The same structure was found in a biochemical survey of other mygalomorphs such as *Atrax*, *Atypus* and *Nemesia* (Markl 1986) and is mirrored in the ESTs of *A. gomesiana* (Lorenzini et al. 2006).

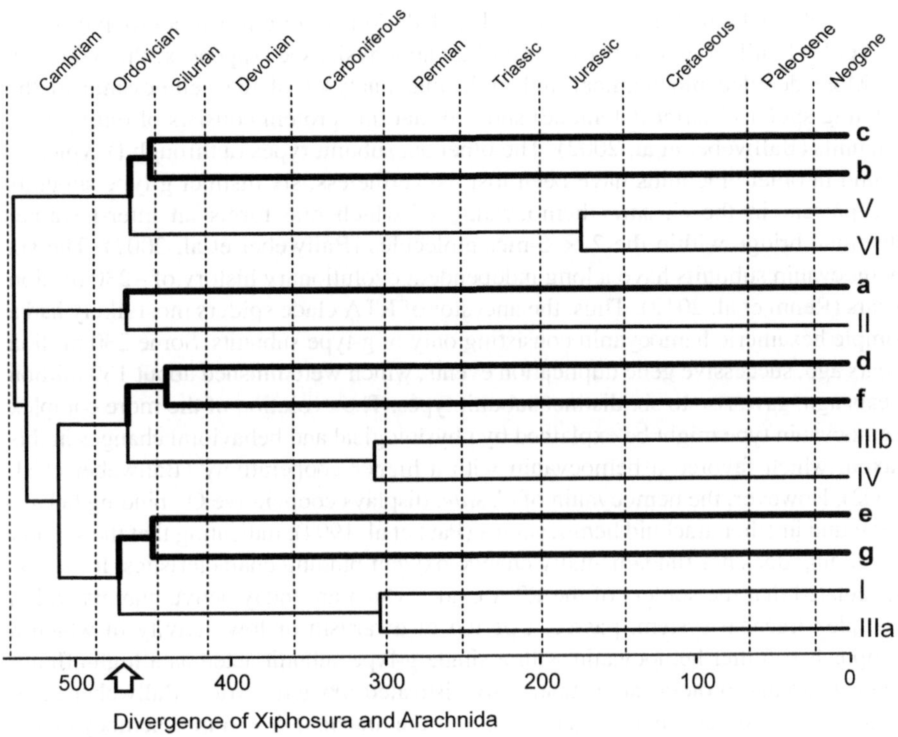

Fig. 1.3 Relationships among chelicerate hemocyanin subunit types. The timescale was deduced from a molecular clock approach using amino acid sequences of hemocyanin subunits. The *thick lines* represent the arachnid subunit types a through g. Adapted from Rehm et al. (2012)

Within the suborder Araneomorphae (Labidognatha), this 4 × 6 hemocyanin is present in the Araneidae (orb weavers), Nephilidae, Pholcidae (cellar spiders), Oecobiidae (disc web spiders), Linyphiidae (dwarf spiders), Theridiidae (cobweb weavers) and Tetragnathidae (long-jawed orb weavers) (Markl 1986; Markl et al. 1986; Averdam et al. 2003). Six hemocyanin subunits of the golden orb web spider *Nephila inaurata* (Nephilidae) have been sequenced, which are orthologous to *A. hentzi* hemocyanin subunits a, b, d, e, f and g (Averdam et al. 2003). Subunit c may be absent in *N. inaurata*, although the 4 × 6-mer structure is conserved.

While many Entelegynae, such as *N. inaurata* (Averdam et al. 2003), have 4 × 6-mer hemocyanins, in other Entelegynae the pattern is more complex (Markl 1986; Markl et al. 1986). *Cupiennius salei* has a mixture of 2 × 6-mer and 1 × 6-mer hemocyanins, which occur in a ratio of about 2:1 (Markl 1980). A similar hemocyanin structure was found in the Agelenidae (funnel weavers), Clubionidae (sac spiders), Ctenidae (wandering spiders), Gnaphosidae (ground spiders), Lycosidae (wolf spiders), Oxyopidae (lynx spiders), Philodromidae (running crab spiders), Pisauridae (nursery web spiders), Salticidae (jumping spiders), Sparassidae (huntsman spiders) and Thomisidae (crab spiders) (Markl 1986; Markl et al. 1986), which all belong to the RTA clade (this term refers to a group of some 37 spider families with a retrolateral tibial apophysis; see Appendix, this volume).

More detailed biochemical and molecular analyses of the hemocyanin of the hunting spider *C. salei* (Ctenidae) showed that this protein consists of only g-type subunits (Ballweber et al. 2002). The other six subunit types (a through f) typically found in other arachnids have been lost. Nevertheless, six distinct g-type subunits are present in the *C. salei* hemocyanin, of which one forms an inter-hexamer disulfide bridge within the 2 × 6-mer molecules (Ballweber et al. 2002). The six hemocyanin subunits have a long independent evolutionary history of ~230 million years (Rehm et al. 2012). Thus, the ancestor of RTA clade spiders most likely had a simple hexameric hemocyanin consisting only of g-type subunits. Some 230 million years ago, successive gene duplication events, which were finished about 130 million years ago, gave rise to six distinct subunit types. The evolution of the more complex hemocyanin type might be explained by physiological and behavioral changes in this taxon, which favored a hemocyanin with a higher cooperativity (Ballweber et al. 2002). However, the hemocyanin of *C. salei* displays cooperative O_2-binding behavior found in other arachnid hemocyanins (Paul et al. 1994), indicating that the subunit rebuilding does not fundamentally change oxygen-binding characteristics. It may be speculated that the spiders of the RTA clade, which are today active hunters with a complex tracheal system, passed a period of dwarfism or low activity in which a simple 1 × 6-mer hemocyanin with a single g-type subunit acted as a high-affinity oxygen storage protein rather than a sophisticated oxygen carrier (Ballweber et al. 2002). It is also conceivable that the loss of the arachnid 4 × 6-mer hemocyanin is actually linked to the evolution of trachea (see Schmitz 2013). However, later in evolution an increase in body size might have rendered simple tracheal respiration ineffective to sustain the metabolic requirements of an active hunter. Thus, the more complex 2 × 6 hemocyanin evolved, which allows allosteric interaction and more efficient O_2 transport. It is also conceivable that the rebuilding of the hemocyanin oligomer was caused by the decrease in atmospheric oxygen levels in the Permian period (Berner et al. 2007).

Another deviation from the standard scheme of arachnid hemocyanin structure was found in the haplogyne spider *Filistata insidiatrix*, which possesses a hexamer of only five subunits, with subunits b and c being absent (Markl 1986). Because b- and c-type subunits are required for making the inter-hexamer contacts, no 4 × 6-mer structure can form. Surprisingly, the haplogyne spider *Dysdera* sp., which was originally reported to have a simple hexamer (Markl 1986; Markl et al. 1986), actually lacks hemocyanin (Rehm et al. 2012).

1.2.3 Loss of Hemocyanin in Some Chelicerates

Hemocyanin is the principal oxygen transport protein of the Chelicerata. No other type of respiratory protein has been identified in this taxon. A hemoglobin-like protein found in the expressed sequence tags (ESTs) of some arachnids is unlikely to have any oxygen transport function (Ertas et al. 2011). Because hemocyanin was present in the early chelicerates, taxa without this protein must have lost it during evolution. Hemocyanins are in fact absent in Opiliones, Pseudoscorpiones, Solifugae and Acari (Markl 1986; Markl et al. 1986; Rehm et al. 2012). It is, however, uncertain whether the loss of hemocyanin hints to a common evolutionary origin of some of these taxa. The lack of hemocyanin in these arachnid taxa may be explained by distinct morphological or physiological features that assist oxygen supply and render a respiratory protein redundant. Such structure may be the well-developed tracheal system, which transports oxygen from the atmosphere to the metabolically active organs. In contrast, the other arachnids possess book lungs, which may be less efficient. Here, hemocyanin is required for efficient oxygen uptake and distribution. Likewise, the lack of an oxygen transport protein in the haplogyne spiders of the genus *Dysdera*, which are—according to the present knowledge—the only Araneae without hemocyanin, may also be explained by the possession of a highly developed tracheal system (Schmitz 2013).

1.3 Synthesis of Spider Hemocyanins

Hemocyanin accumulates in the hemolymph of chelicerates to concentrations of up to 120 mg/ml (Mangum 1985). Spider hemocyanins are released from hemocytes (Kempter 1983). Surprisingly, the chelicerate hemocyanin sequences do not contain any N-terminal signal peptide that targets the protein to the secretory pathway (Voit et al. 2000). Rather the chelicerate hemocyanins are synthesized at free ribosomes and subsequently released by cell rupture (Kempter 1983). Thus, despite of the presence of putative N-glycosylation sites in their amino acid sequences, the native spider hemocyanins do not pass the Golgi apparatus and do not carry any carbohydrate moiety (Markl et al. 1992). Both the absence of the signal peptide and the release from hemocytes by rupture are likely a reminiscence of the origin of the hemocyanin from a prophenoloxidase, which displays the same characteristics (Kusche and Burmester 2001). However, during the evolution of the hemocyanins of Mandibulata (Myriapoda, Crustacea and Hexapoda), a signal peptide to guide the nascent protein to the endoplasmic reticulum was acquired.

1.4 Regulation of Oxygen Transport Function in Spider Hemocyanin

During evolution, the environmental conditions have not only influenced the structures of the hemocyanin oligomers but also shaped their O_2-binding kinetics. The evolutionary advantage of the large hemocyanin oligomers presumably lies in a higher O_2-carrying capacity per mole combined with a lower viscosity and colloid

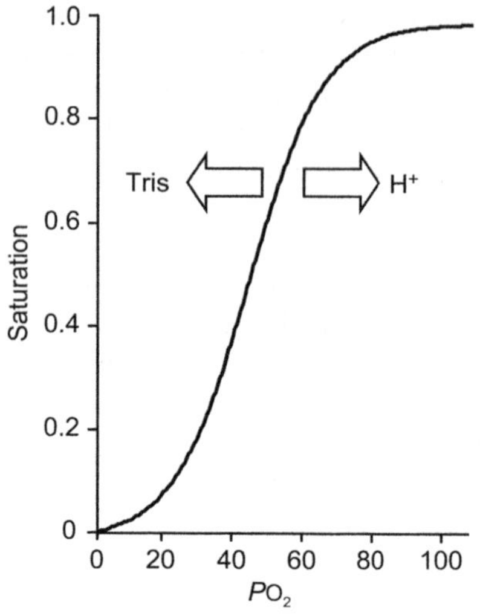

Fig. 1.4 Oxygen-binding kinetics of *Aphonopelma hentzi* hemocyanin at pH 7.5. Data were taken from Paul et al. (1994). Effectors that influence O_2 affinity of spider hemocyanin are indicated

osmotic pressure of the hemolymph. The oligomeric nature of the arthropod hemocyanin also permits cooperativity and allosteric regulation of O_2 binding (van Holde and Miller 1995), resulting in a sigmoid curve (Fig. 1.4). Allostery of hemocyanin enhances the oxygen transport capacity from the respiratory organs to the inner tissues. Cooperativity is based on the interaction between the different subunits and allostery is therefore increased with their number. Spider hemocyanins typically display an allosteric behavior that exceeds the allosteric properties of a typical vertebrate hemoglobin (Mangum 1985; Markl 1986; Savel-Niemann et al. 1988).

Oxygen-binding properties of hemocyanins strongly depend on the particular species but also on the temperature, pH and allosteric effectors. For example, half-saturation pressures (P_{50}) reported for *A. hentzi* hemocyanin vary between 9 and 54 Torr (mm Hg) (Decker and Sterner 1990; Paul et al. 1994). Employing a crude hemolymph preparation, Paul et al. (1994) reported P_{50} values of ~40–50 Torr for *C. salei*, *A. hentzi* and *P. imperator* hemocyanins at the physiological pH of 7.5. Thus, the O_2-binding characteristics of different arachnid hemocyanins are largely similar but also depend on various effectors.

The best characterized effect that modulates the oxygen affinity of spider hemocyanin is the pH dependence (Bohr effect). An acidic pH lowers the oxygen affinities (higher P_{50}) of spider hemocyanins, thereby enhancing O_2 release (Paul et al. 1994). At low oxygenation levels, a reverse Bohr effect has been reported for *A. hentzi* hemocyanin (Sterner and Decker 1994). Such a reverse Bohr effect has also been found in *L. polyphemus*, which has a hemocyanin that increases its O_2 affinity at low pH independently from the oxygenation level. Ions such as Ca^{2+} and Cl^- have pronounced effects on the hemocyanins of decapod crustaceans but only minor influence on the O_2-binding properties of *A. hentzi* hemocyanin. A somewhat unusual allosteric effector is Tris (hydroxy-methyl-amino-methane), which is a common component of chemical buffers (Sterner et al. 1994).

1.5 Non-respiratory Functions of Chelicerate Hemocyanins

While the main role of hemocyanins is the transport of O_2 in the hemolymph, these respiratory proteins may have additional roles. Hemocyanin fragments were found as a component of the cuticle of *A. hentzi* (Paul et al. 1994). Notably, the integration into the cuticle has been demonstrated also for hexamerins and pseudo-hemocyanins, proteins that are related to hemocyanin (Peter and Scheller 1991; Terwilliger et al. 2005). *A. hentzi* hemocyanin may also bind to ecdysone with low affinity ($k_D = 0.5$–4.4 mM), which is, however, similar to the ecdysone affinity of some insect hexamerins (Jaenicke et al. 1999). Ecdysone is responsible for the molting of arthropods. Hemocyanin may transport this hydrophobic steroid hormone in the hemolymph, analogous to the function of the albumins in the vertebrate blood.

It is well known that many hemocyanins may exert a phenoloxidase activity in vitro (Decker et al. 2001; Jaenicke and Decker 2008). This enzymatic function may be induced by limited proteolysis or by the detergent SDS (sodium dodecyl sulfate). However, it has also been demonstrated that in the horseshoe crab, *Tachypleus tridentatus*, the clotting enzyme as well as an antimicrobial peptide are able to convert the hemocyanin into a phenoloxidase without any proteolytic cleavage (Nagai and Kawabata 2000; Nagai et al. 2001). This finding links hemocyanin to the chelicerate immune response (see also Kuhn-Nentwig and Nentwig 2013). In contrast to the other arthropod phyla, chelicerates do not have a true phenoloxidase. Thus, it is likely that the role of this enzyme in pathogen defense has been assumed by the hemocyanin. However, it remains to be demonstrated that the same mechanism identified in the horseshoe crab also works in spiders.

1.6 Conclusions

Hemocyanins are copper proteins that function as the oxygen carriers in the hemolymph of most spiders (Araneae). According to our present knowledge, only the genus *Dysdera* lacks hemocyanin or any other respiratory protein. Spider hemocyanins are large oligo-hexameric proteins ($n \times 6$). The first spider probably had a 4×6-mer hemocyanin composed of seven distinct subunit types (a through g), which had separated more than 450 million years ago. The rather high number of subunits in a spider hemocyanin permits highly cooperative binding of oxygen and thus an efficient transport of oxygen. This 4×6-mer hemocyanin type occurs in the Mygalomorphae (Orthognatha) and is also conserved in many Araneomorphae (Labidognatha). However, the more advanced entelegyne spiders of the RTA clade have 2×6-mer hemocyanins. These spiders have lost the subunits a through f, and six distinct subunits evolved from the g-type subunit 230–120 million years ago. Nevertheless, spider hemocyanins show only few modifications in terms of evolutionary rate and subunit compositions when compared to their crustacean paralogs and are less susceptible to allosteric effectors. The rather conservative nature of the chelicerate hemocyanins may partly be explained by the rather constant terrestrial conditions of the modern arachnids, whereas many crustaceans live in changing aquatic or semiaquatic environments.

References

Averdam A, Markl J, Burmester T (2003) Subunit sequences of the 4 × 6-mer hemocyanin from the golden orb-web spider, *Nephila inaurata*. Eur J Biochem 270:3432–3439. doi:10.1046/j.1432-1033.2003.03730.x

Ballweber P, Markl J, Burmester T (2002) Complete hemocyanin subunit sequences of the hunting spider *Cupiennius salei*: recent hemocyanin remodeling in entelegyne spiders. J Biol Chem 277:14451–14457. doi:10.1074/jbc.M111368200

Berner RA, Vandenbrooks JM, Ward PD (2007) Oxygen and evolution. Science 316:557–558. doi:10.1126/science.1140273

Burmester T (1999a) Evolution and function of the insect hexamerins. Eur J Entomol 96:213–225

Burmester T (1999b) Identification, molecular cloning, and phylogenetic analysis of a non-respiratory pseudo-hemocyanin of *Homarus americanus*. J Biol Chem 274:13217–13222. doi:10.1074/jbc.274.19.13217

Burmester T (2001) Molecular evolution of the arthropod hemocyanin superfamily. Mol Biol Evol 18:184–195

Burmester T (2002) Origin and evolution of arthropod hemocyanins and related proteins. J Comp Physiol B 172:95–107. doi:10.1007/s00360-001-0247-7

Decker H, Sterner R (1990) Nested allostery of arthropodan hemocyanin (*Eurypelma californicum* and *Homarus americanus*). The role of protons. J Mol Biol 211:281–293. doi:10.1016/0022-2836(90)90027-J

Decker H, Ryan M, Jaenicke E, Terwilliger N (2001) SDS-induced phenoloxidase activity of hemocyanins from *Limulus polyphemus*, *Eurypelma californicum*, and *Cancer magister*. J Biol Chem 276:17796–17799. doi:10.1074/jbc.M010436200

Durstewitz G, Terwilliger NB (1997a) cDNA cloning of a developmentally regulated hemocyanin subunit in the crustacean *Cancer magister* and phylogenetic analysis of the hemocyanin gene family. Mol Biol Evol 14:266–276

Durstewitz G, Terwilliger NB (1997b) Developmental changes in hemocyanin expression in the Dungeness crab, *Cancer magister*. J Biol Chem 272:4347–4350. doi:10.1074/jbc.272.7.4347

Ertas B, Kiger L, Blank M, Marden MC, Burmester T (2011) A membrane-bound hemoglobin from gills of the green shore crab *Carcinus maenas*. J Biol Chem 286:3185–3193. doi:10.1074/jbc.M110.160341

Jaenicke E, Decker H (2008) Kinetic properties of catecholoxidase activity of tarantula hemocyanin. FEBS J 275:1518–1528. doi:10.1111/j.1742-4658.2008.06311.x

Jaenicke E, Föll R, Decker H (1999) Spider hemocyanin binds ecdysone and 20-OH-ecdysone. J Biol Chem 274:34267–34271. doi:10.1074/jbc.274.48.34267

Kempter B (1983) Site of hemocyanin biosynthesis in the tarantula *Eurypelma californicum*. Naturwissenschaften 70:255–256

Kuhn-Nentwig L, Nentwig W (2013) The immune system of spiders. In: Nentwig W (ed) Spider ecophysiology. Springer, Heidelberg (this volume)

Kusche K, Burmester T (2001) Diplopod hemocyanin sequence and the phylogenetic position of the Myriapoda. Mol Biol Evol 18:1566–1573

Lamy J, Bijlholt MC, Sizaret PY, van Bruggen EF (1981) Quaternary structure of scorpion (*Androctonus australis*) hemocyanin. Localization of subunits with immunological methods and electron microscopy. Biochemistry 20:1849–1856

Lamy J, Billiald P, Sizaret PY, Cave G, Frank J, Motta G (1985) Approach to the direct intramolecular localization of antigenic determinants in *Androctonus australis* hemocyanin with monoclonal antibodies by molecular immunoelectron microscopy. Biochemistry 24:5532–5542

Lorenzini DM, Da Silva PI Jr, Soares MB, Arruda P, Setubal J, Daffre S (2006) Discovery of immune-related genes expressed in hemocytes of the tarantula spider Acanthoscurria gomesiana. Dev Comp Immunol 30:545–556. doi:10.1016/j.dci.2005.09.001

Mangum CP (1985) Oxygen transport in invertebrates. Am J Physiol 248:505–514

Markl J (1980) Hemocyanins in spiders. XI. The quaternary structure of Cupiennius hemocyanin. J Comp Physiol B 140:199–207

Markl J (1986) Evolution and function of structurally diverse subunits in the respiratory protein hemocyanin from arthropods. Biol Bull (Woods Hole, MA) 171:90–115

Markl J, Decker H (1992) Molecular structure of the arthropod hemocyanins. Adv Comp Environ Physiol 13:325–376

Markl J, Savel A, Linzen B (1981) Hemocyanins in spiders. XIV. Subunit composition of dissociation intermediates and its bearing on quarternary structure of Eurypelma hemocyanin. Hoppe Seylers Z Physiol Chem 362:1255–1262

Markl J, Gebauer W, Runzler R, Avissar I (1984) Immunological correspondence between arthropod hemocyanin subunits. I. Scorpion (*Leiurus*, *Androctonus*) and spider (*Eurypelma*, *Cupiennius*) hemocyanin. Hoppe Seylers Z Physiol Chem 365:619–631

Markl J, Stöcker W, Runzler R, Precht E (1986) Immunological correspondences between the hemocyanin subunits of 86 arthropods: evolution of a multigene protein family. In: Linzen B (ed) Invertebrate oxygen carriers. Springer, Heidelberg

Markl J, Burmester T, Decker H, Savel-Niemann A, Harris JR, Süling M, Naumann U, Scheller K (1992) Quaternary and subunit structure of *Calliphora* arylphorin as deduced from electron microscopy, electrophoresis, and sequence similarities with arthropod hemocyanin. J Comp Physiol B 162:665–680

Nagai T, Kawabata S (2000) A link between blood coagulation and prophenol oxidase activation in arthropod host defense. J Biol Chem 275:29264–29267. doi:10.1074/jbc.M002556200

Nagai T, Osaki T, Kawabata S (2001) Functional conversion of hemocyanin to phenoloxidase by horseshoe crab antimicrobial peptides. J Biol Chem 276:27166–27170. doi:10.1074/jbc.M102596200

Nentwig W (2012) The species referred to as *Eurypelma californicum* (Theraphosidae) in more than 100 publications is likely to be *Aphonopelma hentzi*. J Arachnol 40:128–130

Paul R, Bergner B, Pfeffer-Seidl A, Decker H, Efinger R, Storz H (1994) Gas transport in the haemolymph of arachnids – oxygen transport and the physiological role of haemocyanin. J Exp Biol 188:25–46

Peter MG, Scheller K (1991) Arylphorins and the integument. In: Retnakaran A, Binnington K (eds) The physiology of insect epidermis. CSIRO, Canberra

Rehm P, Pick C, Borner J, Markl J, Burmester T (2012) The diversity and evolution of chelicerate hemocyanins. BMC Evol Biol 12:19. doi:10.1186/1471-2148-12-19

Roeding F, Borner J, Kube M, Klages S, Reinhardt R, Burmester T (2009) A 454 sequencing approach for large scale phylogenomic analysis of the common emperor scorpion (*Pandinus imperator*). Mol Phylogenet Evol 53:826–834. doi:10.1016/j.ympev.2009.08.014

Savel-Niemann A, Markl J, Linzen B (1988) Hemocyanins in spiders. XXII. Range of allosteric interaction in a four-hexamer hemocyanin. Co-operativity and Bohr effect in dissociation intermediates. J Mol Biol 204:385–395. doi:0022-2836(88)90583-9

Schmitz A (2013) Tracheae in spiders—respiratory organs for special functions. In: Nentwig W (ed) Spider ecophysiology. Springer, Heidelberg (this volume)

Sterner R, Decker H (1994) Inversion of the Bohr effect upon oxygen binding to 24-meric tarantula hemocyanin. Proc Natl Acad Sci U S A 91:4835–4839. doi:10.1073/pnas.91.11.4835

Sterner R, Bardehle K, Paul R, Decker H (1994) Tris: an allosteric effector of tarantula haemocyanin. FEBS Lett 339:37–39. doi:0014-5793(94)80379-X

Sugita H, Shishikura F (1995) A case of orthologous sequences of hemocyanin subunits for an evolutionary study of horseshoe crabs: amino acid sequence comparison of immunologically identical subunits of *Carcinoscorpius rotundicauda* and *Tachypleus tridentatus*. Zoolog Sci 12:661–667

Terwilliger NB, Dangott L, Ryan M (1999) Cryptocyanin, a crustacean molting protein: evolutionary link with arthropod hemocyanins and insect hexamerins. Proc Natl Acad Sci U S A 96:2013–2018. doi:10.1073/pnas.96.5.2013

Terwilliger NB, Ryan MC, Towle D (2005) Evolution of novel functions: cryptocyanin helps build new exoskeleton in Cancer magister. J Exp Biol 208:2467–2474. doi:10.1242/jeb.01667

van Holde KE, Miller KI (1995) Hemocyanins. Adv Protein Chem 47:1–81

Voit R, Feldmaier-Fuchs G, Schweikardt T, Decker H, Burmester T (2000) Complete sequence of the 24-mer hemocyanin of the tarantula *Eurypelma californicum*. Structure and intramolecular evolution of the subunits. J Biol Chem 275:39339–39344. doi:10.1074/jbc.M005442200

Weber RE, Vinogradov SN (2001) Nonvertebrate hemoglobins: functions and molecular adaptations. Physiol Rev 81:569–628

Chapter 2
The Circulatory System of Spiders

Christian S. Wirkner and Katarina Huckstorf

2.1 Introduction

Spiders are a fascinating group of animals which exhibit a range of different lifestyles. Despite this variety, however, the spider body is fairly uniform, consisting of a prosoma which mainly serves sensory and locomotory functions and an opisthosoma which mainly serves the remaining vegetative functions. As of the main organ systems, the circulatory system is greatly affected by this bipartition. Although its importance in homeostasis, immune defense, transportation, and hydraulics is known, our understanding of this integrative organ system is still in the early stages. In the following chapter the main structural and functional aspects of the circulatory system are reviewed.

2.2 Comparative Morphology of the Circulatory Organs

Spiders exhibit an open circulatory system made up of the hemolymph vascular system in combination with a complex system of sinuses and lacunae (hemolymph lacunar system). The hemolymph vascular system is composed of the central pumping organ, the heart, and arteries which emanate from it and open out into the hemolymph lacunar system (Wirkner and Richter 2010).

The tubular heart (h; Fig. 2.1) lies dorso-medially in the anterior part of the opisthosoma within a pericardial sinus. It curves convexly in line with the shape of the anterior opisthosoma. The unpaired anterior aorta emanates from the anterior end of the heart and runs through the pedicel to supply the entire prosoma (aa; Fig. 2.1). Posteriorly, the heart narrows and passes into the posterior aorta

C.S. Wirkner (✉) • K. Huckstorf
Department of Zoology, Institute of Biosciences, University of Rostock, Rostock, Germany
e-mail: christian.wirkner@web.de; k.huckstorf@live.de

W. Nentwig (ed.), *Spider Ecophysiology*,
DOI 10.1007/978-3-642-33989-9_2,

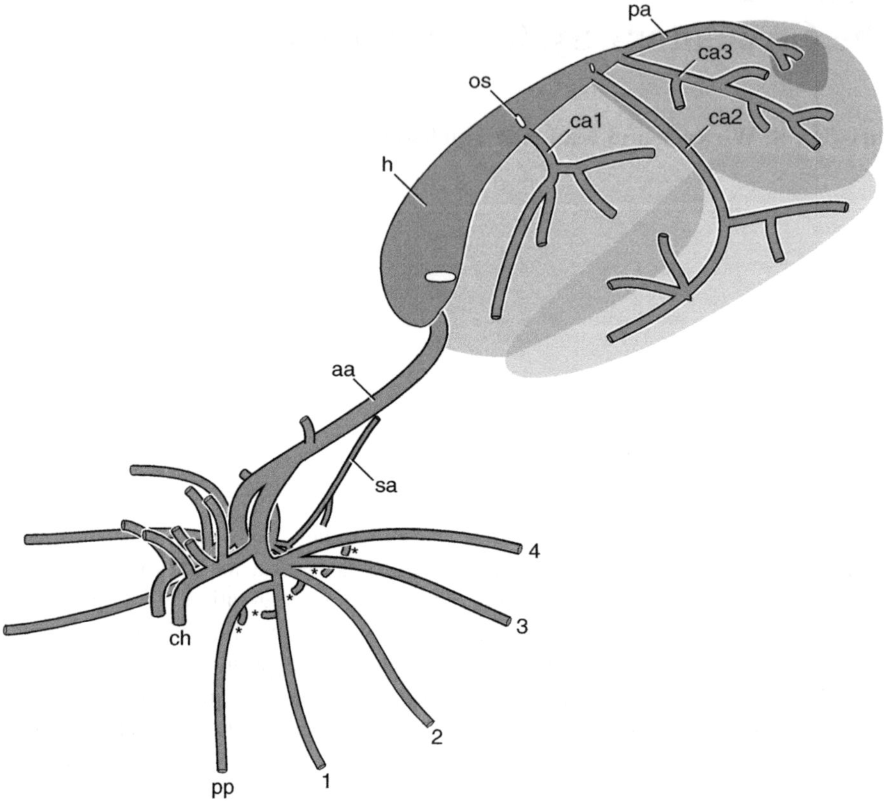

Fig. 2.1 Schematic representation of the main structures of the hemolymph vascular system of a spider. Abbreviations see text

(pa; Fig. 2.1). The heart wall is made up of a thin outer layer, the epicardium, and a thick muscle layer, the myocardium. The muscle layer is comprised of cardiomyocytes which form lunated lamellae that are positioned transversely to the longitudinal axis and project deeply into the lumen of the heart (Seitz 1972; Huckstorf et al. 2013). In the dorsal and ventral part of the heart the lamellae interconnect alternately. Also embedded within the transverse muscle layer are longitudinal muscle bands. The heart is completely surrounded by a pericardial sinus and is suspended via dorsal, ventral, and lateral ligaments which fan out before fusing with the outer layer of the heart. Four (Mygalomorphae, Mesothelae, Palaeocribellatae, and some Austrochiloidea) or two pulmo-pericardial sinuses (remaining Austrochiloidea, Araneoclada) connect the pericardial sinus with the book lungs, enabling oxygen-enriched hemolymph to travel from the book lungs into the pericardial sinus (see also Burmester 2013 and Appendix, this volume). These sinuses have often misleadingly been termed "lung veins" although they do not form vessels directly connected to the hemolymph vascular system.

Fig. 2.2 Micro-computed tomography of a resin cast of the hemolymph vascular system in the opisthosoma of *Cupiennius salei*. Note that only resin filling the arteries is visualized. Modified after Huckstorf et al. (2013). Compare Fig. 2.1

In addition to the anterior and posterior aortae, paired cardiac arteries emanate ventrolaterally from the heart, ramifying strongly to supply all parts of the opisthosoma (ca; Fig. 2.1). The liphistiid *Liphistius malayanus* displays the highest number of cardiac arteries with five pairs (Millot 1933). A stepwise reduction in pairs down to three (e.g., in the ctenid *Cupiennius salei*; Huckstorf et al. 2013) and two (e.g., the cybaeid *Argyroneta aquatica*; Crome 1953) occurred within derived spiders. It was recently possible in *C. salei* to visualize the entire arterial system of the opisthosoma (Fig. 2.2; Huckstorf et al. 2013). The antero-dorsal part of the opisthosoma is supplied by the anterior-most pair of cardiac arteries (ca1; Fig. 2.1), which ramifies just beyond its origin. The middle pair of cardiac arteries (ca2; Fig. 2.1) supply the greater portion of the ventral part of the opisthosoma, i.e. the gonads, spinning glands, and parts of the midgut. After emanating from the heart, they run a long distance ventrally before bifurcating into two strong branches each. The anterior branch runs in an antero-median direction and splits several times. The posterior pair of cardiac arteries (ca3) runs postero-ventrally into the anal region and supplies the postero-dorsal part of the opisthosoma.

The heart is equipped with slit-like openings, the ostia, through which the hemolymph can enter the lumen of the heart (os; Fig. 2.1). Ostia are formed by specialized cardiomyocytes which form lips that can close during systole to prevent hemolymph from flowing out of the heart. As a result of reduction in some spider lineages, the number of pairs of ostia varies (Fig. 2.3). The plesiomorphic state is five pairs as found in liphistiids (e.g. *L. malayanus*). This number was reduced to four in mygalomorphs (e.g., the theraphosid *Aphonopelma hentzi*, often mentioned sub *Eurypelma californicum*, Nentwig 2012) and basal aranaeomorphs (e.g., the austrochilid *Austrochilus* sp.), then three (e.g., the filistatid *Filistata hibernalis*; also basal aranaeomorphs). Within higher spiders a reduction down to two pairs is evident in several groups (Petrunkevitch 1933).

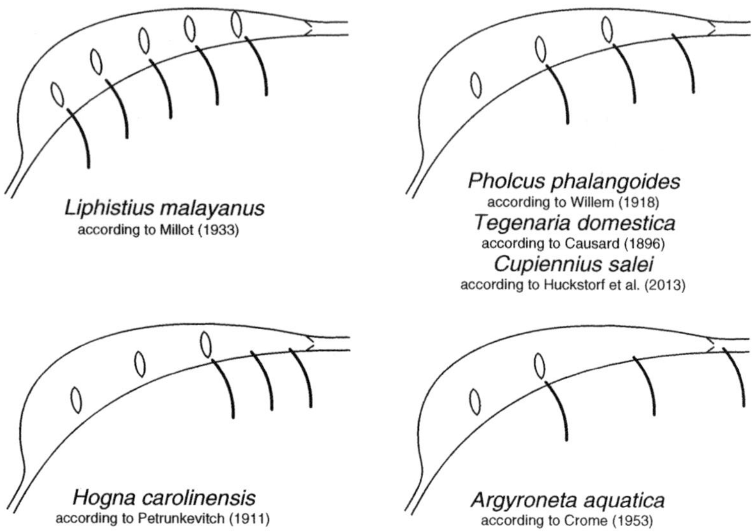

Fig. 2.3 Schematic representation of various spider hearts showing the position of pairs of cardiac arteries and pairs of ostia. Modified after Huckstorf et al. (2013)

The anterior aorta and its branches supply the entire prosoma. From the anterior tip of the heart the anterior aorta runs through the pedicel into the prosoma (aa; Fig. 2.1). It divides into two branches just posterior to the sucking stomach. Flanking the stomach laterally, these two branches run in an anterior direction and give off several small arteries.

Above the subesophageal ganglion, each branch of the anterior aorta bends in a ventral direction to rest on the dorsal side of the subesophageal ganglion. At this bend, one cheliceral artery branches off each aortic branch and runs in an anterior direction, flanking the supraesophageal ganglion dorsolaterally. On their way into the chelicerae the cheliceral arteries (ch; Fig. 2.1) give rise to several other arteries, some of which supply the supraesophageal ganglion. Two pairs of these latter arteries are present in the agelenids *Tegenaria* sp. (Schneider 1892) and *Agelena labyrinthica* (Causard 1896) while in *Cupiennius salei* three pairs occur (Huckstorf et al. 2013). Furthermore, dorsal branches emanating from the cheliceral arteries supply the dorsal musculature of the anterior prosoma. The optical arteries, which also arise from the cheliceral arteries, run dorsally to the eyes and branch repeatedly. The supply of the upper and lower lips varies between spider taxa. In most spiders a branch of the cheliceral arteries supplies the upper and lower lips together (e.g., *Tegenaria* sp.; Schneider 1892). In others, the upper and lower lips are each supplied by arteries which split off from the cheliceral arteries. In *Araneus diadematus*, an artery emanating from the anterior aorta supplies the lower lip (Schneider 1892; Runge and Wirkner, unpublished data). In addition, the cheliceral arteries give off dorsolateral branches to supply the venom glands and the dorsolateral muscles of the anterior prosoma. The cheliceral arteries then run ventrally into the chelicerae, where a number of small arteries branch off.

After having bent, the branches of the anterior aorta run ventrally to form an arch along the dorsal side of the prosomal ganglion. It is from these arches that the arteries for the four pairs of legs emanate (1–4; Fig. 2.1). The pedipalpal arteries either branch off proximally from the arteries supplying the first leg (e.g., *Cupiennius salei*; Huckstorf et al. 2013; pp; Fig. 2.1) or from the branches of the anterior aorta directly (e.g., *Agelena labyrinthica*; Causard 1896; *Argyroneta aquatica*; Crome 1953). The coxae of the pedipalps are supplied by a ventro-anterior branch off the pedipalpal artery. Several small arteries emanate from the leg arteries on their way into the legs. The branches of the anterior aorta are interconnected via several small arteries just above the subesophageal ganglion (connecting arteries). The number of connecting arteries varies between and within spider taxa. The last connecting artery gives rise to the unpaired supraneural artery, which extends above the opisthosomal nerve to the pedicel (sa; Fig. 2.1).

The central nervous system, i.e. the supraesophageal ganglion (brain) and the subesophageal ganglion, is well supplied by vessels (for a review of central nervous system architecture in *Cupiennius salei*, see Seyfarth 2002; Barth 2002). This is achieved by a number of smaller vessels permeating the nerve mass after having branched off the main arteries in the prosoma. The subesophageal neuromeres are supplied medially by unpaired transganglionic arteries and laterally by paired arteries. The latter branch off from the leg and pedipalpal arteries. On each side, one artery emanates from the proximal end of each appendage artery and runs in an antero-ventral direction into the septum between two neighboring neuromeres (*; Fig. 2.1). The septum between the cheliceral and pedipalpal neuromeres is supplied by an artery which branches off the pedipalpal artery. The transganglionic arteries take their origin from the connecting arteries and the proximal part of the supraneural artery. They run through the subesophageal ganglion from the dorsal to the ventral side. From the transganglionic arteries, a number of fine arteries extend into neighboring neuromeres, which they penetrate to over a third of their width. The center of the neuromeres remains largely free of arteries.

Although the vascularization of the central nervous system has long been the object of investigation (e.g., Schneider 1892), it is best understood in *Cupiennius salei* (Huckstorf et al. 2013) where it has been described as a form of capillarization in which supply is provided by a system of afferent and efferent vessels. A similar system of irrigation of the central nervous system is also found in crustaceans (Sandemann 1967; Abbott 1971; Brown and Sherwood 1981). In both crustaceans and spiders, certain areas of the central nervous system seem to be better supplied than others. In *C. salei*, these include the arcuate body and the third optic neuropil of the brain, which are especially rich in synaptic contacts. Hemolymph vessels and loops in the subesophageal ganglion (Fig. 2.4) ensure that supply to the peripheral layers of each neuromere is particularly abundant.

These are the regions where the majority of the neuronal somata are situated and where metabolic activity is high. Along the subesophageal midline, areas with a high oxygen requirement—those with dense synaptic contacts in which intense neuronal activity takes place—are also well supplied. Immunohistochemical studies have shown that octopamine- and serotonin-immunoreactive neurons project

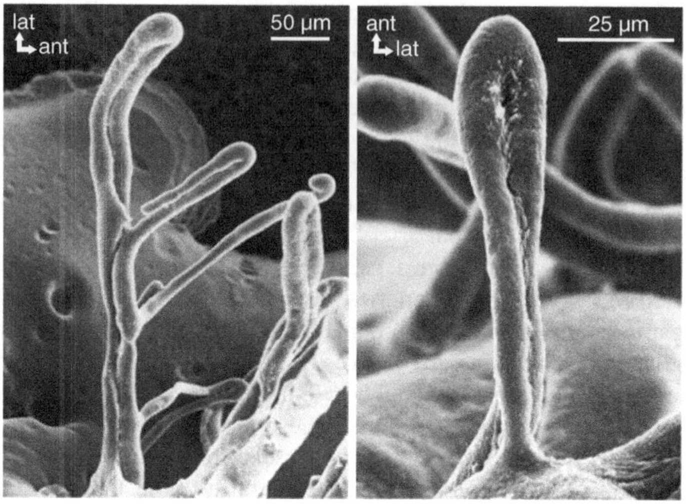

Fig. 2.4 Electron micrograph of a resin cast in *Cupiennius salei* showing loops supplying neuropiles in the subesophageal ganglion. Modified after Huckstorf et al. (2013)

into the close vicinity of hemolymph vessels in the subesophageal neuromeres, suggesting that these neuroactive substances are released into the hemolymph here to supply more peripheral organs (Seyfarth et al. 1990, 1993).

2.3 Functional Morphology of the Circulatory System

On the whole, the circulatory system in spiders has not been studied extensively. Although some aspects (e.g., hydraulic leg movement) are well explored, comparative morphological and functional studies are scarce, with the latter remaining fairly superficial and focusing on the relatively large and thus more easily accessible mygalomorph spiders. The following passages on functional morphology and physiology are therefore based on these few studies.

2.3.1 The Split/Bipartitioned Body

Paul et al. (1989) came up with a hydraulic model for explaining major body functions in spiders. After this model, the spider body can be described as being partitioned into two functional units, an anterior and a posterior region. In the latter, the vegetative/ visceral region, i.e. the opisthosoma, the main parts of the circulatory, respiratory, excretory, and reproductive organs are situated, while the anterior region, the prosoma, houses the main sensory and locomotory functions (apart from food uptake and defense). The regions are connected by the rather thin pedicel, which drastically limits the exchange of body fluids. In terms of function, the two groups of organs display quite different time patterns. Vegetative systems have to perform to the needs of the whole animal/body. They work relatively continuously and have comparatively slow reaction times. Sensory and locomotory functions, on the other hand, require much faster reaction times. In accordance with its vegetative function, the opisthosoma displays long-term volume changes (degree of hydration, nutritional condition, degree of fertility/maturity, etc.), while in the prosoma, except for short-term changes during locomotion, a constant volume is maintained. The relative stiffness of the exoskeleton also reflects this division of physiological functions. The comparatively soft opisthosoma is—to differing degrees—expandable. The prosoma, on the other hand, is fairly rigid, providing the mechanical strength required for both the mechanical (muscular) locomotion system and the peculiar hydraulic system (see Kropf 2013) serving leg extension. The bipartitioning of the body is therefore of double advantage in this regard. On the one hand, only half of the body's cuticle requires reinforcement, and on the other, as this is the same half which necessitates higher internal pressure levels, the effort needed to build up this pressure can be greatly reduced. In terms of the vascular system, however, the partitioning poses a major challenge. First of all, despite differences in the level of supply required by the prosoma and the opisthosoma according to their different functions, hemolymph must be supplied to both regions. The mechanisms behind this shifting of hemolymph supply according to the requirements of the animal and vegetative regions are still not properly understood. A further complication is caused by differences in the configuration of respiratory organs. In species with more than one pair of lungs the contribution of each lung pair to the reoxygenation of the hemolymph and the release of carbon dioxide must be different to that with a complex tracheal system. There also seem to be more obvious "problems" with the bipartition of the spider body. First of all, the pumping action of the heart into the prosoma has to operate against the prosomal pressure generated during leg extension. This might in some cases lead to a severe lack of oxygen supply to the anterior body region, resulting in the complete exhaustion of the spider. In summary it should be clear that the circulatory system in spiders has to compromise between a large number of different functional and structural limitations and requirements.

2.3.2 The Circulatory Cycle

As described above, hemolymph is distributed via a varying number of cardiac arteries and one unpaired posterior aorta into the opisthosoma and via an unpaired anterior aorta into the prosoma. The arteries end openly in the hemolymph lacunar system by means of which hemolymph irrigates the tissues and supplies them with oxygen, nutrients and various other molecules such as hormones. Oxygen-poor hemolymph is conducted through the book lungs and there replenished with oxygen

(Burmester 2013). Oxygen-rich hemolymph flows dorsally through the pulmo-pericardial sinuses into the pericardial sinus. Hemolymph reenters the lumen of the heart via the paired ostia. Measurements have shown, however, that hemolymph is unevenly distributed in the spider body, with the prosoma and opisthosoma not always receiving the same amount (Paul and Bihlmayer 1995). There are three possible ways in which this may be achieved. Firstly, the bidirectional movement of hemolymph could be caused by peristaltic heart contractions, as is known to be the case in insects (e.g., Pass 2000). In spiders, however, peristaltic contractions of this nature seem not to occur (at least not in *A. hentzi*; Paul and Bihlmayer 1995). Secondly, flow distribution may be controlled and regulated by arterial valves. In measurements again performed in *A. hentzi* (Paul and Bihlmayer 1995), the pressures in the anterior aorta and the heart were recorded to be the same, leading the authors to rule out this hypothesis. As arterial valve control has a strong influence on hemolymph distribution in some decapod crustaceans, however (e.g., McMahon and Burnett 1990), there is a strong case for it to be tested more comprehensively in spiders. Last but not least, changes in the resistance of peripheral vessels caused by a passive or active decrease in arterial diameters may also lead to an uneven distribution of hemolymph. Changes in resistance of this nature are known from decapod crustaceans, in which group they are mainly triggered by aminergic and peptidergic substances (McMahon et al. 1997), but spiders have not, as yet, been tested for such substances. The unevenness of the distribution of cardiac output also varies with in accordance with the activity level of the spider. On the basis of simultaneous measurements taken at a number of circulatory structures (e.g., the pulmo-pericardial sinuses and the anterior aorta), Paul and Bihlmayer (1995) were able to show that in *A. hentzi*, forward flow came to a halt during locomotion while flow in a posterior direction was kept up. Flow through both pairs of pulmo-pericardial sinuses was maintained during this phase. The opposite was measured shortly after locomotory activity. In some cases, changes in hemolymph flow direction were measured even during late recovery.

2.3.3 The Cardiac Cycle

The action of the muscular heart is the main motor of hemolymph movement. In arthropods, there are two basic ways in which the heart is activated to pump (Sherman and Pax 1968; Sherman 1985). Hearts which are stimulated by automatically beating myocardial cells are termed myogenic and occur in some crustaceans and hexapods (McMahon et al. 1997). Hearts which are mainly controlled by the central nervous system (though various other chemicals can provide additional stimulation) are termed neurogenic. One indicator of neurogenic control is the presence of nervous structures directly associated with the heart, i.e. cardiac ganglia and nerves. Cardiac nerves were first described in spiders by Wilson (1967), with Sherman and Pax (1969) later demonstrating their presence in a further 28 spider species. A cardiac ganglion lies on the mid-dorsal surface of the heart and extends

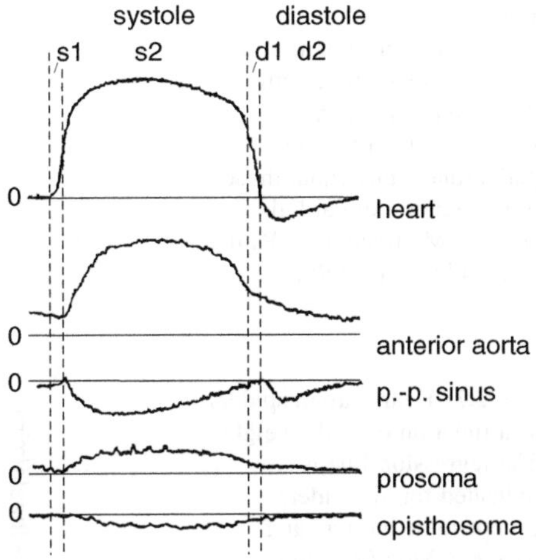

Fig. 2.5 Pressure distribution in parts of the structures involved in circulation in *Aphonopelma hentzi*. Pressure baseline represents the noncardiac body pressure. Modified after Paul and Bihlmayer (1995)

its entire length (Wilson 1967; Sherman and Pax 1968, 1969; Gonzalez-Fernandez and Sherman 1984), initiating and coordinating the heartbeat (Bursey and Sherman 1970) and sending out motor nerves which innervate the myocardial cells (Sherman 1973; Ude and Richter 1974). In *Aphonopelma marxi*, Gonzalez-Fernandez and Sherman (1984) actually demonstrated the presence in spiders of cardioregulatory nerves containing both inhibitory and acceleratory neurons (Gonzalez-Fernandez and Sherman 1984). Other studies involving pharmacological experiments all come to the conclusion that the basic form of heart control in spiders is neurogenic (Sherman and Pax 1970a, b; Martin 1974).

The movement of the heart can be divided into a phase of contraction, the systole, and a phase of relaxation, the diastole. Each of these phases can be again subdivided into two different phases (Paul and Bihlmayer 1995). In the first phase of the systole the myocardium contracts more or less isovolumetrically (Fig. 2.5) and no hemolymph flow occurs through the arterial systems. When pressure in heart surpasses pressures in the arteries, valves between the heart and the arteries are forced open and hemolymph is pumped into the arterial systems. In this first phase of the systole, pressure increase is rapid, then followed by a slow decrease (Fig. 2.5). After this, the myocardium relaxes slowly (the first phase of the diastole), resulting in a rapid decrease in hemolymph pressure to below that in the arterial systems and a closure of the arterial valves. In the second phase of diastole the pressure inside the heart becomes negative. During both the systole and the diastole, pressures inside the pulmo-pericardial sinuses are negative, allowing a continuous hemolymph flow to be kept up for the duration of the whole cardiac cycle (Paul and Bihlmayer 1995).

In mygalomorph spiders at least, the heart and the pericardial sinus act together as a pressure-suction-pump (Paul et al. 1994; Paul and Bihlmayer 1995). While the heart is suspended elastically, the pericardial membrane seems to be fairly fixed, which means that volume changes are significant but minor (Fig. 2.6a). Cardiac

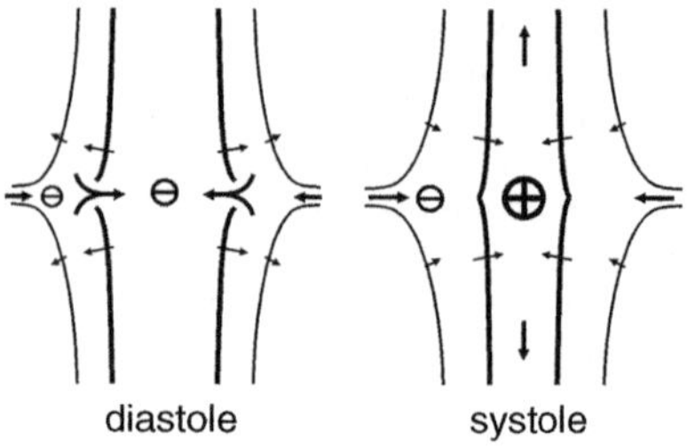

Fig. 2.6 Schematic functional representation of the pressure-suction pump in *Aphonopelma hentzi*. Pressures given for the pericardial sinus equal those in the pulmo-pericardial sinuses. Modified after Paul and Bihlmayer (1995)

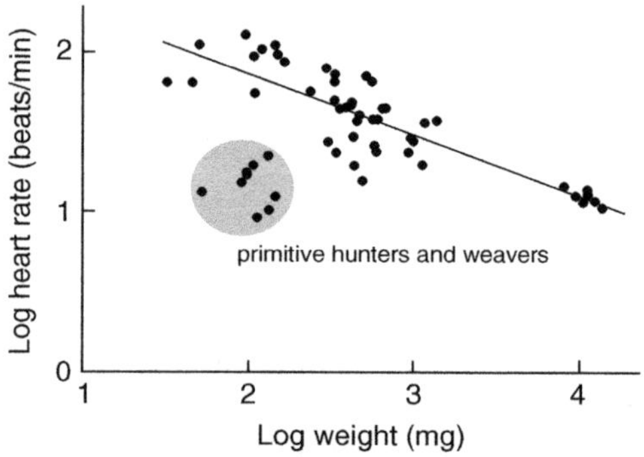

Fig. 2.7 Heart rate in spiders as a function of body weight. The regression line is indicated for all spiders except some hunters and weavers. Modified after Carrel and Heathcote (1976)

systole, i.e. the contraction of the myocardium, results in a change in volume in the pericardial sinus which affects a negative pressure which causes hemolymph to be sucked out of the pulmo-pericardial sinuses at the same time as it is pumped through the heart. The pressure remains negative in the pericardial sinus even during the diastole of the heart (Fig. 2.6b) because as the myocardium is widened through the action of the alary muscles and the pericardial membrane also moves back slightly into its original position, hemolymph continues to be sucked out of the pulmo-pericardial sinuses even as it is sucked into the heart.

2.3.4 Heart Rate

Carrel and Heathcote (1976) measured the resting heart beat in 18 species of spiders using a laser trans-illumination technique. All the animals were unrestrained and had no further equipment attached to their body. Heart rates ranged from 9 to 125 beats per minute. Resting heart rates correlate with body size and, except in the case of some hunting and orb-weaving spiders, with body weight (Fig. 2.7). Naturally, activity leads to an increase in heart beat frequency. Experiments carried out by Paul and Bihlmayer (1995) in *A. hentzi* showed that activity can cause the heart to beat up to ten times faster than in resting individuals, and that it takes more than 50 min for the frequency to drop down to resting values following activity. Similar results were presented by Bristowe (1932) for *Liphistius desulator*. During a short burst of activity the heart rate in this spider went up to 120 beats per minute. In the first 5–10 min after this, a rapid decrease was observed, but it took more than 40 min for the resting heart beat of 48 beats per minute to be regained. Carrel (1987) reported on the various factors that can influence spider heart rate, listing, apart from activity, correlations with body size, metabolic rate and temperature and the effects of arthropod venoms such as those from wasps and millipedes. The reader is referred to that chapter for more detailed information on heart rate.

2.3.5 Hemolymph Pressures

The other main function of the circulatory system is the generation of blood—or in this case hemolymph pressure. The cardiac pressures generated in mygalomorphs at rest are relatively low: 1.2/0.8 kPa systolic/diastolic in *Aphonopelma* sp. (Stewart and Martin 1974) and 0.5–1.5/−0.1 kPa in *A. hentzi* (Paul et al. 1994; Paul and Bihlmayer 1995). Cardiac pressures can change in accordance with alertness, water uptake, and hemolymph loss. The pressure gradients that occur during rest are solely the result of the action of the heart-pericardial system, i.e. the pressure-suction pump. Paul and Bihlmayer (1995) discovered that in *A. hentzi* heart output, i.e. the volume pumped by the heart over a certain time interval, had the capacity to increase without an increase in heart beat frequency. The authors interpret this, on the basis of a series of experiments and measurements, as being the product of an increase in cardiac stroke volume. Apart from the heart there are two further hemolymph pressure generators in spiders. These are certain groups of prosomal and opisthosomal muscles, which, due to their contraction, produce a steady regime of above ambient pressure in the whole spider body (Wilson 1970; Anderson and Prestwich 1975; Paul and Bihlmayer 1995). All three hemolymph generating systems interact on various functional levels such as the hydraulic leg extension, etc. (Kropf 2013).

2.4 Conclusions

The spider circulatory system is a structurally complex and elaborately regulated system with a broad range of functionalities. Apart from its well-known role in metabolism and, plesiomorphically at least, in oxygen distribution, its main functions are hydraulic. Though we know much about the structure and function of this important organ system, it is highly desirable that comparative studies on both the morphological and functional/physiological levels be carried out to shed more light on the complex interactions of this system. The fact that Araneae are a monophyletic group makes them an ideal object to study the different pathways taken during the evolutionary development of this uniform yet uniquely diverse group of chelicerates.

References

Abbott NJ (1971) The organization of the cerebral ganglion in the shore crab. Carcinus maenas. II. The relation of intracerebral blood vessels to other brain elements. Z Zellforsch 120:401–419

Anderson JF, Prestwich KN (1975) The fluid pressure pumps of spiders. Z Morphol Tiere 81: 257–277

Barth FG (2002) A spider's world: senses and behavior. Springer, Berlin

Bristowe WS (1932) The Liphistiid spiders. With an appendix on their internal anatomy by J. Millot. Proc Zool Soc Lond 102:1015–1057

Brown SK, Sherwood DN (1981) Vascularization of the crayfish abdominal nerve cord. J Comp Physiol A 143:93–101

Burmester T (2013) Evolution and adaptation of hemocyanin within spiders. In: Nentwig W (ed) Spider ecophysiology. Springer, Heidelberg (this volume)

Bursey CR, Sherman RG (1970) Spider cardiac physiology. I. Structure and function of the cardiac ganglion. Comp Genet Pharmacol 1:160–170

Carrel JE (1987) Heart rate and physiological ecology. In: Nentwig W (ed) Ecophysiology of spiders. Springer, Berlin

Carrel JE, Heathcote RD (1976) Heart rate in spiders: influence of body size and foraging energetics. Science 193:148–150

Causard M (1896) Recherches sur l'appareil circulatoire des aranéides. Bull Sci France Belgique 29:1–109

Crome W (1953) Die Respirations- und Circulationsorgane der *Argyroneta aquatica* Cl. (Araneae). Wiss Z Humboldt-Universität Berlin 2:53–83

Gonzalez-Fernandez F, Sherman RG (1984) Cardioregulary nerves in the spider *Eurypelma marxi* Simon. J Exp Zool 231:27–37

Huckstorf K, Kosok G, Seyfarth E-A, Wirkner CS (2013) The hemolymph vascular system in *Cupiennius salei* (Araneae: Ctenidae). Zool Anz (in press)

Kropf C (2013) Hydraulic system of locomotion. In: Nentwig W (ed) Spider ecophysiology. Springer, Heidelberg (this volume)

Martin AW (1974) Circulation in invertebrates. Annu Rev Physiol 36:171–186

McMahon BR, Burnett LE (1990) The crustacean open circulatory system: a reexamination. Physiol Zool 63:35–71

McMahon BR, Wilkens JL, Smith PJS (1997) Invertebrate circulatory systems. In: Danzler WH (ed) Handbook of physiology, vol 2. Comparative physiology. American Physiological Society, New York

Millot J (1933) Notes complementaires sur l'anatomie des liphistiides et des hypochilides, a propos d'un travail recent de A. Petrunkevitch. Bull Soc Zool France 58:217–235

Nentwig W (2012) The species referred to as *Eurypelma californicum* (Theraphosidae) in more than 100 publications is likely to be *Aphonopelma hentzi*. J Arachnol 40:128–130

Pass G (2000) Accessory pulsatile organs: evolutionary innovations in insects. Annu Rev Entomol 45:495–518

Paul RJ, Bihlmayer S (1995) Circulatory physiology of a tarantula (*Eurypelma californicum*). Zoology 98:69–81

Paul RJ, Tiling K, Focke P, Linzen B (1989) Heart and circulatory functions in a spider (*Eurypelma californicum*): the effects of hydraulic force generation. J Comp Physiol B 158: 673–687

Paul RJ, Bihlmayer S, Colmorgen M, Zahler S (1994) The open circulatory system of spiders (*Eurypelma californicum, Pholcus phalangoides*). Physiol Zool 67:1360–1382

Petrunkevitch A (1933) An inquiry into the natural classification of spiders, based on a study of their internal anatomy. Trans Conn Acad Arts Sci 31:299–389

Sandemann DC (1967) The vascular circulation in the brain, optic lobes and thoracic ganglia of the crab *Carcinus*. Proc R Soc Lond B 168:82–90

Schneider A (1892) Sur le systeme artériel du scorpion. Tabl Zool 3:157–198

Seitz K-A (1972) Zur Histologie und Feistruktur des Herzens und der Hämocyten von *Cupiennius salei* KEYS. (Araneae, Ctenidae) I. Herzwandung, Bildung und Differenzierung der Hämocyten. Zool Jahrb Anat 89:351–384

Seyfarth E-A (2002) Tactile body raising: neuronal correlates of a 'simple' behavior in spiders. In: Toft S, Scharff N (eds) European Arachnology 2000: Proceedings of the 19th European Colloquium of Arachnology. Aarhus University Press, Aarhus

Seyfarth E-A, Hammer K, Grünert U (1990) Serotonin-like immunoreactive cells in the CNS of spiders. Verh Deut Z 83:640

Seyfarth E-A, Hammer K, Spörhase-Eichmann U, Hörner M, Vullings HG (1993) Octopamine-like immunoreactive neurons in the fused central nervous system of spiders. Brain Res 611: 197–206

Sherman RG (1973) Ultrastructural features of cardiac muscle cells in a tarantula spider. J Morphol 140:215–242

Sherman RG (1985) Neural control of the heartbeat and skeletal muscle in spiders and scorpions. In: Barth FG (ed) Neurobiology of arachnids. Springer, Berlin

Sherman RG, Pax RA (1968) The heart-beat of the spider *Geolycosa missouriensis*. Comp Biochem Physiol 26:529–536

Sherman RG, Pax RA (1969) Electrical activity in single muscle cells of a spider heart. Comp Biochem Physiol 28:487–489

Sherman RG, Pax RA (1970a) Spider cardiac physiology. II. Responses of a tarantula heart to cholinergic compounds. Comp Gen Pharmacol 1:171–184

Sherman RG, Pax RA (1970b) The spider heart. In: Kerkut GA (ed) Experiments in physiology and biochemistry. Academic Press, London

Stewart DM, Martin AW (1974) Blood pressure in the tarantula, *Dugesiella hentzi*. J Comp Physiol 88:141–172

Ude J, Richter K (1974) The submicroscopic morphology of the heart ganglion of the spider *Tegenaria atrica* (C.L. KOCH) and its neuroendocrine relations to the myocard. Comp Biochem Physiol A 48:301–308

Wilson RS (1967) The heartbeat of the spider, *Heteropoda venatoria*. J Insect Physiol 13: 1309–1326

Wilson RS (1970) Some comments on the hydrostatic system of spiders (Chelicerata, Araneae). Z Morphol Tiere 68:308–322

Wirkner CS, Richter S (2010) Evolutionary morphology of the circulatory system in Peracarida (Malacostraca; Crustacea). Cladistics 26:143–167

Chapter 3
Tracheae in Spiders: Respiratory Organs for Special Functions

Anke Schmitz

3.1 Short Description of Respiratory Systems: Lungs or Tracheae

Many people do not know that spiders have a tracheal system. Even biologists often think that in spiders, lungs play the most important role in respiration. But nevertheless in most modern spiders a tracheal system has an important role, and it is still not exactly known why they have evolved tracheae and which purpose they are good for. This chapter wants to give a short overview over the morphology of tracheal systems in spiders and tries to give some explanations for evolution.

Spiders comprise a multifold equipment with respiratory organs. Basal spiders (Mesothelae, Mygalomorphae, and some Araneomorphae like Palaeocribellatae and some Austrochiloidea) possess two pairs of book lungs in the second and third opisthosomal (posterior) segments (see also Appendix, this volume). Most modern spiders (Araneomorphae), however, are bimodal breathers using a combination of lungs and tracheae (Weygoldt and Paulus 1979). Tracheae are invaginations of the body wall or originate from lungs or from other internal structures. Most often the second lung pair is reduced and replaced by tracheae. Moreover, the third opisthosomal segment is in many families stretched in the ventral part, whereby the tracheal spiracles lay at the end of the opisthosoma. This elongation happened as spiders developed their sophisticated webs which required special morphological adaptations. Mesothelae, in which original segmentation of the opisthosoma is still reflected by dorsal cuticular plates, possess the spinnerets directly behind the lung spiracles. Mygalomorph and araneomorph spiders, however, shifted the spinnerets at the end of the body to the tip of the opisthosoma. This is the best solution for web building, as far as possible distance from the legs, together with a high movability between pro- and opisthosoma because of the petiolus. As the spinnerets belong to

A. Schmitz (✉)
Institute for Zoology, Rheinische Friedrich-Wilhelms University Bonn, Bonn, Germany
e-mail: unb30011@uni-bonn.de

W. Nentwig (ed.), *Spider Ecophysiology*,
DOI 10.1007/978-3-642-33989-9_3, © Springer-Verlag Berlin Heidelberg 2013

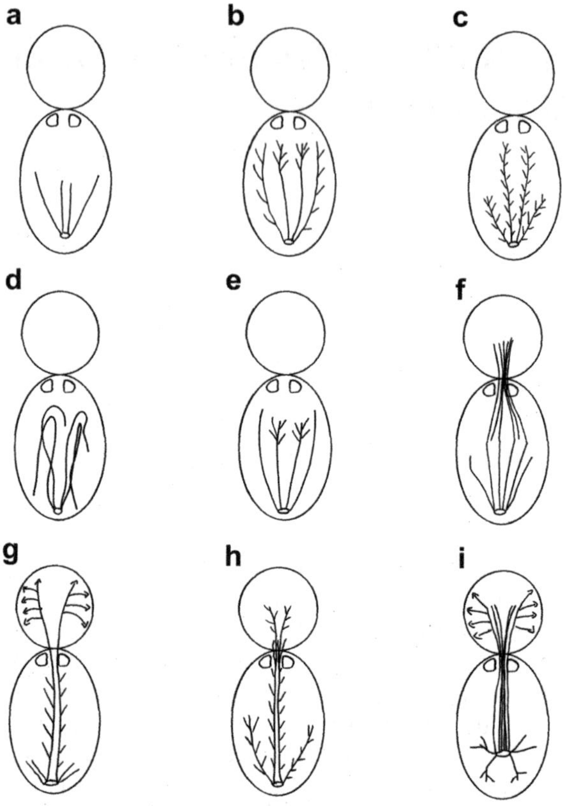

Fig. 3.1 (continued)

the opisthosomal segments 4 and 5, spiders elongated sternite 4 on the ventral side of the body. Most often also segment 3 is elongated thus repositioning the tracheal spiracle (Fig. 3.1).

Therefore in most modern spiders we have the situation with one pair of lungs in the second opisthosomal segment with an elongated third segment and one tracheal spiracle in this segment. Most often, the tracheal spiracles are single openings from which four tube-like tracheae originate. The outer two (primary or lateral) tracheae are remnants of lungs and can be connected to the lung extended lung atria. The inner two (secondary or median) tracheae, however, are new structures and are hollowed apodemes (entapophyses) elongated along the long axis of the spider (Purcell 1895; Lamy 1902; Purcell 1909, 1910; Levi 1967; Forster 1980; Ramirez 2000). True median tracheae are a synapomorphy of Entelegynae, as is the extreme posterior displacement and narrowing of the tracheal spiracle (Ramirez 2000).

In haplogyne spiders, tracheal spiracles are situated behind lung spiracles (e.g., Dysderoidea) (Ramirez 2000). In some species of this group also the first pair of respiratory organs is replaced by tracheae (e.g., Caponiidae, Telemidae). These are the so-called “sieve tracheae”. They look like a bundle of tubes, are probably

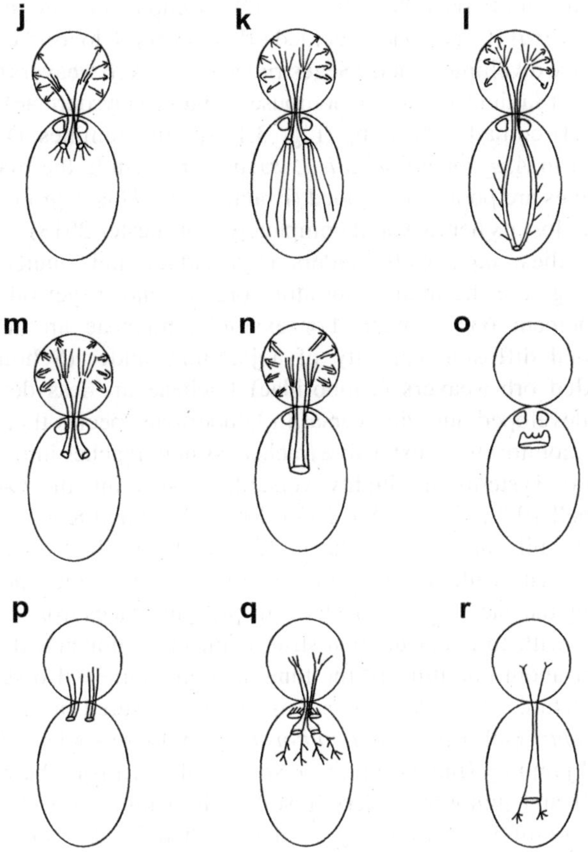

Fig. 3.1 Tracheae in different spider families. Bodies are schematized as two compartments. Lungs are given as simple *circles* in the frontal part of the ophistosoma. Legs are not given. (**a**) Lycosidae, Agelenidae, (**b**) Philodromidae, (**c**) Thomisidae, (**d**) Gnaphosidae, (**e**) Linyphiidae (Linyphiinae), (**f**) Linyphiidae (Erigoninae), (**g**) Uloboridae, (**h**) Salticidae, (**i**) Anyphaenidae, (**j**) Dysderidae, (**k**) Segestriidae, (**l**) Dictynidae, (**m**) *Argyroneta aquatica*, (**n**) Hahniidae, (**o**) Filistatidae, (**p**) Symphytognatidae, (**q**) Caponiidae, (**r**) Oonopidae

rounded derivatives of lung lamellae and occur exclusively in the second opisthosoma segment. In Oonopidae lungs are mainly reduced and tracheae take over the entire respiratory function. In other species, tracheae are completely lacking (e.g., Tetrablemmidae, Pholcidae, Diguetidae, Plectreuridae and Sicariidae) and the one pair of lungs is the only respiratory organ.

Also within the entelegyne spiders the first lung pair may be replaced by tube tracheae: in Symphognathidae spiders, the first lung pair may be replaced by tube tracheae while the third segment lacks respiratory organs. In some species, primary and secondary tracheae remain as tubes restricted to the opisthosoma. In others, however, highly branched tracheae also enter the prosoma via the petiolus and may penetrate the nervous system, muscles, or gut (Kästner 1929; Millot 1949; Bromhall

1987a, b; Schmitz and Perry 2000, 2001, 2002), examples include the water spider (*Argyroneta aquatica*) (Cybaeidae), hackled orb-weavers (Uloboridae), crab spiders (Thomisidae), tube dwelling spiders (Segestriidae), sheetweb weavers (Linyphiidae), jumping spiders (Salticidae) and woodlouse hunters (Dysderidae) (Blest 1976; Millidge 1986; Bromhall 1987a, b) (Fig. 3.1). In the families Dysderidae and Segestriidae and in *Argyroneta aquatica*, tracheae constitute the main respiratory system while lungs are poorly developed (Braun 1931). This is probably the reason why *Dysdera* sp. lost its hemocyanin completely (Burmester 2013).

In other families, e.g., wolf spiders (Lycosidae) and funnel web spiders (Agelenidae) lungs are the main respiratory organs, and in yet others lungs and tracheae complement one another. For example, tracheae are responsible for 25–30 % of total diffusing capacity of a jumping spider (Schmitz and Perry 2001). In hackled orb-weavers (Uloboridae) tracheae are less developed when lungs are well developed and vice versa. In Uloboridae species that use their legs actively for net monitoring an extensive tracheal system reaches into the legs. Thus, spider respiratory systems are highly versatile in meeting the O_2 demands of locomotion (Opell 1979, 1987, 1989, 1990, 1998; Opell and Konur 1992).

Lung lamellae pile up like the sheets of a book in which layers filled with haemolymph alternate with layers containing air. Epidermis lines the inner surface of the lung facing the haemolymph. Pillar cells prevent spaces from collapsing. The air space is lined with cuticle. Cuticular struts connect the dorsal and ventral part of the lamella over about one third of the lung area and spines also stabilize the air space. Ultrastructure of spider lungs has been measured in the theraphosid *Aphonopelma hentzi* (sub *Eurypelma californica*), in house spiders (*Tegenaria* sp.) and in wolf and jumping spiders. In house spiders, the diffusion barrier is twice as thick as in wolf and jumping spiders, resulting in a much lower morphological diffusing capacity (4–9 μl min^{-1} g^{-1} kPa^{-1}) than in the other two species (12–16 μl min^{-1} g^{-1} kPa^{-1}) (Strazny and Perry 1984; Schmitz and Perry 2001, 2002). Thus the flexibility of lung morphometrics in different spiders becomes clear. In *Aphonopelma hentzi*, epidermal and cuticular layers have similar thickness as in wolf spiders and jumping spiders (*Pardosa lugubris* and *Salticus scenicus*, each 0.05–0.5 μm thick) (Schmitz and Perry 2000, 2002). But dimensions of air (5–7 μm) and haemolymph spaces (60–70 μm) are much larger in *Aphonopelma hentzi* than in araneomorph spiders and the linear lung dimensions (length, width, interlamellar width) scale to the 0.2–0.35 power of body mass, i.e., a four- to eightfold increase in lung size over a 900-fold body size range (Moore 1976; Reisinger et al. 1990, 1991). Dimensions are 10–20 μm for the haemolymph space and 1–2 μm for the air space in the wolf spider *Pardosa lugubris* and the jumping spider *Salticus scenicus*, both about 20 mg in body mass (Schmitz and Perry 2000, 2002).

Ultrastructure of tracheae consists of an epidermal outer layer and a cuticular inner layer that builds hoop or spiral thickenings (taenidia) for stabilization (Fig. 3.2). Therefore spider tracheae look very similar to insect tracheae. Both cuticle and epidermis constitute about the same proportion of the tracheal walls: the walls of the smallest tracheae have about the same thickness as the lungs (Schmitz and Perry 2001, 2002).

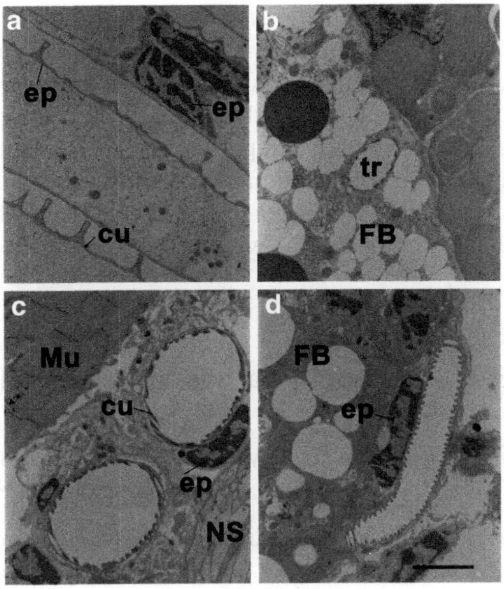

Fig. 3.2 Examples of lungs (**a**) and tracheae (**b–d**) in *Salticus scenicus*, a salticid spider. In (**b**) and (**d**) tracheae (tr) can be found in the fat body (FB) while in (**c**) they are situated between muscles (Mu) and nervous system (NS). *ep* epidermis, *cu* cuticle

3.2 Functional Principles and the Evolution of Respiratory Organs

All lungs are so-called book lungs and they are diffusion lungs, convection is done by the haemolymph alone and the spiracles are diffusion regulators (Paul et al. 1987). Tracheae in spiders, however, can function as tracheal lungs or use terminal diffusion. Transitions between the two types are blurred and even within one animal, tracheae may function in the two general kinds. As tracheal lungs, their entire surface may be used as gas exchanger with haemocyanin within the haemolymph. Therefore, they considerably increase the effectiveness of respiration, last but not least because gas exchange is no longer restricted to the second and third opisthosomal segment. In terminal diffusion, proximal tracheae branches are grouped into bundles that run parallel through the body to create a thick diffusion barrier. O_2 loss is limited to the outer surface of those bundles. The terminal branches (tracheoles) reach into the epithelia of organs in which gas exchange takes place. Therefore the gas exchange is restricted to several places in the body.

Terrestrialization of arachnids especially influenced the evolution of the respiratory organs. Recent studies investigated the ultrastructure of lungs in arachnids (Scholtz and Kamenz 2006). It was concluded that arachnid lungs developed only once and therefore a common arachnid aquatic ancestor, at least for the lung breathers must be postulated. These ancestors might be the scorpions that in turn possibly stem from eurypterid-like ancestors (Weygoldt 1998). In this scenario, lungs were the first respiratory organs for air breathing and they have to be lost before tracheae could develop. Spiders, however, may have evolved from trigonotarbid-like ancestors, which were contemporaneous with aquatic scorpions in the Devonian. First fossil records of terrestrial Mesothelae exist from late Carboniferous to early Permian around 295 million years ago (Selden 1996). Also other tracheal breathers might have evolved separately thus indicating that tracheae developed independently in all arachnid groups. One hint supporting this hypothesis is that number and position of tracheal spiracles being different in each group. First fossil records of tracheate arachnids are the harvestmen (Opiliones) appearing in the early Devonian 410 MYA, without any hint that this arachnid group stands in any evolutionary line with the lung breathers (Dunlop et al. 2003).

3.3 Potential Reasons to Develop Tracheal Systems

In the literature four hypotheses for the evolution of spider tracheae can be found (Ellis 1944; Levi 1967, 1976; Anderson and Prestwich 1975):

1. The first theory indicates that tracheae are an adaptation for the reduction of water loss. During evolution, spiders reduced body size and therefore are threatened by evaporation in low humidities. Unlike insects, spiders possess a relatively thin body cuticle and are threatened by evaporation in arid habitats. This hypothesis may not explain all tracheated spiders, as not all tracheal breathers are active during the day or in full sun. For example, the six-eyed spider *Dysdera erythrina* has highly developed tracheae and reduced lungs but is active only at dawn and night and rests under leaves or stones during the day.
2. The second theory indicates that spiders developed tracheae in response to increased local O_2 demand. Tracheae supply organs with high O_2 needs relative to surrounding organs. Two examples support this hypothesis. In jumping spiders that rely mainly on their eyes for prey capture and protection against predation, tracheae supply predominantly the prosomal nervous system. It was hypothesized that tracheae supply the highly energy consuming signal processing of the eyes by allowing a constant aerobic metabolism (Schmitz 2004, 2005). Orb-weavers (Uloboridae) actively monitor their nets with the third leg pair during prey capture, tracheal supply to the muscles of the third leg is better than that to the other legs (Opell 1987, 1990; Opell and Konur 1992).
3. Moreover, higher mass specific O_2 demands occur, for example, in free ranging hunters. Literature partly supports this hypothesis as jumping spiders have higher metabolic rates and better aerobic capabilities than spiders with a less developed tracheal system, e.g., wolf spiders (Prestwich 1983a, b; Schmitz 2004, 2005). However, also spiders breathing mainly with lungs may have increased metabolic rates compared to the average metabolism in spiders, e.g., araneid spiders that develop over only 1 year. Free ranging hunters such as wolf spiders and jumping spiders with different respiratory configurations may have similar metabolic rates, which are greater than in web hunters. Moreover, spiders possess haemocyanin, which is most effective with lung breathing. Highly effective tracheae serving increased metabolic rates would need a modified use of haemocyanin.

Comparison among tracheated spiders revealed a higher O_2 affinity and lower haemocyanin concentration in spiders with a well-developed tracheal system (Schmitz, unpublished). This point needs further investigation but indicates that haemocyanin is used for O_2 storage in tracheated spiders, even if it is better suited for O_2 transport and release at the tissues in non-tracheated spiders.

4. The fourth theory is related to the hydraulic type of leg extension in spiders (see Kropf 2013). This process is accomplished by high pressure in the prosoma generated by muscle contraction and supported by the separation of the prosoma from the opisthosoma. Therefore, breathing with book lungs alone will cause an O_2 lack in the prosoma, but tracheae reaching into the prosoma would solve that problem. As the tracheae are situated in the prosoma and in some species also in the legs, the O_2 support for the muscles is guaranteed. This hypothesis has been questioned because most spiders run in short spurts, which allows haemolymph exchange between pro- and opisthosoma during resting phases. Therefore, separation of those two body compartments is not complete. Moreover, not all walking or running spiders possess tracheae in the prosoma (e.g., wolf spiders) and vice versa not all spiders with prosomal tracheae are free ranging runners (e.g., crab spiders, Thomisidae).

3.4 Having a Double Respiratory System: Advantage or Disadvantage?

Resting metabolic rates of spiders are 50–80 % of that expected in poikilotherms (Anderson 1970, 1996; Greenstone and Bennett 1980; Anderson and Prestwich 1982) according to Hemmingsen's equation $V_{O_2} = 0.82\,M^{0.75}$ (V_{O_2} in µl h^{-1} and M in mg) (Hemmingsen 1960). This is because spiders are ‚sit-and-wait' predators that have irregular access to prey and therefore a great ability to starvation accompanied with a reduction of metabolism (Anderson 1974), low energy needs in prey capture because of venom use and a high anaerobic capacity (Prestwich 1983a, b). In general, resting metabolic rates correlate with life style. Spiders that live longer or use webs for prey capture have lower metabolic rates than prey-stalking spiders or species that complete their life cycle within 1 year; the latter groups include araneid or theridiid spiders that have similar or higher metabolic rates than many poikilotherms (Anderson and Prestwich 1980; Anderson 1994). Bimodal breathers, e.g., jumping spiders, have higher resting rates than pure lung breathers, e.g., mygalomorph spiders (Anderson 1970). In spiders that lack tracheae or have poorly developed tracheae, e.g., mygalomorph and wolf spiders, resting rate is proportional to the respiratory surface area (Anderson 1970; Anderson and Prestwich 1982; Prestwich 1983a, b).

During food deprivation, metabolism is low but aerobic. During low activity, e.g. web building or egg production, the aerobic–anaerobic partition depends on ATP needs, species-specific respiratory and muscle capacities. During short phases of high activity, anaerobic metabolism predominates. D-Lactate is the major anaerobic by-product, and the legs and prosoma are the main site of lactate accumulation (Prestwich 1983a, b). After anaerobic activity, O_2 debt is repaid during recovery. Most spiders are completely exhausted after 1–2 min of maximum activity, e.g. after being chased. The length of recovery depends on the duration of anaerobiosis and body mass. Complete lactate removal requires 30–45 min in small spiders and several hours in large species, e.g., mygalomorph spiders. Even free hunting species (such as wolf spiders or jumping spiders) use a sit-and-wait strategy and are dependent on anaerobic capacities for running in short spurts or jumping after slowly sneaking up on the prey. Such behaviour does not require prolonged high metabolic rates. In *Cupiennius salei*, mitochondria comprise only 0.1 % of leg muscle mass (Linzen and Gallowitz 1975) compared to 10 % in mammalian locomotive muscle. Therefore, anaerobic metabolism is the standard strategy of a spider and this seems to be not generally changed when animals developed tracheae.

Therefore, one of the most interesting questions regarding the respiratory biology of spiders is why these animals reduced the existing and well functioning lungs and developed tracheae. This evokes a conflict in gas exchange as lungs are designed to work together with the haemolymph and the haemocyanin herein and tracheae are most effective when used for terminal diffusion. In four- and two-lunged species, peak O_2 uptake occurs at submaximal activities that require lower haemolymph pressures and permit continuous circulation and O_2 exchange (Prestwich 1983a, b). Experiments that measured maximum O_2 uptake of spiders running on a treadmill show an aerobic scope of 3–10 in most species, but it can reach 17.8 times (McQueen 1980; Culik and McQueen 1985). Running increases the aerobic metabolic rate to its maximum and ATP needs are complemented by anaerobic pathways. At high or exhaustive speeds this leads to a considerable O_2 debt. This suggests that tracheae may have evolved in conjunction with higher aerobic needs (Prestwich 1988; Schmitz 2005). As already stated, the anaerobic partition in running not only depends on the velocity and the equipment with respiratory organs but also depends on the aerobic capabilities of the muscle tissue. For jumping spiders it was shown that tracheae support aerobic metabolism at highly intense activity (Prestwich 1988; Schmitz 2005).

Spiders have an open circulatory system without capillaries in which at least in the legs gas exchange has to take place along the open portion of the circulatory system. In four-lunged spiders, anterior and posterior circulation is separated, thus that haemolymph from the prosoma passes only through the anterior lung pair and haemolymph from the opisthosoma passes through the posterior lungs (Paul et al. 1989; Wirkner and Huckstorf 2013). Moreover prosomal perfusion is interrupted during fast locomotion, most likely caused by a muscular valve at the anterior end of the pedicel (Paul and Bihlmayer 1995). It was also measured in *Aphonopelma hentzi* that in resting animals the arterial O_2 pressure (P_{aO_2}) is 3.7 kPa, stays constant during walking, increases during the recovery phase and is maximum at the end of this phase (about 9.8 kPa) (Angersbach 1978). The crucial variable for O_2 transport in the haemolymph is the arterious-venous P_{CO_2} difference (P_{avO_2}). During rest, spiracles are nearly closed and the P_{avO_2} is small. During recovery after an exhaustive run, spiracles are open and the P_{avO_2} increases because of an increase in P_{aO_2}. Together with an increase in heart rate, this results in a more intensive use of haemocyanin in respiration. In araneomorph spiders the correlation of heart rates and equipment with tracheae was tested. Spiders with prosomal tracheae have significantly lower maximum heart rates than spiders with tracheae limited to the opisthosoma. Return to normal heart rates in recovery phases after running are faster in spiders with prosomal tracheae (Bromhall 1987a, b). Other authors correlated the heart rate with life style and not with tracheal supply to the prosoma. In this study resting heart rates of spiders were found to be primarily a function of body size and can be used as a measure for metabolism. The less active an animal is, the lower is the resting heart rate, thus reflecting fundamental differences in foraging strategies among spiders. For example spitting spiders (Scytodidae) and brown spiders (Sicariidae) (both with low activity ranges) do not have lower metabolic rates but lower heart rates compared with salticid spiders which are more active for hunting prey (Carrel and Heathcote 1976; Greenstone and Bennett 1980; Carrel 1987). Sometimes both types of interpretation overlap as it is the case in salticid spiders that are more active, have prosomal tracheae and have a higher heart rate than other species.

3.5 Conclusions

Tracheae enable spiders to become more flexible in their respiratory behaviour. They can breathe more adapted to their individual needs even if this differs among all species. Spiders can use tracheae as a general increase in respiratory surface, but as most of them have lost their second lung pair this has not to be the main reason for developing tracheae. More convincing is the idea of using tracheae for local oxygen demands. Every family, or even species, can use the tracheae in an individual way. Some use them for delivering O_2 to their nervous system, e.g., jumping spiders, others to deliver O_2 to their legs, e.g., Uloboridae. So a tracheal system is a very flexible structure and so it is used in spiders.

References

Anderson JF (1970) Metabolic rates in spiders. Comp Biochem Physiol 33:51–72

Anderson JF (1974) Responses to starvation in the spiders *Lycosa lenta* (Hentz) und *Filistata hibernalis* (Hentz). Ecology 55:576–585

Anderson JF (1994) Comparative energetics of comb-footed spiders (Araneae: Theridiidae). Comp Biochem Physiol A 109:181–189

Anderson JF (1996) Metabolic rates of resting salticid and thomisid spiders. J Arachnol 24:129–134

Anderson JF, Prestwich KN (1975) The fluid pressure pumps of spiders (Chelicerata, Araneae). Z Morph Tiere 81:257–277

Anderson JF, Prestwich KN (1980) Scaling of subunit structures in book lungs of spiders (Araneae). J Morphol 165:167–174

Anderson JF, Prestwich KN (1982) Respiratory gas exchange in spiders. Physiol Zool 55:72–90

Angersbach D (1978) Oxygen transport in the blood of the tarantula *Eurypelma californicum*: pO2 and pH during rest, activity and recovery. J Comp Physiol 123:113–125

Blest AD (1976) The tracheal arrangement and the classification of linyphiid spiders. J Zool (Lond) 180:185–194

Braun F (1931) Beiträge zur Biologie und Atmungsphysiologie der *Argyroneta aquatica* Cl. Zool Jb Syst 62:175–262

Bromhall C (1987a) Spider heart-rates and locomotion. J Comp Physiol B 157:451–460

Bromhall C (1987b) Spider tracheal systems. Tissue Cell 19:793–807

Burmester T (2013) Evolution and adaptation of hemocyanin within spiders. In: Nentwig W (ed) Spider ecophysiology. Springer, Heidelberg (this volume)

Carrel JE (1987) Heart rate and physiological ecology. In: Nentwig W (ed) Ecophysiology of spiders. Springer, Berlin

Carrel JE, Heathcote RD (1976) Heart rate in spiders: influence of body size and foraging strategies. Science 193:148–150

Culik BM, McQueen DJ (1985) Monitoring respiration and activity in the spider *Geolycosa domifex* (Hancock) using time-lapse televison and CO_2-analysis. Can J Zool 63:843–846

Dunlop JA, Anderson LI, Kerp H, Hass H (2003) Preserved organs of Devonian harvestmen. Nature 425:916

Ellis CH (1944) The mechanism of extension in the legs of spiders. Biol Bull 86:41–50

Forster RR (1980) Evolution of the tarsal organ, the respiratory system and the female genitalia in spiders. In: Gruber J (ed) International Congress of Arachnology, vol 8, Vienna, pp 269–284

Greenstone MH, Bennett AF (1980) Foraging strategy and metabolic rates in spiders. Ecology 61:1255–1259

Hemmingsen AM (1960) Energy metabolism as related to body size and respiratory surfaces, and its evolution. Rep Steno Mem Hosp 9:1–110

Kästner A (1929) Bau und Funktion der Fächertracheen einiger Spinnen. Z Morphol Tiere 13:463–558

Kropf C (2013) Hydraulic system of locomotion. In: Nentwig W (ed) Spider ecophysiology. Springer, Heidelberg (this volume)

Lamy E (1902) Les trachées des araignées. Ann Sci Natur Zool 15:149–280

Levi HW (1967) Adaptations of respiratory systems of spiders. Evolution 21:571–583

Levi HW (1976) On the evolution of tracheae in Arachnids. Bull Br Arachnol Soc 3:187–188

Linzen B, Gallowitz P (1975) Enzyme activity patterns in muscles of the lycosid spider, *Cupiennius salei*. J Comp Physiol 96:101–109

McQueen DJ (1980) Active respiration rates for the burrowing wolf spider *Geolycosa domifex* (Hancock). Can J Zool 58:1066–1074

Millidge AF (1986) A revision of the tracheal structures of the Linyphiidae (Araneae). Bull Br Arachnol Soc 7:57–61

Millot J (1949) Ordre des Aranéides (Araneae), systeme respiratoire. In: Grassé P (ed) Traité de zoologie, vol VI., pp 637–646

Moore SJ (1976) Some spider organs as seen by the scanning electron microscope, with special reference to the book-lung. Bull Br Arachnol Soc 3:177–187

Opell BD (1979) Revision of the genera and tropical american species of the spider family Uloboridae. Bull Mus Comp Zool 148:443–549

Opell BD (1987) The influence of web monitoring tactics on the tracheal systems of spiders in the family Uloboridae (Arachnida, Araneida). Zoomorphology 107:255–259

Opell BD (1989) Centers of mass and weight distribution in spiders of the family Uloboridae. J Morphol 202:351–359

Opell BD (1990) The relationships of book lung and tracheal systems in the spider family Uloboridae. J Morphol 206:211–216

Opell BD (1998) The respiratory complementary of spider book lung and tracheal systems. J Morphol 236:57–64
Opell BD, Konur DC (1992) Influence of web-monitoring tactics on the density of mitochondria in leg muscles of the spider family Uloboridae. J Morphol 213:341–347
Paul R, Bihlmayer S (1995) Circulatory physiology of a tarantula (*Eurypelma californicum*). Zool-Anal Complex Syst 98:69–81
Paul R, Fincke T, Linzen B (1987) Respiration in the tarantula *Eurypelma californicum*: evidence for diffusion lungs. J Comp Physiol B 157:209–217
Paul R, Tiling K, Focke P, Linzen B (1989) Heart and circulatory functions in a spider (*Eurypelma californicum*): the effects of hydraulic force generation. J Comp Physiol B 158:673–687
Prestwich KN (1983a) Anaerobic metabolism in spiders. Physiol Zool 56:112–121
Prestwich KN (1983b) The roles of aerobic and anaerobic metabolism in active spiders. Physiol Zool 56:122–132
Prestwich KN (1988) The constraints on maximal activity in spiders. I. Evidence against the fluid insufficiency hypothesis. J Comp Physiol 158:437–447
Purcell F (1895) Note on the development of the lungs, entapophyses, tracheae and genital ducts in spiders. Zool Anz 486:1–5
Purcell WF (1909) Development and origin of the respiratory organs in Araneae. Q J Microsc Sci 54:1–110
Purcell WF (1910) The phylogeny of tracheae in Araneae. Q J Microsc Sci 54:519–563
Ramirez MJ (2000) Respiratory system morphology and the phylogeny of haplogyne spiders (Araneae, Araneomorphae). J Arachnol 28:149–157
Reisinger PWM, Focke P, Linzen B (1990) Lung morphology of the tarantula, *Eurypelma californicum*, Ausserer, 1871 (Araneae: Theraphosidae). Bull Br Arachnol Soc 8:165–170
Reisinger PWM, Tutter I, Welsch U (1991) Fine structure of the gills of the horseshoe crabs *Limulus polyphemus* and tachypleus tridentatus and of the book lungs of the spider *Eurypelma californicum*. Zool Jb Anat 121:331–357
Schmitz A (2004) Metabolic rates during rest and activity in differently tracheated spiders (Arachnida, Araneae): *Pardosa lugubris* (Lycosidae) and *Marpissa muscosa* (Salticidae). J Comp Physiol B 174:519–526
Schmitz A (2005) Spiders on a treadmill: influence of running activity on metabolic rates in *Pardosa lugubris* (Araneae, Lycosidae) and *Marpissa muscosa* (Araneae, Salticidae). J Exp Biol 208:1401–1411
Schmitz A, Perry SF (2000) Respiratory system of arachnids I: Morphology of the respiratory system of *Salticus scenicus* and *Euophrys lanigera* (Arachnida, Araneae, Salticidae). Arthr Struct Develop 29:3–12
Schmitz A, Perry SF (2001) Bimodal breathing in jumping spiders: morphometric partitioning of lungs and tracheae in *Salticus scenicus* (Arachnida, Araneae, Salticidae). J Exp Biol 204:4321–4334
Schmitz A, Perry SF (2002) Respiratory organs in wolf spiders: morphometric analysis of lungs and tracheae in *Pardosa lugubris* (L.) (Arachnida, Araneae, Lycosidae). Arthr Struct Dev 31:217–230
Scholtz G, Kamenz C (2006) The book lungs of Scorpiones and Tetrapulmonata (Chelicerata, Arachnida): evidence for homology and a single terrestrialisation event of a common arachnid ancestor. Zoology 109:2–13
Selden PA (1996) Fossil mesothele spiders. Science 379:498–499
Strazny F, Perry SF (1984) Morphometric diffusing capacity and functional anatomy of the book lungs in the spider *Tegenaria* spp. (Agelenidae). J Morphol 182:339–354
Weygoldt P (1998) Evolution and systematics of the Chelicerata. Exp Appl Acarol 22:63–79
Weygoldt P, Paulus HF (1979) Untersuchungen zur Morphologie, Taxonomie und Phylogenie der Chelicerata. II. Cladogramme und die Entfaltung der Chelicerata. Z Zool Syst Evol-Forsch 17:177–200
Wirkner C, Huckstorf K (2013) The circulatory system of spiders. In: Nentwig W (ed) Spider ecophysiology. Springer, Heidelberg (this volume)

Part II
Locomotion and Dispersal

Spiders move with a combined muscles/hydraulic pressure system. Some leg segments only possess flexor muscles, and the hydraulic hemolymph pressure, fine-tuned by a system of valves, provides the necessary back-pressure. This reduces the muscle system in major parts of the long leg tubes of a spider and allows at the same time larger flexors, so that spiders in general are more powerful than comparable insects. Moreover, many spiders modified the third leg claw to scopulae with exciting ultrastructural features which enable them to stick to smooth surfaces and even to walk, upside down on a glass panel. Since many spiders, at least in specific developmental stages, can also balloon, a process which would perhaps better be called "flying with the wind," they are an extremely agile group, in space and time, able to reach all spots of their habitat and beyond.

Chapter 4
Hydraulic System of Locomotion

Christian Kropf

4.1 Introduction

Spiders can be seen from a functional viewpoint as semi-hydraulic machines (Manton 1958). Just to name a few functions of a spider's hydraulic system, walking, running, climbing and jumping, grasping of prey, moulting, leg-autotomy, extrusion of silk or expanding the male palpal organ are performed under locally increased pressure of body fluid (hemolymph). This is made possible by the open circulatory system of spiders where hemolymph leaves the arteries and flows back to the lungs between the internal organs (Foelix 2011, with references therein). Increased hemolymph pressure is generated by activity of the heart in the opisthosoma, and by certain muscles, situated both in the prosoma and in the opisthosoma. While the heart, being able to pump hemolymph both towards anterior and posterior (Paul et al. 1989), is mainly responsible for the "normal" circulation of hemolymph (see Wirkner and Huckstorf 2013), prosomal and opisthosomal muscles generate locally increased hemolymph pressure enabling the spider to perform various hydraulic movements in different body parts. However, we are far from understanding which muscles in which body parts exactly produce this pressure and how these mechanisms work in detail.

Spider locomotion is of special interest in this context. It has been known for a long time that spiders lack extensor muscles in the main joints of their walking legs, i.e. the femur–patella and tibia–metatarsus joint. Contrary to assumptions of earlier authors who considered elasticity of joint membranes or a local rise of hemolymph pressure inside the legs as responsible for leg joint extension, it is now generally acknowledged that extension of these leg joints is based mainly on an increase of prosomal hemolymph pressure (Shultz 1989, with references therein). Hydraulic

---

C. Kropf (✉)
Department of Invertebrates, Natural History Museum, Bern, Switzerland

Institute of Ecology and Evolution, University of Bern, Bern, Switzerland
e-mail: christian.kropf@iee.unibe.ch

W. Nentwig (ed.), *Spider Ecophysiology*,
DOI 10.1007/978-3-642-33989-9_4, © Springer-Verlag Berlin Heidelberg 2013

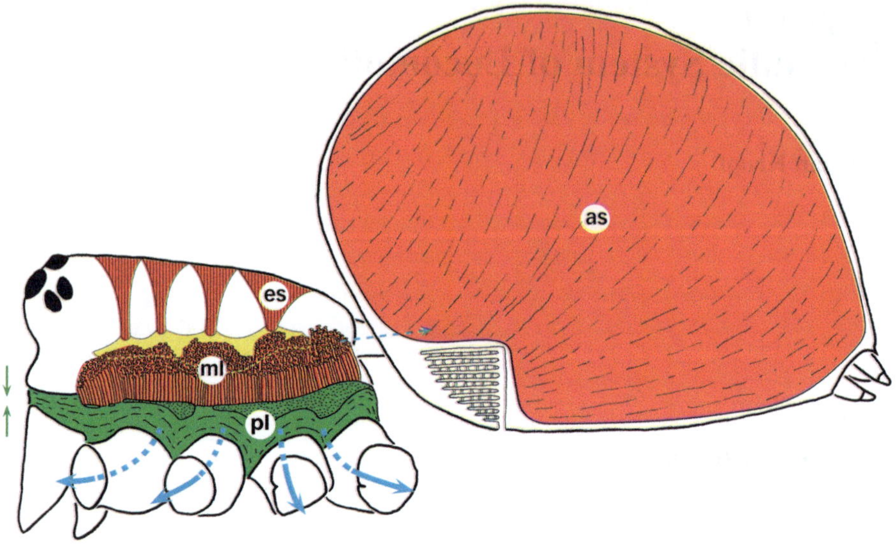

Fig. 4.1 Simplified schematic illustration of the putative main elements of a spider's hemolymph pressure producing system. It is not yet clear, if the musculi laterales, or the endosternal suspensor muscles, or both generate hemolymph pressure. Muscles in *red*, endosternite in *yellow*, pleura in *green*, hemolymph flow during activity in *blue*. Note that only a minor hemolymph flow from prosoma to opisthosoma occurs during activity. *as* abdominal sac, *es* fourth dorsolateral endosternal suspensor muscle, *ml* musculi laterales, *pl* pleura

leg extension saves space in the cuticular leg tubes (as no extensor muscles are required) and enables slender leg segments that are almost exclusively filled with flexor muscles, which in turn may be adaptive with respect to prey capture (Anderson and Prestwich 1975; Rovner 1980).

A dorsoventral compression of the prosoma is made possible by certain dorsoventrally running muscles and by soft and flexible lateral pleurae (Fig. 4.1) as shown in most spiders. Prosomal compression at the onset of activity (Wilson 1970) leads to a volume decrease in the prosoma, increase of hemolymph pressure, hemolymph flow into the legs, and to extension of leg joints (Anderson and Prestwich 1975; Blickhan and Barth 1985; Paul et al. 1989). In this review, I will discuss functional morphology and physiology of the main structures involved in hydraulic leg movements and present new data on spiders that probably use a different hydraulic system for generating hemolymph pressure.

4.2 Muscles Generating Hemolymph Pressure in the Prosoma

It is not yet clear what muscles in the spider's prosoma actually generate hemolymph pressure. Wilson (1970) attributed the function of hemolymph pressure production to the so-called "musculi laterales" (Brown 1939; "carapace

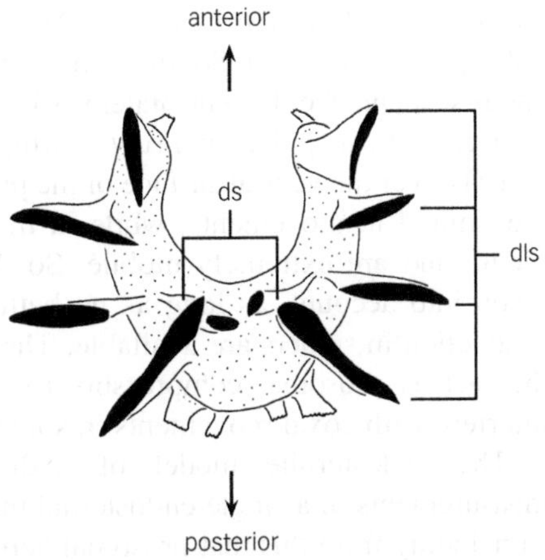

Fig. 4.2 Endosternite of *Steatoda bipunctata* (Theridiidae) in dorsal view. Areas of muscle attachments on the carapace indicated in *black*. *ds* dorsal endosternal suspensor muscles, *dls* dorsolateral endosternal suspensor muscles

compressor" of Whitehead and Rempel 1959). These muscles are running from the dorsolateral region of the carapace to the pleura where they insert at sclerites in the pleural membrane (Fig. 4.1) or possibly also on cuticular folds of the lateral borders of the carapace (Brown 1939; Whitehead and Rempel 1959 and others).

Most subsequent authors followed Wilson's suggestion (e.g., Anderson and Prestwich 1975; Palmgren 1978; Prestwich 1988a; Paul et al. 1989, 1994; Paul and Bihlmayer 1995). Later, Shultz (1991, 1993) proposed that suspensor muscles of the endosternite (a mesodermal skeletal plate in the prosoma of most arachnids; Fig. 4.1) generate most of the prosomal pressure in "hydraulic" arachnids. Shultz' hypothesis is based on detailed morphological and electromyographical investigations in a whipscorpion, *Mastigoproctus giganteus*. The endosternal suspensor muscles connect the endosternite with the carapace (dorsal and dorsolateral suspensors) or (in whipscorpions, see below) with sternum and coxal processes (ventral suspensors). It was found that during unrestrained locomotion of the whipscorpion, the third portion of the musculi laterales contracts in a step-coupled pattern while the activity of one dorsal endosternal suspensor muscle correlates with changes in the prosomal pressure baseline. Therefore, endosternal suspensor activity should provide the general level of pressure in the prosoma during activity. The function of the musculi laterales was ascribed to "controlling of coxal movements by acting as an antagonist to prosomal pressure at the prosoma–coxa joint" (Shultz 1991: 29).

Unfortunately, no such data exist for spiders. It may be problematical to transfer the conclusions from whipscorpions to spiders without restrictions, because the construction of the prosoma is considerably different in the two groups: first, only one dorsal endosternal suspensor (out of four pairs connecting endosternite and carapace) was tested (Shultz 1991, 1993), but these muscles are mostly reduced to a single and often tiny pair in spiders (Fig. 4.2) or are even totally absent (e.g., in *Harpactea*, Dysderidae, Palmgren 1978). Second, the endosternite of whipscorpions is attached to the sternum by a pair of endosternal suspensors; this is not the case in the vast majority of araneomorph spiders where ventral suspensors are totally lacking (Brown 1939; Whitehead and Rempel 1959; Firstman 1973; Palmgren 1978, 1980). So, their endosternite hangs "freely" in the prosomal cavity and is suspended only ventrolaterally by extrinsic coxal muscles. Third, the sternum is large in most spiders and tiny in whipscorpions; the leg coxae (especially coxae II–IV) cover the ventral surface of the prosoma of whipscorpions almost totally and are limited in movements, while in most spiders the coxae originate in the soft pleura and are extremely mobile. So, Wilson's (1970) argument should still be taken into account, as long as no better data on prosomal hemolymph pressure production in spiders are available: The activity of endosternal suspensor muscles that act as prosomal compressors for fluid pressure generation would strongly interfere with coxal movements in spiders and thus be dysfunctional.

The endosternite model of hydraulic pressure generation is based on measurements of a single endosternal muscle in a single species of whipscorpions. Concluding from this that prosomal hemolymph pressure is produced by all dorsal and dorsolateral endosternal suspensors (Shultz 1993) not only in whipscorpions, but also in all arachnids lacking extensor muscles in certain leg joints seems premature. As an example, in Ricinulei this cannot be true as they totally lack contractile endosternal suspensors (Firstman 1973). Finally, the possibility should be taken into consideration, that prosomal hemolymph pressure in spiders may be generated by both muscle sets and even extrinsic and intrinsic leg muscles could play a substantial role (Anderson and Prestwich 1975; Whitehead and Rempel 1959; Palmgren 1981; Shultz 1991). Because of these concerns, musculi laterales and endosternal suspensors are both pictured in Fig. 4.1.

4.3 Muscles Generating Hemolymph Pressure in the Opisthosoma

In almost all spiders a thin subepidermal muscle sheet, the so-called "abdominal sac," lies as a continuous layer beneath the epidermis of the opisthosoma (Millot 1949; Fig. 4.1). Noticeably, it is composed of smooth musculature (Vosseler 1891). In its physiological properties, this muscle sheet should resemble to the tonically active smooth muscles of vertebrates (Anderson and Prestwich 1975). Contraction of the abdominal sac leads to a decrease of opisthosomal volume and to an increase of hydrostatic pressure there (Wilson 1970; Anderson and Prestwich 1975; Paul et al. 1994). In addition, up to four pairs of dorsoventrally running muscles can occur in the opisthosoma that fully or at least partly consist of smooth muscle fibres (Millot 1936; Geiler and Beier 1971) and are probably also involved in hydraulic pressure production. Furthermore, it seems obvious that several other muscles are able to compress the opisthosoma and so may contribute to increase its fluid pressure (Robinson and Paim 1969; Wilson and Bullock 1973; Stewart and Martin 1974; Anderson and Prestwich 1975).

4.4 Hemolymph Pressure in Legs and Prosoma

In an inactive theraphosid (*Aphonopelma hentzi* sub *Eurypelma californicum*, Nentwig 2012), the hemolymph pressure in legs is slightly higher than in the prosoma and pressure in the prosoma slightly higher than in the opisthosoma, but the differences are small (Stewart and Martin 1974; Paul and Bihlmayer 1995). These differences are obviously necessary and sufficient for circulation of hemolymph and its reflow to the heart. During activity, the prosoma gets flattened, its volume decreases and the internal hemolymph pressure increases. During "normal" walking, prosomal pressures reach roughly 2–3 times the resting values (Anderson and Prestwich 1975; Paul et al. 1989; Paul and Bihlmayer 1995) and may reach peaks during maximum activity of almost 50 times the resting pressure (i.e., 64 kPa; Table 4.1; Stewart and Martin 1974). In legs, the maximum values may be even higher (Table 4.1; Blickhan and Barth 1985). It seems that many of the published values for "resting pressures" in fact represent pressures of motionless, but more or less alert spiders, probably due to experimental conditions (Table 4.1; Wilson 1970; Stewart and Martin 1974; Anderson and Prestwich 1975). Different levels of alertness could explain the variation in published "resting pressures"; the variation in activity pressures could be due to different levels of activity (Table 4.1). Other factors influencing pressures may be feeding status, water uptake, or blood loss (Paul and Bihlmayer 1995).

Parry and Brown (1959b) roughly calculated the activity pressures in the femur–patella joint and the tibia–metatarsus joint of the fourth leg of the salticid *Sitticus pubescens* before a jump (i.e., before the leg got fully stretched). The unusually high values varied between 65.3 and 144.0 kPa for the femur–patella joint and between 17.3 and 46.7 kPa for the tibia–metatarsus joint, respectively. The differences between the two joints were attributed to flexor muscle tension (Parry and Brown 1959b). Blickhan and Barth (1985) measured a pressure of 130 kPa in a leg of the ctenid *Cupiennius salei* during autotomy.

4.5 Leg Joint Extension

Hemolymph enters the legs by two different pathways. First, it flows into the leg arteries and flows back into the prosoma via venous return channels (see Wirkner and Huckstorf 2013). Two of these channels were found in the femur of *Pholcus phalangioides* (Pholcidae), a smaller one (ventro-anteriorly situated) and a larger one dorso-posteriorly (Paul et al. 1994). Second, when prosomal pressure increases, hemolymph enters the legs via channels ("lacunae") between the leg muscles. The best-studied leg joint is the hydraulic tibia–metatarsus joint, a hinge joint with a dorsal axis of rotation and a dorsoventral plane of movement (Blickhan and Barth 1985; see also Zentner 2013, Figs. 34.1 and 34.2). Due to the dorsal axis of rotation, no extensor muscles can occur and therefore hemolymph pressure and flexor

Table 4.1 Hemolymph pressure recordings in kPa (kN/m²) from legs and prosoma, taken from the literature

Leg at "rest"	Leg in activity	Prosoma at "rest"	Prosoma in activity	Species	Reference
5.3–14.7	14.7–60.0			*Tegenaria atrica* (Agelenidae)	Parry and Brown (1959a)
≤1.3	20.0–30.7	<1.3	>20.0–64.0	*Aphonopelma hentzi* (Theraphosidae)	Stewart and Martin (1974)
0.8–10.6 (mean 5.1)	10.2–61.3 (mean 26.0)			*Kukulcania hibernalis* (Filistatidae)	Anderson and Prestwich (1975)
	≤70			*Cupiennius salei* (Ctenidae)	Blickhan and Barth (1985)
1.3–4.0	26.7–63.3 (and below)			*Kukulcania hibernalis*	Prestwich (1988a)
		2.7–10.7 (mean 5.5)	18.0–38.7 (mean 28.9)	*Aphonopelma hentzi*	Paul et al. (1989)

Original values given in mm or cm Hg are converted and rounded to 0.1 kPa. Therefore and because of different accuracy of data, the values shown are more or less approximate values. The pressure values for "rest" probably do not always represent values during rest, but values of alert and motionless spiders (see text). Maximum values probably represent "startle" responses

muscle forces act antagonistically in the tibia–metatarsus joint. The diameters of the hemolymph channels (and thus their hemodynamic resistances) are strongly influenced by intrinsic muscle activity, showing the interplay of internal forces developed both by muscles and by hemolymph pressure. The tibia is almost completely filled with flexor muscles. A dorsal hemolymph channel supplies the adjacent leg segments, while several dorsoventral channels branch off laterally from the dorsal channel and run towards distal, where they empty at the tibia–metatarsus joint (Fig. 4.3). A muscular ring at the proximal end of the metatarsus can be closed and so prevents hemolymph flow from the dorsoventral channels into the dorsal channel or into the metatarsus, in this way enabling extension of the tibia–metatarsus joint independently from the hemolymph filling of the more distal segments. Muscle rings and joint membrane are able to store both hemolymph and energy (Blickhan and Barth 1985).

The joint membrane deserves special interest. By raising hemolymph pressure, the membrane gets inflated. During pressure load, axial tensions develop that would lead to high torques opposing joint extension if the membrane were isotropic (black arrows in Fig. 4.4b). The articular membrane in the tibia–metatarsus joint, however, is bellows-like folded. By the special design of this membrane, axial tensions are resolved radially in tangential stresses applying at the dorsal edge of the tibia, in this way preventing unfavourable opposite torques during joint extension (black arrows in Fig. 4.4a; Blickhan and Barth 1985).

Indicators of high hemolymph pressures are erectile leg spines as occurring in many freely hunting spiders (Parry and Brown 1959b; Rovner 1980; Weihmann et al. 2010). This erection may not occur during slower movements and needs ca. 4.6 ms to complete in jumping *C. salei* spiders. Spine erection always occurs

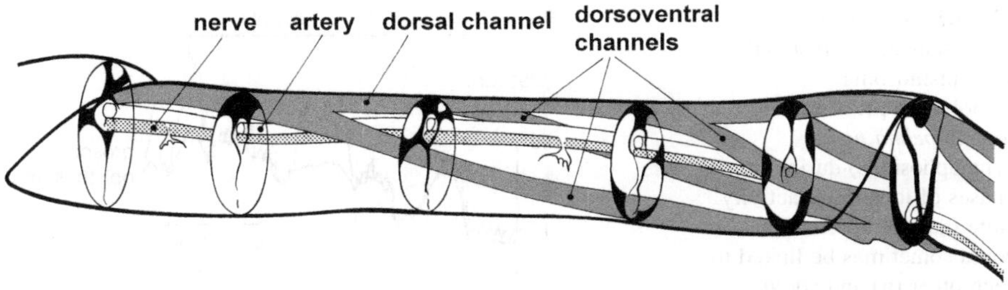

Fig. 4.3 Simplified illustration of hemolymph channels in the tibia of a spider. Cross sections with hemolymph spaces (*black*) at different locations are indicated (modified from Blickhan and Barth 1985)

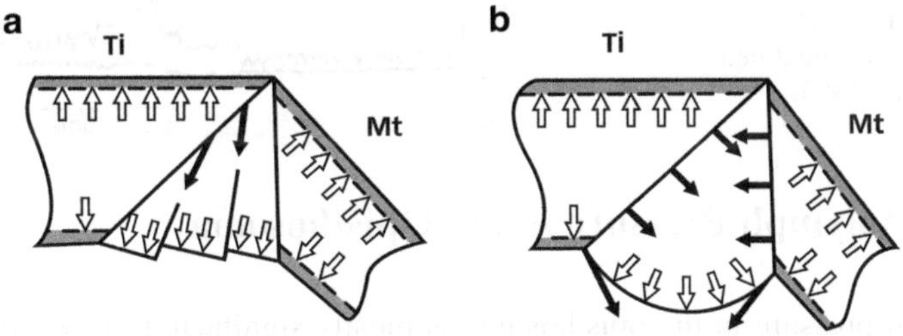

Fig. 4.4 Mechanical design of the articular membrane at the tibia–metatarsus joint. (**a**) The bellows-like folding of the membrane resolves axial tensions, developed under pressure load, in tangential stresses (see text). (**b**) A hypothetical isotropic membrane would lead to high torques preventing joint extension (modified from Blickhan and Barth 1985)

simultaneously in all segments of all legs (evidence for a simultaneous increase in hemolymph pressure in all legs) and is coupled with the first visible movements. However, spine erection may take place between 10 ms before and 14 ms after the first visible movement (Weihmann et al. 2010).

Hemolymph pressure within the tibia also serves other functions. For example, it compensates (by absorbing pressure loads) for potential weaknesses of the thin-walled tibial cuticle that could otherwise easily fail by buckling. Torques sufficient for buckling of the tibia in the region of the tibia–metatarsus joint can be compensated by a hemolymph pressure of ca. 20 kPa that lies well within the physiological range of hemolymph pressure inside the leg (Table 4.1). Hemolymph pressure probably also participates in developing torques in other joints than the femur–patella and tibia–metatarsus joint (Blickhan and Barth 1985).

Recently, Weihmann et al. (2010, 2012) proposed that in large spiders (with a body weight of >3 g) at least propulsion of the hind leg is generated by both hydraulics and activity of proximal flexors in the three basal leg joints (body–coxa, coxa–trochanter and trochanter–femur joint, respectively). By pushing the leg against the substrate, flexor muscle contribution to torques in any distal joint (including the two hydraulic joints) was shown to be dominant over hydraulic pressure contribution in hind leg propulsion (Weihmann et al. 2012).

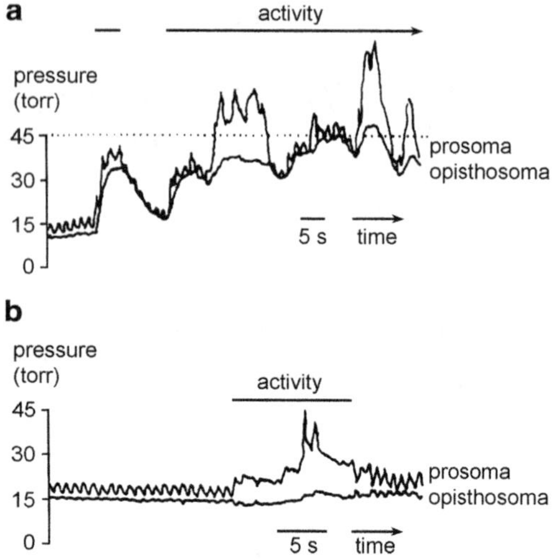

Fig. 4.5 Simultaneous measurement of prosomal and opisthosomal hemolymph pressures of *Aphonopelma hentzi* (Theraphosidae) during phases of locomotor activity. Note that both pressures could sometimes be linked to each other (**a**) and could sometimes not be linked (**b**). Pressure oscillations caused by the heartbeat are superimposed, which is seen most clearly on the prosomal traces during rest and moderate activity (modified from Paul et al. 1995)

4.6 Hemolymph Pressure in the Opisthosoma

Hemolymph pressure in the opisthosoma is mostly significantly lower than in the prosoma during high activity, while during moderate activity or rest the pro- and opisthosomal values may come very close (Fig. 4.5; Wilson and Bullock 1973; Stewart and Martin 1974; Anderson and Prestwich 1975; Paul et al. 1989, 1994; Paul and Bihlmayer 1995). The circulation pathways in the opisthosoma are complex and depend on the activity level: During rest and recovery, two separate hemolymph circulations exist, at least in theraphosids, see Wirkner and Huckstorf 2013): (1) an anterior circulation loop in the hemolymph spaces of the pedicel and the most anterior part of the opisthosoma transporting hemolymph from the prosoma to the anterior book lungs and further to the heart and (2) a posterior one more behind, transporting hemolymph from the opisthosoma to the heart via the posterior lungs (Paul et al. 1989). During activity, hemolymph from the opisthosoma returns to the heart via both the anterior and posterior lung veins (Paul et al. 1989, 1994; Paul and Bihlmayer 1995). There are no studies about spiders with only one pair of lungs or with tracheae only, in this respect. The high activity values (Table 4.2) shown in the study of Stewart and Martin (1974) must be seen as exceptional at the moment.

4.7 Hydraulic Interaction Between Prosoma and Opisthosoma

The narrow pedicel (representing the first opisthosomal segment) connects the two main body parts of a spider. It contains various muscles, the gut, nerve cord and aorta passing through but does not show any visible valves that could separate

Table 4.2 Hemolymph pressure recordings in kPa (kN/m^2) from the anterior and posterior part of the opisthosoma (see text), taken from the literature

Anterior at "rest"	Anterior in activity	Posterior at "rest"	Posterior in activity	Species	Reference
3.3	22.7–51.3			*Aphonopelma hentzi*	Stewart and Martin (1974)
		1.0–1.9	2.3–4.0	*Kukulcania hibernalis*	Anderson and Prestwich (1975)
	≤10.7	2.7	5.3–8.0	*Aphonopelma hentzi*	Paul et al. (1989)

Original values given in mm Hg are converted and rounded to 0.1 kPa. The pressure values for "rest" probably do not always represent values during rest, but values of alert and motionless spiders (see text). The data from Wilson (1962) are excluded here because of probable methodological shortcomings (Anderson and Prestwich 1975)

hydrostatically the prosoma from the opisthosoma. Therefore, Wilson (1965) concluded that no such separation exists. In addition, Wilson and Bullock (1973) calculated hemolymph volume decrease in the prosoma during high activity and, associated with this, hemolymph volume increase in the opisthosoma, from recording vertical linear movements of the spider's (*Amaurobius ferox*, Amaurobiidae) cuticle. Opisthosomal volume increase during activity could not be shown when the pedicel was ligatured. Based on these and other findings (Anderson and Prestwich 1975), it was concluded that the locomotory exhaustion shown by many spiders after maximum activity should be due to loss of hemolymph from the prosoma into the opisthosoma and to lack of oxygen supply for the prosoma through interruption of the normal hemolymph flow. Hemolymph backflow into the prosoma during recovery should be rather slow. The opisthosomal pressure-generating muscles should attenuate the "hemolymph-loss" into the opisthosoma during activity and enhance recovery from the resulting imbalance of body fluid (Wilson and Bullock 1973; Anderson and Prestwich 1975).

However, the "fluid insufficiency hypothesis" did not withstand later results of various authors. In active *Aphonopelma*, only a marginal hemolymph flow (<0.1 ml, about 3–3.5 % of the total hemolymph volume) into the opisthosoma at the beginning of activity (that was possibly recorded by Wilson and Bullock 1973) can be observed (Fig. 4.1) that quickly returns into the prosoma after activity. A high prosomal-to-opisthosomal pressure difference can be maintained for the whole activity period with no further shifting of hemolymph into the opisthosoma and hemocytes in the anterior ventral part of the opisthosoma stop flowing during activity (Paul et al. 1989, 1994; Paul and Bihlmayer 1995). Prestwich (1988a, b) presented evidence that the locomotory collapse is not related to unbalanced body fluid but to accumulation of anaerobic metabolic by-products and depletion of energy-rich phosphagen sources. Taken together, these data strongly contradict the fluid insufficiency hypothesis. Hemolymph flow between pro- and opisthosoma also depends on the frequency of carapace depression: In relation to hemolymph flow, the pedicel and the structures behind it (e.g., the book lungs) behave like a low-pass filter: Sinusoidal carapace compression at high frequencies (>8 Hz) resulted in <10 % of hemolymph flow (back and forth) as compared to low frequency (i.e. 0.3 Hz) compression (Paul et al. 1989). Finally, 15 araneomorph species of 11 different families were able to maintain forced running for at least 90 s and voluntary running for more than 15 min, without showing the expected locomotory exhaustion (Bromhall 1987).

At the onset of activity, a small prosomal volume increase in *Aphonopelma* can be observed, probably caused by opisthosomal muscle contraction (Paul et al. 1989). It is followed by a strong prosomal volume decrease caused by prosomal muscle contraction. In this phase, anteriorly directed hemolymph flow generated by the heart has to cope with increasing prosomal pressure. Therefore, the perfusion of the prosoma of *Aphonopelma* gets reduced during activity and stops between 6.7 and 9.3 kPa, when prosomal pressure exceeds systolic heart pressure (Paul et al. 1989); however, heartbeat is still maintained, causing posteriorly directed perfusion (Fig. 4.5; Paul et al. 1989, 1994), as is also the case in the above-mentioned araneomorph spiders (Bromhall 1987; see also Wirkner and Huckstorf 2013).

Opisthosomal pressures often follow prosomal ones during moderate activity, but the two may also be unlinked or opisthosomal pressure increase can even be absent (Fig. 4.5). These and other results obtained from injecting casting resin into hemolymph spaces indicate an occlusion mechanism at the anterior end of the pedicel, probably consisting of muscles having a valve function (Paul et al. 1994; Paul and Bihlmayer 1995). These muscles, however, have not yet been identified

4.8 Spiders with an Opisthosomal Hemolymph Pressure Pump

The data presented above stem mainly from a handful of species. It seems probable that the demands on the pressure-generating system of spiders vary tremendously between different spider taxa, depending on their lifestyle, ecological demands, favoured prey capture strategy, anatomical constraints, etc. An example for such an anatomically constrained taxon is given here and concerns spiders with a strongly sclerotized body, especially with a sclerotized pleura.

In spiders of the family Tetrablemmidae (and in species of different other families), the pleura is strongly sclerotized and immovably connected to both carapace and sternum (Fig. 4.6a). In this way the prosoma forms an extremely stiff capsule. Such a capsule seems unable for hemolymph pressure generation, as one precondition for the necessary volume decrease, a soft and flexible pleura, is missing. Simple deformation tests in freshly killed specimens of the tetrablemmid *Perania nasuta* by using a forceps show, that a dorsoventral compression of the prosoma is practically impossible because of the hardiness of the cuticle. Not even minuscule leg movements can be elicited by doing so and the cuticle rather breaks than to deform even slightly. However, already a minor dorsoventral compression of the opisthosoma immediately resulted in well-defined extension of all eight legs in the femur–patella joint and the tibia–metatarsus joint.

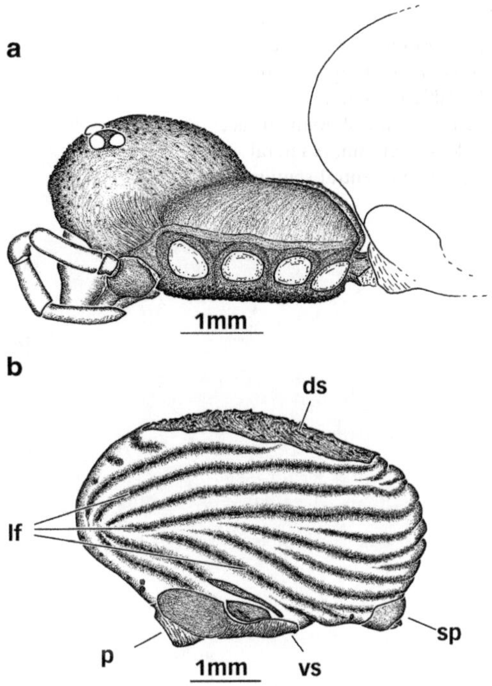

Fig. 4.6 Female of *Perania nasuta* (Tetrablemmidae) in lateral view. (**a**) Prosoma, legs and hairs omitted. Note the inflexible and strongly sclerotized pleura. (**b**) Opisthosoma. Note the cuticular folds with lateral longitudinal rows of sclerites. *ds* dorsal scutum, *lf* lateral folds with sclerites, *p* pedicel tube, *sp* spinnerets, *vs* ventral scutum

The opisthosoma shows a dorsal and a ventral scutum and several longitudinal furrows laterally (Fig. 4.6b). In these furrows, sclerites are situated that serve as muscle attachments. The smooth subepidermal muscle layer in the opisthosoma ("abdominal sac") is strongly modified in *Perania* and in another species of tetrablemmid, *Indicoblemma lannaianum*. The abdominal sac is unusually strongly developed and partitioned into powerful dorsoventrally running muscles that attach at the lateral sclerites and at the dorsal or ventral scutum, respectively (Fig. 4.7). Obviously, a contraction of these muscles leads to a volume decrease in the opisthosoma and to an increase of hemolymph pressure. I propose that in Tetrablemmidae and in other armoured spider taxa with a similar sclerotisation of the body, it is the modified abdominal sac that mainly generates the hemolymph pressure necessary for leg joint extension. Detailed work about these problems will be published elsewhere.

4.9 Conclusions

Spiders lack extensor muscles in the main joints of their walking legs (femur–patella and tibia–metatarsus joint, respectively). These leg joints are extended by a rise of hemolymph pressure in the prosoma, caused by contraction of prosomal dorsoventrally running muscles (musculi laterales or endosternal

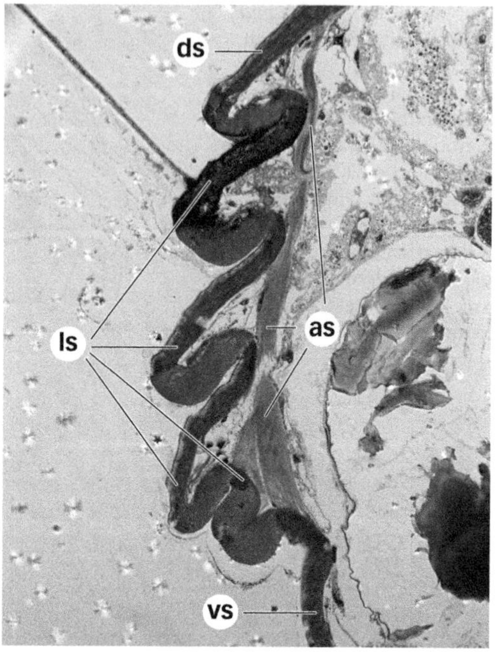

Fig. 4.7 Cross section of the opisthosoma of a female *Indicoblemma lannaianum* (Tetrablemmidae). *as* portions of the abdominal sac, *ds* dorsal scutum, *ls* lateral sclerites, *vs* ventral scutum

suspensor muscles, or both). This leads to volume decrease in the prosoma (made possible by soft and flexible pleurae) and to hemolymph flow into the legs. Inside the legs, hemolymph flows in channels ("lacunae") whose cross sections are strongly influenced by intrinsic muscle activity. In hydraulic leg joints, only flexor muscles occur that act antagonistically to hemolymph pressure. The articular membranes in leg joints are bellows-like folded, in this way preventing unfavourable torques opposing joint extension. During maximum activity, perfusion of the prosoma stops because prosomal pressure exceeds systolic heart pressure. In the opisthosoma, a smooth subepidermal muscle-sheet ("abdominal sac") and several other muscles generate hemolymph pressure that is considerably lower than prosomal pressure during high activity, but may be very close to it during rest or moderate activity. Opisthosomal pressures are sometimes linked to prosomal ones, but sometimes not. This implies (contrary to previous opinions) a temporary hydrostatic separation between pro- and opisthosoma, probably due to a muscular valve at the anterior end of the pedicel that is not yet defined anatomically. Only few spider species have been studied with respect to body hydraulics so far, and various deviations of the here described hydraulic system may occur. For example, in some spiders the prosoma forms an extremely stiff capsule that is almost impossible to deform and therefore unable to generate hemolymph pressure. It is suggested that these spiders produce hemolymph pressure for leg extension mainly in the opisthosoma by contraction of a strongly modified abdominal sac.

Acknowledgements Thanks go to P. Schwendinger (Muséum d'Histoire naturelle, Genève, Switzerland) for kindly providing living specimens of *P. nasuta*, M. Burger (Honolulu, Hawaii) for the serial sections of *I. lannaianum*, T. Sonderegger and M. Hohn (Bern, Switzerland) for the graphics design.

References

Anderson JF, Prestwich KN (1975) The fluid pressure pumps of spiders (Chelicerata, Araneae). Z Morph Tiere 81:257–277

Blickhan R, Barth FG (1985) Strains in the exoskeleton of spiders. J Comp Physiol A 157:115–147

Bromhall C (1987) Spider heart-rates and locomotion. J Comp Physiol B 157:451–460

Brown RB (1939) The musculature of *Agelena naevia*. J Morphol 64:115–166

Firstman B (1973) The relationship of the chelicerate arterial system to the evolution of the endosternite. J Arachnol 1:1–54

Foelix RF (2011) Biology of spiders. Oxford University Press, New York

Geiler H, Beier R (1971) Nachweis glatter Muskulatur im Opisthosoma von *Araneus diadematus* Clerck. Zool Anz 187:434–438

Manton SM (1958) Hydrostatic pressure and leg extension in arthropods, with special reference to arachnids. Ann Mag Nat Hist 13:161–182

Millot J (1936) Métamérisation et musculature abdominale chez les Araneomorphes. Bull Soc Zool France 61:181–204

Millot J (1949) Ordre des Aranéides (Araneae). In: Grassé P (ed) Traité de Zoologie 6. Masson et Editeurs, Paris

Nentwig W (2012) The species referred to as *Eurypelma californicum* (Theraphosidae) in more than 100 publications is likely to be *Aphonopelma hentzi*. J Arachnol 40:128–130

Palmgren P (1978) On the muscular anatomy of spiders. Acta Zool Fenn 155:1–41

Palmgren P (1980) Some comments on the anatomy of spiders. Ann Zool Fenn 17:161–173

Palmgren P (1981) The mechanism of the extrinsic coxal muscles of spiders. Ann Zool Fenn 18:203–207

Parry DA, Brown RHJ (1959a) The hydraulic mechanism of the spider leg. J Exp Biol 36:423–433

Parry DA, Brown RHJ (1959b) The jumping mechanism of salticid spiders. J Exp Biol 36:654–664

Paul RJ, Bihlmayer S (1995) Circulatory physiology of a tarantula (*Eurypelma californicum*). Zoology 98:69–81

Paul RJ, Tiling K, Focke B, Linzen B (1989) Heart and circulatory functions in a spider (*Eurypelma californicum*): the effects of hydraulic force generation. J Comp Physiol B 158: 673–687

Paul RJ, Bihlmayer S, Colmorgen M, Zahler S (1994) The open circulatory system of spiders (*Eurypelma californicum*, *Pholcus phalangioides*): a survey of functional morphology and physiology. Physiol Zool 67:1360–1382

Prestwich KN (1988a) The constraints of maximal activity in spiders I. Evidence against the fluid insufficiency hypothesis. J Comp Physiol B 158:437–447

Prestwich KN (1988b) The constraints of maximal activity in spiders. II. Limitations imposed by phoshagen depletion and anaerobic metabolism. J Comp Physiol B 158:449–456

Robinson GL, Paim U (1969) Opisthosomal musculature of female *Araneus diadematus* (Araneae: Argiopidae). Can Entomol 101:337–352

Rovner J (1980) Morphological and ethological adaptations for prey capture in wolf spiders (Araneae, Lycosidae). J Arachnol 8:201–215

Shultz JW (1989) Morphology of locomotor appendages in Arachnida: evolutionary trends and phylogenetic implications. Zool J Linn Soc 97:1–56

Shultz JW (1991) Evolution of locomotion in Arachnida: the hydraulic pressure pump of the Giant Whipscorpion, *Mastigoproctus giganteus* (Uropygi). J Morphol 210:13–31

Shultz JW (1993) Muscular anatomy of the giant whipscorpion *Mastigoproctus giganteus* (Lucas) (Arachnida: Uropygi) and its evolutionary significance. Zool J Linn Soc 108:335–365

Stewart DM, Martin AW (1974) Blood pressure in the tarantula, *Dugesiella hentzi*. J Comp Physiol 88: 141–172

Vosseler J (1891) Untersuchungen über glatte und unvollkommen quergestreifte Muskeln der Arthropoden. Laupp, Tübingen

Weihmann T, Karner M, Full RJ, Blickhan R (2010) Jumping kinematics in the wandering spider *Cupiennius salei*. J Comp Physiol A 196:421–438

Weihmann T, Günther M, Blickhan R (2012) Hydraulic leg extension is not necessarily the main drive in large spiders. J Exp Biol 215:578–583

Whitehead WF, Rempel JG (1959) A study of the musculature of the Black Widow Spider, *Latrodectus mactans* (Fabr). Canad J Zool 37:831–870

Wilson RS (1962) The control of dragline spinning in the garden spider. Q J Microsc Soc 104: 557–571

Wilson RS (1965) The pedicel of the spider *Heteropoda venatoria*. J Zool 147:38–45

Wilson RS (1970) Some comments on the hydrostatic system of spiders (Chelicerata, Araneae). Z Morph Tiere 68:308–322

Wilson RS, Bullock J (1973) The hydrostatic interaction between prosoma and opisthosoma in *Amaurobius ferox* (Chelicerata, Araneae). Z Morph Tiere 74:221–230

Wirkner C, Huckstorf K (2013) The circulatory system of spiders. In: Nentwig W (ed) Spider ecophysiology. Springer, Heidelberg (this volume)

Zentner L (2013) Modelling and application of the hydraulic spider leg mechanism. In: Nentwig W (ed) Spider ecophysiology. Springer, Heidelberg (this volume)

Chapter 5
Functional Aspects of Spider Scopulae

Senta Niederegger

5.1 Introduction: Attachment Systems in Insects, Geckos and Spiders

Two types of systems for the attachment to variable surfaces evolved within arthropods. The smooth attachment system (Fig. 5.1a) consisting of soft pads located on the pretarsi can be found in many insects, e.g. arolia in Hymenoptera (bees, wasps, ants) or euplantulae in Ensifera (crickets) (Gorb 2012). Few arachnids exhibit a smooth attachment system such as Solifugae and Pseudoscorpiones with suctorial organs. The hairy attachment system (Fig. 5.1b) consists of surfaces covered with simple or complex attachment structures. The simple system can exemplary be found in pretarsi of Diptera where the attachment pads pulvilli are unilaterally covered with single setae terminating in spatulae (as in Fig. 5.1b) (Niederegger and Gorb 2003). A more complex formation with setae branching into microtrichia before terminating in spatula-shaped endings can be found in the scopulae located on tarsi and occasionally on metatarsi of many spiders and in geckos (Arzt et al. 2003). Surface structure such as setae and microtrichia on attachment devices leads to an increase of the tenacity (adhesion per unit of the measured contact area) compared to a nonstructured pad with the same area. Furthermore, structured attachment systems show higher reliability of contact on various surface profiles and an increased tolerance to individual contact defects (Peressadko and Gorb 2004).

S. Niederegger (✉)
Department of Forensic Medicine, University Hospital Jena, Jena, Germany
e-mail: senta.niederegger@med.uni-jena.de

W. Nentwig (ed.), *Spider Ecophysiology*,
DOI 10.1007/978-3-642-33989-9_5, © Springer-Verlag Berlin Heidelberg 2013

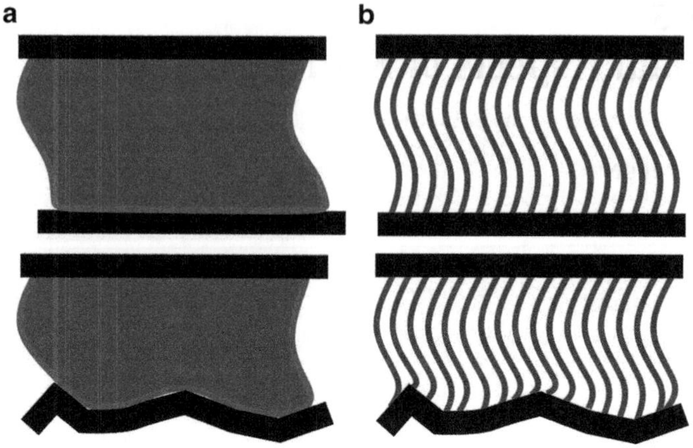

Fig. 5.1 (**a**) Smooth attachment system on smooth surface (*top*) and on rough surface (*bottom*). (**b**) Hairy attachment system on smooth surface (*top*) and on rough surface (*bottom*) (from Gorb 2012)

5.2 Scopulae in Spiders

Scopulae are located at the tip of the tarsus ventral of the two claws (Fig. 5.2), and some spiders have additional tarsal and metatarsal scopulae. They can be found in a large variety of spider families, ranging from mygalomorph spiders (Barychelidae, Ctenizidae, Nemesiidae, Theraphosidae) to haplogyne spiders (Dysderidae, Caponiidae, Orsolobidae) and to higher non-araneoid entelegyne spiders (Ammoxenidae, Anyphaenidae, Clubionidae, Ctenidae, Gnaphosidae, Liocranidae, Lycosidae, Salticidae, Selenopidae, Homalonychidae, Philodromidae, Sparassidae, Thomisidae, Zoridae) (nonexclusive listing). Scopulae enable spiders to climb smooth, steep surfaces such as large leaves, bark or stones and in an artificial environment even window panes. As a rule of thumb, web-building spiders do not have scopulae, but many wandering spiders do.

5.2.1 Morphology

Claw scopulae as well as tarsal and metatarsal scopulae (Fig. 5.3a) located on the ventral side of tarsi and metatarsi bear thousands of setae which are distally bent at their tips (Fig. 5.3b). The concave part of the bent seta is covered with fine cuticular outgrowths called microtrichia (Fig. 5.3c), terminating in small spatula-shaped endings, the spatulae (Fig. 5.3d). Tarsal and metatarsal scopulae are composed of setae with almost equal length, whereas the claw scopulae are built from setae with different lengths, resulting in a paw-like shape of the claw tuft (Fig. 5.2) (Niederegger and Gorb 2006). Scopulae can also consist of setae with a round

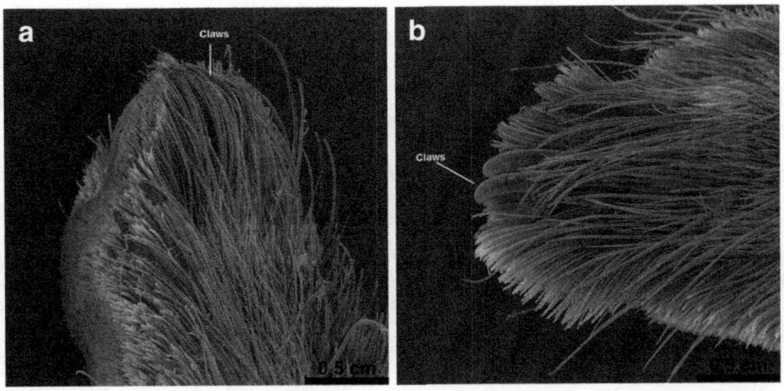

Fig. 5.2 Claw scopula of *Aphonopelma seemanni* (Theraphosidae). (**a**) Dorsal view, (**b**) lateral view

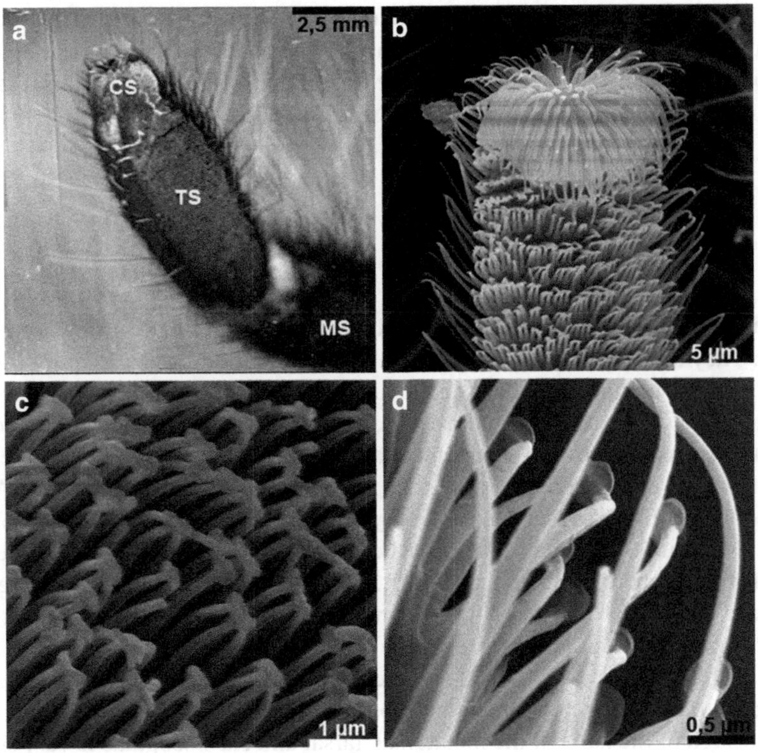

Fig. 5.3 (**a**) Tarsus of the right hind leg of *Aphonopelma seemanni* (Theraphosidae) in contact with glass surface during vertical walking. (**b**) Seta, SEM picture; (**c**) microtrichia and spatulae, SEM picture; (**d**) spatulae in contact to smooth surface, SEM picture. *CS* claw scopula, *TS* tarsal scopula, *MS* metatarsal scopula

Fig. 5.4 *Aphonopelma seemanni* (Theraphosidae). (**a**) Standing horizontally on a glass surface. (**b**) Metatarsus and tarsus of the second left leg in horizontal walking. *P* pedipalps, *MT* metatarsus, *MS* metatarsal scopula, *T* tarsus, *TS* tarsal scopula, *C* claws, *CS* claw scopula

cross section, ventrally and laterally covered with microtrichia without spatulae as found in Thomisidae. The claw scopula is a bipartite structure originating from two plates flanking the central claw lever articulated with the claws (Hill 1977). Claws and scopulae can be moved independently by means of hydraulic inflation or deflation of the tarso-pretarsal joint or by movement of a dorsal pretarsal levator and an opposing ventral pretarsal depressor (Hill 2006).

5.2.2 Theraphosidae

Theraphosidae are hairy and often very large spiders hunting on trees or on the ground. They are able to produce silk, but most do not build webs for prey capture, as their primary hunting method is ambush. As an example, *Aphonopelma seemanni* (Fig. 5.4a) possesses three scopulae: the claw scopula which is located under the claws, the tarsal scopula and the metatarsal scopula on each leg (Fig. 5.4b) except for the hind leg in which the scopulae are restricted to the claws and tarsus. As the claw scopula hairs are densely packed, only the tips of the claws are visible from the ventral side. Due to different length in the hairs, an almost even but very flexible profile for attachment is generated (Fig. 5.2). When walking on glass, however, the spider uses only the claw scopula to contact the surface (Fig. 5.4a). Even on a vertical surface the animal never uses the whole area of the leg scopulae to build contact with the surface. The tarsal and metatarsal scopulae are most likely used in hunting (Dunlop 1995) to grab and hold on to prey.

The design of the hairs of the metatarsal, tarsal and claw scopulae is very similar as described in the previous section, although the hairs of the claw scopula have a wider variation in length. The hairs in the middle of the metatarsal and tarsal scopulae are oriented at a slight angle to the leg axis, and the lateral hairs run fan-shaped to the borders of the leg (Niederegger 2003). This arrangement is probably advantageous for retaining the captured prey and not for attachment purposes in walking. In the connection of the tarsal scopula to the claw scopula, the hairs have almost no slope to the leg axis but are oriented in the direction of the walking surface. The spatulae of *Aphonopelma seemanni* are round (Fig. 5.3c), whereas spatulae in other spiders might be of a more triangular shape (Sect. 5.2.3).

Divided scopulae have a longitudinal band of long thick setae, whereas entire tarsal scopulae have homogeneous adhesive hairs. A possible division of the tarsal scopula in Theraphosidae is related to spider size. All theraposids have divided scopulae in juvenile states, but they become entire in adults of several species (Pérez-Miles 1994).

5.2.3 Salticidae

The mostly small jumping spiders are active hunters and usually stalk and jump their prey rather than building webs. For safety reasons, jumping spiders produce silky tether lines before each jump enabling them to leap onto their prey even on vertical or inverted surfaces (Foelix 2011). *Phidippus audax* possesses claw scopulae on each leg as well as two ventrolateral scopulae on the distal portion of the tarsus. These scopulae are found particularly on forelegs of older spiders. No metatarsal scopulae are present on the legs of jumping spiders. In the mid-ventral tract of the distal tarsus, plume setae can be found between the scopulae. Each plume seta bears a single large apical filament. The specific function of these setae is unknown, but they may facilitate thrust in jumping and rapid locomotion. It was found in *Phidippus rimator* that the scopulae are not present in the first instar but appear in the second instar. The setae associated with the pretarsus subsequently increase in size and number with every moult and the increase in size of the spider (Hill 1977). In *Evarcha arcuata,* no additional attachment structures other than claws and the very characteristic and tuft-like claw scopulae were found on any leg (Fig. 5.5a, b). The spatulae of jumping spiders have a triangular appearance (Fig. 5.5c) (Kesel et al. 2003).

5.2.4 Ctenidae

Wandering spiders are ambush hunters and attack their prey with high velocity as soon as it is close by. Silky tether lines are produced for safety, and the prey is retained by the tarsal and metatarsal scopulae (Niederegger 2003). *Cupiennius salei* (Fig. 5.6a) possesses morphologically very similar scopulae (Fig. 5.6b) as the theraphosid *Aphonopelma seemanni* presented earlier (Sect. 5.2.2). Claw scopulae as well as tarsal and metatarsal scopulae are present. The microtrichia of *C. salei* (Fig. 5.6c) have a very similar appearance as the microtrichia of *A. seemanni*; interestingly, however, the spatulae of the considerably smaller spider

Fig. 5.5 *Evarcha arcuata* (Salticidae). (**a**) Lateral view of the tarsal adhesive apparatus, showing both claws and the scopula. (**b**) Ventral view of the scopula. Scopula area = 3.2×10^4 mm^2; mean: 3.7×10^4 mm^2. (**c**) Triangular shape of spatulae. *C* claws, *CS* claw scopula (from Kesel et al. 2003)

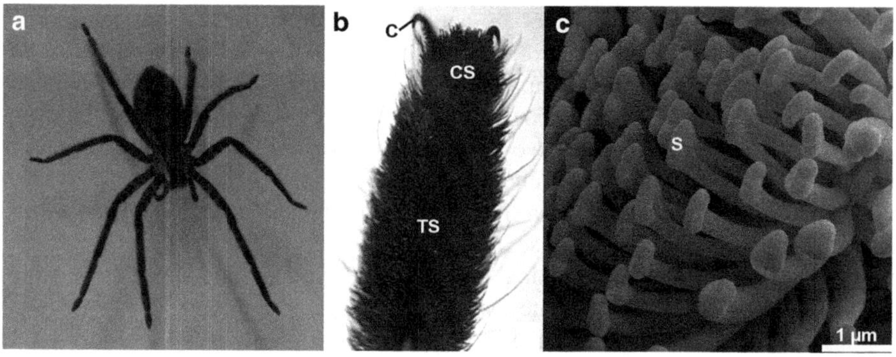

Fig. 5.6 *Cupiennius salei* (Ctenidae). (**a**) Hanging slightly vertically on a glass surface. (**b**) One middle leg in contact to glass surface. (**c**) Setae of *Cupiennius salei*. *C* claw, *CS* claw scopula, *TS* tarsal scopula *S* spatula

C. salei are about twice the size (0.30 ± 0.05 µm^2) of the spatulae of *A. seemanni* (0.13 ± 0.02 µm^2) (Niederegger and Gorb 2006). This is quantitatively explained by applying the principles of contact mechanics, according to which splitting up the contact into finer subcontacts increases adhesion. Arzt et al. (2003) found a strong inverse scaling effect in attachment devices as heavier animals exhibit finer adhesion structures.

5.3 Forces

Scopulae of spiders were reported as dry adhesive devices relying on the molecular van der Waals interactions between spatulae and substrate, similar to the hairy system of geckos (Kesel et al. 2003; Autumn et al. 2000). Additionally, the

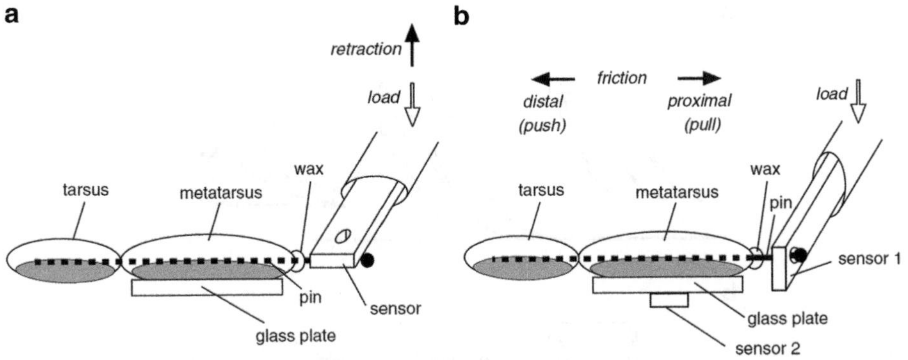

Fig. 5.7 Experimental set-up for force measurements. (**a**) Adhesion test: the force transducer (sensor) registered both load and adhesive forces while pressing onto (*white arrow*) and retracting the segments from (*black arrow*) a glass surface without sliding. (**b**) Friction test: two force transducers simultaneously registered normal (sensor 2) and friction (sensor 1) forces generated after applying a load (*white arrow*) and sliding the segments over a glass plate in both distal and proximal directions (*black arrows*) (from Niederegger and Gorb 2006)

humidity of the environment can have strong influence on the adhesion of spider scopulae (Wolff and Gorb 2012). Attachment forces should therefore be similar for fresh and dried samples. Force measurements using a micro force meter on fresh and dried samples of severed tarsal and metatarsal scopulae of *Aphonopelma seemanni* and *Cupiennius salei* revealed that no adhesion forces were generated when the scopulae were simply lowered to and lifted from a surface (Fig. 5.7a). An axial movement (Fig. 5.7b) on the surface on the other hand yielded considerable forces. Friction forces produced by pulling the spider leg proximally over the surface were much lower than forces produced in distal pushing movements of the spider leg. Measurements further showed that forces for *A. seemanni* were usually higher in tarsi than in metatarsi but for *C. salei* the forces were higher in metatarsi than in tarsi. Overall, however, both spiders generated almost equal forces even though the spatulae of the much larger *A. seemanni* were only about half as large as the spatulae of *C. salei*. Friction experiments additionally showed that desiccation of the foot leads to lower forces (Niederegger and Gorb 2006).

The distal part of the seta is ventrally covered by the spatulae bearing microtrichia and slightly bent in distal direction (see Sect. 5.2.1). When the tarsal or metatarsal scopulae are pulled towards the body (proximal shear), the setae are strongly bent in the opposite direction and contact the substrate with their spatulae-lacking dorsal part (Fig. 5.8a). No adhesive contact can be established between the setae and the substrate. When the tarsal or metatarsal scopulae are pushed away from the body (distal shear), however, the spatulae contact the substrate, and adhesion forces as well as molecular van der Waals forces can be generated (Fig. 5.8a) (Niederegger and Gorb 2006).

According to atomic force microscopy measurements in the jumping spider *Evarcha arcuata,* a single triangular spatula can produce a force of 38.12 nN.

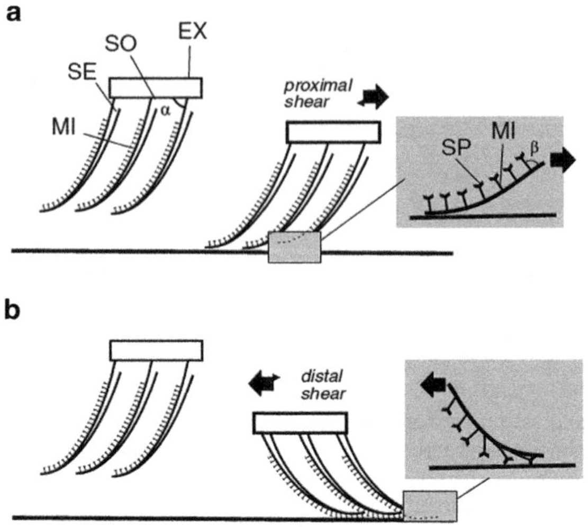

Fig. 5.8 Effect of the leg-sliding direction on the orientation of setae and spatulae in tarsal and metatarsal scopulae. (**a**) Proximal pulls result in the alignment of setae with their proximal sides to the substrate. This side does not bear microtrichia with spatula-shaped tips. In this case, the contact area is reduced because of the cylindrical shape of the setal shaft contacting smooth glass. (**b**) Distal pushes result in the orientation of setae with their distal sides to the substrate. This side bears microtrichia with spatula-shaped tips. Thin flattened spatulae provide contact formation between setae and substrate. *EX* exoskeleton, *MI* microtrichia, *SE* seta, *SO* socket, *SP* spatula, α indicates slope of setae relative to the exoskeleton surface, β indicates slope of microtrichia relative to the seta (from Niederegger and Gorb 2006)

With an estimated total number of 78.000 spatulae per claw scopula, eight spider feet can generate a maximal force of 2.97×10^{-3} N. Maximal adhesion perpendicular to the substrate would measure 2.38×10^{-2} N. With a body mass of 15 mg, corresponding to 1.47×10^{-4} N, the adhesive force is 160 times the weight of the animal (Kesel et al. 2003).

5.4 Fluid Secretion

It was postulated for a long time that the branched, hairy attachment structures found in spiders and geckos work by dry intermolecular interaction (van der Waals forces) enhanced solely by capillary forces generated by the very thin film of water naturally present on most surfaces (Foelix 2011).

Video recordings of a living *Aphonopelma seemanni* adhering to a vertical glass surface, however, suggest filament-like secretion from the spider scopulae while sliding downwards. The substance had a liquid appearance when secreted, but the filaments hung over the edge of the glass or even intertwined which would not be possible if it was a fluid. In the footprints of severed legs, drop-like secretions were found additionally to filament-like (Gorb et al. 2006). The findings were supported by a study on three distantly related theraphosid species in which nozzle-like setae on the tarsi were found to be responsible for silk depositions (Rind et al. 2011). Silk-extruding spigots arrayed on the spinnerets and connected to internal silk glands are assumed to be modified setae on modified appendages (e.g. Selden et al. 2008). Silk-secreting tarsal structures were therefore interpreted as missing link to the ancestral condition (Gorb et al. 2006; Rind et al. 2011).

Criticism to the hypothesis arose as it was postulated the nozzle-like structures were not spigots but thermosensors (Pérez-Miles et al. 2009) or chemoreceptors (Foelix et al. 2012). While Pérez-Miles et al. (2009) were not able to find any secretion after sealing the spinnerets with paraffin, Foelix et al. (2012) observed secretions directly correlated with the ribbed hairs found on the theraphosid tarsus. As chemosensory hairs are known to be filled with proteinaceous fluid (e.g. Foelix and Chuwang 1973), it was suggested that the substance might be "receptor lymph" (Foelix et al. 2012), although the composition of this fluid was not determined.

Hydrophobic fluid footprints however were found in many spiders, solifugids and mites. An approach utilizing interference reflection microscopy revealed persistent, hydrophobic footpad secretions in every species studied by Peattie et al. (2011). As opposed to the hairy attachment system in flies, however, fluid secretion in spider scopulae was not continuous. It was therefore postulated by the authors that the scopulae can function in both wet and dry modes.

Given that the hairy attachment system of spiders can function in wet and dry condition and the cohesive forces of a thin water film naturally occurring on most surfaces suffice for adhesion, humidity should influence attachment abilities of spiders. Wolff and Gorb (2012) found 50 % higher traction forces for *Philodromus dispar* (Philodromidae) in an environment with a relative humidity of 70 % than at a relative humidity of 15 %. Very low forces were detected at a relative humidity of 99 %, especially if water condensation could be observed on the substrate which then became slippery to the spider.

5.5 Conclusions

Spider scopulae are powerful attachment devices for locomotion on smooth and/or steep surfaces as well as for prey capture. Forces can be generated to hold up to 160 times the body weight, and the system can, in its most advanced state, probably be run in dry and wet modes according to necessity.

References

Arzt E, Gorb S, Spolenak R (2003) From micro to nano contacts in biological attachment devices. Proc Natl Acad Sci USA 100:10603–10606. doi:10.1073/pnas.1534701100

Autumn K, Liang YA, Hsieh ST, Zesch W, Chan WP, Kenny TW, Fearing R, Full RJ (2000) Adhesive force of a single gecko foot-hair. Nature 405:681–685

Dunlop JA (1995) Movements of scopulate claw tufts at the tarsus tip of a tarantula spider. Neth J Zool 45:513–520

Foelix RF (2011) Biology of spiders. University Press, Oxford

Foelix RF, Chuwang IW (1973) Morphology of spider sensilla. 2. Chemoreceptors. Tissue Cell 5:461–478

Foelix R, Rast B, Peattie AM (2012) Silk secretion from tarantula feet revisited: alleged spigots are probably chemoreceptors. J Exp Biol 215:1084–1089

Gorb SN (ed) (2012) Adhesion and friction in biological systems. Biologically-inspired systems, vol 3. Springer, London

Gorb SN, Niederegger S, Hayashi CY, Summers AP, Voetsch W, Walther P (2006) Silk-like secretion from tarantula feet. Nature 443:407. doi:10.1038/443407a

Hill DE (1977) The pretarsus of salticid spiders. Zool J Linn Soc 60:319–338

Hill DE (2006) Jumping spider feet (Araneae, Salticidae). Peckhamia Epublications 3:1–41

Kesel AB, Martin A, Seidl T (2003) Adhesion measurements on the attachment devices of the jumping spider *Evarcha arcuata*. J Exp Biol 206:2733–2738

Niederegger S (2003) The hairy attachment system in arthropods. PhD thesis, Eberhard Karls University, Tübingen

Niederegger S, Gorb S (2003) Tarsal movements in flies during leg attachment and detachment on a smooth substrate. J Ins Physiol 49:611–620. doi:10.1016/S0022-1910(03)00048-9

Niederegger S, Gorb SN (2006) Friction and adhesion in the tarsal and metatarsal scopulae of spiders. J Comp Physiol A 192:1223–1232. doi:10.1007/s00359-006-0157-y

Peattie AM, Dirks JH, Henriques S, Federle W (2011) Arachnids secrete a fluid over their adhesive pads. PLoS One 6(5). doi:10.1371/journal.pone.0020485

Peressadko A, Gorb SN (2004) When less is more: experimental evidence for tenacity enhancement by division of contact area. J Adhesion 80:247–261. doi:10.1080/00218460490430199

Pérez-Miles F (1994) Tarsal scopula division in Theraphosinae (Araneae, Theraphosidae) – its systematic significance. J Arachnol 22:46–53

Pérez-Miles F, Panzera A, Ortiz-Villatoro D, Perdomo C (2009) Silk production from tarantula feet questioned. Nature 461:E9

Rind FC, Birkett CL, Duncan BJA, Ranken AJ (2011) Tarantulas cling to smooth vertical surfaces by secreting silk from their feet. J Exp Biol 214:1874–1879. doi:10.1242/Jeb.055657

Selden PA, Shear WA, Sutton MD (2008) Fossil evidence for the origin of spider spinnerets, and a proposed arachnid order. Proc Natl Acad Sci USA 105:20781–20785. doi:10.1073/pnas.0809174106

Wolff JO, Gorb SN (2012) The influence of humidity on the attachment ability of the spider *Philodromus dispar* (Araneae, Philodromidae). Proc Biol Sci 279:139–143. doi:10.1098/rspb.2011.0505

Chapter 6
Cost–Benefit Balance of Dispersal and the Evolution of Conditional Dispersal Strategies in Spiders

Dries Bonte

6.1 Introduction

Movement is a central life history trait in organisms inhabiting a world that is spatially heterogeneous. From birth to death, organisms need to displace themselves to fulfil essential life functions related to foraging, escaping predators, avoiding competition and inbreeding. According to the general framework proposed by Nathan et al. (2008), the movement pattern can be described by four basic components that all depend on the organisms physiological conditions: (1) the organism's internal state, which defines its intrinsic motivation to move; (2) the motion and (3) navigation capacities representing, respectively, the organism's basic ability to move and affect where and when to move; and (4) the broad range of external factors affecting physiological processes related to movement. While movement describes any change in location, dispersal refers only to the movement of an individual from its natal site to (possibly several) sites at which it reproduces (Clobert et al. 2009). Consequently, many of the movements undertaken by an organism may not be relevant for dispersal, such as long-distance migration of birds and other movements of highly active animals within their home range for foraging. Furthermore, dispersal can comprise a single event through a lifetime (e.g. in most sessile organisms) or can be a very complicated compound process including, for example, many bouts of settlement and reproduction.

In spiders, dispersal is generally conducted by means of two distinct locomotory strategies. Ambulatory locomotion allows short-distance displacement relevant for foraging or mate searching and can eventually result in dispersal away from the place of birth (Bonte et al. 2004; Lambeets and Bonte 2009). In contrast, spiders have developed silk-related dispersal strategies that allow displacements over longer distances. Such dispersal strategies are generally referred to as ballooning

D. Bonte (✉)
Department of Biology, Terrestrial Ecology Unit, Ghent University, Ghent, Belgium
e-mail: dries.bonte@ugent.be

W. Nentwig (ed.), *Spider Ecophysiology*,
DOI 10.1007/978-3-642-33989-9_6, © Springer-Verlag Berlin Heidelberg 2013

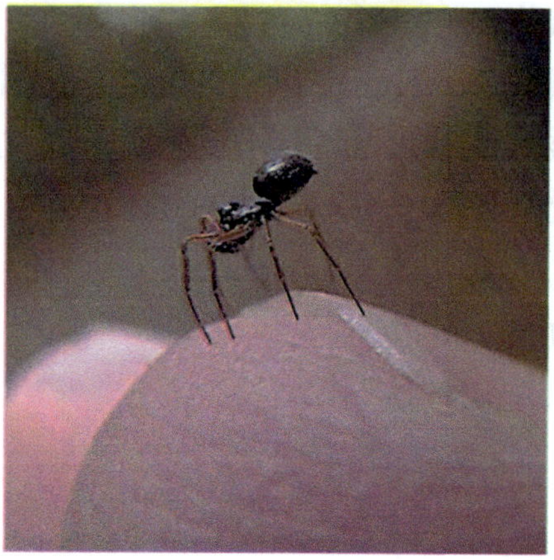

Fig. 6.1 *Erigone atra* or *E. dentipalpis* (Linyphiidae) adopting a typical tiptoeing position before the onset of aerial dispersal. Picture credit by ARABEL image bank/© R. Louvigny

when meteorology and behaviour facilitate drag-induced lift, allowing the "body" to become airborne (Bell et al. 2005), but I prefer to use the term aerial dispersal since alternative displacement strategies using silk are possible.

The production of silk is consequently central to dispersal in spiders. Spiders have up to six silk glands producing a multitude of threads which emerge from spigots (Vollrath and Knight 2001). These spigots are placed on spinnerets terminating at the spider's opisthosoma and are commonly found on most spiders except the Anyphaenidae, Clubionidae and Salticidae in which they are absent (Bell et al. 2005). Presumably, silk lines for aerial dispersal are classical draglines that are produced in such a way that they do not get immediately attached to the surface, but instead are released in the atmosphere. According to Richter (1970), six or eight silk threads are produced prior to take-off, pointing at the possible involvement of the eight glandulae ampullaceae (glands in the anterior spinnerets, associated with the production of draglines) in aerial dispersal. Araneomorph spiders do succeed in this by displaying distinct behaviours preceding transfer, so-called tiptoe behaviour (i.e. stretching of legs, raising opisthosoma and producing silk threads; see Fig. 6.1) after climbing up in the vegetation to exposed locations. By means of this behaviour, spiders initiate either long-distance dispersal when silk threads are transported in the air (ballooning) or short-distance dispersal when silk threads are used as bridges (rappelling). Due to aerodynamic constraints, ballooning is restricted to smaller individuals. A mathematical model revealed that ballooners are predicted to have little control over their aerodynamic drag and their dispersal within the atmospheric boundary layer and that the length of the silk line is only crucial at lift-off (Reynolds et al. 2006). While this strategy following tiptoe behaviour appears to be the most commonly used technique, some spiders use alternative methods, for instance, dropping in mygalomorphs by which spiderlings drop from an elevated point in the vegetation thereby remaining attached to a dragline and catch upward velocities while continuously producing silk till the line breaks at a sensitive point (Eberhard 1987). Silk-assisted aerial dispersal is known from almost all spider families, and phylogenetic analysis suggests that it likely evolved in parallel with the raise of grasses in the Cretaceous era (135–65 million years ago) and the subsequent temporal and spatial heterogeneities in habitat structure due to, for example, herbivory and trampling. I refer to the excellent review of James Bell and colleagues (Bell et al. 2005) for a comprehensive review on the evolution of ballooning from a phylogenetic perspective. Because ballooning has been demonstrated to be a heritable trait in a selection of species (Bonte et al. 2003, 2009; Bonte and Lens 2007) and because aerial-dispersal strategies have been shown to be strongly dependent on the local context and individual condition, I will outline more into detail how the cost–benefit balance of dispersal may influence aerial dispersal strategies in spiders through selection and phenotypic plasticity.

6.2 Benefits and Costs of Dispersal in Spiders

6.2.1 Benefits

Widely acknowledged benefits of dispersal are increased fitness in spatio-temporally changing environments due to bet hedging, avoidance of competition with conspecifics and kin and inbreeding avoidance (Cressman and Krivan 2006; Gandon 1999; Gandon and Michalakis 1999; Krivan et al. 2008; Leturque and Rousset 2002; Palmqvist et al. 2000). In spiders, benefits for aerial dispersal are likely to be attributed to survival in disturbed habitat (Bell et al. 2005) but also avoidance of competition, that is, avoiding cannibalistic events and competition for resources with relatives (Bonte et al. 2007b, 2011a). While aerial dispersal is largely associated with habitat disturbance in spiders both among and within species (Entling et al. 2011) and/or the avoidance of resource competition (De Meester and Bonte 2010), natal dispersal strategies of hatchlings are mostly beneficial for avoiding kin competition. Because of the huge costs of aerial dispersal in reaching unsuitable habitat, such natal dispersal often confers short-distance displacements by ambulatory displacements (Mestre-Arias and Bonte 2012) or even maternally assisted displacements which we referred to as hitchhiking dispersal (Bonte et al. 2007b). While all these factors are simultaneously shaping the process of dispersal, each of them may influence dispersal in different ways (Lecomte et al. 2004; Stenseth and Lidicker 1992) depending on the associated costs.

6.2.2 Costs

As recently outlined in a review, costs of dispersal can be categorised into four major categories: energetic, time, risk and opportunity costs (Bonte et al. 2012). In short, energetic costs are due to lost metabolic energy related to the development of specific dispersal attributes; time costs refer to costs by time not invested in other primary activities during dispersal; risk costs are costs related to both mortality risks and attrition, while opportunity costs are incurred by giving up prior residence and familiarity-related advantages. While we do not have a proper understanding on the magnitude of the costs associated with spider dispersal, it is likely that costs related to silk production (Craig et al. 1999) can be substantial, especially if multiple threads are simultaneously or sequentially produced. Spiders climb elevated structures in the vegetation to prepare for ballooning by tiptoeing, thereby spending energy on nonroutine movements, but they also expose themselves to predators and experience mortality or damage costs before and during the departure preparation phase (Young and Lockley 1988). The observation that spiders display antipredator behaviour when preparing for dispersal by taking a position that allows for a fast escape by jumping when attacked additionally suggests that predation during dispersal preparation may be quite common. Similar risk costs are likely to be incurred during transfer, but no data so far are available on predation risks in the air.

Settlement costs during aerial dispersal (i.e. failure to reach suitable habitat) are likely high when habitat availability is low and spatially uncorrelated, with possibilities of reaching a suitable habitat proportional to the availability of habitats. This is expected because of any absence of control over the aerodynamic drag within the atmospheric boundary layer (Reynolds et al. 2006). Although no attempts have been made to quantify these costs, selection pressures arising from such potentially area-related mortality risks can be deduced indirectly from decreased ballooning responses in spiders inhabiting small habitat islands or heavily fragmented landscapes (Bonte et al. 2006, 2007a). Finally, social spiders are likely to incur substantial integration costs of aerial dispersal due to increased Allee effects when founding new colonies at reduced group sizes (Lubin and Bilde 2007). These costs are therefore thought to have led to the evolution of group dispersal strategies in social spiders through budding or fission, thereby additionally maintaining genetic relatedness within groups (Lubin and Bilde 2007).

6.3 Information Use and Phenotypic Plasticity of Dispersal

The balance between costs and benefits of dispersal will determine the eventual dispersal strategy. This balance is expected to be determined by both biotic and abiotic environmental factors like the strength of wind velocity and habitat area determining successful immigration in suitable habitat, local densities and sex ratio determining future reproductive success (context dependency) or intrinsic factors related to body condition directly determining the amount of energy that can be spend on dispersal or avoiding predation risks during the preparation (condition dependency). Because of such context- and condition-dependent cost–benefit balances, dispersal is expected to be a plastic trait, allowing a quick and opportunistic adoption of the strategy in fast-changing environment (Stamps 2001). Such plastic responses can, however, only occur when individuals detect information from the environment and use it for dispersal decision making (Clobert et al. 2009). This information can be perceived during development through changes in internal state or it can be perceived by cognitive processes, so by actively collecting information.

6.3.1 Context-Dependent Dispersal

Information on the prevailing environmental conditions is detected by sensory mechanisms for visual, olfactory or wind stimuli but potentially also through its impact on development when starvation or different temperatures during development directly impact physiological conditions. In the latter case, changes in the environment affect body condition. Such conditional effects on dispersal in spiders are discussed below. Information can also be genetically acquired when natural selection has generated adaptive changes in gene frequency that determine the eventual phenotype. Such adaptive changes (sensu natural selection) comprise the adaptation of aerial dispersal strategies in relation to changes in landscape structure (Bonte et al. 2006, 2007a). Evidently, information on the landscape structure cannot be acquired during an individual's life span, especially not when aerial dispersal takes place in the juvenile life phases. However, spiders with well-developed visual capacities might adopt emigration strategies by using cues on the local habitat structure for homing when incidentally leaving the local habitat. Such strategies have been observed in wolf spiders (Lycosidae) from riverine (Lambeets and Bonte 2009; Lambeets et al. 2010) or grassland systems (Bonte et al. 2004), respectively, swept by floods or leaving optimal habitat by increased ambulatory movements.

For aerial dispersal, information on the turbulence and wind speed is essential for successful displacements. When wind velocities are too high, upward draughts may be limited and silk threads may become entangled thereby hindering effective displacements (Bell et al. 2005; Reynolds et al. 2006). Alternatively, species inhabiting small habitat islands, like isolated salt marshes along the North Sea, may experience higher probabilities of being blown in the hostile sea. Aerial dispersal strategies are therefore adjusted to the prevailing wind velocities as censored by trichobothria on the legs (Bell et al. 2005), with maximal take-off frequencies when wind currents are below 3 m/s and when upward thermal winds are present (Greenstone 1990; Vugts and Vanwingerden 1976). Because the consequences of differences in wind speed and eventual dispersal distance will largely depend on the local habitat configuration, local adaptation in the use of this source of information has been observed in the wolf spider *Pardosa purbeckensis* with individuals from small populations showing decreased emigration behaviour with increased wind strength relative to those individuals originated from large-area salt marshes (Bonte et al. 2007a).

In agrobiont spiders, the availability and configuration of habitat is seasonally changing. Species like *Erigone* typically inhabit large crop areas from spring till harvest in late summer and are then forced to retreat to small strips of natural habitat in road verges or other more permanent vegetation which is typically sparsely distributed in intensively managed agricultural landscapes. This implies that costs of dispersal, that is, possibilities of reaching such habitat, change seasonally with low probabilities of finding suitable habitat in late summer and autumn, while nearly risk-free in spring. Given the limited movement possibilities and visual sensory capacities, agrobiont spiders are not able to explore the condition of the landscape. Instead, adaptive strategies have evolved in function of the experienced temperature during development (Bonte et al. 2008b). While temperature has an overall effect on body condition (see below), dispersal strategies are independently adopted in relation to these thermal conditions, with those individuals developing at low temperatures (mimicking early spring conditions) performing long-distance ballooning strategies, thereby optimising colonisation of crop habitat in the vicinity, and those having experienced warm summer temperature conditions more likely to show repeated short-distance rappelling movement, thereby increasing possibilities of reaching patches of hibernation habitat (Bonte et al. 2008b).

While changes in habitat quality determine fitness at longer time intervals, local population structure may have more immediate impacts on fitness through resource competition. Dispersal is a prime mechanism to avoid competition when densities are high and will, under the condition that individuals are able to perceive information related to local crowding, result in positive density-dependent emigration strategies (Poethke and Hovestadt 2002). Equally, males and females do not share the same interest. In spiders, but also other arthropods (Anholt et al. 2001), males are trying to optimise mating opportunities, while females need to optimise food intake for further investment in reproduction. In the spider *Erigone atra*, emigration strategies have been demonstrated to follow this prediction (Fig. 6.2), with higher dispersal in an artificial metapopulation when densities of individuals from the same sex increase and, in the case of males, when the proportion of males to females increased (De Meester and Bonte 2010). Interestingly, such density-dependent dispersal strategies were only recorded for short-distance dispersal strategies, that is, rappelling. Because of the aggregated distribution of spiders, for instance, in crops, such strategies are consequently expected to maximise resource use while avoiding costs of ending up in unfavourable habitat, where resources are totally lacking. Results of this experiment indicate that spiders are able to perceive information on the local population structure, but it remains an open question how this is done. Note that these density-dependent strategies emerged only by the fact of interactions with conspecifics and not through the action of food shortage since prey availability was not limiting (see Sect. 6.3.2 for effects of food limitation).

While information received from direct interactions during development is likely, our experiments additionally pointed at the possibility of using silk-related information, possibly the presence of specific associated volatiles. Indeed, long-distance dispersal strategies are not triggered by changes in local population structure, but in contrast strongly determined by the presence of dispersal-associated silk in the environment. Conversely, *Erigone* spiders were found to be more engaged in performing costly long-distance dispersal when some conspecifics

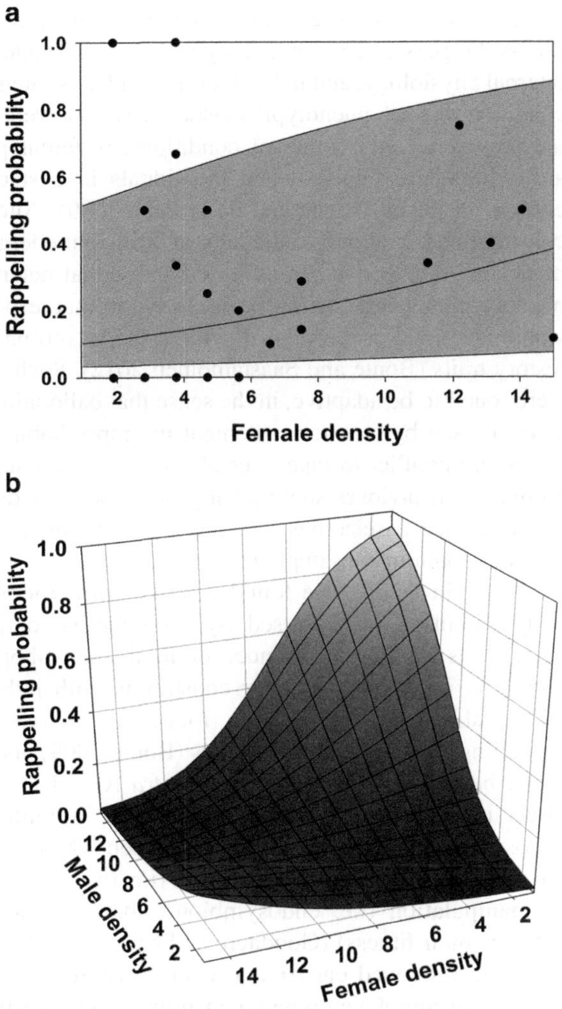

Fig. 6.2 Density-dependent rappelling in the spider *Erigone atra* (Linyphiidae). Females adjust their strategy only in response to female density (*top panel*), while males (*bottom panel*) use information on both the number of direct competitors (males) and the availability of mating partners (females). Adapted from De Meester and Bonte (2010)

have done so previously (De Meester and Bonte 2010). Such cues do indicate that, for instance, meteorological conditions are suitable and are a source of cheap public information on the suitability of the dispersal window. Of course, some individuals should engage in sampling the environmental conditions first. Relatively few dispersing individuals are therefore expected to induce a cascade of emigration events, eventually leading to synchronised, mass-ballooning events.

6.3.2 Condition-Dependent Dispersal

In the previous paragraph, we showed how spiders use information from their environment for dispersal. In all these cases, adaptive decisions are made in the sense that they are expected to be beneficial for the individual, thereby maximising fitness. Dispersal decision making is, however, equally expected to be influenced by internal physiology, and individuals may adapt strategies according to their own body condition through phenotypic plasticity. We showed, for instance, that dependent on the prevailing environmental conditions, different phenotype-dependent strategies are expected to evolve when individuals in best condition are able to bear the costs of dispersal (Bonte and de la Pena 2009). Alternatively, body condition may constrain dispersal due to the lack of sufficient energetic reserves, thereby inducing nonadaptive responses. In all cases, individual heterogeneity in body condition, or phenotype in general, will give rise to syndromal responses (Dingemanse et al. 2010), where dispersal is correlated with specific morphological, physiological or life history traits (Bonte and Saastamoinen 2012). Such phenotype-dependent strategies were found to be adaptive, in the sense that ballooning phenotypes were additionally characterised by a faster settlement in empty habitat and an increased willingness to go into conflict to take over already occupied sites (Bonte et al. 2011a). These correlated behaviours subsequently point out that dispersive phenotypes are likely to experience a selective advantage at settlement and that dispersive phenotypes are not a random subsample from the population.

Again, in *E. atra* we found positive correlations between aerial dispersal and body condition (as assessed by reproductive output) in an experiment where individuals were reared under optimal and suboptimal temperature conditions (Bonte et al. 2008b). Similar constraints with individuals in poor condition not willing/able to disperse were found for inbreeding on both ballooning and rappelling in three *Erigone* species (Bonte 2009) and for the presence of endosymbiotic bacteria on ballooning in *E. atra* (Goodacre et al. 2009). Interestingly, no direct impact on fitness was observed in individuals infected with the specific endosymbionts, leaving the question open whether such physiological constraints are due to negative impacts on health (i.e. the endosymbionts behaving as a disease) or manipulation (i.e. endosymbionts manipulating host dispersal phenotype to enhance their fitness) (Goodacre and Martin 2013).

While I showed earlier that overcrowding induced dispersal, immediate food shortage during development constrained the investments in silk threads produced for taking off in *Erigone arctica* and *Erigone dentipalpis* (Bonte et al. 2008a), suggesting that a lowered body condition decreases the eventual distance that can be dispersed. This study did, however, not separate rappelling from ballooning and differences in the length of the produced silk thread were likely to be related to a shift in dispersal strategy (with ballooning silk threads being shorter than those produced for rappelling). In a more recent experiment, we therefore assessed the impact of starvation on the expression of dispersal in *E. dentipalpis* more into detail, thereby including the impact of maternal effects on both natal ambulatory and adult aerial dispersal strategies (Mestre-Arias and Bonte 2012). We found that food supply had a pervasive influence on all aspects of *E. dentipalpis* dispersal behaviour. Food stress induced ambulatory emigration during the juvenile stage, and only males decreased rappelling under current food shortage. Short-distance dispersal

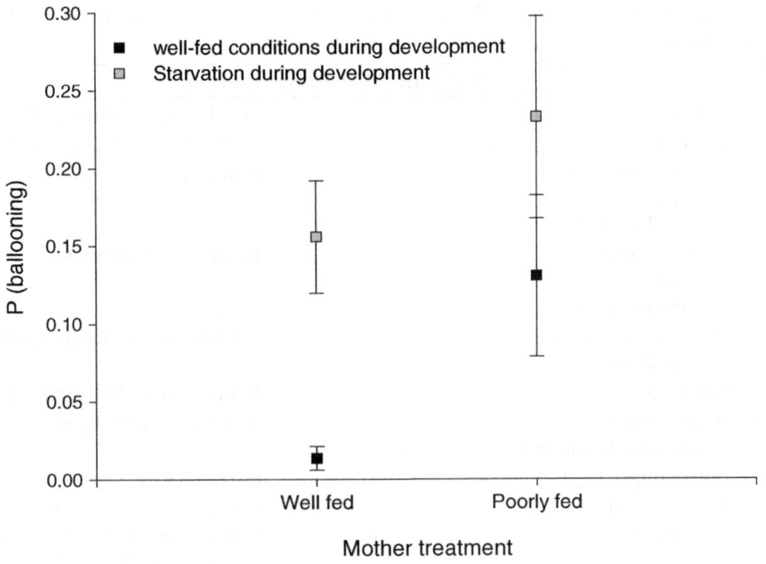

Fig. 6.3 The relative contribution of food conditions of the mother and during own development on ballooning strategies in *Erigone dentipalpis* (Linyphiidae). Adapted from Mestre-Arias and Bonte (2012). P (ballooning) denotes the probability that spiders performed ballooning under standardised laboratory conditions

strategies were consequently plastic in relation to food shortage during development and only constraining adult dispersal. Interestingly, the interaction between maternal and current food deprivation caused a marked increase in long-distance ballooning, with lowest ballooning rates in cases where food was excessively available during the development of mother and her offspring and highest when it was limiting during the two subsequent generations (Fig. 6.3). This factorial experiment consequently demonstrated that reactions of spiders to food supply are more complex than previously assumed and corroborates that the influence of an ecological factor on dispersal is not confined to a single point in time but acts through its temporal variation across generations (Ronce 2007). Maternal effects are increasingly being recognised as key determinants of population dynamics, since they not only affect offspring life history traits, such as survival and reproduction, but also dispersal behaviour and gene flow. As such, both immediate and past environmental conditions are expected to have important consequences for metapopulation cohesiveness and persistence by inducing lag responses in population dynamics.

6.4 Conclusions and Perspectives

Aerial dispersal in spiders is a complex trait, subject to selection and phenotypic plasticity in relation to both environmental parameters and body condition. We refer to these as, respectively, context- and condition-dependent effects

Table 6.1 An overview of the different context- and condition-dependent aerial-dispersal strategies recorded by reaction norms in *Erigone* spiders

		Rappelling	Ballooning	Reference
Context	Increased density	+	=	De Meester and Bonte (2010)
	Sex ratio	+	=	De Meester and Bonte (2010)
	High temperatures during development	+	–	Bonte et al. (2008a, b)
	Low temperature during development	–	+	Bonte et al. (2008a, b)
	Presence of dispersal silk threads	=	+	De Meester and Bonte (2010)
	Patch size	=	–	Bonte et al. (2006, 2008a, b)
Condition	Body condition (general fecundity)	–	–	Bonte et al. (2008a, b)
	Inbreeding	–	–	Bonte (2009)
	Endosymbionts	=	–	Goodacre et al. (2009)
	Starvation	–	+	Bonte et al. (2008a, b) and Mestre-Arias and Bonte (2012)
	Maternal effects	=	+	Mestre-Arias and Bonte (2012)

It is clear that changes in the environmental and body conditions affect short-distance dispersal (rappelling) and long-distance dispersal (ballooning) in a different and sometimes opposite direction (– decrease, + increase, = no effect)

(see Table 6.1 for an overview). By using standardised experiments, we showed that such context- and condition-dependent strategies are not in the same direction for, respectively, long-distance dispersal through ballooning and short-distance dispersal strategy by rappelling, in line with predictions from theory (Bonte et al. 2010; Clobert et al. 2009; Travis et al. 2012). While many observed changes are likely to be adaptive, a poor body condition can constrain dispersal, thereby inducing nonadaptive behavioural changes. Our experiments also point at the existence of maternal effects that additively determine long-distance dispersal decision making.

References

Anholt BR, Vorburger C, Knaus P (2001) Mark-recapture estimates of daily survival rates of two damselflies (*Coenagrion puella* and *Ischnura elegans*). Can J Zool 79:895–899

Bell JR, Bohan DA, Shaw EM, Weyman GS (2005) Ballooning dispersal using silk: world fauna, phylogenies, genetics and models. Bull Entomol Res 95:69–114

Bonte D (2009) Inbreeding depresses short and long distance dispersal in three congeneric spiders. J Evol Biol 22:1429–1434

Bonte D, de la Pena E (2009) Evolution of body condition-dependent dispersal in metapopulations. J Evol Biol 22:1242–1251

Bonte D, Lens L (2007) Heritability of spider ballooning motivation under different wind velocities. Evol Ecol Res 9:817–827

Bonte D, Saastamoinen M (2012) Dispersal syndromes in spiders and butterflies. In: Clobert J, Baguette M, Benton TG, Bullock JM (eds) Dispersal and spatial evolutionary ecology. Oxford University Press, Oxford 161–170

Bonte D, Deblauwe I, Maelfait JP (2003) Environmental and genetic background of tiptoe-initiating behaviour in the dwarfspider *Erigone atra*. Anim Behav 66:169–174

Bonte D, Lens L, Maelfait JP (2004) Lack of homeward orientation and increased mobility result in high emigration rates from low-quality fragments in a dune wolf spider. J Anim Ecol 73:643–650

Bonte D, Vanden Borre J, Lens L, Maelfait JP (2006) Geographical variation in wolf spider dispersal behaviour is related to landscape structure. Anim Behav 72:655–662

Bonte D, Bossuyt B, Lens L (2007a) Aerial dispersal plasticity under different wind velocities in a salt marsh wolf spider. Behav Ecol 18:438–443

Bonte D, Van Belle S, Maelfait JP (2007b) Maternal care and reproductive state-dependent mobility determine natal dispersal in a wolf spider. Anim Behav 74:63–69

Bonte D, Lukac M, Lens L (2008a) Starvation affects pre-dispersal behaviour of *Erigone* spiders. Basic Appl Ecol 9:308–315

Bonte D, Travis JMJ, De Clercq N, Zwertvaegher I, Lens L (2008b) Thermal conditions during juvenile development affect adult dispersal in a spider. Proc Natl Acad Sci USA 105:17000–17005

Bonte D, De Clercq N, Zwertvaegher I, Lens L (2009) Repeatability of dispersal behaviour in a common dwarf spider: evidence for different mechanisms behind short- and long-distance dispersal. Ecol Ent 34:271–276

Bonte D, Hovestadt T, Poethke HJ (2010) Evolution of dispersal polymorphism and local adaptation of dispersal distance in spatially structured landscapes. Oikos 119:560–566

Bonte D, De Meester N, Matthysen E (2011a) Selective integration advantages when transience is costly: immigration behaviour in an agrobiont spider. Anim Behav 81:837–841

Bonte D, van Dyck H, Bullock JM et al (2012) Costs of dispersal. Biol Rev. 87:290–312. doi:10.1111/j.1469-185X.2011.00201.x

Clobert J, Le Galliard JF, Cote J, Meylan S, Massot M (2009) Informed dispersal, heterogeneity in animal dispersal syndromes and the dynamics of spatially structured populations. Ecol Lett 12:197–209

Craig CL, Hsu M, Kaplan D, Pierce NE (1999) A comparison of the composition of silk proteins produced by spiders and insects. Int J Biol Macromol 24:109–118

Cressman R, Krivan V (2006) Migration dynamics for the ideal free distribution. Am Nat 168:384–397

De Meester N, Bonte D (2010) Information use and density-dependent emigration in an agrobiont spider. Behav Ecol 21:992–998

Dingemanse NJ, Kazem AJN, Reale D, Wright J (2010) Behavioural reaction norms: animal personality meets individual plasticity. Trends Ecol Evol 25:81–89

Eberhard WG (1987) How spiders initiate airborne lines. J Arachnol 15:1–9

Entling MH, Stampfli K, Ovaskainen O (2011) Increased propensity for aerial dispersal in disturbed habitats due to intraspecific variation and species turnover. Oikos 120:1099–1109

Gandon S (1999) Kin competition, the cost of inbreeding and the evolution of dispersal. J Theor Biol 200:345–364

Gandon S, Michalakis Y (1999) Evolutionarily stable dispersal rate in a metapopulation with extinctions and kin competition. J Theor Biol 199:275–290

Goodacre SL, Martin OY (2013) Endosymbiont infections in spiders. In: Nentwig W (ed) Spider ecophysiology. Springer, Heidelberg (this volume)

Goodacre SL et al (2009) Microbial modification of host long-distance dispersal capacity. BMC Biol 7:32

Greenstone MH (1990) Meteorological determinants of spider ballooning – the roles of thermals vs the vertical wind-speed gradient in becoming airborne. Oecologia 84:164–168
Krivan V, Cressman R, Schneider C (2008) The ideal free distribution: a review and synthesis of the game-theoretic perspective. Theor Popul Biol 73:403–425
Lambeets K, Bonte D (2009) Between-population variation in homeward orientation behaviour in two riparian wolf spiders. Behav Processes 82:62–66
Lambeets K, Van Ranst J, Bonte D (2010) Is movement behavior of riparian wolf spiders guided by external or internal information? J Arachnol 38:313–318
Lecomte J, Boudjemadi K, Sarrazin F, Cally K, Clobert J (2004) Connectivity and homogenisation of population sizes: an experimental approach in *Lacerta vivipara*. J Anim Ecol 73:179–189
Leturque H, Rousset F (2002) Dispersal, kin competition, and the ideal free distribution in a spatially heterogeneous population. Theor Popul Biol 62:169–180
Lubin Y, Bilde T (2007) The evolution of sociality in spiders. In: Brockmann HJ, Roper TJ, Naguib M, WynneEdwards KE, Barnard C, Mitani J (eds) Advances in the study of behavior, vol 37. Elsevier, San Diego
Mestre-Arias L, Bonte D (2012) Food stress during juvenile and maternal development shapes natal and breeding dispersal in a spider. Behav Ecol 23:759–764
Nathan R et al (2008) A movement ecology paradigm for unifying organismal movement research. Proc Natl Acad Sci USA 105:19052–19059
Palmqvist E, Lundberg P, Jonzen N (2000) Linking resource matching and dispersal. Evol Ecol 14:1–12
Poethke HJ, Hovestadt T (2002) Evolution of density-and patch-size-dependent dispersal rates. Proc Biol Sci 269:637–645
Richter CJJ (1970) Aerial dispersal in relation to habitat in eight wolf spider species (Par- dosa, Araneae, Lycosidae). Oecologia 5: 200–214
Reynolds AM, Bohan DA, Bell JR (2006) Ballooning dispersal in arthropod taxa with convergent behaviours: dynamic properties of ballooning silk in turbulent flows. Biol Lett 2:371–373
Ronce O (2007) How does it feel to be like a rolling stone? Ten questions about dispersal evolution. Annu Rev Ecol Evol Syst 38:231–253
Stamps JA (2001) Habitat selection by dispersers: integrating proximate and ultimate approaches. In: Clobert J, Danchin E, Dhondt AA, Nichols JD (eds) Dispersal. Oxford University Press, Oxford
Stenseth NC, Lidicker WZ (eds) (1992) Animal dispersal: small mammals as a model. Chapman & Hall, London
Travis JMJ et al (2012) Modelling dispersal: an eco-evolutionary framework incorporating emigration, movement, settlement behaviour and the multiple costs involved. Methods Ecol Evol. 3:628–641. doi:10.1111/j.2041-210X.2012.00193.x
Vollrath F, Knight DP (2001) Liquid crystalline spinning of spider silk. Nature 410:541–548
Vugts HF, Vanwingerden W (1976) Meteorological aspects of aeronautic behavior of spiders. Oikos 27:433–444
Young OP, Lockley TC (1988) Dragonfly predation upon *Phiddipus audax* (Araneae, Salticidaae). J Arachnol 16:121–122

Part III
Immune System and Pathogens

Hemocytes contain the immune system of spiders with a cellular and a humoral response. Different hemocyte types react very fast to an invasion of various pathogens by releasing a variety of defensive compounds into the hemolymph. Combined with such a defence is a very potent repair mechanism which probably via a clotting cascade and coagulin production stops hemolymph loss, in principle resembling the human blood clotting cascade. With its combination of ancestral elements and modern parts, spiders obviously can defend rather successfully against bacterial and fungal invaders. Since spiders can be affected by a variety of fungal pathogens but infected spiders nevertheless are rare, one could ask how resistant spiders are against fungi. Given the substantial negative or positive effects on immunity, behaviour and reproduction, bacterial pathogens may have on spiders, the recent debate also focuses on their parasitic or endosymbiotic nature.

Chapter 7
The Immune System of Spiders

Lucia Kuhn-Nentwig and Wolfgang Nentwig

7.1 Introduction

Spiders, as all other arthropods, have an open circulatory system, and their body fluid, the hemolymph, freely moves between lymphatic vessels and the body cavities (see Wirkner and Huckstorf 2013). The hemolymph can be considered as a multifunctional organ, central for locomotion (Kropf 2013), respiration (Burmester 2013) and nutrition, and it amounts to approximately 20 % of a spider's body weight. Any injury includes not only immediate hemolymph loss but also pathogen attacks and subsequent infections. Therefore spiders have to react to injuries in a combined manner to stop fluid loss and to defend against microbial invaders. This is achieved by an innate immune system which involves several host defence systems such as hemolymph coagulation and the production of a variety of defensive substances (Fukuzawa et al. 2008).

In spiders, the immune system is localised in hemocytes which are derived from the myocardium cells of the heart wall where they are produced as prohemocytes and from where they are released as different cell types into the hemolymph (Seitz 1972). They contribute to the defence against pathogens by phagocytosis, nodulation and encapsulation of invaders. The humoral response includes mechanisms which induce melanin production to destroy pathogens, a clotting cascade to stop hemolymph loss and the constitutive production of several types of antimicrobial peptides, which are stored in hemocyte granules and released into the hemolymph (Fukuzawa et al. 2008) (Fig. 7.1).

The immune system of spiders is an innate immune system. It is hemolymph-based and characterised by a broad but not very particular specificity. Its advantage is a fast response within minutes to a few hours. This is in contrast to the adaptive immune system of vertebrates which can react to very specific pathogens, thus

L. Kuhn-Nentwig (✉) • W. Nentwig
Institute of Ecology and Evolution, University of Bern, Bern, Switzerland
e-mail: lucia.kuhn@iee.unibe.ch

W. Nentwig (ed.), *Spider Ecophysiology*,
DOI 10.1007/978-3-642-33989-9_7, © Springer-Verlag Berlin Heidelberg 2013

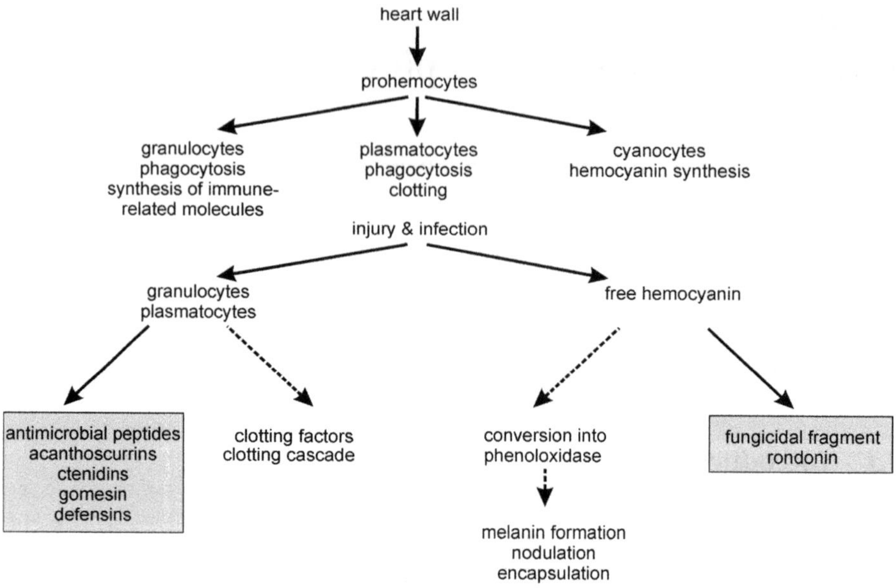

Fig. 7.1 Scheme of the immune system of spiders: origin and function of hemocytes (*upper part*) and reaction to injury and infection by microbial invaders (*lower part*). The reaction of spiders to injury and infection has not been well investigated and only the boxed components have been detected in spider hemocytes or hemolymph, whereas the clotting cascade and melanin formation pathway are adaptations from xiphosurans. For further details, see text

resulting in much more specific responses. Moreover, it creates an immunological memory during the lifetime of the species. The disadvantage is that it needs more time to react with antibody production, usually many hours to a few days, and needs to be built up during early ontogenesis.

7.2 Hemocytes

Seitz (1972) described the development of hemocytes in the heart wall of *Cupiennius salei* (Ctenidae), but did not recognise their immunological function. He rather thought that they were mainly storage cells, involved in moulting and yolk production. Sherman (1973) analysed the hemocytes of the theraphosid *Aphonopelma marxi* and realised the context to immune defence and clotting. He also reviewed the confusing nomenclature of blood cells (Sherman 1981) and distinguished three mature hemocyte types and a precursor cell type.

- Granulocytes are the most important hemocyte cell type. They are filled with granules and the most prominent feature of this hemocyte type, which contain several antimicrobial peptides to be released into the hemolymph (degranulation), is to kill pathogens. The nucleus is smaller than in plasmatocytes and they

also eliminate invaders via phagocytosis (Fukuzawa et al. 2008). Granulocytes vary in shape, including irregular, trapezoidal, triangular, oval and spindle forms, and their size is about 10 × 25 μm.
- Plasmatocytes are the most common hemocyte type and they have no granular inclusions, i.e. their cytoplasm is generally agranular or finely granular. They are polymorphic and show spindle, oval to ellipsoidal size, some are triangular or irregular. Plasmatocytes have approximately the same size as granulocytes (5–10 × 20–30 μm) and they are involved in phagocytosis of pathogens and in clotting (Sherman 1981).
- Cyanocytes represent the least common hemocyte type in the hemolymph and correspond to oenocytes (Sherman 1981). They are rare in the hemolymph (<5 % according to Millot 1949) but are frequently found in the heart (Sherman 1973). Their cytoplasma is dense and contains many ribosomes and protein crystals which were identified as hemocyanin in the case of *Aphonopelma marxi*. Cyanocytes are usually rounded or elongated, and they are the largest hemocytes with 20 × 50 μm (Sherman 1981). Cyanocytes synthesise hemocyanin which then is released into the hemolymph (Kemper 1983).
- Prohemocytes are small, mostly round cells with a large, compact nucleus and thin homogenous cytoplasm. As precursor cells of the other hemocytes, they are stem cells (Sherman 1981) and are mainly found in the heart wall (Fukuzawa et al. 2008; Seitz 1972).

Leberidocytes have also frequently been mentioned among blood cells. They are obviously only involved in the moulting process (Sherman 1981) and correspond to the moulting hemocytes described by Seitz (1976).

After an injury occurs, hemocytes migrate to the infection site. They directly fight against pathogens by phagocytosis, the formation of nodules to entrap bacteria (nodulation), and by the formation of capsules to entrap larger objects (encapsulation). Most importantly, hemocytes release components of the coagulation cascade to induce clotting and antimicrobial peptides to kill the invading microorganisms. The direct importance of phagocytosis has probably been overestimated but presumably it is important for the removal of cellular debris and for remodelling damaged tissues (Fukuzawa et al. 2008). While the overall functions of hemocytes in spiders are understood, it is unclear which hemocyte type is responsible for which part and up to which degree they are specialised for specific tasks.

7.3 The Humoral Response

The humoral response starts with the recognition of invading microorganisms, triggered by compounds from the bacterial surface, and leads to the initialisation of complex processes. They include (1) the start of cascades to induce clotting, (2) the start of cascades to induce melanisation and (3) the release of several antimicrobial peptides. Clotting is an important step to reduce further hemolymph loss and to enable subsequent wound healing. Melanin production is an unspecific but, nevertheless, very efficient system to destroy bacterial invaders. Antimicrobial peptides, also described as killing factors, are effective against a wide spectrum of pathogens. They are known from most major taxonomic groups of organisms and they differ considerably in structure and specificity. In spiders, so far four groups of antimicrobial peptides have been identified: defensins, glycine-rich peptides, a small open-ended cyclic peptide, and a hemocyanin fragment.

7.3.1 The Clotting and Agglutination System

Hemocytes circulating in the hemolymph react very sensitively to contact with pathogens and release the content of their granules into the hemolymph immediately via rapid exocytosis. This process has not as yet been investigated in spiders but well investigated in the horseshoe crab *Tachypleus tridentatus* (Xiphosura) (Kawabata 2010), which represents the most ancestral extant chelicerate group, and thus might be a model which could at least be partially valid for spiders.

The clotting process is triggered by pathogen-associated molecular patterns, resulting in a proteolytic cascade and hemolymph coagulation (Iwanaga and Lee 2005). Compounds from the bacterial surface, such as lipopolysaccharides (major components of the cell wall of Gram-negative bacteria) and β-1,3 glucans (typical for fungi), initiate the start of a coagulation cascade in horseshoe crabs and probably also in spiders. In the presence of lipopolysaccharides, all the components of the cascade are released from hemocytes by degranulation. After binding of lipopolysaccharides to factor C, the serine protease zymogen is autocatalytically activated to factor C′, which itself activates factor B to B′. Activation of the proclotting enzyme occurs through factor B′ and results in the conversion of coagulogen into noncovalent coagulin homopolymers (Osaki and Kawabata 2004).

In a cDNA library of hemocytes from the theraphosid *Acanthoscurria gomesiana*, immune related transcripts have been identified as factor C precursor, coagulation factor B precursor and proclotting enzyme precursor-like components (Lorenzini et al. 2006). For *Cupiennius salei*, these three components of the clotting system have also been identified in a hemocyte cDNA library (Kuhn-Nentwig, unpublished). Transcripts concerning coagulogen precursor or stablin and proxin precursors involved in crosslinking of the coagulin homopolymers via hemocyte-derived transglutaminase, as verified for horseshoe crabs, have not been detected so far in spiders (Kawabata 2010; Kawabata et al. 2002; Matsuda et al. 2007). However, ESTs concerning a protein-glutamine gamma-glutamyltransferase (EC2.3.2.13) are known from *A. gomesiana* and *C. salei* (Lorenzini et al. 2006; Kuhn-Nentwig, unpublished results). No information is available concerning the β-1,3 glucan initiated clotting cascade via activation of factor G, as reported for horseshoe crabs (Kawabata 2010).

In horseshoe crabs, pathogens can also be recognised and agglutinated by a series of compounds present in the hemolymph and circulating hemocytes. Besides several agglutinins, five types of lectins and three types of C-reactive proteins have been described (Kawabata 2010). Lorenzini and co-workers identified for *A. gomesiana* a transcript encoding tachylectin 5a and 5b-like components, hinting to similarities in these systems (Iwanaga and Lee 2005; Lorenzini et al. 2006).

7.3.2 The Phenoloxidase System and Melanin

During the early evolution of arthropods, hemocyanins evolved from a phenoloxidase-like enzyme. Hemocyanin, one of the major components of spider hemolymph, is usually known as an oxygen carrier (see Burmester 2013), while phenoloxidases are key enzymes in the melanin-producing pathway of arthropods that plays an important role in the sclerotisation of the cuticle. Since melanin is also toxic to microorganisms, additional functions in wound healing and in the humoral immune system evolved at the same time, and it has been argued that the simultaneous appearance of the arthropod exoskeleton and of an effective immune system obviously fostered the evolution of arthropods remarkably (Burmester 2002).

In contrast to other arthropods, chelicerates do not have a true phenoloxidase and hemocyanin itself acts also as a phenoloxidase after limited proteolysis with trypsin or chymotrypsin, as shown for the theraphosid *Aphonopelma hentzi* (sub *Eurypelma californicum*, see Nentwig 2012) (Decker and Rimke 1998). During the evolution of spider hemocyanin, two hemocyanin subunits not only conserved the original properties of phenoloxidases but also gained the property of the other subunits to bind oxygen reversibly (Decker and Tuczek 2000; Jaenicke and Decker 2004). For horseshoe crabs, the above described coagulation cascade is linked to prophenoloxidase activation in which the clotting enzyme transforms hemocyanin to phenoloxidase without proteolytic cleavage (Nagai and Kawabata 2000). Furthermore, the functional conversion of hemocyanin to phenoloxidase is also mediated by the binding of the antimicrobial peptides tachyplesin and tachystatins to the hemocyanin α-subunit (Nagai et al. 2001).

In arthropods in general, the activity of phenoloxidase follows a similar pattern as in the above described clotting process with recognition of compounds from bacterial membranes by recognition molecules. They react on serine proteases which activate a prophenoloxidase to a phenoloxidase (Söderhäll and Cerenius 1998). This is performed by hemocyanin and it leads to the production of melanin, a polymer of dihydroxyindole carboxylic acids, the monomers of which are derived from the amino acid tyrosine. The intermediary compounds of this process and melanin itself are toxic to microorganisms. Within minutes after infection, melanin is deposited either onto the surface of the microorganisms or onto a capsule of microorganisms and this process is called melanisation. Microorganisms are engulfed in a hardened gel and this leads finally to their destruction (Söderhäll and Cerenius 1998; Iwanaga and Lee 2005).

7.3.3 Antimicrobial Peptides

The contact of hemocytes with bacterial membrane compounds leads to an immediate release of a variety of antimicrobial peptides. The main location of these peptides seems to be the granulocytes but they contain different granule types which, in turn, may each contain different sets of antimicrobial peptides (Fukuzawa et al. 2008). While several compounds with antimicrobial activity were found for the horseshoe crab *Tachypleus tridentatus* (Iwanaga and Lee 2005), only a few substances have been described so far for spiders (Silva et al. 2000; Lorenzini et al. 2003; Baumann et al. 2010a, b).

7.3.3.1 Defensins

Spider defensins are small, amphipathic cysteine-rich peptides and consist of 37 amino acid residues. The peptides exhibit six conserved cysteines and follow most likely the cysteine bonding pattern of arthropod defensins (C1–C4, C2–C5, C3–C6), which characterises them as members of the "ancestral group" of invertebrate defensins (Froy and Gurewitz 2003). Their precursors reveal a putative signal peptide of 23–24 amino acid residues, followed by the peptide precursor of 37 amino acid residues and the stop signal. Defensins are primarily active against Gram-positive bacteria.

Defensins are known from several groups of araneomorph spiders such as Araneidae (*Argiope* sp.), Tetragnathidae (*Meta menardi*), Agelenidae (*Tegenaria atrica*), Sparassidae (*Polybetes pythagoricus*), and Ctenidae (*Cupiennius salei* and *Phoneutria reydyi*). Within spiders, defensins are rather similar with 75–100 % similarity at amino acid level. The overall similarity with defensins from ticks and scorpions is in the range of up to 70 %, while the similarity with insects is much lower (Baumann et al. 2010a).

In *C. salei*, it could be shown that defensins are constitutively expressed not only in hemocytes but also in other organs such as ovaries, subesophageal nerve mass, hepatopancreas and muscles of uninfected spiders. This underlines the important role defensins play, not only for defence in the hemolymph, but also in various tissues. Despite analysis of a hemocyte cDNA library of the theraphosid *Acanthoscurria gomesiana* (Lorenzini et al. 2006), up to now no defensin could be identified on the mRNA or peptide level for mygalomorph spiders.

7.3.3.2 Glycine-Rich Peptides

These antimicrobially acting peptides are cationic and characterised by an unusually high glycine content of 71–73 % and only 6–8 further amino acids. This results in repeats of 3–6 glycines, mostly interrupted by a different amino acid. Two very similar peptide families are known so far, acanthoscurrins from mygalomorph and ctenidins from araneomorph spiders.

Two acanthoscurrin isoforms have been isolated from the hemocytes of *Acanthoscurria gomesiana*. They are 130 and 132 amino acids long, with molecular masses of 10.1 and10.2 kDa. Acanthoscurrins have a positive net charge of +8 at physiological pH and are characterised by a three-fold repeat of 26 amino acids: GGGLGGGGLGGGGLGGGKGLGGGGLG. These peptides are active against the Gram-negative bacterium *Escherichia coli* SBS363 (Minimal inhibitory concentration MIC = 2.3–5.6 μM) and the yeast *Candida albicans* (MIC = 1.15–2.3 μM), but however no activity could be measured against the Gram-positive *Micrococcus luteus* up to a concentration of 5.6 μM. Acanthoscurrins are constitutively expressed in hemocytes, stored in their granules and released into the hemolymph upon infection (Fukuzawa et al. 2008; Lorenzini et al. 2003).

In *Cupiennius salei* three ctenidin isoforms have been isolated, which act against Gram-negative *E. coli* (MIC = 2.5–5 μM) but not against the Gram-positive bacteria *Staphylococcus aureus*. Up to a tested concentration of 5 μM, no activity against *C. albicans* was detectable. Growth patterns of bacteria suggest that ctenidins act bacteriostatically rather than bactericidally. The three mature peptides are 109, 119 and 120 amino acid residues long, with molecular masses between 8.8 and 9.6 kDa. Ctenidins are distinguished from acanthoscurrins by a central sequence of 10 amino acids which interrupts the glycine repeat: VIDGKDDVGL. The function of this sequence is still unknown. Ctenidins are constitutively expressed mainly in hemocytes, to a small amount also in the subesophageal nerve mass, but not in six other tissues so far investigated (Baumann et al. 2010b).

7.3.3.3 Small Open-End Cyclic Peptide

Gomesin is a small, open-ended cyclic peptide consisting of 18 amino acids (2.3 kDa) isolated from hemocytes of *Acanthoscurria gomesiana* (Silva et al. 2000). This peptide forms two internal disulphide bridges and adopts a β-hairpin like fold, thus strongly resembling antimicrobial peptides known from other arachnids (tachyplesin and polyphemusin from horseshoe crabs, androctin from scorpions). Gomesin is constitutively expressed in hemocytes, it is stored in the granules and released into the hemolymph after lipopolysaccharide stimulation in a concentration-dependent manner (Fukuzawa et al. 2008). Gomesin has been tested against a variety of microorganisms and it strongly affects the growth of 24 out of 27 tested strains of Gram-positive and Gram-negative bacteria, as well as the development of nine filamentous fungi species and five yeast species. Also eukaryotic cells, such as *Leishmania amazonensis* cells or human blood cells, are affected (Silva et al. 2000). Although gomesin is so far only known from one theraphosid spider, it can be assumed that comparable peptides are more widely spread among spiders, especially since related substances are found in other arachnid groups. Gomesin also exhibits in vitro and in vivo cytotoxicity on tumour cells (Rodrigues et al. 2008), and it is supposed that the cytotoxic effect of this peptide involves L-type calcium channel ion influx, intracellular signalling as well as the generation of reactive oxygen species (Soletti et al. 2010).

7.3.3.4 Hemocyanin Fragments

Very recently, the isolation of rondonin, an antifungal peptide from the hemolymph of the theraphosid *Acanthoscurria natalensis* (sub *A. rondoniae*) has been reported. The amino acid sequence of this small peptide (IIIQYEGHKH) corresponds to a C-terminal fragment of the 'd' subunit of hemocyanin identified from *Acanthoscurria gomesiana* and *Aphonopelma hentzi*. This protein fragment is not active against several Gram-negative and Gram-positive bacteria at a tested concentration up to 67 μM, but exerts a fungicidal activity against fungi and yeasts in micromolar concentrations (fungi: MIC = 1.1–2.1 μM; yeasts: MIC = 8.37–33.5 μM) (Riciluca et al. 2012).

7.4 Hemocytes Derived Compounds Involved in the Immune Response

Mygalin is an acylpolyamine isolated from the hemocytes of mygalomorph and araneomorph spiders (Baumann 2009; Pereira et al. 2007). The compound was identified as bis-acylpolyamine *N*1,*N*8-bis(2,5-dihydroxybenzoyl) spermidine with a molecular mass of 417 Da. Mygalin is only active against *E. coli* in a concentration of 85 μM, but not active against the Gram-positive bacteria *M. luteus* and the yeast *C. albicans*. Pereira et al. supposed an antibacterial activity mediated through the production of hydrogen peroxide. Contrary to this, mygalin isolated from the hemocytes of *Cupiennius salei* exhibits only weak activity in the millimolar range against *E. coli* and *S. aureus* (MIC = 0.5–1 mM). This discrepancy between the results from both publications is not yet understood and needs further investigations (Baumann 2009). Recently, mygalin was also identified as a potent modulator of vertebrate innate immune responses (Mafra et al. 2012).

A single insulin-like growth factor binding domain protein (SIBD-1) was isolated from the hemocytes of *C. salei*. SIBD-1 (8.7 kDa) is characterised by six disulfide bridges and carries a sugar moiety at Thr 2 (Fuc-GlcA-GalNAc-Thr). The peptide is mainly expressed in hemocytes. After infection, the peptide content in the hemocytes decreases and also the temporal SIBD-1 expression seems to be downregulated. While SIBD-1 itself does not affect pathogens, these results indicate that it may be involved in regulatory processes of the immune system (Kuhn-Nentwig et al. 2011; Trachsel et al. 2012).

From the hemocytes of the theraphosid *Lasiodora* sp. the first elastase inhibitor showing antibacterial activity against *Enterococcus faecalis* has been described (Soares et al. 2011).

7.5 Conclusions

The immune system of spiders is localised in their hemocytes and reacts very fast to an invasion by pathogens and to injuries by releasing several compounds into the hemolymph. The immune response comprises (1) phagocytosis, nodulation and encapsulation of invaders, (2) the regulation of hemocyanin/phenoloxidase to produce melanin which destroys pathogens, (3) a clotting cascade to stop hemolymph loss and immobilise invaders and (4) the constitutive production of antimicrobial peptides. So far four different antimicrobial peptide groups have been identified in spiders. Firstly, six similar defensins have been detected in different spider groups. A second group of antimicrobial peptides comprises several glycine-rich peptides (acanthoscurrins and ctenidins), a third group of small cysteine-rich peptides contains gomesin, an 18 amino acid residue peptide, and as a fourth group, a hemocyanin fragment has been identified.

References

Baumann T (2009) Immune defence of *Cupiennius salei*. PhD thesis, University of Bern

Baumann T, Kämpfer U, Schürch S, Schaller J, Largiadèr C, Nentwig W, Kuhn-Nentwig L (2010a) Ctenidins: antimicrobial glycine-rich peptides from the hemocytes of the spider *Cupiennius salei*. Cell Mol Life Sci 67:2787–2798

Baumann T, Kuhn-Nentwig L, Largiadèr C, Nentwig W (2010b) Expression of defensins in non-infected araneomorph spiders. Cell Mol Life Sci 67:2643–2651

Burmester T (2002) Origin and evolution of arthropod hemocyanins and related proteins. J Comp Physiol B 172:95–107. doi:10.1007/s00360-001-0247-7

Burmester T (2013) Evolution and adaptation of hemocyanin within spiders. In: Nentwig W (ed) Spider ecophysiology. Springer, Heidelberg (this volume)

Decker H, Rimke T (1998) Tarantula hemocyanin shows phenoloxidase activity. J Biol Chem 273:25889–25892

Decker H, Tuczek F (2000) Tyrosinase/catecholoxidase activity of hemocyanins: structural basis and molecular mechanism. Trends Biochem Sci 25:392–397

Froy O, Gurewitz M (2003) Arthropod and mollusk defensins-evolution by exon-shuffling. Trends Genet 19:684–687

Fukuzawa AH, Vellutini BC, Lorenzini DM, Silva PI Jr, Mortara RA, da Silva JMC, Daffre S (2008) The role of hemocytes in the immunity of the spider *Acanthoscurria gomesiana*. Dev Comp Immunol 32:716–725

Iwanaga S, Lee BL (2005) Recent advances in the innate immunity of invertebrate animals. J Biochem Mol Biol 38:128–50

Jaenicke E, Decker H (2004) Functional changes in the family of type 3 copper proteins during evolution. Chembiochem 5:163–169. doi:10.1002/cbic.200300714

Kawabata S (2010) Immunocompetent molecules and their response network in horseshoe crabs. Adv Exp Med Biol 708:122–136

Kawabata S, Osaki T, Iwanaga S (2002) Innate immunity in the horseshoe crab. In: Ezekowitz RAB, Hoffmann JA (eds) Infectious disease: innate immunity. Humana, New York

Kemper B (1983) Site of hemocyanin biosynthesis in the tarantula *Euryplema californicum*. Naturwissenschaften 70:255–256

Kropf C (2013) Hydraulic system of locomotion. In: Nentwig W (ed) Spider ecophysiology. Springer, Heidelberg (this volume)

Kuhn-Nentwig L, Largiadèr CR, Streitberger K, Chandru S, Baumann T, Kämpfer U, Schaller J, Schürch S, Nentwig W (2011) Purification, cDNA structure and biological significance of a single insulin-like growth factor-binding domain protein (SIBD-1) identified in the hemocytes of the spider *Cupiennius salei*. Insect Biochem Mol Biol 41:891–901

Lorenzini DM, da Silva PI, Fogaça AC Jr, Bulet P, Daffre S (2003) Acanthoscurrin: a novel glycine-rich antimicrobial peptide constitutively expressed in the hemocytes of the spider *Acanthoscurria gomesiana*. Dev Comp Immunol 27:781–791

Lorenzini DM, Da Silva PI Jr, Soares MB, Arruda P, Setubal J, Daffre S (2006) Discovery of immune-related genes expressed in hemocytes of the tarantula spider *Acanthoscurria gomesiana*. Dev Comp Immunol 30:545–556

Mafra DG, Da Silva PI Jr, Galhardo CS, Nassar R, Daffre S, Sato MN, Borges MM (2012) The spider acylpolyamine mygalin is a potent modulator of innate immune responses. Cell Immunol 275:5–11

Matsuda Y, Osaki T, Hashii T, Koshiba T, Kawabata S (2007) A cysteine-rich protein from an arthropod stabilizes clotting mesh and immobilizes bacteria at injury sites. J Biol Chem 282: 33545–33552

Millot J (1949) Appareil circulatoire. In: Grasse P (ed) Traité de Zoologie, vol VI. Masson, Paris, pp 639–646

Nagai T, Kawabata S (2000) A link between blood coagulation and prophenol oxidase activation in arthropod host defense. J Biol Chem 275:29264–29267

Nagai T, Osaki T, Kawabata S (2001) Functional conversion of hemocyanin to phenoloxidase by horseshoe crab antimicrobial peptides. J Biol Chem 276:27166–27170

Nentwig W (2012) The species referred to as *Eurypelma californicum* (Theraphosidae) in more than 100 publications is likely to be *Aphonopelma hentzi*. J Arachnol 40:128–130

Osaki T, Kawabata S (2004) Structure and function of coagulogen, a clottable protein in horseshoe crabs. Cell Mol Life Sci 61:1257–1265

Pereira LS, Silva PI Jr, Miranda MT, Almeida IC, Naoki H, Konno K, Daffre S (2007) Structural and biological characterization of one antibacterial acylpolyamine isolated from the hemocytes of the spider *Acanthocurria gomesiana*. Biochem Biophys Res Commun 352:953–959

Riciluca KCT, Sayegh RSR, Melo RL, Silva PI Jr (2012) Rondonin an antifungal peptide from spider (*Acanthoscurria rondoniae*) haemolymph. Res Immunol 2:66–71

Rodrigues EG, Dobroff AS, Cavarsan CF, Paschoalin T, Nimrichter L, Mortara RA, Santos EL, Fazio MA, Miranda A, Daffre S, Travassos LR (2008) Effective topical treatment of subcutaneous murine B16F10-Nex2 melanoma by the antimicrobial peptide gomesin. Neoplasia 10: 61–68

Seitz KA (1972) Zur Histologie und Feinstruktur des Herzens und der Hämocyten von *Cupiennius salei* KEYS. (Araneae, Ctenidae). Zool Jb Anat 89:351–384

Seitz KA (1976) Zur Feinstruktur der Häutungshämocyten von *Cupiennius salei* KEYS (Araneae, Ctenidae). Zool Jb Anat 96:280–292

Sherman RG (1973) Ultrastructurally different hemocytes in a spider. Can J Zool 51:1155–1159

Sherman RG (1981) Chelicerates. In: Ratcliffe NA, Rowley AF (eds) Invertebrate blood cells, vol 2. Academic, London

Silva PI Jr, Daffre S, Bulet P (2000) Isolation and characterization of gomesin, an 18-residue cysteine-rich defense peptide from the spider *Acanthoscurria gomesiana* hemocytes with sequence similarities to horseshoe crab antimicrobial peptides of the tachyplesin family. J Biol Chem 275:33464–33470

Soares T, Ferreira FRB, Gomes FS, Coelho LCBB, Torquato RJS, Napoleao TH, Cavalcanti MSM, Tanaka AS, Paiva PMG (2011) The first serine protease inhibitor from *Lasiodora* sp. (Araneae: Theraphosidae) hemocytes. Prog Biochem 46:2317–2321

Söderhäll K, Cerenius L (1998) Role of the prophenoloxidase-activating system in invertebrate immunity. Curr Opin Immunol 10:23–28

Soletti RC, del Barrio L, Daffre S, Miranda A, Borges HL, Moura-Neto V, Lopez MG, Gabilan NH (2010) Peptide gomesin triggers cell death through L-type channel calcium influx, MAPK/ERK, PKC and PI3K signaling and generation of reactive oxygen species. Chem Biol Interact 186:135–143

Trachsel C, Widmer C, Kämpfer U, Buhr C, Baumann T, Kuhn-Nentwig L, Schürch S, Schaller J, Baumann U (2012) Structural and biochemical characterization of native and recombinant single insulin-like growth factor-binding domain protein (SIBD-1) from the Central American hunting spider *Cupiennius salei* (Ctenidae). Proteins 80:2323–2329

Wirkner C, Huckstorf K (2013) The circulatory system of spiders. In: Nentwig W (ed) Spider ecophysiology. Springer, Heidelberg (this volume)

Chapter 8
Endosymbiont Infections in Spiders

Sara L. Goodacre and Oliver Y. Martin

8.1 Introduction: Occurrence of Endosymbiont Infections in Spiders

Spiders are known to be infected with a diverse range of intracellular bacteria belonging to the Rickettsiales, Bacteroidetes, Mollicutes and the Gammaproteobacteria (reviewed in Goodacre and Martin 2012; Martin and Goodacre 2009). In other invertebrates these bacteria, which are termed "endosymbionts," are associated with a wide range of effects on their hosts' reproductive biology. They include the widely reported *Wolbachia pipientis,* which is estimated to infect more than 60 % of insect species (Hilgenboecker et al. 2008). These symbionts are thought primarily to act as selfish genetic elements, promoting their own transmission by increasing the reproductive success of infected females. They do not survive outside the host and their mechanism of transmission is thus almost exclusively vertical, from mother to offspring, although incongruities between host and bacterial phylogenies indicate that occasional horizontal transmission events must also occur.

8.1.1 The Enemies Within: Symbionts or Reproductive Parasites?

Studies of the interaction between invertebrate hosts and their obligate intracellular bacteria, such as *Hamiltonella defensa* in the pea aphid (Degnan et al. 2009),

S.L. Goodacre (✉)
School of Biology, University of Nottingham, Nottingham, UK
e-mail: sara.goodacre@nottingham.ac.uk

O.Y. Martin
ETH Zurich, Experimental Ecology, Institute for Integrative Biology, Zürich, Switzerland
e-mail: oliver.martin@env.ethz.ch

W. Nentwig (ed.), *Spider Ecophysiology*,
DOI 10.1007/978-3-642-33989-9_8,

indicate that these kinds of associations have persisted over millions of years and that they are mutualistic with each partner gaining from the other. In this sense use of the word "symbiosis" is an appropriate descriptor. In contrast, bacteria such as *Wolbachia*, *Rickettsia* and *Spiroplasma*, which are the focus of the current review, do not fall into the same category. They are not considered to be strictly in symbiosis with their hosts because they act as reproductive parasites and persist only through manipulating their hosts' biology. Their association with the host is thought in many cases to have existed over a relatively short (ecological) time frame.

It is now known, however, that there may be an extraordinarily diverse range of benefits conferred by endosymbionts such as *Wolbachia*. For example it is shown that their presence can be advantageous to the host by conferring protection against viruses or heat shock (reviewed in Goodacre and Martin 2012). Studies of infections in arthropods and other groups of invertebrates as a whole also indicate that such subtle advantages may be more commonly associated with infection than was originally thought to be the case (reviewed in Jaenike 2012).

The length of time during which there has been an association between endosymbionts and their spider hosts is not straightforward to estimate, but inferences can be made from the molecular identities of the bacteria that they carry. For instance *Rickettsia* endosymbionts that are found in spiders appear to be phylogenetically distinct from those found in other arthropod groups (Goodacre et al. 2006; Weinert et al. 2009) and distributed widely throughout the araneomorph spiders studied to date. This finding is in contrast with the observation that the *Wolbachia* and *Spiroplasma* found in spiders are not spider specific but come from lineages known to infect a diverse range of other arthropod hosts. This observation is consistent with more recent infection. Mechanisms through which the latter might have been acquired are not well understood but potentially include horizontal transmission to spider hosts from infected prey species.

The possibility that transitions from initial parasitism/pathogenicity to mutualistic relationships have occurred between endosymbionts and their spider hosts has yet to be fully investigated. Such transitions could potentially be rapid if they only involve small numbers of genetic changes, and thus both *Rickettsia* and *Wolbachia* might provide benefits to their host that earn them the title "symbiont," regardless of the amount of evolutionary time that they have had to co-evolve. Genome evolution has been particularly well characterised in *Rickettsia,* where rapid "shuffling" of genomic content occurs through mechanisms such as horizontal gene transfer (e.g. Weinert et al. 2009) or recombination between different bacterial strains (Goodacre et al. 2006). Recombination, which is also thought to occur amongst *Wolbachia* (e.g. Jiggins 2002), rapidly creates novel bacterial strains that potentially have different phenotypes in their host than either of the parental host strains. The nature of the relationship between a bacterium and its host may also change rapidly through evolution of the host genome. For instance it is known that the evolution of male-killer suppression can be swift, occurring over small numbers of generations (Hornett et al. 2006).

8.1.2 *Prevalence in Araneae*

All four of the most commonly reported reproductive parasites of invertebrates have been documented in spiders (Table 8.1): *Wolbachia* and *Rickettsia* (from the Rickettsiales), *Spiroplasma* (a Mollicute) and *Cardinium* (a member of the Bacteroidetes group). A recently discovered, less well-documented gamma-proteobacterium, *Arsenophonus* has also been reported to infect spiders, but in only one species to date (Duron et al. 2008a, Table 8.1). Infections in spiders are found in both entelegyne and haplogyne species, and it is of note that most species studied thus far have been found to carry at least one endosymbiotic type. An exception to this is the study of subsocial and social *Stegodyphus* spiders where initial investigations have not detected *Rickettsia, Wolbachia, Cardinium* or *Spiroplasma* (Goodacre, unpublished data). These species were studied because they show a marked primary sex-ratio bias and endosymbiotic bacteria are known in some species to cause such a skew (discussed below). Further study of *Stegodyphus* is necessary to rule out the possibility that they carry an as yet unidentified bacterial infection.

Studies of endosymbionts in spiders thus far have largely been through detection of bacterial DNA through the use of PCR-based molecular methods, with few attempts to characterise where within specific tissues the bacteria are distributed. Studies of other arthropods demonstrate that bacteria can be highly localised to specific ranges of tissue type and that more than one localisation pattern can be observed within the same host. For instance, *Rickettsia* may either be found widely distributed throughout the hemocoel of particular insects or localised to bacteriocytes, and there is evidence to suggest that the bacterial phenotype is dependent upon which of these is the case (Caspi-Fluger et al. 2011). Microbes that resembled *Rickettsia* were reported from the nervous tissue of a jumping spider, *Salticus* spp. as early as 1923 (Cowdry) and data from the spider *Erigone atra* indicate that *Rickettsia* are found in regions of nervous tissue but not throughout the spider (Goodacre et al. 2009; Goodacre, unpublished data). Detailed study of precise tissue or cellular localisations appears not yet to have been carried out in any other spider.

8.2 Consequences of Endosymbiont Infection

8.2.1 *Classic Bacterial Phenotypes in Invertebrate Hosts*

Extensive studies of *Wolbachia* and other bacterial infections of insect hosts have uncovered four classic reproductive phenotypes that these reproductive parasites can cause (reviewed in Duron et al. 2008a). These are: male-killing (at early or late stages of development), feminization, thelytokous parthenogenesis and cytoplasmic incompatibility. Through these manipulations, the predominantly

Table 8.1 Recorded endosymbiont infections in spiders

Families	Species	Symbiont(s)	Reference(s)
Agelenidae	*Tegenaria duellica*	*Cardinium, Rickettsia, Wolbachia*	Martin and Goodacre (2009) and Goodacre et al. (2006)
	Agelenopsis sp.	*Spiroplasma, Wolbachia*	Goodacre et al. (2006) and Baldo et al. (2008)
Araneidae	*Araneus diadematus*	*Arsenophonus, Cardinium, Rickettsia, Spiroplasma, Spiroplasma ixodetis*	Martin and Goodacre (2009), Duron et al. (2008a) and Goodacre et al. (2006)
	Araneus ventricosus	*Wolbachia*	Wang et al. (2010)
	Araneus sp.	*Wolbachia*	Rowley et al. (2004)
	Cyclosa conica	*Cardinium*	Duron et al. (2008a, b)
	Eriovixia cavaleriei	*Wolbachia*	Wang et al. (2010)
	Larinia argiopiformis	*Wolbachia*	Wang et al. (2010)
	Larinia tabida	*Wolbachia*	Rowley et al. (2004)
	Linyphia triangularis	*Cardinium, Wolbachia*	Duron et al. (2008a, b)
	Nereine clathrata	*Spiroplasma ixodetis*	Duron et al. (2008a)
	Poecilopachys australasiae	*Wolbachia*	Rowley et al. (2004)
	Zygiella x-notata	*Cardinium, Rickettsia, Spiroplasma*	Martin and Goodacre (2009) and Goodacre et al. (2006)
	gen. sp.	*Cardinium*	Chang et al. (2010)
Clubionidae	*Clubiona terrestris*	*Rickettsia*	Goodacre et al. (2006)
Cybaeidae	*Cybaeus* sp.	*Cardinium*	Perlman et al. (2010)
Desidae	*Badumna longinqua*	*Wolbachia*	Rowley et al. (2004)
Dysderidae	*Dysdera crocata*	*Wolbachia*	Cordaux et al. (2001)
	Dysdera erythrina	*Wolbachia*	Cordaux et al. (2001)
Gnaphosidae	*Zelotes latreillei*	*Spiroplasma*	Goodacre et al. (2006)
Linyphiidae	*Baryphma trifrons*	*Rickettsia*	Goodacre et al. (2006)
	Bathyphantes gracilis	*Wolbachia*	Goodacre et al. (2006)
	Centromerita bicolor	*Spiroplasma*	Goodacre et al. (2006)
	Diplocephalus cristatus	*Wolbachia*	Goodacre et al. (2006)
	Diplocephalus latifrons	*Spiroplasma*	Goodacre et al. (2006)
	Diplocephalus picinus	*Spiroplasma*	Goodacre et al. (2006)
	Entelecara flavipes	*Wolbachia*	Goodacre et al. (2006)
	Erigone arctica	*Spiroplasma*	Goodacre et al. (2006)
	Erigone atra	*Cardinium, Rickettsia, Spiroplasma, Wolbachia*	Goodacre et al. (2006, 2009)
	Erigone dentipalis	*Rickettsia, Wolbachia*	Goodacre et al. (2006)

(continued)

Table 8.1 (continued)

Families	Species	Symbiont(s)	Reference(s)
	Erigone longipalpis	*Rickettsia, Wolbachia*	Goodacre et al. (2006)
	Erigone promiscua	*Wolbachia*	Goodacre et al. (2006)
	Gonatium rubens	*Wolbachia*	Goodacre et al. (2006)
	Hylaphantes graminicola	*Rickettsia*	Goodacre et al. (2006)
	Hypomma bituberculatum	*Rickettsia, Wolbachia*	Goodacre et al. (2006)
	Hypomma cornutum	*Rickettsia*	Goodacre et al. (2006)
	Lepthyphantes minutus	*Cardinium, Rickettsia, Wolbachia*	Goodacre et al. (2006) and Martin and Goodacre (2009)
	Macrargus rufus	*Wolbachia*	Goodacre et al. (2006)
	Megalepthyphantes collinus	*Wolbachia*	Goodacre et al. (2006)
	Meioneta rurestris	*Wolbachia*	Goodacre et al. (2006)
	Micrargus herbigadus	*Rickettsia*	Goodacre et al. (2006)
	Microlinyphia imprigra	*Rickettsia*	Goodacre et al. (2006)
	Microlinyphia pusilla	*Rickettsia*	Goodacre et al. (2006)
	Microneta viaria	*Rickettsia*	Goodacre et al. (2006)
	Nereine clathrata	*Wolbachia*	Goodacre et al. (2006)
	Nereine montana	*Rickettsia, Wolbachia*	Goodacre et al. (2006)
	Nereine peltata	*Rickettsia, Wolbachia*	Goodacre et al. (2006)
	Oedothorax apicatus	*Wolbachia*	Goodacre et al. (2006)
	Oedothorax fuscus	*Rickettsia, Wolbachia*	Goodacre et al. (2006)
	Oedothorax gibbosus	*Rickettsia, Wolbachia*	Goodacre et al. (2006) and Vanthournout et al. (2011)
	Oedothorax retusus	*Wolbachia*	Goodacre et al. (2006)
	Pityohyphantes phrygianus	*Rickettsia, Wolbachia*	Goodacre et al. (2006) and Gunnarsson et al. (2009)
	Peponcranium ludicrum	*Spiroplasma, Wolbachia*	Goodacre et al. (2006)
	Pocadicnemus juncea	*Wolbachia*	Goodacre et al. (2006)
	Silometopus elegans	*Wolbachia*	Goodacre et al. (2006)
	Stemonyphantes lineatus	*Wolbachia*	Goodacre et al. (2006)
	Tenuiphantes tenuis	*Wolbachia*	Goodacre et al. (2006)
	Tenuiphantes zimmermanni	*Rickettsia*	Goodacre et al. (2006)
	Tiso vagans	*Wolbachia*	Goodacre et al. (2006)
	Troxochrus scabriculus	*Rickettsia, Spiroplasma, Wolbachia*	Goodacre et al. (2006)
	Walckenaeria antica	*Wolbachia*	Goodacre et al. (2006)

(continued)

Table 8.1 (continued)

Families	Species	Symbiont(s)	Reference(s)
	Walckenaeria atrotibialis	*Wolbachia*	Goodacre et al. (2006)
	Walckenaeria obtusa	*Wolbachia*	Goodacre et al. (2006)
Liocranidae	gen. sp.	*Cardinium*	Martin and Goodacre (2009)
Lycosidae	*Alopecosa pulverulenta*	*Cardinium, Wolbachia*	Duron et al. (2008a, b)
	Pardosa agrestis	*Spiroplasma*	Goodacre et al. (2006)
	Pardosa agricola	*Spiroplasma*	Goodacre et al. (2006)
	Pardosa bifasciata	*Spiroplasma*	Goodacre et al. (2006)
	Pardosa hyperborea	*Spiroplasma*	Goodacre et al. (2006)
	Pardosa lugubris	*Spiroplasma, Spiroplasma poulsonii*	Duron et al. (2008a) and Goodacre et al. (2006)
	Pardosa monticola	*Spiroplasma*	Goodacre et al. (2006)
	Pardosa pullata	*Spiroplasma, Wolbachia*	Duron et al. (2008a, b) and Goodacre et al. (2006)
	Pardosa riparia	*Spiroplasma*	Goodacre et al. (2006)
	Pardosa sphagnicola	*Spiroplasma*	Goodacre et al. (2006)
Nephilidae	*Nephila clavata*	*Wolbachia*	Oh et al. (2000) and Wang et al. (2010)
	Nephila plumipes	*Wolbachia*	Rowley et al. (2004)
Oxyopidae	*Oxyopes sertatus*	*Wolbachia*	Wang et al. (2010)
Pholcidae	*Holocnemus pluchei*	*Cardinium*	Duron et al. (2008a, b)
	Pholcus crypticolens	*Wolbachia*	Wang et al. (2010)
	Pholcus phalangioides	*Cardinium, Rickettsia, Wolbachia*	Martin and Goodacre (2009), Goodacre et al. (2006), Rowley et al. (2004) and Duron et al. (2008a, b)
Salticidae	*Evarcha falcata*	*Cardinium*	Duron et al. (2008a, b)
	Ghelna canadensis	*Cardinium*	Martin and Goodacre (2009)
	Habrocestum pulex	*Cardinium*	Martin and Goodacre (2009)
	Holoplatys sp.	*Wolbachia*	Rowley et al. (2004)
	Maevia inclemens	*Cardinium*	Martin and Goodacre (2009)
	Marpissa lineata	*Cardinium*	Martin and Goodacre (2009)
	Neon neli	*Cardinium*	Martin and Goodacre (2009)
	Pelegrina proterva	*Cardinium*	Martin and Goodacre (2009)
	Phidippus audax	*Cardinium*	Martin and Goodacre (2009)
	Salticus scenicus	*Cardinium*	Martin and Goodacre (2009)
Tetragnathidae	gen. sp.	*Cardinium*	Chang et al. (2010)
	Leucauge dromedaria	*Wolbachia*	Rowley et al. (2004)

(continued)

Table 8.1 (continued)

Families	Species	Symbiont(s)	Reference(s)
	Meta mengei	*Cardinium, Spiroplasma ixodetis, Rickettsia, Wolbachia*	Martin and Goodacre (2009), Goodacre et al. (2006) and Duron et al. (2008b)
	Metellina segmenta	*Spiroplasma ixodetis, Wolbachia*	Duron et al. (2008a, b)
	Pachygnatha degeeri	*Cardinium, Wolbachia*	Duron et al. (2008a, b)
	Pachygnatha listeri	*Wolbachia*	Duron et al. (2008a, b)
	Tetragnatha montana	*Cardinium, Spiroplasma ixodetis, Wolbachia*	Martin and Goodacre (2009) and Duron et al. (2008a, b)
Theridiidae	*Coleosoma octomaculatum*	*Wolbachia*	Wang et al. (2010)
	Enoplognatha ovata	*Cardinium, Wolbachia*	Martin and Goodacre (2009) and Duron et al. (2008a, b)
	gen. sp.	*Cardinium*	Chang et al. (2010)
	gen. sp.	*Wolbachia*	Goodacre et al. (2006)
	gen. sp.	*Rickettsia*	Goodacre et al. (2006)
Thomisidae	*Diaea* sp.	*Wolbachia*	Rowley et al. (2004)
	Diaea circumlita	*Wolbachia*	Rowley et al. (2004)
	gen. sp.	*Cardinium*	Martin and Goodacre (2009)
Titanoecidae	*Nurscia albofasciata*	*Wolbachia* A	Yun et al. (2011)
Uloboridae	*Zosis geniculatus*	*Wolbachia*	Rowley et al. (2004)

Information collected from the cited references, or in case of Goodacre et al. (2006) the original data (as paper does not provide species-level information)

maternally inherited microbe favours its own transmission either by skewing the sex ratio towards females or by causing incompatibilities between infected and uninfected hosts, such that uninfected females are less reproductively successful than their infected counterparts. Examples of each of these four phenotypes are known to be caused by *Wolbachia* and both male-killing and parthenogenesis are known to be caused by *Ricketsia*. Several other bacterial strains, such as *Arsenophonus, Flavobacterium* and *Spiroplasma* species, are known to be male-killers.

There are few studies to date testing spiders infected with endosymbionts for the classically observed phenotypes described above. Cytoplasmic incompatibility, feminization and parthenogenesis driven by endosymbiont infections have been found in other arachnids such as mites (Weeks et al. 2001; Chigira and Miura 2005; Gotoh et al. 2007; Groot and Breeuwer 2006), but the hypothesis that there are similar bacterial phenotypes in spiders remains largely untested. It seems likely, however, that spiders too are vulnerable to such reproductive manipulation by intracellular bacteria (Goodacre 2011). A microbial cause for parthenogenesis in spiders is an avenue worth investigating given the large number of candidate parthenogenetic species in this group (Martin and Goodacre 2009; reviewed by

Goodacre 2011). Similarly, cytoplasmic incompatibility in particular species has been suggested as a potential explanation for observed instances where not all crosses are equally fertile (Goodacre et al. 2006).

In contrast to the circumstantial case for endosymbiont-driven parthenogenesis and cytoplasmic incompatibility in spiders there is good evidence for a microbial involvement in sex-ratio bias in two particular species. The first species is the linyphiid *Pityohyphantes phrygianus* where it appears that *Wolbachia* influences the sex ratio through an alteration in behavioural traits linked to female control over offspring sex ratio (Gunnarsson et al. 2009). The second species is the linyphiid *Oedothorax gibbosus* where there is evidence to suggest that *Wolbachia* infection causes a skewed sex ratio through male-killing (Vanthournout et al. 2011). In both *Pityohyphantes* and *Oedothorax* curing females of their infections using antibiotic treatment restores a more even sex ratio. It should be noted, however, that in neither case does the presence of *Wolbachia* itself appear to fully explain the variation amongst females in the sex ratio of their offspring.

8.2.2 Impacts on Host Behaviour and Fitness

Recent evidence indicates that the symbiont *Rickettsia* hampers long distance dispersal in the spider *Erigone atra* without any other detectable effects on fitness (Goodacre et al. 2009). Specifically, infected females were found to be less likely to adopt long-range dispersal behaviour (ballooning). This represents a highly interesting and novel finding, as the effect in question relates to a non-reproductive behaviour with potential impacts on spatial patterns of gene flow. It is not clear, however, how this behavioural change increases the reproductive fitness of infected individuals relative to their uninfected counterparts. One hypothesis (yet to be tested) is that the population carries many bacterial strains that are both incompatible with one another and geographically localised. If this were the case, reduced dispersal might increase the chances of finding a mate with a compatible strain. This argument has been proposed to explain the findings of Vala et al. (2004) who showed that *Wolbachia* infected *Tetranychus urticae* females aggregate their offspring to a greater extent than uninfected individuals. The authors proposed that this behavioural change is driven by *Wolbachia* infection and that it promotes matings between siblings, which are likely to have inherited the same, compatible bacterial variant, thereby ensuring that the cross is fertile.

It has also recently been shown that *Wolbachia* can influence offspring sex ratio via an alteration of female post-copulatory position in the linyphiid spider *Pityohyphantes phrygianus*. In this species females can adopt a variety of positions post-mating; a ventral-side-up position leads to proportionally more sons than a dorsal-side-up equivalent (Gunnarsson et al. 2009). This is particularly noteworthy as it represents a *behavioural* manipulation with the potential to skew sex ratio towards females, i.e. with results akin to the classic phenotypes outlined above. Of further interest is the observation that there is a relationship between female body

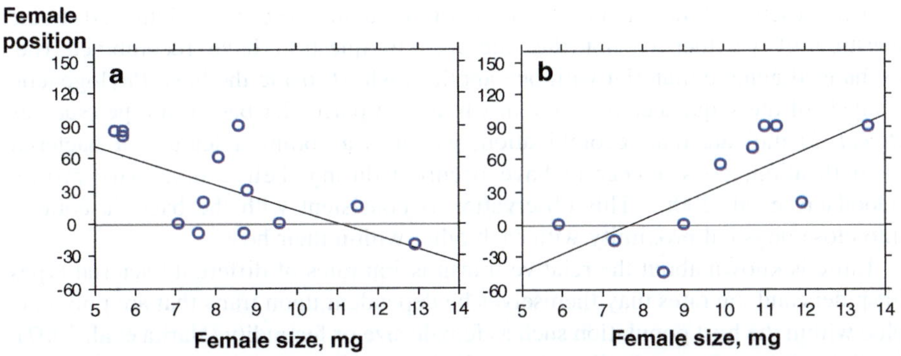

Fig. 8.1 Opisthosomal position of infected female *Pityohyphantes* (0° ventral side up; 180° dorsal side up) after mating with (**a**) infected and (**b**) uninfected male (data taken from Gunnarsson et al. 2009)

size in *P. phrygianus* and the sex ratio of offspring, but that the nature of this relationship is altered by the presence of *Wolbachia* (Fig. 8.1). Large females are shown to adopt a more ventral-side-up position post mating and hence produce greater number of sons than smaller females. However, this was only true if both the male and female were infected with *Wolbachia*. Infected females mating with uninfected males showed precisely the opposite trend: large females adopt a more dorsal-side up position and hence produce proportionally fewer males than smaller individuals. No such relationship between female size and sex ratio has been observed in females that do not carry *Wolbachia*. This may be because no such relationship exists or because none was detected in the small number of individuals included in the study thus far. Further experiments are ongoing to determine how *Wolbachia* favours its own transmission in *Pityohyphantes* (Cotterill and Goodacre, unpublished data).

8.2.3 Multiple Infections

It is not uncommon for multiple infections with different strains or species of symbiotic bacteria to occur even within individuals. Spiders from diverse groups such as the araneid *Araneus diadematus* and the agelenid *Tegenaria duellica* both carry at least four endosymbiont types (Table 8.1). Individual spiders may even carry more than one genetic variant of a particular type of bacterium. As an example, multiple copies of the *Rickettsia* citrase synthase gene were recovered from individual spiders by Goodacre et al. (2006) using conventional PCR and sequencing. This strongly supports the view that at least two (and potentially many more) variants of *Rickettsia* bacteria were present within the same individual spider although the method used means it is not possible to ascertain whether or not these occupied the same host tissue type.

The presence of more than one variant of the same type of bacterium leads to the question of whether or not the microbes in question have recombined (i.e. exchanged genetic material with one another) whilst inside the host. Phylogenetic analysis of the sequences of different variants of particular bacterial types such as *Rickettsia* indicate that recombination, which is a common feature of bacterial diversification, does appear to have occurred during their evolutionary history (Goodacre et al. 2006). This observation is consistent with the bacteria coming into close physical proximity with each other within their host.

Little is known about the relative transmission rates of different bacterial types in spiders and the rates may themselves be dependent upon traits that are polymorphic within the host population such as female size or fecundity (Narita et al. 2007). Spatial patterns of genetic diversity of the host, which are common in arthropods, may also determine differences in the response of populations of the same species to a particular endosymbiont infection. Recent study of the pholcid spider *Holocnemus pluchei* showing the presence of divergent mitochondrial clades (Stefanini and Duron 2012) illustrates one such potential case. Transmission may also be influenced by environmental factors such as external temperature (van Opijnen and Breeuwer 1999). Microbial species compete with each other for vertical transmission, thus the presence of multiple infections might also influence the transmission rate of the individual competing bacterial strains. The importance of disentangling the different phenotypes of potentially competing bacteria is illustrated by studies such as that of *Bryobia* spider mites, which have shown that phenotypes of one bacterial type are modified by the presence of other strains. For example, the presence of *Wolbachia* modifies the phenotype exerted by co-infecting *Cardinium* (Ros and Breeuwer 2009).

8.3 Conclusions and Directions for Future Research

A main focus of studies on endosymbiont infections has been to determine their effect on host reproduction. In this context it is worth noting that whilst some classic bacterial phenotypes, such as cytoplasmic incompatibility, have not yet been observed in spiders, they have been detected in other arachnids and their current absence might therefore reflect lack of study of the group rather than that this phenotype is absent. The recent work on dispersal behaviour (Goodacre et al. 2009) indicates, however, that effects need not be limited to traits associated with reproduction. It would hence be worthwhile when evaluating effects on host biology to assess a more comprehensive and diverse range of host traits.

When considering the effects on a host of a bacterial endosymbiont it is also important to account for both spatial heterogeneity and temporal changes in infection frequency. Spatial variation in the distribution of particular bacteria is known to be common and has been found within the wider metapopulation of species such as the linyphiid spider *Erigone atra* (Goodacre et al. 2009). Rapid temporal changes are known from studies of other invertebrates, such as male-killing bacteria in *Hypolimnas* butterflies (Charlat et al. 2007). They certainly also occur in spiders, for example a twofold decrease in *Wolbachia* prevalence was observed in *Pityohyphantes phrygianus* over a 4-year period (Cotterill and Goodacre, unpublished data). The phenotypic effects observed in a study population thus are neither predicted to be constant throughout the wider metapopulation nor necessarily currently to be at equilibrium.

In the case of *E. atra* where *Rickettsia* infections are associated with reduced dispersal of females only and where infection frequency varies amongst populations it is expected that the net effects of the infection are (1) fluctuations in the local sex ratio and (2) localised restrictions in female-mediated gene flow. Alterations in the population sex ratio driven by endosymbiont infections can influence processes involved in sexual selection and sexual conflict. For instance, the intensity of male–male competition and opportunities for mate choice will be determined by the number of each sex present with consequences for the strength of selection on reproductive traits in both sexes.

The absolute change in the number of individuals following altered female dispersal rate may further contribute to density or frequency-dependent processes. Again, local population density will have a major effect on reproduction, and could further shape reproductive isolation between populations (Martin and Hosken 2003). Restricted gene flow is further predicted to increase the rate at which localised adaptations can be fixed through selection and to reduce effective population size thus increasing the level of inbreeding and the chance of fixation of non-adaptive traits through random genetic drift. Spiders represent a tractable system for studying the potentially subtle ways in which endosymbionts can affect their hosts' biology at both local and wider scales.

References

Baldo L, Ayoub NA, Hayashi CY, Russell JA, Stahlhut JK, Werren JH (2008) Insight into the routes of *Wolbachia* invasion: high levels of horizontal transfer in the spider genus *Agelenopsis* revealed by *Wolbachia* strain and mitochondrial DNA diversity. Mol Ecol 17:557–569. doi:10.1111/j.1365-294X.2007.03608.x

Caspi-Fluger A, Inbar M, Mozes-Daube N, Mouton L, Hunter MS, Zchori-Fein E (2011) *Rickettsia* 'in' and 'out': two different localization patterns of a bacterial symbiont in the same insect species. PLoS One 6:e21096. doi:10.1371/journal.pone.0021096

Chang J, Masters A, Avery A, Werren JH (2010) A divergent *Cardinium* found in daddy long-legs (Arachnida: Opiliones). J Invertebr Pathol 105:220–227. doi:1016/j.jip.2010.05.017

Charlat S, Hornett EA, Fullard JH, Davies N, Roderick GK, Wedell N, Hurst GDD (2007) Extraordinary flux in sex ratio. Science 317:214. doi:10.1126/science.1143369

Chigira A, Miura K (2005) Detection of '*Candidatus* Cardinium' bacteria from the haploid host *Brevipalpus californicus* (Acari: Tenuipalpidae) and effect on the host. Exp Appl Acarol 37:107–116. doi:10.1007/s10493-005-0592-4

Cordaux R, Michel-Salzat A, Bouchon D (2001) *Wolbachia* infection in crustaceans: novel hosts and potential routes for horizontal transmission. J Evol Biol 14:237–243. doi:10.1046/j.1420-9101.2001.00279.x

Cowdry EV (1923) The distribution of *Rickettsia* in the tissues of insects and arachnids. J Exp Med 37:431–456

Degnan P, Yub Y, Sisneros N, Wing RA, Moran NA (2009) *Hamiltonella defensa,* genome evolution of protective bacterial endosymbiont from pathogenic ancestors. Proc Natl Acad Sci U S A 106:9063–9068. doi:10.1073/pnas.0900194106

Duron O, Bouchon D, Boutin S, Bellamy L, Zhou LQ, Engelstädter J, Hurst GDD (2008a) The diversity of reproductive parasites among arthropods: *Wolbachia* do not walk alone. BMC Biol 6:27. doi:10.1186/1741-7007-6-27

Duron O, Hurst GDD, Hornett EA, Jostling JA, Engelstädter J (2008b) High incidence of the maternally inherited bacterium *Cardinium* in spiders. Mol Ecol 17:1427–1437. doi:10.1111/j.1365-294X.2008.03689.x

Goodacre SL (2011) Endosymbiont infections in spiders. Adv Ins Physiol 40:137–153. doi:10.1016/B978-0-12-387668-3.00003-9

Goodacre SL, Martin OY (2012) Modification of insect and arachnid behaviours by vertically transmitted endosymbionts: infections as drivers of behavioural change and evolutionary novelty. Insects 3:246–261. doi:10.3390/insects3010246

Goodacre SL, Martin OY, Thomas CFG, Hewitt GM (2006) *Wolbachia* and other endosymbiont infections in spiders. Mol Ecol 15:517–527. doi:10.1111/j.1365-294X.2005.02802.x

Goodacre SL, Martin OY, Bonte D, Hutchings L, Woolley C, Ibrahim K, Thomas CFG, Hewitt GM (2009) Microbial modification of host long-distance dispersal capacity. BMC Biol 7:32. doi:10.1186/1741-7007-7-32

Gotoh T, Noda H, Ito S (2007) *Cardinium* symbionts cause cytoplasmic incompatibility in spider mites. Heredity 98:13–20. doi:10.1038/sj.hdy.6800881

Groot T, Breeuwer JAJ (2006) *Cardinium* symbionts induce haploid thelytoky in most clones of three closely related *Brevipalpus* species. Exp Appl Acarol 39:257–271. doi:10.1007/s10493-006-9019-0

Gunnarsson B, Goodacre SL, Hewitt GM (2009) Sex ratio, mating behaviour and *Wolbachia* infections in a sheetweb spider. Biol J Linn Soc 98:181–186. doi:10.1111/j.1095-8312.2009.01247.x

Hilgenboecker K, Hammerstein P, Schlattmann P, Telschow A, Werren JH (2008) How many species are infected with *Wolbachia*? – a statistical analysis of current data. FEMS Microbiol Lett 281:215–220. doi:10.1111/j.1574-6968.2008.01110.x

Hornett EA, Charlat S, Duplouy AMR, Davies N, Roderick GK, Wedell N, Hurst GDD (2006) Evolution of male-killer suppression in a natural population. PLoS Biol 4:1643–1648. doi:10.1371/journal.pbio.0040283

Jaenike J (2012) Population genetics of beneficial heritable symbionts. Trends Ecol Evol 27: 226–232. doi:10.1016/j.tree.2011.10.005

Jiggins FM (2002) The rate of recombination in *Wolbachia* bacteria. Mol Biol Evol 19:1640–43

Martin OY, Goodacre SL (2009) Widespread infections by the bacterial endosymbiont *Cardinium* in arachnids. J Arachnol 37:106–108. doi:10.1636/SH08-05.1

Martin OY, Hosken DJ (2003) The evolution of reproductive isolation through sexual conflict. Nature 423:979–982. doi:10.1038/nature01752

Narita S, Nomura M, Kageyama D (2007) Naturally occurring single and double infection with *Wolbachia* strains in the butterfly *Eurema hecabe*: transmission efficiencies and population density dynamics of each *Wolbachia* strain. FEMS Microbiol Ecol 61:235–245. doi:10.1111/j.1574-6941.2007.00333.x

Oh HW, Kim MIG, Shin SW, Bae KS, Ahn YJ, Park HY (2000) Ultrastructural and molecular identification of a *Wolbachia* endosymbiont in a spider, *Nephila clavata*. Insect Mol Biol 9: 539–543. doi:10.1046/j.1365-2583.2000.00218.x

Perlman SJ, Magnus SA, Copley CR (2010) Pervasive associations between *Cybaeus* spiders and the bacterial symbiont *Cardinium*. J Invertebr Pathol 103:150–155. doi:10.1016/j.jip.2009.12.009

Ros VID, Breeuwer JAJ (2009) The effects of, and interactions between, *Cardinium* and *Wolbachia* in the doubly infected spider mite *Bryobia sarothamni*. Heredity 102:413–422. doi:10.1038/hdy.2009.4

Rowley SM, Raven RJ, McGraw EA (2004) *Wolbachia pipientis* in Australian spiders. Curr Microbiol 49:208–214. doi:10.1007/s00284-004-4346-z

Stefanini A, Duron O (2012) Exploring the effect of the *Cardinium* endosymbiont on spiders. J Evol Biol. doi:10.1111/j.1420-9101.2012.02535.x

Vala F, Egas M, Breeuwer JA, Sabelis MW (2004) *Wolbachia* affects oviposition and mating behaviour of its spider mite host. J Evol Biol 17:692–700. doi:10.1046/j.1420-9101.2003.00679.x

van Opijnen T, Breeuwer JA (1999) High temperatures eliminate *Wolbachi*a, a cytoplasmic incompatibility inducing endosymbiont, from the two-spotted spider mite. Exp Appl Acarol 23:871–881. doi:10.1023/A:1006363604916

Vanthournout B, Swaegers J, Hendrickx F (2011) Spiders do not escape reproductive manipulations by *Wolbachia*. BMC Evol Biol 11:15. doi:10.1186/1471-2148-11-15

Wang Z, Deng C, Yun Y, Jian C, Peng Y (2010) Molecular detection and the phylogenetics of *Wolbachia* in Chinese spiders (Araneae). J Arachnol 38:237–241. doi:10.1636/JOA_B09-69.1

Weeks AR, Marec F, Breeuwer JA (2001) A mite species that consists entirely of haploid females. Science 292:2479–2482. doi:10.1126/science.1060411

Weinert L, Werren J, Aebi A, Stone G, Jiggins FM (2009) Evolution and diversity of *Rickettsia* bacteria. BMC Biol 7:6. doi:10.1186/1741-7007-7-6

Yun Y, Peng Y, Liu FX, Lei C (2011) *Wolbachia* screening in spiders and assessment of horizontal transmission between predator and prey. Neotrop Entomol 40:164–169. doi:10.1590/S1519-566X2011000200002

Chapter 9
Fungal Pathogens of Spiders

Harry C. Evans

9.1 Introduction

This has been a somewhat neglected topic: typically, a no-man's land between mycologists and arachnologists, where few venture to tread and even fewer attempt to negotiate the pitfalls that surround the host–pathogen associations. Invariably, the host is completely overgrown by the pathogen, making further identification problematic at best and contentious at worse, even in the unlikely scenario that the fungal forayer has access to a "friendly" spider systematist. Equally, fungal-colonised spiders would hold no interest for the ecologically- or taxonomically inclined arachnologist who would probably view them merely as mouldy specimens: if, indeed, they could be recognised as spiders. More importantly, however, the association would tend to be overlooked in the context of spider ecology. This is evidenced by the predecessor of this book (Nentwig 1987), in which no mention is made of fungal pathogens despite the many references to natural enemies of spiders, as well as similar omissions in contemporary publications on the biology of spiders (Foelix 1982). Indeed, in the most recent edition (Foelix 2011), the only natural enemies discussed are insects (wasps, flies), nematodes and vertebrates with the over-riding statement that: "The main enemies of spiders are spiders themselves." Perhaps this poor interdisciplinary collaboration should come as no surprise since, even in mycology and pathology circles, the study of fungal pathogens associated with arthropods, in general, is viewed as an esoteric by-water (Samson et al. 1988). Unfortunately, this has been perpetuated to the present day: a review chapter on fungal entomopathogens in a book on insect pathology (Vega et al. 2012) has a section entitled "Host range beyond insects" in which no mention is made of spider pathogens. Evans and Samson (1987) attempted to address this shortfall in a mini-review on the fungal pathogens of spiders. However, because it was published

H.C. Evans (✉)
E-UK, CAB International, Egham, Surrey, UK
e-mail: h.evans@cabi.org

W. Nentwig (ed.), *Spider Ecophysiology*,
DOI 10.1007/978-3-642-33989-9_9, © Springer-Verlag Berlin Heidelberg 2013

in a relatively obscure mycological journal, it would have been lost to arachnologists, in particular, and to arthropod ecologists, in general; and, seemingly, to entomopathologists (Vega et al. 2012). This chapter will broaden the scope and update the content of the original review, since much has changed in the intervening 25 years, no more so than the spectacular advances in molecular taxonomy.

9.2 History

One of the first references to parasitism of spiders by fungi was in a pioneering treatise by the entomologist R.C. Gray who—according to Petch (1932b)—included the first illustration dating from 1817 of a fungus on a spider host, despite the fact that the book title suggested it dealt exclusively with insect hosts (Gray 1856). This fungus, *Isaria arachnophila*, was later redescribed by the renowned Victorian mycologist M.C. Cooke—who gave it the name, European Spider Isaria, and reported it as common—in the first comprehensive review of entomogenous or entomopathogenic fungi (Cooke 1892). The latter term is considered the most suitable designation for the fungi discussed here, since entomogenous is applied to any fungal association with arthropods. As argued previously (Evans and Samson 1987), a more precise descriptor for the spider pathogens would be araneopathogenic fungi—and, Kobayasi (1941) even proposed that they should be assigned to a distinct grouping within the genus *Cordyceps* and its allies, under the name *Arachnidicolae* to distinguish them from the *Entomogenae* group attacking insect hosts. However, the better-known term entomopathogenic is considered more appropriate for the fungi implicated in arthropod mortality. Indeed, the old confusing terms of vegetable wasps and plant worms (Cooke 1892) were still being used nearly a century later (Kobayasi and Shimizu 1983). These were early Chinese and European interpretations of the fungus-arthropod association, pre-dating the germ theory of disease, which was believed to be the transformation of an insect into a plant.

Cooke (1892) devoted a chapter to the fungi on Arachnida and went on to describe other *Isaria* species on spiders from the New World, including *I. aranearum*—"parasitic on some species of *Aranea* and *Phalangium*"—and, *I. gigantea*—on an indigenous tarantula species in Cuba. This set the scene for a succession of publications that recorded, described or reviewed the fungal pathogens associated with spiders: Massee (1895), Johnston (1915), Petch (1923, 1932a, b, 1948), Kobayasi (1941), Mains (1939, 1950a, 1954), Petch and Bisby (1950), Evans (1974), Kobayasi and Shimizu (1976, 1982a, b, 1983).

9.3 Biology

Because there are no studies on the infection process in spiders per se, the assumption is that the mechanisms involved are essentially similar to those in insect hosts (Evans and Samson 1987). However, closer examination suggests that perhaps this is not the case. For example, there is very little overlap between the fungi attacking insects and those recorded on spiders (Sect. 9.4), offering tantalising evidence that fungal pathogens of spiders are highly specific and that this specificity must operate at the exoskeleton level. In contrast to insects, spiders possess a mesocuticle, whilst the spider opisthosoma lacks an exocuticle, and is a non-hardened structure composed predominantly of mesocuticle (Foelix 2011). Certainly, as some of the images included here clearly demonstrate (Figs. 9.1d, f, g and 9.2d), this appears to be the part of the spider body most vulnerable to colonisation by pathogenic fungi. Thus, it could be conjectured that fungal pathogens of spiders have evolved specific mechanisms to penetrate the unique epicuticle–mesocuticle structure, rather than the more complex cuticular layers (epi-, exo-, endo-) which make up the insect exoskeleton (Samson et al. 1988).

Studies of insect pathogens show that the spores land on and adhere to the outer epicuticle—a highly complex structure appearing as a series of electron dense and translucent layers and composed of lipoproteins and wax—using both physical (electrostatic) and chemical (mucilage) forces. Often, a limpet-like structure (appressorium), formed by the germinating spore, offers further adhesion whilst serving as a "beach-head" for the infection hypha to penetrate the exoskeleton, employing both mechanical pressure and chemical (enzymes and toxins) degradation (Charnley 2003). Once through the outer host defences, the infection hypha must overcome the cellular and humoral defences: for example, by confronting and disrupting haemocyte activity or even by-passing the defences completely. In effect, the fungus is not recognised as a threat because of a failure to distinguish between self and non-self: the high point of host-parasite coevolution (Evans 1988). In the *Ascomycota*—the fungal phylum to which the overwhelming majority of the recognised spider pathogens belong—the fungi are pleomorphic and invasion of the haemocoele is via a parasitic, budding yeast-like phase. When invasion is complete, toxins are produced to kill the host and the pathogen switches to a saprophytic phase, manifested by rapid hyphal growth, characteristic of most fungi. Absorption of haemocoele fluids by the mycelial network causes dehydration of the host. Undoubtedly, a cascade of antibiotics and toxins is employed in protecting the cadaver from secondary invasion by opportunistic micro- and macro-organisms.

The resultant mummified cadaver—full of fungal food reserves, typically in the form of lipid-rich hyphal bodies in the opisthosoma—produces external structures, with mycelium growing from the body joints, sutures and orifices. Subsequently, and depending on climatic conditions, more co-ordinated hyphal growth appears: typically, in the form of stalks or outgrowths on which the asexual and sexual stages are borne. Thus, with the release of their spores from these structures—elevated above the host for more efficient dissemination—the cycle is complete.

There is every reason to believe that the fungi pathogenic on spiders share a similar biology; perhaps, with the added rider that there may be critical host-recognition factors operating at the cuticular level which separate them from insect pathogens. Similarly, the method of growth from and external colonisation of the opisthosoma is probably unique to spider hosts because of the predominantly mesocuticular nature of this structure.

Fig. 9.1 (**a**) *Cordyceps cf. singeri* on a small trapdoor spider (Ctenizidae), Rio Napo, Eastern Ecuador. The host is completely covered in *white* mycelium and the two fleshy fruit-bodies (stromata) emerge from the burrow, bearing the perithecia towards the apex. (**b**) *Cordyceps cylindrica* on a large trapdoor spider, Rio Napo, Eastern Ecuador. The distinctly capitate, yellow fruit-body has pushed aside the trapdoor, revealing the base of the lid. (**c**) *Cordyceps cylindrica* on a trapdoor spider, Rio Negro, Amazonas, Brazil. The greenish-yellow fruit-body (~10 cm in length) is cylindrical, rather than capitate, and the well-camouflaged trapdoor is only slightly displaced. (**d**) *Ophiocordyceps cf. engleriana* on free-living spiders attached to the underside of shrub leaves, Ecuador. The spider opisthosoma is covered by dark brown mycelium (subiculum) on which the majority of fruit-bodies develop. On the *right*, the host bears mainly the sexual (teleomorph) stage: the exposed flask-shaped perithecia, containing the ascospores, are borne aloft. In contrast, the host on the *left* has predominantly the asexual (anamorph) fruit-bodies (synnemata), typical of *Akanthomyces*, with powdery conidia formed laterally towards the apical region. (**e**) *Ophiocordyceps caloceroides* on large mygalomorph spider (~11 cm in length), showing the fruit-bodies (>50 cm in length) which are brightly coloured in the upper quarter; most of the fruit-body lies beneath the soil inside the burrow with the mummified spider at the bottom. (**f**) *Torrubiella* sp. on spider host attached to shrub leaf in Atlantic rainforest of SE Brazil, showing the yellow subiculum covering the opisthosoma from which the yellow perithecia, clothed in white mycelium, form directly. (**g**) Another *Torrubiella* sp. on a different spider host from Atlantic rainforest, with the perithecia borne on a creamish-coloured subiculum producing a cryptic *Granulomanus* anamorph. Photos H.C. Evans and C.A. Ellison (e)

Fig. 9.2 (**a**) Smaller, salticid spider on a shrub leaf in Atlantic rainforest of SE Brazil, bearing *Torrubiella* perithecia on the opisthosoma and the distinctive white sporulating heads of the *Gibellula* anamorph over the entire body. (**b**) *Torrubiella* sp. in evergreen forest in eastern Ghana: the pale yellow subiculum completely obscures the spider host and the white perithecia form around the periphery; whilst, centrally, a solitary cylindrical fruit-body (synnema) of the *Gibellula* anamorph arises. (**c**) Salticid spider in Atlantic rainforest with the opisthosoma enveloped in a white subiculum bearing the *Granulomanus* synanamorph directly, with the dominant *Gibellula* anamorph on synnemata; *Gibellula* conidiogenous heads cover the legs. (**d**) Another *Gibellula* sp. producing numerous lilac-coloured synnemata on a large spider in Atlantic rainforest: the pale yellow subiculum forms mostly on the opisthosoma. (**e**) A spider host attached to a shrub leaf in Atlantic rainforest with another species of *Gibellula*: the stout, creamish-white synnemata are densely produced on the opisthosoma to which the white subiculum is confined. Photos H.C. Evans

9.4 Taxonomy

Evans and Samson (1987) noted that all the confirmed or purported fungal pathogens of spiders pertain to the phylum *Ascomycota*, specifically to the order *Clavicipitales* (now synonymised in the *Hypocreales*), and that there are no representatives in the other phyla of fungi. Most notably, the absence of taxa in the *Entomophthorales* (*Zygomycota*)—ubiquitous pathogens of terrestrial arthropods, including other arachnids, such as mites (Acari) and harvestmen (Opiliones) (Leatherdale 1958)—was highlighted. This still appears to hold true, with a literature search recording only the occurrence of the zygomycete *Mucor hiemalis*—a common soil fungus—attacking lab cultures of *Cupiennius salei* (Ctenidae) and *Ischnothele guianensis* (Dipluridae) (Nentwig and Prillinger 1990). In all probability, this would seem to be an opportunistic pathogen of spiders predisposed to infection rather than an obligate parasite: as would the report of a *Mucor* sp. on a *Modisimus* spider (Pholcidae) from a cave system in Cuba (Mercado et al. 1988). Similarly, the discovery of an adult female of *Pardosa amentata* (Lycosidae) in a pitfall trap in Switzerland "infected" with an un-named *Conidiobolus* (*Anclystaceae*, *Entomophthorales*)—a genus containing many plurivorous species—must be treated, for the moment, as a doubtful pathogenic record (Keller and Wegensteiner 2010).

Some explanation is needed here concerning the intricacies of fungal taxonomy and nomenclature for the benefit of non-mycologists. The greater majority of fungi are pleomorphic, in that they have more than one independent form or spore stage in the life cycle, and most produce both asexual (imperfect) and sexual (perfect) reproductive stages, now called anamorph and teleomorph, respectively. Nevertheless, for many fungi the teleomorph is undescribed, or unconnected with the anamorph, or rarely produced; whilst others have taken separate evolutionary pathways and lost sexuality completely—replacing it by alternative mechanisms, such as parasexuality. Under the code governing the scientific names of fungi, it is possible to treat both anamorph and teleomorph as species with separate Latinised binomials. Now that DNA sequencing is routinely being used in fungal systematics, anamorph–teleomorph relationships are becoming clearer and, in those cases where such associations are established, the teleomorph name takes priority. In order to avoid ambiguity, however, and for easier access to the literature, the fungi are listed here under both their teleomorph (*Ascomycota*) and anamorph (*Hyphomycetes*) names, and cross-indexed. Nevertheless, as demonstrated below, the taxonomy of fungal pathogens of spiders can still be said to be in a state of transition and, therefore, the groupings listed here should be regarded as provisional.

9.4.1 Ascomycota, Hypocreales, Clavicipitaceae Sensu Lato

9.4.1.1 *Cordyceps* Sensu Lato

This is by far the best known entomopathogenic genus, containing an estimated 400 species, but it has long been recognised as being polyphyletic (Sung et al. 2007).

Kobayasi (1941) delimited three sub-genera in his pioneering monograph on the genus *Cordyceps*—maintaining this in a later revision (Kobayasi 1982)—whilst, much more recently, Sung et al. (2007) further refined the classification of *Cordyceps* and its allies (*Clavicipitaceae* sensu lato) using multi-gene phylogeny. They recognised three families (*Cordycipitaceae*, *Ophiocordycipitaceae* and *Clavicipitaceae* sensu stricto) and four genera (*Cordyceps*, *Ophiocordyceps*, *Metacordyceps* and *Elaphomyces*): the latter being predominantly parasitic on other fungi. Significantly, in terms of host specificity, the anamorphs of *Metacordyceps* are some of the most commonly reported pathogens of insects—as well as ubiquitous soil inhabitants—but *Metarhizium* has never been recorded from spider hosts. Similarly, the common plurivorous white muscardine fungus (*Beauveria bassiana*) is rarely recorded on spiders (Petch 1931, 1948) and, because host identification appears to be less than robust, the genus is not included here.

Kobayasi (1941) listed five species of *Cordyceps* in his section *Arachnidicolae*, and more have been added since from North America and East Asia (Mains 1954; Kobayasi and Shimizu 1976, 1982a): these were beautifully illustrated in a subsequent publication (Kobayasi and Shimizu 1983). However, only one is retained in the genus *Cordyceps* sensu stricto, whilst others (5 spp.) are of uncertain affinity in *Cordyceps* sensu lato, and the remainder (3 spp.) have been transferred to the genus *Ophiocordyceps* (Sung et al. 2007). It is possible that these pathogens are evolutionary distinct and may prove to belong in a separate clade specific to spider hosts, judging from the phylogenetic tree constructed by Johnson et al. (2009).

Nevertheless, there remains considerable confusion over the taxonomy of all the species in *Cordyceps* sensu lato, which are well represented on larger, ground-dwelling mygalomorph spiders, especially in South America. For example, *Cordyceps* cf. *singeri* (Mains 1954)—with delicate, reddish, clavate fruit-bodies (Fig. 9.1a)—is relatively common on trapdoor spiders (Ctenizidae) in the Amazon region of both Brazil and Ecuador; whilst *C. cylindrica* occurs on similar hosts in the same habitats, but the spiders are considerably bigger, as are the yellowish-white, stout fruit-bodies (Fig. 9.1b, c). *C. cylindrica*, in fact, epitomises the original concept of the genus: the buried, mummified host, enveloped in white mycelium, sprouting a stout fleshy stalk (clava), and raising the club-shaped "payload" (ascostroma) into the air—in this case, pushing through the trapdoor (Fig. 9.1c)—to facilitate spore dispersal. Embedded in the ascostroma are flask-shaped structures (perithecia) filled with cylindrical sacs (asci) containing the filiform sexual spores (ascospores), which are forcibly discharged through the specialised thickened apex (cap)—diagnostic for and unique to the *Clavicipitaceae* sensu lato. In *C. cylindrica*, the ascospores break into part-spores after release, or on contact, the process being analogous to a scatter gun.

Cordyceps cylindrica is of particular interest because it demonstrates and emphasises our lack of knowledge of not only the taxonomy but also the biology and ecology of fungal pathogens of spiders. Seemingly, this species has a pantropical distribution but, outside of the Neotropics (Petch 1937; Mains 1954; Evans 1982), it occurs mainly in its anamorph state, *Nomuraea atypicola*, in both Africa (Samson 1974; Samson and Evans 1977; Rong and Grobbelaar 1998) and Asia

(Kobayasi 1941; Kobayasi and Shimizu 1983; Hywel-Jones and Sivichai 1995; Tzean et al. 1997b). However, Evans (1982) showed that ascospore isolations from Amazonian specimens produced *N. atypicola* in culture—although the anamorph is never seen in the field situation (Fig. 9.1b, c)—confirming the suspicions of Petch (1937) that these are linked. Later, Kobayasi and Shimizu (1976) reported the teleomorph in Japan on a trapdoor spider, as did Li et al. (2005) in China, although the anamorph—with fruit-bodies reminiscent of the teleomorph, and said to be common "in the gardens and roadsides" (Kobayasi 1941)—appears to be the dominant form in Japan (Kobayasi and Shimizu 1983). The host of the holotype is a trapdoor spider of the genus *Latouchia* (sub *Kishinouyeus*, Petch 1939; Kobayasi 1941). Most other records of *N. atypicola* describe the anamorph as growing directly on the hosts, which, invariably, are spiders attached to foliage (Samson 1974; Hywel-Jones and Sivichai 1995; Tzean et al. 1997a, b). It is possible that this represents a species complex, with one or more taxa on trapdoor spiders and at least one other that attacks free-living spiders (Greenstone et al. 1987).

9.4.1.2 *Ophiocordyceps*

Sung et al. (2007) transferred three *Cordyceps* species found on spiders to this genus—*Ophiocordyceps arachneicola, O. caloceroides* and *O. engleriana*—based only on original descriptions. One of the distinguishing generic features is that the fruit-bodies are rarely fleshy or brightly coloured, typically, being darkly pigmented and wiry. Whilst *Ophiocordyceps engleriana* fits this description (Fig. 9.1d), *O. arachneicola*—described by Kobayasi (1941) on the orb-web spider, *Araneus ventricosus*, from Japan—and *O. caloceroides* (a species resembling and named after the soft fleshy basidiomycete genus *Calocera*, Fig. 9.1e), do not conform to the genus description.

9.4.1.3 *Torrubiella*

The genus is typified by the perithecia forming directly on the host body (Figs. 9.1f, g and 9.2a), and not in or on aerial structures, thus separating it from *Cordyceps* sensu lato (Petch 1923). Kobayasi (1982) provided a key to the known taxa, whilst Kobayasi and Shimizu (1982b) monographed the genus and described 34 species on spiders, many of them new and collected in Japan. This had risen to 40 species in their next publication (Kobayasi and Shimizu 1983), but others have been added since or not included in their publications (O'Donnell et al. 1977; Humber and Rombach 1987; Tzean et al. 1997b, 1998), pushing up still further the total number of known species attacking spiders. Anamorphs, when present, belong to *Gibellula–Granulomanus* (Fig. 9.2a, b), as well as to *Akanthomyces* and *Lecanicillium*. Recently, the phylogeny of the genus has been investigated (Johnson et al. 2009) and, of the nine *Torrubiella–Gibellula* species from spiders sequenced, all fall in a unique clade within the *Cordycipitaceae*. Since they form this distinct evolutionary group, and the type species of the genus, *Torrubiella aranacida*, was described on a spider host (Boudier 1885), *Torrubiella* should be reserved for spider pathogens.

9.4.2 Anamorphic Fungi, Hyphomycetes

These used to be placed in the phylum *Deuteromycota*, but this is no longer accepted as a formal taxonomic category since it is an artificial assemblage of polyphyletic groups (Kirk et al. 2008). All the spider pathogens can be placed in the *Hyphomycetes*, comprising genera producing their spores (asexual conidia) on aerial hyphae rather than enclosed within fruit-bodies (conidiomata).

9.4.2.1 *Akanthomyces*

Typically, the spider host is covered by a white to pale brown mycelium from which arise multiple stalks (synnemata) composed of and organised into hyphal strands, producing the spore-bearing structures (conidiogenous cells or phialides) laterally (Fig. 9.1d). Hywel-Jones (1996) reported numerous specimens of "small hunting spiders" infected with *Akanthomyces* on leaves in the forests of Thailand and described three new species, together with another three confirmed species in the genus: also, recorded in North America (Mains 1950b), Ghana (Samson and Evans 1974) and Papua New Guinea (Samson and Brady 1982). Later, Hsieh et al. (1997) recorded four species—including one new taxon—from spiders in Taiwan, which were illustrated by Tzean et al. (1997b), whilst others have been added from China (Huang et al. 2000). *Akanthomyces arachnophilus* has been linked with *Torrubiella flava* (Petch 1923; Samson and Evans 1974), and DNA sequencing seems to confirm this association since both of the *Akanthomyces* species included in the phylogenetic tree came out in the unique *Torrubiella* clade (Johnson et al. 2009).

9.4.2.2 *Clathroconium*

This was described as a new monotypic genus collected on a free-living spider in Ghana, with no known teleomorph (Samson and Evans 1982). A second species from Cuba on the cobweb spider, *Nesticodes rufipes* (Theridiidae), was described subsequently (Mercado et al. 1988). The distinctive spiral-shaped (clathroid or helicoid) conidia form a powdery covering on the host

9.4.2.3 *Gibellula*

This is a highly conserved, spider-specific genus: past records from insect hosts (Petch 1932a; Mains 1950a) have proven to be erroneous (Samson and Evans 1973). The South American species have been monographed (Samson and Evans 1992), and eight species were delimited. However, with the additional species described by Kobayasi and Shimizu (1976, 1982b, 1983), Tzean et al. (1997a, b) and Huang et al. (1998b) from Asia, as well as undescribed species from Ghana and Brazil (Fig. 9.2b–e), the number of species should rise significantly. The characteristic *Aspergillus*-like heads are borne directly on the spider (Fig. 9.2a), or on solitary (Fig. 9.2b) to multiple (Fig. 9.2c–e) brightly-coloured synnemata, sometimes together with the *Torrubiella* teleomorph stage (Fig. 9.2a, b).

9.4.2.4 *Granulomanus*

Associated with *Gibellula* as a synanamorph (Samson and Evans 1977, 1992; Evans and Samson 1987), and characteristically borne on curled hyphal masses growing on or around the spider body. However, it can also occur directly on or replace the *Gibellula* sporulating structures (Petch 1944; de Hoog 1978).

9.4.2.5 *Hirsutella*

Typified by long, hair-like synnemata on which the narrow-cylindrical conidiogenous cells produce solitary conidia, usually in mucus. *Hirsutella darwinii* was described on a spider host enclosed in a silk cocoon from the Galápagos Islands (Evans and Samson 1982).

9.4.2.6 *Hymenostilbe*

Similar to *Akanthomyces* but the conidia are borne singly on short, tooth-like projections (denticles): at least three species, reported from the USA (Mains 1950b), Ghana (Samson and Evans 1975) and China (Huang et al. 1998a).

9.4.2.7 *Isaria*

Two species, formerly placed in the genus *Paecilomyces*, have been recorded on spiders (Samson 1974), forming short, white to cream, powdery synnemata on the host. In Ghana, *Isaria farinosa* is said to be common on spiders and pseudoscorpions (Chelonithidae) (Samson and Evans 1977), but with a wider host range elsewhere (Samson 1974; Evans and Samson 1982; Tzean et al. 1997b).

9.4.2.8 *Lecanicillium*

Three species associated with spiders have been assigned recently to this genus (Zare and Gams 2001); all previously described in disparate genera, such as *Acremonium*, *Cephalosporium*, *Engyodontium*, *Sporotrichum* and *Verticillium* (Petch 1931, 1935, 1937, 1948; Gams et al. 1984), because of the simple or verticillate arrangement of conidiogenous cells produced directly on the mycelium. *Lecanicillium tenuipes* (formerly, *Engyodontium aranearum*) has been linked mainly with opilionid hosts (Gams et al. 1984), and its occasional occurrence on insect hosts raises questions as to the specificity of this taxon. However, *Lecanicillium aranearum* recorded on spiders in Ghana and Sri Lanka—where it is associated with *Torrubiella alba* (Petch 1932b)—would appear to be spider-specific.

9.4.2.9 *Nomuraea*

Anamorph of *C. cylindrica* has been confirmed as *N. atypicola*, but probably represents a species complex (see above).

9.5 Ecology

Until recently, the statement that most of the examples of spiders infected with entomopathogenic fungi are to be found in the sub-tropics and tropics—especially in forest ecosystems—held true (Evans and Samson 1987). There have been reports, of course, from both North America and Europe—as discussed in previous sections—and, exemplified in the lists of entomogenous fungi from the British Isles, produced initially for mycologists (Petch 1948), and later modified for entomologists as a host catalogue (Leatherdale 1958). Essentially, however, they were sporadic ad hoc collections made by foraying mycologists and amateur entomologists. Nevertheless, these scattered records could prove to be just the tip of the iceberg and spiders in cold temperate regions may be more prone to fungal attack than previously supposed: as, for example, the discovery of a new *Torrubiella* (*T. falklandica*) on a spider host on a mountain summit in the Falkland Islands (O'Donnell et al. 1977). Another more recent example demonstrates how better-targeted collecting can reveal unexpectedly high levels of spider mortality in the most unlikely of habitats and during, what would appear to be, the most unfavourable time of year for infection: an upland lake in Wales in the depths of winter (D. McNeil, personal communication). Collections made around the lake, during a chilly day in December 2011, yielded a total of 97 specimens of a fungus—identified from images as being close *Gibellula pulchra*—on what appears to be the same host—confirmed as the orb stretch spider, *Metellina merianae* (Tetragnathidae).

High numbers of spiders infected with this ubiquitous pathogen had also been reported much earlier in Ghana from cocoa farms—80 specimens from 10 trees—and from forest sites—120 specimens from a 20 m^2 quadrat over a 12-month sampling period—which led to the conclusion that *G. pulchra* "may well be an important mortality factor" (Samson and Evans 1973; Evans 1974). Nentwig (1985) voiced a similar opinion concerning fungal mortality in web-building spiders during surveys in Panama. Much earlier, Mains (1939) reported on the occurrence of ca. 70 spiders infected with *Cordyceps thaxteri* attached to leaves, collected between July and August 1887 in the mountains of North Carolina. There appear to have been few reports since from the USA, which begs the question: was this a "one-off" event or has nobody been looking and, in fact, does the fungus exert a significant control of spiders in this ecosystem? However, without knowing the actual populations of spiders in these habitats, it is impossible to assess the ecological impact of pathogenic fungi and, therefore, their role in ecosystem functioning. Equally, we know little about their specificity and infection strategies—especially, the function of the various spore types and how spores reach their targets. This is particularly intriguing in the case of Ctenizidae hosts since: "Most trapdoor spiders never leave their burrow during their whole life" (Buchli 1969).

9.6 Conclusions

As shown here, the diversity and common occurrence of fungal pathogens of spiders in both temperate and tropical regions merits more in-depth studies to clarify the considerable gaps in our knowledge. At the very least, arachnologists should recognise them as potential players in the populations dynamics of spiders: from the equatorial rainforests of the Amazon basin and West Africa to the bleak South Atlantic in the Southern Hemisphere; and, from the temperate climes of North America to those of Japan in the Northern Hemisphere. Their exclusion from previous publications on the natural enemies and the general biology of spiders would seem to be a serious oversight.

References

Boudier JLE (1885) Note sur un nouveau genre et quelques nouvelles espèces de Pyrénomycètes. Rev Mycol 7:224–227

Buchli HHR (1969) Hunting behavior in Ctenizidae. Am Zool 9:175–193

Charnley AK (2003) Fungal pathogens of insects: cuticle degrading enzymes and toxins. Adv Bot Res 40:241–321

Cooke MC (1892) Vegetable wasps and plant worms. Society for Promoting Christian Knowledge, London
de Hoog GS (1978) Notes on some fungicolous hyphomycetes and their relatives. Persoonia 10:33–81
Evans HC (1974) Natural control of arthropods, with special reference to ants (Formicidae), in the tropical high forest of Ghana. J Appl Ecol 11:37–49
Evans HC (1982) Entomogenous fungi in tropical forest ecosystems: an appraisal. Ecol Entomol 7:47–60
Evans HC (1988) Coevolution of entomogenous fungi and their insect hosts. In: Pirozynski KA, Hawksworth DL (eds) Coevolution of fungi with plants and animals. Academic, London, pp 149–171
Evans HC, Samson RA (1982) Entomogenous fungi from the Galápagos Islands. Can J Bot 60:2325–2333
Evans HC, Samson RA (1987) Fungal pathogens of spiders. Mycologist 21:152–159
Foelix RF (1982) Biology of spiders. Harvard University Press, Cambridge
Foelix RF (2011) Biology of spiders. Oxford University Press, Oxford
Gams W, de Hoog GS, Samson RA, Evans HC (1984) The hyphomycete genus *Engyodontium*: a link between *Verticillium* and *Aphanocladium*. Persoonia 12:135–147
Gray RC (1856) Notices of insects that are known to form the bases of fungoid parasites. Privately Printed, London
Greenstone MH, Ignoffo CM, Samson RA (1987) Susceptibility of spider species to the fungus *Nomuraea atypicola*. J Arachnol 15:266–268
Hsieh LS, Tzean SS, Wu WJ (1997) The genus *Akanthomyces* on spiders from Taiwan. Mycologia 89:319–324
Huang B, Fan MZ, Li ZZ (1998a) A new species of *Hymenostilbe*. Mycosystema 17:198
Huang B, Ding DG, Fan MZ, Li ZZ (1998b) A new entomopathogenic fungus on spiders. Mycosystema 17:109–113
Huang B, Wang SB, Fan MZ, Li ZZ (2000) Two species of *Akanthomyces* from Dabieshan mountains. Mycosystema 19:172–174
Humber RA, Rombach MC (1987) *Torrubiella ratticaudata* sp. nov. (Pyrenomycetes: Clavicipitales) and other fungi from spiders on the Solomon Islands. Mycologia 79:375–382
Hywel-Jones NL (1996) The genus *Akanthomyces* on spiders in Thailand. Mycol Res 100:1065–1070
Hywel-Jones NL, Sivichai S (1995) *Cordyceps cylindrica* and its association with *Nomuraea atypicola* in Thailand. Mycol Res 99:809–812
Johnson D, Sung G-H, Hywel-Jones NL, Luangsa-ard JJ, Bischoff JF, Kepler RM, Spatafora JW (2009) Systematics and evolution of the genus *Torrubiella* (*Hypocreales*, *Ascomycota*). Mycol Res 113:279–289
Johnston JR (1915) The entomogenous fungi of Porto Rico. Insular Exp Sta Porto Rico 10:1–33
Keller S, Wegensteiner R (2010) A species of the fungus genus *Conidiobolus* as a pathogen of a lycosid spider. Mitt Schweiz Entomol Ges 83:227–231
Kirk PM, Cannon PF, Minter DW, Stalpers JA (2008) Dictionary of the fungi, 10th edn. CAB International, Wallingford
Kobayasi Y (1941) The genus *Cordyceps* and its allies. Sci Rep Tokyo Bunrika Daig B, Koishikawa, Japan 84:53–260
Kobayasi Y (1982) Keys to the taxa of the genera *Cordyceps* and *Torrubiella*. Trans Jap Mycol Soc 23:329–364
Kobayasi Y, Shimizu D (1976) Some species of *Cordyceps* and its allies on spiders. Kew Bull 31:557–566
Kobayasi Y, Shimizu D (1982a) *Cordyceps* species from Japan 4. Bull Natn Sci Mus Tokyo B 8:79–91
Kobayasi Y, Shimizu D (1982b) Monograph of the genus *Torrubiella*. Bull Natn Sci Mus Tokyo B 8:43–78

Kobayasi Y, Shimizu D (1983) Iconography of vegetable wasps and plant worms. Hoikusha Publ Co, Osaka
Leatherdale D (1958) A host catalogue of British entomogenous fungi. Entomol Mon Mag 94:103–105
Li CR, Chen AH, Wang M, Fan MZ, Li ZZ (2005) *Cordyceps cylindrica* and its anamorph *Nomuraea atypicola*. Mycosystema 24:4–18
Mains EB (1939) *Cordyceps* from the mountains of North Carolina and Tennessee. J Elisa Mitchell Sci Soc 55:117–129
Mains EB (1950a) The genus *Gibellula* on spiders in North America. Mycologia 42:306–321
Mains EB (1950b) Entomogenous species of *Akanthomyces*, *Hymenostilbe* and *Insecticola* in North America. Mycologia 42:566–589
Mains EB (1954) Species of *Cordyceps* on spiders. Bull Torrey Bot Club 81:492–500
Massee G (1895) A revision of the genus *Cordyceps*. Ann Bot 9:1–44
Mercado A, Alayo R, Mena J, De Armas LF (1988) Hongos entomógenos de Cuba: nueva especie de *Clathroconium* sobre araña. Acta Bot Cubana 56:1–5
Nentwig W (1985) Parasitic fungi as a mortality factor of spiders. J Arachnol 13:272–274
Nentwig W (ed) (1987) Ecophysiology of spiders. Springer, Berlin
Nentwig W, Prillinger H (1990) A zygomycetous fungus as a mortality factor of spiders. J Arachnol 18:118–121
O'Donnell KL, Common RS, Imshaug HA (1977) A new species of *Torrubiella* on a spider from the Falkland Islands. Mycologia 69:618–622
Petch T (1923) Studies in entomogenous fungi. III. Torrubiella. Trans Br Mycol Soc 9:108–128
Petch T (1931) Notes on entomogenous fungi. Trans Brit Mycol Soc 21:34–67
Petch T (1932a) *Gibellula*. Ann Mycol 30:386–393
Petch T (1932b) Notes on entomogenous fungi. Trans Br Mycol Soc 16:209–245
Petch T (1935) Notes on entomogenous fungi. Trans Br Mycol Soc 19:161–194
Petch T (1937) Notes on entomogenous fungi. Trans Br Mycol Soc 21:34–67
Petch T (1939) Notes on entomogenous fungi. Trans Br Mycol Soc 23:127–148
Petch T (1944) Notes on entomogenous fungi. Trans Br Mycol Soc 27:81–93
Petch T (1948) A revised list of British entomogenous fungi. Trans Br Mycol Soc 31:286–304
Petch T, Bisby GR (1950) The fungi of Ceylon. Ceylon Government Press, Colombo
Rong IH, Grobbelaar E (1998) South African records of associations between fungi and arthropods. Afr Plant Prot 4:43–63
Samson RA (1974) *Paecilomyces* and some allied Hyphomycetes. Stud Mycol 6:1–119
Samson RA, Brady BL (1982) *Akanthomyces novoguineensis* sp. nov. Trans Br Mycol Soc 79:571–573
Samson RA, Evans HC (1973) Notes on entomogenous fungi from Ghana. I. The genera *Gibellula* and *Pseudogibellula*. Acta Bot Neerl 22:522–528
Samson RA, Evans HC (1974) Notes on entomogenous fungi from Ghana. II. The genus *Akanthomyces*. Acta Bot Neerl 23:28–35
Samson RA, Evans HC (1975) Notes on entomogenous fungi from Ghana. III. The genus *Hymenostilbe*. Proc K Nederl Akad Wetensch C 78:73–80
Samson RA, Evans HC (1977) Notes on entomogenous fungi from Ghana. IV. The genera *Paecilomyces* and *Nomuraea*. Proc K Nederl Akad Wetensch C 80:128–134
Samson RA, Evans HC (1982) *Clathroconium*, a new helicosporous hyphomycete genus from spiders. Can J Bot 60:1577–1580
Samson RA, Evans HC (1992) New species of *Gibellula* on spiders (Araneida) from South America. Mycologia 84:300–314
Samson RA, Evans HC, Latge J-P (1988) Atlas of entomopathogenic fungi. Springer, Berlin
Sung G-H, Hywel-Jones NL, Sung JM, Luangsa-ard JJ, Shreshta B, Spatafora JW (2007) Phylogenetic classification of *Cordyceps* and the clavicipitaceous fungi. Stud Mycol 57:1–63
Tzean SS, Hsieh LS, Wu WJ (1997a) The genus *Gibellula* on spiders from Taiwan. Mycologia 89:309–318

Tzean SS, Hsieh LS, Wu WJ (1997b) Atlas of entomopathogenic fungi from Taiwan. Council of Agriculture, Taipei
Tzean SS, Hsieh LS, Wu JJ (1998) *Torrubiella dimorpha*, a new species of spider parasite from Taiwan. Mycol Res 102:1350–1354
Vega FE, Meyling NV, Luangsa-ard JJ, Blackwell M (2012) Fungal entomopathogens. In: Vega FE, Kaya HK (eds) Insect pathology. Academic, London, pp 172–220
Zare R, Gams W (2001) A revision of *Verticillium* section *Prostrata*. IV. The genera *Lecanicillium* and *Simplicillium* gen nov. Nova Hedwigia 73:1–50

Part IV
Chemical Communication and Reproduction

Besides the well-known world of optical and vibrational communication, spiders also live in a much less known chemical world. Volatile pheromones are of glandular origin or emitted from the body cuticle and play an important role for communication between mother and offspring or in social spiders. Also in solitary spiders, as part of the courtship behaviour, sexual pheromones are of upmost importance for successful reproduction. It is fascinating that spiders evolved a system of unusual multiple sex chromosomes with a considerable diversity in some groups. Astonishingly diverse is also the male reproductive system with respect to its gross morphology, sperm ultrastructure, involved glands and secretions. Further reasons for the high biodiversity of spiders?

Chapter 10
Chemical Communication and Contact Cuticular Compounds in Spiders

Marie Trabalon

10.1 Introduction

The study of animal communication has played an important part in developing methods and paradigms in ethology, sociobiology and animal cognition. Animal communication and the understanding of the animal world in general is a rapidly growing field. Animal communication is any behaviour emitted by one animal (emitter or sender) that, by sending a signal, has an effect on the current or future behaviour of another animal (receiver). The sender and receiver of a communication may be of the same species or of different species, although the majority of animal communication is intraspecific (between two or more individuals of the same species). Communication signals can solicit various canals, but here, we focus on chemical communication.

Many reports highlight the importance for spiders of chemical intraspecific communication in reproduction (sexual and parental behaviour) and social interactions. Spiders present two modes of living: solitary most of the time but also gregarious during the first days of life. In these gregarious or communal groups, spiders must be able to distinguish between members of their own group and others. This is a real problem during the capture of prey, because the attacking spiders have to decide instantly whether they should bite or not. Before engaging in a killing bite, arachnids touch the unknown animal. This contact allows them to identify the animal as a prey or as a conspecific. Contact chemoreceptors seem thus to play a key role in discriminating between prey and conspecifics. Social behaviour shifts linked to their development could be induced by changes in the chemical composition of the cuticle. Nevertheless, our knowledge of the nature of the chemical contact compounds spiders use to communicate is limited.

M. Trabalon (✉)
University of Rennes 1, UMR 6552 CNRS EthoS, Rennes, France
e-mail: marie.trabalon@univ-rennes1.fr

W. Nentwig (ed.), *Spider Ecophysiology*,
DOI 10.1007/978-3-642-33989-9_10,

10.2 Chemical Communication

Spiders produce both volatile and contact pheromones. Female spiders produce volatile chemical signals aimed to communicate with either males or other females. Spider contact pheromones are typically associated with the silk, draglines or substrate, whereas volatile pheromones are emitted from the silk or body cuticle. Spider pheromones play an important role as chemical signals for communication (for social as well as solitary spiders) and reproduction.

10.2.1 Molecular Olfactory Transfer

Chemical communication is considered to be the most primitive mode of communication in animals. The literature customary distinguishes between olfaction and contact chemoreception. For both modes, the stimuli are specific compounds or blends of compounds termed semiochemicals, the distinction being that compounds targeted by olfaction are volatile. The term volatile is polysemous and is often used in different contexts. Here, we use volatile in its strict chemical meaning. The volatility of a compound depends on the size (molecular weight), polarity and structure of the molecule and can be expressed as its vapour pressure. Molecules with high molecular weight and high polarity have a low vapour pressure. Most salts, for example, are not volatile, but some apolar compounds with relatively high molecular weights can be sufficiently volatile to be used in chemical communication. As vapour pressure decreases with increasing size or polarity of the molecule, evaporation rate declines, and the signal can therefore be emitted for a longer period, because the time needed for biosynthesis or translocation is a less limiting factor. Larger molecules also diffuse more slowly, and thus, the signal stays longer in the environment. Such molecules would be suitable to be carried openly all the time, for example, on the cuticle. A low concentration in the air requires highly sensitive receptors and excludes action over long distances, because of diffusional dilution.

In addition to volatilization, chemicals can also be transferred through space by adsorption on small particles released by the emitter. Non-volatile pheromones, which require direct contact of the receiver with the emitter or with a substratum on which they are deposited in order to be perceived, are also frequently used for communication. Signals can remain active for a long time on a surface, where they are better protected from degradation than in the air. They are often embedded in a matrix that also has protective properties.

10.2.2 Production of Compounds Involved in Spider Communication

Semiochemical ectomones are divided into two groups according to their function: allelochemicals (allomones and kairomones) and pheromones. Allelochemicals are compounds used for communication between different species, and pheromones are used for communication between members of the same species. Pheromones serve as critical signals for sex (Uhl 2013) or conspecific recognition, danger warning, territorial marking, social cues in kin and individual recognition, dominance cues, state of health and even mating status crucial for mate assessment (Uetz and Roberts 2002; Wyatt 2003; Johansson and Jones 2007). Spiders emit semiochemical compounds from their body cuticle and/or web silk although no specific glands that could secrete ectomones have been identified.

Spider silk is a well-known vector for vibratory, tactile as well as chemical signals. Using silk or webs for ectomones, transmission allows spiders continuous semiochemical emission without the need to emit pheromones actively from glands (Schulz 1997). Silk ectomones can be produced either in the silk glands (Schulz 2004) or have a cuticular origin and applied to the silk during web construction (Trabalon et al. 2005). Cuticular ectomones are synthesized near the integument and transported to the epicuticle for emission (Trabalon et al. 1996), possibly via modified cuticular wax glands or opisthosomal organs (Pollard et al. 1987).

Spiders' silk and cuticle are covered by a lipid layer. The primary function of cuticular lipids is to form a passive barrier to impede water evaporation through terrestrial arthropods' cuticle. While minimizing transpiration and thus protecting terrestrial insects from desiccation, cuticular lipids are also often involved in various types of chemical communication (Howard 1993; Singer 1998; Pourié et al. 2005). The main components of this layer are generally long-chain aliphatic hydrocarbons and fatty acids, but this layer can also include smaller amounts of methyl esters, long-chain aliphatic alcohols and aldehydes, glycerides and cholesterol. Similar compounds are present on the web and spider cuticle. For example, the web and cuticle from two agelenid and amaurobiid (Prouvost et al. 1999b; Trabalon et al. 1996, 1997; Trabalon and Assi-Bessekon 2008) and one theraphosid species (Trabalon 2011) comprised a complex mixture of fatty acids, alcohols and long-chain aliphatic hydrocarbons. Hydrocarbons form complex and varied mixtures with unsaturated hydrocarbons. In those species, the hydrocarbon fraction consists of *n*-, monomethyl- and dimethylalkanes, containing a relatively high proportion of even-numbered carbon chain components. Neither species exhibited unsaturated hydrocarbons. The abundance of even-numbered carbon chain alkanes and odd-numbered carbon chain fatty acyl groups, along abundant methyl branches, suggests that the propionyl-CoA and its carboxylated product, methyl-malonyl-CoA, play important roles in the biosynthesis of these unique waxes. The origin of the alkanes is at present unknown; the possibility that they come from the insect diet has not been investigated.

These lipid compounds can be transported from biosynthetic tissues to tissues that do not synthesize them (prosoma, legs) either by physical translocation to the cuticular surface or by haemolymph. The importance of the role of the haemolymph in transporting cuticular lipids is fully considered and is supported by the fact that, in several insect species, the cuticle and haemolymph fatty acid and hydrocarbon profiles are similar (Schal et al. 1998). For example, the variations of the qualitative composition of cuticular lipids of *Brachypelma albopilosum* (Theraphosidae) (fatty acids, methyl esters and hydrocarbons) are similar to the variations of their haemolymphatic lipids (fatty acids, methyl esters and hydrocarbons). This was the first time that the presence of circulating methyl esters in the haemolymph of this spider was established, as in the haemolymph of insects (Trabalon 2011).

10.2.3 Chemoreceptors

Spiders' olfactory and contact-chemoreception systems are roughly analogous to what we call a sense of smell and a sense of taste, respectively. Spiders use their legs and palps for smelling and tasting, as this is where their chemoreceptors tend to be concentrated, especially on the distal segments. However, their most important chemoreceptors are contact chemoreceptors or taste hairs. These chemosensitive hairs are curved, with blunt tips that are open to the exterior. Each taste hair is usually innervated by 21 sensory cells: two nerve endings with tubular bodies terminate at the hair base, while the other 19 dendrites traverse the hair shaft to the opening near the hair tip (Foelix and Chu-Wang 1973). Chemoreceptive hairs are important for both prey and conspecific recognition, and in particular recognition of potential mates. In addition to salt and sugar solutions, many amino acids are effective stimulants (Vallet et al. 1998). The sensilla of *Tegenaria atrica* (Agelenidae) are capable of encoding five concentration levels ranging from 0.01 to 1 mol, and they are functional immediately after these spiders leave their cocoons (Vallet et al. 1998). Recently extremely high concentrations of chemosensitive hairs were observed on male *Liphistius* spiders, with up to 2,000 hairs on a single tarsus. The fact that females and juveniles do not possess such hairs suggests that these hairs are probably specialized in the detection of female pheromone (Foelix et al. 2010). Prouvost et al. (1999a) observed significant differences in the distributions of these sensilla on the anteroposterior and dorsoventral areas of their legs (pretarsus and tarsus) and on their pedipalps in relation to the sex of *Tegenaria atrica*. Both on the front leg and the pedipalp, chemoreceptors are located on the tip of the tarsus. However, most of the hairs on the pedipalp are on its dorsal surface, whereas most of the hairs on the front leg tarsus are on its ventral surface (Fig. 10.1). These distributions appear compatible with the fact that these sensilla are intended to receive tactochemical cues. Indeed, the analysis of the positions of these two appendages in relation to the substrate shows that the ventral surface of the locomotor limbs' tarsi rests on the substrate, whereas the tip of the dorsal surface of the tarsus of the pedipalp comes into contact with the substrate. The different distributions of the sensilla on these appendages are therefore consistent with their function as contact chemoreceptors.

Another chemoreceptor is the so-called tarsal organ, which usually forms a spherical pit on the dorsal surface of each tarsus (Foelix 2011). Tarsal organs may also appear rod shaped, like a hair (Fig. 10.1e), but only in very rare cases (Foelix 2011). Like hairs they are multiply innervated (by around 20 neurons) and are connected to the external environment by seven small pores (Foelix and Chu-Wang

Fig. 10.1 Scanning electron microscope views of hair sensilla of young *Brachypelma albopilosum* (Theraphosidae) spiders. (**a**) Ventral view of the tarsus; (**b**) ventral view of the pedipalp; (**c** and **d**) chemosensitive hair on the tarsus; (**e**) tarsal organ

1973). Electrophysiological investigations showed that different volatile substances can stimulate the tarsal organs (Tichy et al. 2001). The tarsal organ responds to certain pungent odours such as acids or ammonia vapour, but it appears to be unresponsive to alcohols, aldehydes, esters and so on, and to natural scents coming from other spiders. Thus, we are still left with the question of where spiders' odours receptors are actually located.

10.3 Social Tactochemical Communication

Spiders are predatory and cannibalistic and are not generally known to be social. Only about 25 out of the more than 42,000 species are considered to be permanently social. Non-territorial social spiders present sophisticated forms of cooperation including collective building of a silky structure (irregular webs), group hunting and cooperative brood care. Permanent social spiders also exhibit developed parental behaviour, and their young are gregarious. Plausibly the most important mechanisms maintaining the social structure are the different forms of chemical communication. The ability to distinguish between a group member and an unfamiliar conspecific is a critical element of social behaviour. Furthermore, a number of studies suggest that important traits in social spider evolution, namely, tolerance and interattraction among group members, could be based on cuticular chemical cues, as for social insects. Krafft (1971) suggested that tactochemical communication plays a role in colony cohesion and organization of the social spider *Agelena consociata* (Agelenidae). Kullmann (1972) found that chemosensory perception alone allows social *Stegodyphus* species (Eresidae) to discriminate between prey and conspecifics. Cuticular chemical profiles varied quantitatively between *Anelosimus eximius* colonies (Theridiidae), suggesting that cuticular chemical composition in social spiders may contain information about colony identity, despite a lack of obvious discrimination against non-colony members in this species (Pasquet et al. 1997).

These differences appear to be linked to both genetic variations and ecological factors (Fig. 10.2). The cuticular chemical compounds are hydrocarbons, fatty acids, methylesters and, particularly, novel *n*-propyl esters of long-chain methyl-branched fatty acids (Bagnères et al. 1997). We hypothesized that these propyl esters played a part in social interactions (Trabalon 2000). Indeed, how could volatile molecules be involved in communication between two individuals within a large community, when the "conversation" between these two individuals would be "drowned" and imperceptible in an "odorous cloud" emitted by the whole colony? The different interactions within the group would eventually be blurred, creating cacophony. The question that remains, for the time being, is whether spiders' non-volatile cuticular compounds are used in social interactions. To our knowledge, the recognition systems and cues that spiders use for kin discrimination have not yet been investigated, and no precise data concerning spiders' social pheromones exist.

10.3.1 Chemical Communication During Temporary Gregarious Groups

After mating, some spider species exhibit parental behaviour, often followed by a time-limited period when the young are gregarious (Fig. 10.3). This gregarious period begins inside the cocoon and persists for several days after emergence of spiderlings. For example, young *Tegenaria atrica* remain together immediately after leaving their cocoon, form a dense ball nearby and express little locomotor activity for one week (Trabalon et al. 1996). During this period, the young moult simultaneously. They react neither to the weakening of the web produced by their mother's movements nor to the vibrations provoked by prey. Mothers stay close to their cocoons, but no corporal contact between mothers and young occurs. One week after emergence, the locomotor activity of the young increases. Their aggregation becomes less dense, and after another moult, the young progressively run all over their maternal web. The 20-day-old young move quickly and the resulting weakening of the web alerts the mother. The young that do not yet disperse out of

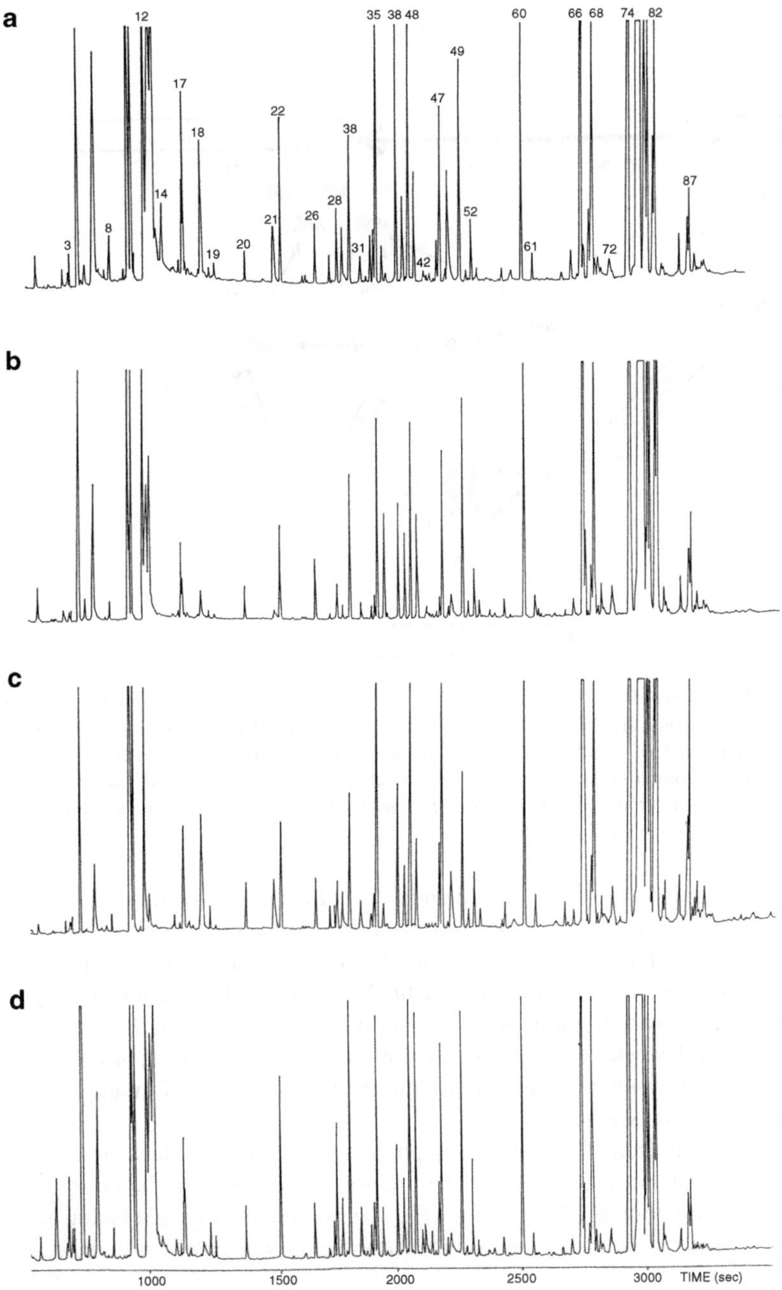

Fig. 10.2 Gas chromatographs of cuticle total extracts of adult females *Anelosimus eximius* (Theridiidae) from four colonies in French Guyana. (**a**) Colony FRG 29; (**b**) colony FRG 70. These two colonies were situated along the same road in a secondary forest 5 km apart. (**c**) Colony R143 was situated along another road about 40 km west of the first road in a secondary forest; (**d**) colony Ste Elie was in a primary forest 60 km north of the other colonies. These gas chromatographs show that the cuticle extracts of females from different colonies present the same lipid compounds

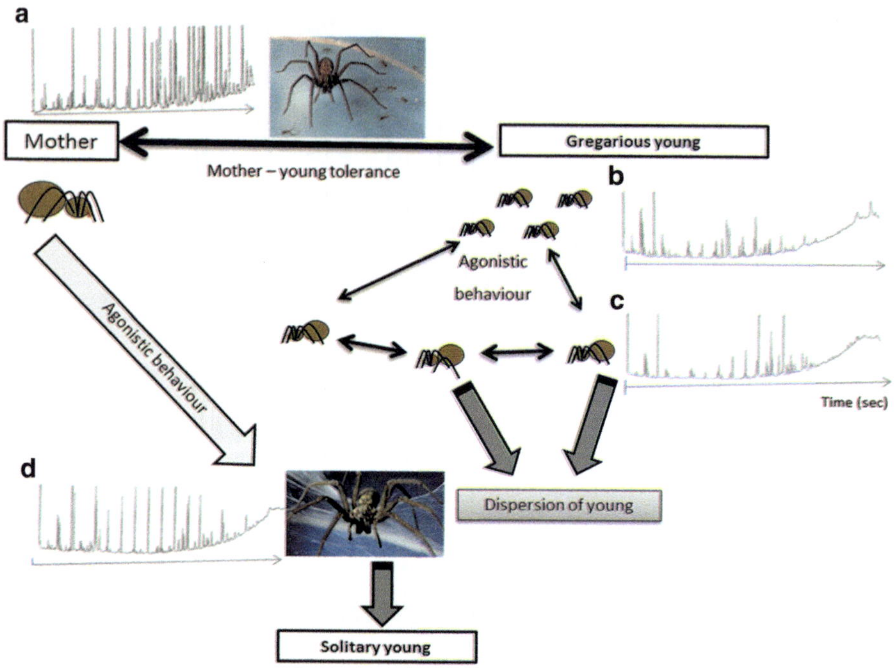

Fig. 10.3 Variations of mother-young social interactions and variations of cuticular lipids during temporary gregarious grouping in *Tegenaria atrica* (Agelenidae). (**a**) Gas chromatograph of cuticle total extract of mother; (**b**) gas chromatograph of cuticle total extract of 5-day-old young; (**c**) gas chromatograph of cuticle total extract of 20-day-old young; (**d**) gas chromatograph of cuticle total extract of 80-day-old young. These gas chromatographs reveal the increasing complexity of cuticular lipid extracts with age

the maternal web feed on prey killed and left by the female. During this period, the young frequently make contact with one another, in fortuitous encounters or when feeding. These contacts are accompanied by agonistic interactions which tend to increase with age, whereas contacts with the mother do not elicit any agonistic reactions, either by the young or by the adult (Trabalon et al. 1996, 1998). At the end of the third week of development, the young start catching small prey, and during the fourth week, after another moult, the young leave their maternal web and settle on individual webs.

The transition from the gregarious phase to the solitary phase is linked to a change in the tolerance of the mother in relation to changes in the composition of the cuticular compounds of the young (Fig. 10.3). Five compounds characteristic of mother females (*n*-eicosane, 2-methylhexacosane, *n*-octacosane, *n*-nonacosane and *n*-tritriacontane) are present on the cuticle of tolerated gregarious young. When the young are 20 days old and begin to avoid one another, their cuticular chemical profiles begin to change. Consequently, two new chemicals (5-methylhentriacontane and methyloctadecadienoate) emerge and four different compounds are variously synthesized/released (*n*-octadecane, *n*-octacosane, octadecadienoic and octadecenoic acids). Nevertheless, these changes do not appear sufficient to cause tolerance to cease completely as the young continue to tolerate one another. The compounds that vary during the dispersal phase, as well as the occurrence of cannibalism in young, are linked to four compounds that are characteristic of solitary virgin females (methyloctadecanoate, *n*-tricosane, *n*-pentacosane and *n*-heptacosane). After dispersal, young are no longer tolerated by other conspecifics and their cuticular chemical profiles are modified: *n*-heneicosane, 3-methylpentacosane and 14- + 12- + 10-methyltriacontane emerge and the synthesis/release of three compounds (*n*-heptadecane, methyltetradecanoate and *n*-octadecane) changes. So tactochemical information plays an important role in modulating agonistic behaviour after close contact between female spiders and their young. Females' cannibalism appears to correspond to an increase in polar compound levels (methylesters and fatty acids) and a decrease in apolar compound levels (hydrocarbons) in young with age.

The cuticular chemical composition of subsocial *Stegodyphus lineatus* spiders also changes during the first 50 days of development especially around dispersal (Grinsted et al. 2011). The relative proportions of longer alkanes increase with age. This increase of longer hydrocarbons could occur mainly to prepare the cuticle for the risky dispersal stage, when spiders become exposed to environmental conditions in their arid habitat different from those in their maternal nest. Developmental changes in cuticular compound composition could possibly decrease acceptance and tolerance of nest mates, inducing their dispersal. Evolutionary modifications of developmental changes in chemical composition like these could have been important in extending the cooperative stage of tolerance towards siblings from temporary, in subsocial spiders, to permanent in social spider species.

In *Coelotes terrestris* (Agelenidae), postmating changes in female pheromone emission are associated with changes of cuticular and web chemical compounds and increased mother behaviour (Trabalon and Assi-Bessekon 2008). For example, the postmating webs of female *Coelotes terrestris* contain significantly less palmitic acid, 1-octadecanol, 13- + 11-methylpentacosane and 3-methylheptacosane and significantly more 1-docosanol, *n*-pentacosane, *n*-hentriacontane and 17- + 15- + 13- + 11-methylpentatriacontane than the webs of virgins. The web therefore seems to be an attracting factor essential for the emergence of maternal behaviour. The web also plays a major role during the gregarious stage, attracting the young. Silk could therefore be an essential factor in spider socialization. Our results show that the lipids deposited on the web by "incubating" and "parental" females form a contact attractant pheromone for mated females and gregarious young and a repulsing pheromone for virgin females and young during the dispersal period. So incubating females' web contains low levels of two fatty acids (palmitic and oleic acids) and one alcohol (1-octadecanol) and high levels of two alcohols (1-eicosanol and 1-docosanol) and one fatty acid (stearic acid). Modifications of chemical profiles after copulation render the web unattractive to males and solitary young spiders, but attractive to mated females. This modification of chemical information appears to be a key factor to induce dispersion of young subsocial spiders and to reduce subsequent competition for the same habitat.

10.3.2 Chemical Communication in Parasocial Groups

Solitary adults can form parasocial groups where they can benefit from sharing silk structures or by joining individual prey-catching webs. This type of aggregation is promoted by abundance of food in the environment. The subsequent reduction of interindividual distances brings increasing advantages, in terms of individual survival and access to reproduction, through greater protection against predators, increased probability of intersexual encounters and cooperation to capture otherwise inaccessible large prey. Buskirk (1981) points out that the behaviour of parasocial spiders differs from that of solitary conspecifics. Intraspecific agonistic behaviour (cannibalism), therefore, decreases markedly. One factor facilitating formation of these groups could be prey density. According to Ruttan (1990), the growing young spiders' increasing need for food drives them to disperse more or less rapidly according to prey availability in that particular area. When young *Tegenaria atrica* (Pourié and Trabalon 1999a, b) are prevented from dispersing and are fed ad libitum, cannibalism rates decrease significantly. Thus, the trophic factor has a considerable influence on maintaining tolerance and inhibiting cannibalism in spider groups. According to these observations, the occurrence of cannibalism could be related to the first signs of "hunting" or predatory behaviour at the time of natural dispersal of young. Under natural conditions, when prey density is low, cannibalism would be an alternative to foraging. Cannibalism rate decreases in experimental groups of young *Tegenaria atrica* coincides with variations in their cuticular chemical signature (Pourié and Trabalon 2001). The development of chemicals of spiders reared in isolation remains unchanged during ontogeny regardless of their level of nutrition, and their chemical signature becomes enriched as individuals grow older, with increasingly heavy compounds. Level of food intake does not influence the synthesis/release of the chemical compounds found on the cuticle. We observed the same development in the cuticular chemical signatures of young that had been reared in groups and fed ad libitum. However, the increasing complexity of their chemical signature does not follow the same developmental pattern as that of isolated individuals (Pourié et al. 2005). More precisely, palmitic acid and 13.17- + 11.17- + 9.17-dimethylhentriacontane are not detected in grouped individuals. Furthermore, grouped individuals secrete methyloctadecenoate and *n*-heptatriacontane only emerge after their fifth moult, contrary to isolated young that start secreting these compounds just before dispersal. These compounds perhaps inhibited cannibalism in our experimental groups of solitary young. We can reasonably believe that the absence of competition for access to food could induce spiders not to differentiate themselves by their chemical signals and thereby enhance their defences against being bitten.

However, another hypothesis must be taken into consideration. The necessary factor for keeping together an experimental group of solitary young of relatively stable size is mutual tolerance among growing individuals. This tolerance is prompted by the absence of competition for food and is perhaps maintained by a change in chemical communication among individuals. The interactions that progressively lead to mutual tolerance among young could be compared to learning processes. Indeed, throughout the eight months we tested our experimental groups, observing repeated interactions in a rich environment, our subjects learned progressively to restrict their agonistic interactions by assessing the advantages and disadvantages of a situation. This allows adults from these groups to display a higher level of tolerance and conversely, very low rates of cannibalism of adult conspecifics. Consequently, very likely if favourable environmental factors incite the young not to disperse, then, when adult, these spiders could possibly preserve the communal lifestyle based on maintaining social interactions (i.e. social tolerance).

Clearly, however, this is not a question of comparing experimental groups of adult *Tegenaria atrica* to spider societies. These results simply demonstrate the pre-existence of a certain behavioural plasticity in some solitary species. Normally solitary spiders are capable of showing mutual tolerance according to one essential environmental factor: food resource availability. This social tolerance appears linked to a very slight change in their cuticular chemical signature during ontogeny.

10.3.3 Solitary Spiders' Chemical Communication for Intraspecific Recognition

Brachypelma vagans (Theraphosidae) females frequently locate and enter the burrows of conspecifics and then attack and consume the other female (Hénaut and Machkour-M'Rabet 2005). A resident female lays down silk in the burrow, and spiders are often attracted to cues (a phenomenon known as sericophily) from the silk of conspecifics (Hodge and Stoffer-Isser 1997). The principal compounds responsible for influencing females' behaviour are nonpolar organic compounds (Dor et al. 2008).

Recently, we studied the social recognition, agonistic behaviour and contact cuticular compounds of the closely related *Brachypelma albopilosum* (Trabalon 2011). Females do not build webs, and therefore, they forage to catch their prey. Field observations indicate that females during a foraging period can locate and enter the burrows of conspecifics, as do *Brachypelma vagans*. However, *Brachypelma albopilosum* females do not bite and consume other females. Our laboratory study showed that their agonistic and foraging behaviour depends on the nutritional state of these females. "Resident" females that ate 24 h before a test did not forage. Only females that had not eaten for several days before a test left their territory and investigated their environment. When one of these foraging females encountered a conspecific, her behaviour could vary. After entering a conspecific's burrow, 60 % of the foraging females did not show any aggressive behaviour; they simply retreated after making contact. However, predation reactions were frequent after contacting a prey item. Only 40 % of foraging females attempted to attack a resident by biting after making corporal contact. The cuticular lipid profiles of resident and aggressive-foraging females are the same. Only the cuticular profiles of "no-contact" foraging spiders differed from those of the other females. The pheromones of these females contained significantly less fatty acids and methyl esters and higher hydrocarbon levels. Levels of free fatty acids (linoleic and oleic acids), of methyl esters (methyl palmitate and methyl stearate), of cholesterol and of six hydrocarbons (tetradecane, pentadecane, hexadecane, tetracosane, nonacosane and triacontane) do not distinguish clearly aggressive females from others. We evidenced that aggressive females are characterized by increased levels of a single compound: stearic acid. More precisely, our results show that the important decline of circulating cholesterol, fatty acids (palmitic, linoleic, and oleic acids) levels and their methyl esters coincides with the manifestation of predatory behaviour and fight avoidance.

The reactions of *Tegenaria atrica* to conspecific intruders are not always of a predator–prey nature. Individuals of this solitary species are capable of adjusting their behaviour according to whether the conspecific intruder is adult or juvenile. This adjustment is linked to the age and physiological state of the individuals (Trabalon et al. 1998). Therefore, whatever their physiological state, adult *Tegenaria atrica* females tolerate gregarious juveniles. Nevertheless, the levels of tolerance of these different types of females decrease as the young develop. This form of tolerance of very recently hatched young has induced females of some spider species to care for their brood to a greater or lesser extent.

10.4 Contact Pheromones and Female Physiological State

Sexual and/or social behaviour (tolerance or cannibalism) depends on a female's physiological state. For example, the chemical cuticular profiles of females' agelenids, ovarian maturation, ecdysteroid levels and sexual receptivity are correlated. When the ovaries of unreceptive virgin females are at a previtellogenic stage, they become cannibalistic towards males (Trabalon et al. 1992, 1998), but when receptive females reach the onset of vitellogenesis, they accept copulation. Fertilization triggers the end of vitellogenesis and appraisal of females' cannibalistic behaviour (Pourié and Trabalon 2003). Only three studies have investigated the internal factors that regulate pheromone emission, reproductive behaviour and ovarian maturation.

The haemolymphatic ecdysteroid levels of *Coelotes terrestris* and *Tegenaria domestica* peaked during transition between previtellogenesis and vitellogenesis (Trabalon et al. 1992). Similarly, the highest levels of *Tegenaria atrica*' total circulating ecdysteroids (20-hydroxyecdysone) were detected in unmated females during the transition between the previtellogenic and the early vitellogenic phases of oocyte development. Pourié and Trabalon (2003), investigating the role of exogenous ecdysteroids on vitellogenesis, reported partial characterization of vitellogenin obtained from the haemolymph, ovaries and eggs of *Tegenaria atrica*. Injections of 20-hydroxyecdysone in unmated female induced vitellogenesis, as in other arthropods. Finally, injections of 20-hydroxyecdysone increase the frequency of sexual receptivity, inhibit cannibalism during sexual activity and alter production of contact sex pheromones (methyl esters and fatty acids) on web and cuticle (Trabalon et al. 2005).

Both genetic and environmental factors can influence the composition and ratios of compounds of the cuticular chemical profiles of arthropods (Liang and Silverman 2000, van Zweden et al. 2009). However, body condition in terms of nutrient storage can be considered as a sign of both an overall healthy state and good reproductive status. Consequently, assessments of body conditions become an essential factor in conservation biology and evolutionary ecology studies (Blanckenhorn and Hosken 2003, Stevenson and Woods 2006). Two of the most commonly used condition-dependent body size estimates are body mass (Jakob et al. 1996; Green 2001), abdomen dimension and mass/size relationships (Jakob et al. 1996). These measures often used as shortcuts, are not direct measurements of the body conditions of *Brachypelma albopilosum* and may sometimes fail to correlate with their nutritional state (Trabalon 2011). Modulation of females' aggressive behaviour is related to their lipidic state rather than their body size. According to our observations, females are capable of modulating their aggressive behaviour in relation to their nutritional state and in particular to their levels of circulating lipids. Females that presented aggressive interactions possessed higher levels of haemolymph circulating lipids, in particular, of methyl esters and hydrocarbons than nonaggressive females that avoided conflict. Females that avoided conspecifics presented the lowest circulating lipid levels. Spiders with high-energy reserves engage more often in aggressive interactions than do those that have eaten less. Presumably, the decision occurs through comparison between the intrinsic state of energy reserves and the perception of a cuticular lipid profile of the opponent. Perception of this chemical information appears to be a key factor inhibiting contact between females, and our results indicate that aggression is influenced by haemolymphatic lipid changes that could alter either the neurochemistry or neurophysiology associated with agonistic behaviour.

10.5 Conclusions

The fact that spiders use chemical communication is well established, although information concerning their chemical communication is far less extensive than for other arthropod taxa. Chemical communication is used in many contexts, including prey detection, predator avoidance, reproduction and social recognition. Chemical compounds detected by spiders by their specialized sensory hairs and pit organs concentrated on their legs and palps. As these chemicals mediate social interactions in solitary as well as social species, they could play a role in the transition from solitary to social life mode.

The chemical composition of the arachnid cuticle, especially the lipid layer is used for information transfer. Production of cuticular compounds is sex-, age- and nutrition-dependent. Cuticular lipids' qualitative and/or quantitative changes play a role in intra- and interspecific relationships in spiders. A better understanding of the functions and modes of action of cuticular compounds would contribute to furthering spider behavioural ecology. The endocrine regulation of cuticular product synthesis and behavioural reproduction is still not fully known, and the role of ecdysteroids is a much neglected field of research. Species specificity of chemical signals is commonly assumed in many of the behavioural studies described above, but only a few studies have tested species specificity of cuticular chemical communication in spiders. This assumption may not be well-founded, as spiders do not co-occur in the same habitat and the use of different chemicals has not necessarily been selected.

Acknowledgments I would like to thank Dr. Ann Cloarec and Dr. Colette Rivault for reading the manuscript.

References

Bagnères AG, Trabalon M, Blomquist G, Schulz S (1997) Waxes of the social spider *Anelosimus eximius* (Araneae, Theridiidae): abundance of novel n-propyl esters of long-chain methyl-branched fatty acids. Arch Insect Biochem Physiol 36:295–314

Blanckenhorn WU, Hosken DJ (2003) Heritability of three condition surrogates in the yellow dung fly. Behav Ecol 14:612–618

Buskirk RE (1981) Sociality in the Arachnida. In: Hermann HR (ed) Social insects. Academic, New York

Dor A, Machkour-M'Rabet S, Legal L, Williams T, Hénaut Y (2008) Chemically mediated burrow recognition in the Mexican tarantula *Brachypelma vagans* female. Naturwissenschaften 95:1189–1193

Dumpert K (1978) Spider odor receptor: electrophysiological proof. Experientia 34:754–756

Foelix RF (2011) Biology of spiders, 3rd edn. Oxford University Press, New York

Foelix RF, Chu-Wang I (1973) The morphology of spider sensilla II. Chemoreceptors. Tissue Cell 5:461–478

Foelix RF, Erb B, Michalik P (2010) Scopulate hairs in male *Liphistius* spiders: probable chemoreceptors. J Arachnol 38:599–603

Green AJ (2001) Mass/length residuals: measures of body condition or generators of spurious results? Ecology 82:1473–1483

Grinsted L, Bilde T, D'Ettorre P (2011) Cuticular hydrocarbons as potential kin recognition cues in a subsocial spider. Behav Ecol 22:1187–1194

Hénaut Y, Machkour-M'Rabet S (2005) Canibalismo y cleptobiosis en la tarantula *Brachypelma vagans*. Entomol Mex 4:30–32

Hodge MA, Stoffer-Isser A (1997) Conspecific and heterospecific attraction: a mechanism of web-site selection leading to aggregation formation by web-building spiders. Ethology 103:815–826

Howard RW (1993) Cuticular hydrocarbons and chemical communication. In: Stanley-Samuelson DW, Nelson DR (eds) Insect lipids: chemistry, biochemistry and biology. University of Nebraska Press, Lincoln

Jakob EM, Marshall SD, Uetz GW (1996) Estimating fitness: comparison of body condition indices. Oikos 77:61–67

Johansson BG, Jones TM (2007) The role of chemical communication in mate choice. Biol Rev 82:265–289

Krafft B (1971) Les interactions entre les individus chez *Agelena consociata* araignée sociale du Gabon. Arachn Congres Inter V: 159–164
Kullmann EJ (1972) Evolution of social behaviour in spiders (Araneae; Eresidae and Theridiidae). Am Zool 12:419–426
Liang D, Silverman J (2000) "You are what you eat": diet modifies cuticular hydrocarbons and nestmate recognition in the Argentine ant, *Linepithema humile*. Naturwissenschaften 87:412–416
Pasquet A, Trabalon M, Bagnères AG, Leborgne R (1997) Does group closure exist in the social spider *Anelosimus eximius*? Behavioural and chemical approaches. Insectes Soc 44:159–169
Pollard SD, Macnab AM, Jackson RR (1987) Communication with chemicals: pheromones and spiders. In: Nentwig W (ed) Ecophysiology of spiders. Springer, Berlin
Pourié G, Trabalon M (1999a) Agonistic behaviour of female *Tegenaria atrica* in the presence of different aged spiderlings. Physiol Entomol 24:1–7
Pourié G, Trabalon M (1999b) Relationships among food and contact signals in experimental group-living young of *Tegenaria atrica*. Arch Insect Biochem Physiol 42:188–197
Pourié G, Trabalon M (2001) Plasticity of agonistic behaviour in relation to diet and contact signals in experimentally group-living of *Tegenaria atrica*. Chemoecology 11:175–181
Pourié G, Trabalon M (2003) The role of 20-hydroxyecdysone on the control of spider vitellogenesis. Gen Comp Endocrinol 131:250–257
Pourié G, Ibarra F, Francke W, Trabalon M (2005) Fatty acids mediate aggressive behavior in the spider *Tegenaria atrica*. Chemoecology 15:161–166
Prouvost O, Keller JM, Trabalon M (1999a) Comportement sexuel et répartition des sensilles chez les mâles et femelles de *Tegenaria atrica* C. Koch (*Agelenidae*). Bull Acad Soc Lorraines Sci 38:6–14
Prouvost O, Trabalon M, Papke M, Schulz S (1999b) Contact sex signals on web and cuticle of *Tegenaria atrica* (Araneae, Agelenidae). Arch Insect Biochem Physiol 40:194–202
Ruttan LM (1990) Experimental manipulations of dispersal in the subsocial spider, *Theridion pictum*. Behav Ecol Sociobiol 27:169–173
Schal C, Sevala VL, Young H, Bachmann JAS (1998) Synthesis and transport of hydrocarbons: cuticle and ovary as target tissues. Am Zool 38:382–393
Schulz S (1997) The chemistry of spider toxins and spider silk. Angew Chem 36:314–326
Schulz S (2004) Semiochemistry of spiders. In Cardé, RT, Miller JG (eds) Advances in Insect Chemical Ecology, Cambridge University Press, USA, 110–150
Singer TL (1998) Roles of hydrocarbons in the recognition systems of insects. Am Zool 38:394–405
Stevenson RD, Woods WA (2006) Condition indices for conservation: new uses for evolving tools. Integr Comp Biol 46:1169–1190
Tichy H, Gingl E, Ehn R, Papke M, Schulz S (2001) Female sex pheromone of a wandering spider (*Cupiennius salei*): identification and sensory reception. J Comp Physiol A 187:75–78
Trabalon M (2000) The forms of communication in spiders. Act Coll Insect Soc 13:1–11
Trabalon M (2011) Agonistic interactions, cuticular and hemolymphatic lipid variations during the foraging period in spider females, *Brachypelma albopilosa*. J Insect Physiol 57:735–743
Trabalon M, Assi-Bessekon D (2008) Effects of web chemical signatures on intraspecific recognition in a subsocial spider, *Coelotes terrestris* (Araneae). Anim Behav 76:1571–1578
Trabalon M, Bautz AM, Morinière M, Porcheron P (1992) Ovarian development and correlated changes in hemolymphatic ecdysteroid levels in two spiders, *Coelotes terrestris* and *Tegenaria domestica* (*Araneae, Agelenidae*). Gen Comp Endocrinol 88:128–136
Trabalon M, Bagnères AG, Hartmann N, Vallet AM (1996) Change in cuticular compounds composition during the gregarious period and after dispersal of the young in *Tegenaria atrica* (Araneae, Agelenidae). Insect Biochem Mol Biol 26:77–84
Trabalon M, Bagnères AG, Roland C (1997) Contact sex signals in two sympatric spider species, *Tegenaria domestica* and *Tegenaria pagana*. J Chem Ecol 23:747–758

Trabalon M, Pourié G, Hartmann N (1998) Relationships among cannibalism, contact signals, ovarian development and ecdysteroid levels in *Tegenaria atrica* (Araneae, Agelenidae). Insect Biochem Mol Biol 28:751–758

Trabalon M, Niogret J, Legrand-Frossi C (2005) Effect of 20-hydroxyecdysone on cannibalism, sexual behavior, and contact sex pheromone in the solitary female spider, *Tegenaria atrica*. Gen Comp Endocrinol 144:60–66

Uetz GW, Roberts JA (2002) Multisensory cues and multimodal communication in spiders: insights from video/audio playback studies. Brain Behav Evol 9:222–230

Uhl G (2013) Spider olfaction: attracting, detecting, luring and avoiding. In: Nentwig W (ed) Spider ecophysiology. Springer, Heidelberg (this volume)

Vallet AM, Marion-Poll F, Trabalon M (1998) Preliminary electrophysiological study of the contact chemoreceptors in a spider. CR Acad Sci Paris 321:463–469

Van Zweden JS, Dreier S, D'Ettorre P (2009) Disentangling environmental and heritable nestmate recognition cues in a carpenter ant. J Insect Physiol 55:158–163

Wyatt TD (2003) Pheromones and animal behaviour: communication by smell and taste. Cambridge University Press, Cambridge

Chapter 11
Spider Olfaction: Attracting, Detecting, Luring and Avoiding

Gabriele Uhl

11.1 Introduction

The sensory world of spiders is composed of a wealth of chemical, tactile, acoustic and visual information that we are only slowly beginning to understand (Uhl and Elias 2011). Each signal modality has specific properties such as speed of propagation, directionality, persistence and effect of obstacles (McGregor 2005). Consequently, the combination and relative importance of the sensory channels used by a given spider species depends largely on its habitat and mode of living. For example, nocturnal or web-building species will depend less on visual signals than diurnal or cursorial species, but even vision-based jumping spiders use and exploit chemical, tactile and acoustic signals (Elias et al. 2005).

In the following, I present and discuss the current state of research on biofunctional molecules that transfer information between individuals, so-called semiochemicals, and I will focus on volatile substances. Pheromones are semiochemicals that mediate interactions between individuals of the same species. Pheromones with low molecular weight are airborne, not limited by environmental barriers, effective day and night and relatively long lasting. Volatile pheromones provide the best means to transmit signals over long distances. Chemicals of higher molecular weight, on the other hand, adhere to a substrate or to the animal's body or silk and are perceived by contacting the respective structures. Contact pheromones in combination with tactile information are thought to provide a more detailed picture of the state of the individual compared to volatile pheromones (see Trabalon 2013). However, there is evidence accumulating showing that volatile substances can signal much more than the mere presence of the emitter.

G. Uhl (✉)
General and Systematic Zoology, Zoological Institute and Museum, University of Greifswald, Greifswald, Germany
e-mail: gabriele.uhl@uni-greifswald.de

W. Nentwig (ed.), *Spider Ecophysiology*,
DOI 10.1007/978-3-642-33989-9_11,

Starting in the late nineteenth century (Peckham and Peckham 1887), many behavioural studies have shown that airborne and contact pheromones are used in information transfer in spiders (Gaskett 2007; Uhl and Elias 2011). Combinations of airborne as well as contact pheromones are suspected to have evolved especially in the context of reproduction. Here, a male spider has to find a female from a distance, assess her reproductive state and willingness to mate, perform courtship displays, make sure that he is not mistaken for prey, find the genital region and eventually mate successfully by exploiting all chemical, but also vibratory, visual and tactile information available to him. Apart from playing an important role in the context of mating, pheromones are also known to have evolved in the context of territorial marking and alarm substances, an aspect which as yet has not been explored in spiders.

If the participants involved in communication are from different species, the semiochemicals are called allelochemicals. Allelochemicals that benefit the sender (allomones) have been found in spiders that lure prey by copying the sex pheromones of the target species, attention (Herberstein and Wignall 2011). Allelochemicals that benefit the receiver (kairomones) are used to find prey or to avoid predators (Nelson and Jackson 2011a, b). Such kairomones are cues that signal the presence of a predator at a distance, early enough to avoid a potentially deadly physical encounter. Kairomones that signal the availability of prey may represent any volatile cue that is released from potential prey. Clearly, the biofunctional molecules that are used as pheromones or allelochemicals must have shaped the evolution of spider mating systems as well as the dynamics between predator and prey.

11.2 Volatile Sex Pheromones: Finding and Assessing Mates

Spiders are predators, most of which lead a solitary life. However, when sexually mature, the potential mating partners have to be located for reproduction. Early behavioural observations suggested that the olfactory sense is the major sensory channel by which mating partners are found. Mate attraction over long distances has been described for many spider species especially in web-building spiders where several adult males can be observed to arrive at a web of a female (Gaskett 2007; Foelix 2011; Uhl and Elias 2011). The wealth of observations that is mainly based on entelegyne spiders suggests that the females are attracting males as virgins and cease to produce the pheromone after mating. For example, in the Australian redback spider, *Latrodectus hasselti* (Theridiidae), males locate webs of adult virgin females from a distance (Kasumovic and Andrade 2004). An amino acid derivative functions as a contact pheromone which may be also responsible for mate attraction over long distances (Jerhod et al. 2010). When web extracts from virgin female *Linyphia triangularis* (Linyphiidae) are sprayed onto webs of juvenile or mated females, males perform an intriguing web reduction behaviour when encountering these manipulated webs. Web reduction by the male is typically performed with virgin females in several *Linyphia* species (Watson 1986; Schulz and Toft 1993). A butyric acid ((3*R*,3*R*′)-3-(3-hydroxybutyryloxy)butyric acid) was found to induce the web reduction behaviour. The acid degrades mainly autocatalytically into two acids (Schulz 2004). If these decay products are responsible for remote mate attraction, they could be produced with no extra costs.

To date, volatile pheromones have been identified and tested in only three species: *Agelenopsis aperta* (Agelenidae: Papke et al. 2001), *Argiope bruennichi* (Araneidae: Chinta et al. 2010) and *Pholcus beijingensis* (Pholcidae: Xiao et al. 2009; see Gaskett 2007).

The courtship behaviour of *Agelenopsis aperta* (Fig. 11.1a) includes a large number of different action patterns during vibratory and visual displays, and males respond to a volatile pheromone (8-methyl-2-aonanone) that is emitted by virgin females (Singer et al. 2000; Papke et al. 2001). At very low dosages, the synthesised pheromone was found to attract males in a choice arena system where it elicited most elements and stages of male courtship behaviour. The full range of male actions was observed only when females responded to male displays, demonstrating that vibratory communication between the sexes takes over at least in the last phase of courtship (Papke et al. 2001). The pheromone was not found in juveniles and in females younger than two weeks past the final moult, and its production subsides following insemination (Papke et al. 2001). The delay in female signalling after the final moult may reflect the time a female needs to produce secretion in which sperms are eventually stored (Uhl 1994). Alternatively, late female pheromone advertisement may be adaptive in the context of female mate choice since female genital structures will be fully hardened by this time, which may reduce the potential of male manipulations of freshly moulted and defenceless females.

In the orb-weaving spider *Argiope bruennichi* (Fig. 11.1b), a trimethyl methylcitrate is used as mate attractant (Chinta et al. 2010). The citrate is emitted in a specific diasteromeric ratio by virgin females and their webs. The citrate was not found in body extracts and headspace analyses of juveniles. After mating females cease emitting the pheromone. The substance naturally occurs in two enantiomers: the major enantiomer being trimethyl (2*R*,3*S*)-methylcitrate and the minor occurs in the 2*S*,3*S* configuration. In natural extracts, these enantiomers were found in a ratio between 6:1 and 25:1. It was demonstrated in a natural habitat of *Argiope bruennichi* that the citrate indeed functions as a sex pheromone (Chinta et al. 2010). When comparing different enantiomer ratios, more males arrived at 6:1 ratio traps compared to 2:1, but the difference was not statistically significant (Chinta et al. 2010). Nevertheless, it is still undetermined as to whether the enantiomer ratio a female produces depends on her condition and may represent a cue for male mate choice. Since males run a high risk of being cannibalised during mating (Schneider et al. 2006), male mate choice may be particularly pronounced in this species. The use of a derivate of citric acid as a sex pheromone already suggests that production costs are not negligible since citric acid constitutes a major primary metabolite. In the Krebs cycle, citric acid undergoes a series of chemical transformation and is eventually regenerated to continue the cycle. Interestingly, another spider species is known to use a derivative of citric acid. Females of the wandering

Fig 11.1 (**a**) *Agelenopsis aperta* (Agelenidae) mating trials in the laboratory (phot. S. Riechert). (**b**) *Argiope bruennichi* (Araneidae): receptive females assume a specific mating posture that allows sufficient space between her venter and the web (phot. G. Uhl). (**c**) *Pholcus beijingensis* (Pholcidae), female carrying the egg sac in her chelicerae (phot. Yue Chen). (**d**) *Evarcha culicivora* (Salticidae) male feeding on *Anopheles gambiae* (phot. R. Jackson). (**e**) *Hogna helluo* (Lycosidae), predator of *Pardosa milvina* (phot. J.M. Schmidt). (**f**) *Portia fimbriata* (Salticidae) feeding on *Jacksonoides queenslandicus* (Salticidae) (phot. R. Jackson). (**g**) *Mastophora stowei* (Araneidae) hunting at night with her bolas (phot. K.F. Haynes and K.V. Yeargan)

spider *Cupiennius salei* (Ctenidae) apply a contact pheromone to their silk. This unsymmetrical (*S*)-dimethyl citrate induces male courtship behaviour and was shown to elicit neural responses when in contact with chemosensitive sensilla

(Tichy et al. 2001). So far no other animals apart from these two spider species are known to use a derivative of citric acid as pheromone. It will need to be studied whether the withdrawal of citric acid for the production of a pheromone is costly and whether the production of a specific enantiomer ratio can be considered a cue that honestly reveals female condition.

The third spider species for which volatile pheromones are known is the haplogyne pholcid *Pholcus beijingensis* (Xiao et al. 2009, Fig. 11.1c). Males are attracted by silk and extracts of both virgin and mated females in a two-choice arena system. Interestingly, mated females remain receptive and signal continuously after mating which corresponds with the high paternity success for last males to mate that was found for other pholcids (Eberhard et al. 1993; Schäfer et al. 2008). Due to the specific haplogyne female genital system, males have access to the female sperm storage site and can displace or remove previous males' sperm. Continuous signalling of the female should therefore not be in the male's interest. It seems therefore that it is the female sex that controls the dynamics of signalling. Females also reduce pheromone production during the time of brood care, when they carry the egg sac in their chelicerae. The pheromone that triggers male searching behaviour was found to consist of a blend of two acetates in a 2:1 ratio ((*E*,*E*)-farnesyl acetate and hexadecyl acetate). This is the first known case of a pheromone that consists of more than one component in spiders (Xiao et al. 2009). Multicomponent pheromones that function only when produced in a specific ratio are highly common in insects and mammals (Wyatt 2003). Since we have only started to work on spider olfaction, future studies will show if spider airborne signals can be similarly complex. The available information on the communication system of *Pholcus beijingensis* points in this direction.

Understanding the dynamics and adaptive value of female signalling is a great challenge for future studies. In *Agelenopsis aperta* and *Argiope bruennichi,* only virgin females were found to produce the respective pheromone, whereas mated female *Pholcus beijingensis* continue signalling. These coarse signalling patterns need to be investigated in more detail. Virgin females clearly need to attract males as sperm donors to secure fertilisation of their eggs. In addition, competition between males over access to the female may serve to filter out males of high quality that eventually gain access to the females. However, why females stop signalling after mating is less clear, since polyandrous females may profit from competing sperm of different males inside her genital tract. Furthermore, multiple mating reduces the risk of mating with a sterile male, reduces the risk of inbreeding, increases genetic diversity of offspring and allows sperm choice by the female (Hosken and Stockley 2003). On the other hand, long-term signalling may be costly from a female's perspective if the signal attracts predators or parasitoids. Also, long-term signalling may be ineffective in species in which the morphology of the female genital tract favours sperm from the first male or in cases in which effective mating plugs are applied to the female genital openings during mating that strongly reduce subsequent mating success as is the case for *Argiope bruennichi* (Uhl et al. 2010). Keeping in mind that most entelegyne female spiders possess two copulatory ducts that lead to separate sperm storage organs; "mated" females may have experienced sperm transfer into one or both spermathecae. If copulation results in only one insertion, only one spermatheca will contain sperm, and a female's reproductive state may therefore be called "half-mated" in contrast to "fully mated" females that experienced sperm transfer into both spermathecae. "Half-mated females" may profit from continued signalling to receive sperm from at least two males, unless the single mate is the male of highest quality available. Female signalling should therefore be continuous or depend on the quality of the males. From a male's perspective, however, a female should cease to be attractive to other males as soon as he has transferred sperm. The selective advantage for male manipulative behaviour is especially obvious in species in which genital plugging did not evolve. However, even in species with effective plugging, only a male that plugs both copulatory openings can expect exclusive paternity as was shown for *Argiope bruennichi* (Nessler et al. 2007). If he is only allowed to use one opening because the female fends him off or cannibalises him, any substance that reduces female remating success with rival males will be advantageous. This raises the question if spider males can transfer manipulative substances that render the female unattractive or cause her to stop signalling, as was found to occur in many animal mating systems and which is especially well studied in *Drosophila* (Simmons 2001; Chapman and Davies 2004).

On the other end of the time spectrum, the onset of female signalling also requires further scrutiny. What explanation is there for the time lag between final moult and pheromone production as was found in *Agelenopsis aperta*? Can we find a similar time lag in other species, or is it more common to find signals shortly after the final moult? Behavioural observations strongly suggest that the majority of spiders start signalling early after the moult to adulthood (Uhl and Elias 2011). Recent observations of the mating behaviour of *Argiope bruennichi* point towards another interesting aspect of the mating system that is probably connected to the timing of signalling. In *Argiope bruennichi*, mating occurs while the female is in the process of moulting (Uhl and Renner, unpublished) as was found for *Argiope aurantia* (Foellmer and Fairbairn 2003). This behaviour is termed "opportunistic mating" since it is considered a male strategy to overcome female aggressive behaviour and mate choice. Mating with moulting and thus defenceless females may be especially advantageous in species with a high probability of sexual cannibalism. Indeed, 80 % of *Argiope bruennichi* males do not survive their first insertion when mating with a virgin female with hardened genitalia since females invariably attack the male during mating (Schneider et al. 2006). When a male mates with a moulting female, however, sexual cannibalism does not occur and males survive to perform a second insertion with the same or another female (Uhl and Renner, unpublished). In terms of survival and mating frequency, opportunistic mating is clearly advantageous to males. However, it is unclear how males find subadult females that preferably are close to the final moult and whether volatile signals or cues are involved. Females in the subadult state did not emit the citrate (Chinta et al. 2010), but volatile substances may only be emitted very shortly before moulting starts. However, if opportunistic mating is disadvantageous for females, we would expect a time lag between the moulting process and the onset of signalling which would allow the female to be under control of mating. Consequently, studies are required that reveal the dynamics of female pheromone advertisement and the mechanisms that are applied by males for finding defenceless females.

To complete the complex picture, male signalling also needs to be taken into account. In two wolf spider species (*Allocosa brasiliensis*, *Allocosa alticeps*) males produce volatile signals that are used by females for finding male burrows where they initiate courtship (Aisenberg et al. 2010). In the salticid *Evarcha culicivora*, both sexes perform courtship and mate assessment. The presence of male odour leads to escalation in agonistic behaviour between females (Cross and Jackson 2009). Females were found to discriminate not only between sexes but also between developmental stages based on as yet unknown odour signals (Cross et al. 2009). In addition, females prefer males that recently fed on blood-carrying mosquitoes over those that did not have a recent indirect blood meal (Jackson et al. 2005, Fig. 11.1d). This case demonstrates that species and sex-specific signals and dietary odour cues can be combined for mate attraction and mate assessment (Cross et al. 2009). Indeed, male volatile substances are used for mate assessment in a spitting spider (*Scytodes* sp., Scytodidae) in which female reproductive output measured as number and size of eggs and hatching success was higher after mating with males whose odour the females preferred during a two-choice trial (Koh et al. 2009). Olfactory choice of mating partners requires that the structure and/or composition of the signal varies depending on the condition or genetic architecture of the potential mate. Current evidence suggests that chemical signals may provide good means of assessing genotypes (Johanson and Jones 2007). Compatibility between mating partners that is assessed by olfactory signals/cues similar to discrimination on MHC-based pheromones in vertebrates (Johnston 2003) has not been considered for spiders but represents the next major step in the investigation of spider olfaction.

11.3 Finding Prey and Avoiding Being Preyed on

The contexts in which spiders use olfaction are not restricted to sexual selection. Since spiders are key predators of terrestrial ecosystems, spiders must be able to distinguish between profitable hunting sites and therefore must exploit cues that transfer information on food availability as well as on the degree of predation risk and competition at the site. Chemical cues are suspected to be highly important in these contexts.

11.3.1 Site Selection

Although spiders generally hatch from the eggs in habitats that are suitable for them, they change habitats during development from a freshly hatched spiderling to an adult individual. On the one end of the extreme, spiderlings that migrate by ballooning must have means to evaluate the quality of the new habitat and decide on staying or leaving accordingly (Bonte et al. 2012; Bonte 2013). Even if spiderlings stay in their native habitat, they often change microhabitats by, e.g. moving from denser to more open vegetation or building webs at different heights above the ground depending on their developmental stage (Kevan and Greco 2001). Generally, spiders should choose foraging patches to which they are best adapted and which offer sufficient prey. However, the cues that are used by spiders to find suitable patches have rarely been investigated. The crab spider *Thomisus spectabilis* which typically waits on flowers to catch flower visitors uses olfactory cues to identify flowers as rewarding foraging patches (Heiling et al. 2004). In contrast, another thomisid *Misumena vatia* showed no preference for flowers, floral extracts or common sesquiterpenoid floral scents (β-caryophyllene and nerolidol) in natural concentrations (Junker et al. 2011). Interestingly, another sit-and-wait predator that does not hunt on flowers, the pisaurid *Pisaura mirabilis*, seems to avoid flowers, extracts and floral scents (Junker et al. 2011) which could be due to a deterrent effect of plant volatiles that was shown for various arthropods (Junker and Blüthgen 2010). Plants seem to have evolved floral compounds that deter predators that exploit the plant-pollinator mutualism and reduce the reproductive success of plants (Irwin et al. 2004). Possibly, flower-visiting spider predators are able to tolerate the defensive floral scents (as may be the case in *Misumena vatia*) or even use them to find the flower host from a distance (as seems to be the case in *Thomisus spectabilis*).

11.3.2 The Smell of Prey

Foraging efficiency may be significantly increased by exploiting volatile odours of prey. Predators may exploit any chemical substance that informs about the presence of potential prey or may have specialised on detecting species-specific substances evolved for intraspecific communication of the prey species. Although prey detection at a distance by means of such kairomones may be highly advantageous, our knowledge on this context is limited. *Cupiennius salei* was shown to use its olfactory sense to detect prey from a distance (Hostettler and Nentwig 2006). Interestingly, it was shown that spiders are able to choose prey according to the availability of venom: spiders with empty venom glands preferred venom-sensitive prey (cockroach species 1) over venom-insensitive prey (cockroach species 2), whereas spiders with full venom glands did not distinguish between these prey types (Hostettler and Nentwig 2006). In most studies available, however, it was tested if spiders react differently to substrates containing chemical cues from potential prey compared to control substrates making it difficult to separate chemotactile and volatile information (Johnson et al. 2011 and references therein). Females of *Latrodectus hesperus* were given the choice of rocks previously housing cricket prey versus control rocks lacking cricket cues. *Latrodectus hesperus* preferentially chose microhabitat patches that contained chemical prey cues (Johnson et al. 2011). In the wolf spider, *Trochosa parthenus*, and in the lynx spider *Oxyopes salticus*, individuals that were taken from the field stayed in areas for longer that contained cues from prey that occurred in the natural habitat relative to unfamiliar prey (Punzo and Kukoyi 1997). However, spiders reared in the laboratory showed no preference for the prey types offered. This suggests that even polyphagous spiders that are generally considered to exhibit little prey preference (Nentwig 1987) exhibit associative learning of chemical cues (Punzo and Kukoyi 1997; Persons and Rypstra 2000). An insight into the complex predator–prey dynamics offers the detailed studies on two North American wolf spiders *Hogna helluo* (Fig. 11.1e) and its syntopic, smaller prey species *Pardosa milvina*. *Hogna helluo* seems to form an olfactory searching image of its prey species and shows a strong response to specific prey even after a long period without food (9 days; Persons and Rypstra 2000). *Hogna helluo* that were fed *Pardosa milvina* preferred areas (measured as longer residency time and decreased mobility) that contained chemical cues from *Pardosa milvina*, whereas if fed crickets they preferred areas containing cues from crickets. A volatile component must be involved in site selection since *Hogna hello* that had fed *Pardosa milvina* moved with high probability directly from the untreated centre of the arena to the area that contained cues from *Pardosa milvina* (Persons and Rypstra 2000). Possibly, *Hogna helluo* exploits the sex pheromone of *Pardosa milvina* (Searcy et al. 1999) apart from exploiting chemical contact cues to increase foraging success. These findings demonstrate that even a spider that is considered a generalist predator can associate chemical cues with particular prey. It remains to be investigated how many different cues a generalist spider can learn, and how easily it can switch prey types by means of associative learning.

Spiders that are more specialised predators clearly respond to olfactory and contact chemical cues of their specific prey. Chemical espionage by intercepting messages is common in these predator–prey systems (Stowe et al. 1995). For example, the myrmecophagic jumping spider *Habrocestum pulex* spends more time and exhibits specific stalking sequences on ant-treated soil compared to control soil (Clark et al. 2000). In olfactometer tests, *Habrocestum pulex* discriminated between the odour of disturbed ants as well as their synthetic alarm pheromone (6-methyl-5-hepten-2-one) and a control. The alarm pheromones that are produced by injured or disturbed ants are used as kairomones not only by myrmecophagic jumping spiders but also in the less vision-based zodariid spiders (Allan et al. 1996). In addition, *Habrocestum pulex* located ants faster when ant-derived cues were present suggesting that olfaction and vision can be highly integrated. Attentional priming by kairomones was also found for the salticid *Portia fimbriata* that mainly feeds on the salticid *Jacksonoides queenslandicus* (Fig. 11.1f). Chemical cues of *Jacksonoides queenslandicus* heighten *Portia fimbriata's* attention to optical cues of the preferred prey (Jackson et al. 2002). Attentional priming seems to be preprogrammed and not based on prior experience (Nelson and Jackson 2011a).

An intriguing case of prey selection occurs in the jumping spider *Evarcha culicivora*. These spiders feed indirectly on vertebrate blood by preferentially attacking female mosquitoes that have had a recent blood meal (Jackson et al. 2005). The study in which confounding effects of prey behaviour were eliminated suggests that *Evarcha culicivora* can identify the preferred prey by odour alone and by sight alone (Jackson et al. 2005). To my knowledge, this is the only study that demonstrates that a predator selects prey on the basis of what the prey has recently eaten.

11.3.3 Luring Prey

Chemical compounds that are involved in interspecific interactions and in which the emitter has an adaptive benefit are termed allomones. In the case of the bolas spiders, the spider takes the role of an illicit signaller that aggressively mimics the chemical signals normally used by their prey in intraspecific communication (Stowe et al. 1995; Vereecken and McNeil 2010). Most information available on chemical aggressive mimicry (Peckham's mimicry) in bolas spiders (Fig. 11.1g) comes from studies on the American *Mastophora* species (Stowe et al. 1995; Gemeno et al. 2000; Haynes et al. 2001). Bolas are sticky globules at the end of a silk thread that spiders produce with no apparent trigger or within two minutes after having received an auditory cue of an approaching moth (Haynes et al. 2001). In hunting position, bolas spiders have their front legs extended and hold the bolas with one leg. To catch prey, *Mastophora hutchinsoni* and *M. cornigera* draw back the front leg and flick the bolas in the direction of the prey, whereas bolas spiders in Africa and Australia whirl the bolas in a circle (Haynes et al. 2001). Prey that gets stuck at the globule is instantly bitten, paralysed and wrapped by the spider. Interestingly, only late instar and adult females produce bolas, and it is not the bolas that contain the pheromone mimic (Stowe et al. 1987). Females in hunting position without bolas emit the pheromone mimics and release more of the substances than when in a resting position or when feeding.

Usually, illicit signallers tend to exploit only one or a limited number of taxa, because the production of signals requires the evolution of specialised signalling mechanisms (Vereecken and McNeil 2010). Interestingly, females of *Mastophora cornigera* capture mainly noctuid moth species from the same family. Nineteen different moth species are recorded for *Mastophora cornigera* with up to nine different species by a single female (Stowe et al. 1995). Variability in the volatiles found in individual females was considerable. Analysis of the compounds involved in prey attraction demonstrated that *Mastophora cornigera* can attract many moth species since it relies on common components found in their pheromone blends (Stowe et al. 1995). The array of moth species as prey is more limited in *Mastophora hutchinsoni*. The two species that compose more than 90 % of its diet, however, use very different two-component pheromone blends with no overlap in the pheromone composition, and their biosynthesis requires different enzymatic reactions (Yeargan 1994). Nevertheless, *Mastophora hutchinsoni* produces compounds that are identical to those of its lepidopteran prey in the appropriate blend ratio (Gemeno et al. 2000). The main moth species differ in periodicity with one early-flying and one late-flying species. *Mastophora hutchinsoni* seems to produce components of both species at all times during the night but drastically decreases the production of the pheromone of the early-flying moth over the course of the night (Haynes et al. 2002). The differential temporal production leads to an intriguingly efficient exploitation of different prey species.

When looking at the foraging habits of early instars and male *Mastophora* species, the story of aggressive mimicry becomes even more complex: early instars of females and all stages of males also attract prey, but they do not use bolas; they sit on leaf edges with their front two pairs of legs stretched up in the air (Yeargan 1994). Their targets are male moth flies (Psychodidae). As in the adults, each of the bolas spiders exploits different psychodid species demonstrating that species-specific chemical cues are produced to attract male prey (Yeargan and Quate 1996). Since the pheromones in the Psychodidae have not yet been identified, it is not known how strongly the pheromones match the female sex pheromones of the various prey species. It would be of interest to determine what triggers the dramatic switch from producing pheromone mimics of flies to moths and how this is physiologically accomplished and to identify the pathways for the synthesis of the allomones. Overall, studies on aggressive chemical mimicry in bolas spiders suffer from the cryptic and nocturnal life of the spiders. Other spider species also have been suspected to attract insects such as the araneid *Kaira alba* (Levi 1993) and the theridiid *Phoroncidia studo* (Eberhard 1981) but have not been studied yet.

11.3.4 Avoiding Predation

Chemical signals further play a role in detecting and avoiding predators (Dicke and Grostal 2001). Several spider species were shown to be able to detect airborne predator-avoidance kairomones (Nelson and Jackson 2011b). The wolf spider *Pardosa milvina* that is detected by its predator *Hogna helluo* by means of kairomones shows antipredator responses to airborne, silk and excreta-based chemical cues from *Hogna helluo* (Schonewolf et al. 2006; Persons et al. 2001, 2002, Lehmann et al. 2004). In pitfall capture experiments, *Pardosa milvina* showed avoidance of traps baited with males or females of the large wolf spider *Hogna helluo* but did not avoid traps containing crickets or empty traps (Schonewolf et al. 2006). *Pardosa milvina* showed graded antipredator behaviour relative to the quantity of kairomones present (Persons and Rypstra 2001) and showed reduced activity in areas containing chemical cues from *Hogna helluo* that had previously been fed *Pardosa milvina* (Persons et al. 2001). These findings suggest that *Pardosa milvina* is not only able to detect chemical cues from predators and their quantity but also distinguish cues from predators fed different diets. Avoidance extends even to the offspring in *Pardosa milvina*: spiderlings remain on the back of the mother with higher probability in containers that were previously occupied by an early instar of *Hogna helluo* (Persons and Lynam 2004). Similarly, the incubation duration of the eggs in the spitting spider *Scytodes pallida* is significantly reduced when airborne cues from the araneophagic jumping spider *Portia labiata* are present (Li and Jackson 2005). Carrying eggs in the chelicerae is typical for spitting spiders and makes the mother more vulnerable to predators. Incubation time was also influenced by the predator's previous diet: incubation time was shorter in the presence of volatile cues from *Portia labiata* that had previously fed on *Scytodes pallida* compared to cues from *Portia labiata* fed on house flies (Li and Jackson 2005). Overall, potential prey can derive information from volatile substances that contain detailed cues not only about the presence of a predator and its previous diets but also about size and hunger level of the predator (Bell et al. 2006).

11.4 Info-disruption

As demonstrated above, chemical signals and cues are generally used in spider communication systems on the intra- and interspecific level. Natural chemical signalling, however, can be impeded or disturbed by anthropogenic chemicals that are used to manage weeds, fungi, insects and bacteria. Studies on the effects of anthropogenic chemicals are highly biased towards aquatic systems (Lürling and Scheffer 2007), but recently it was shown that herbicides can directly influence the behaviour of terrestrial arthropods such as spiders (e.g. Michalkova and Pekár 2009; Benamú et al. 2010, see also Pekár 2013). In a field and in the laboratory, exposure to a glyphosate-based herbicide resulted in reduced mate location and distorted reaction towards female airborne signals in males of the agrobiont wolf spider *Pardosa milvina* (Griesinger et al. 2011). The study suggests that although the herbicide is not acutely toxic, it has strong effects on intraspecific information flow. Exposure to the glyphosate-based herbicide further resulted in decreased activity in *Pardosa milvina* which may impact its ability to forage as well as colonise new habitat patches (Wrinn et al. 2012). If anthropogenic chemicals distort the perception of natural chemical signals and cues, considerable changes at the population and community levels can be expected.

11.5 Where Are Chemical Signals Produced and Perceived?

Despite the importance of semiochemicals for the sensory world of spiders, we still do not know where the chemicals are produced and know little about how they are perceived. Candidate cells for the production of chemical signals have been suspected in the silk apparatus, spinnerets or in the epidermis (overview in Uhl and Elias 2011, see also Townley and Tillinghast 2013). The few spider sex pheromones that were identified are derived from primary metabolites suggesting that the biosynthetic pathways differ between spiders and insects (Francke and Schulz 1999; Jurenka 2004). The production of signals from primary metabolites may involve higher costs compared to the production of signals from secondary metabolites.

The receptors for chemical signals are suspected to be tip pore sensilla (Barth 2002; see also Trabalon 2013). In *Cupiennius salei*, electrophysiological methods and single-cell recordings demonstrated neural responses in tip pore sensilla after contact with female silk or synthetic contact pheromone (Tichy et al. 2001). Tip pore sensilla exhibit a blunt tip with a subterminal single pore and include dendritic nerve endings (Foelix and Chu-Wang 1973). The chemosensitive sensilla are described as having a steeper angle of insertion compared to mechanosensory sensilla, having curved hair shafts and spiral ornamentation. However, much of what we know originates from the study of *Cupiennius salei* only (Barth 2002). The chemosensitive hairs are most numerous in the tarsi and metatarsi of the spider's legs suggesting that they are mainly involved in gustation (Foelix 2011). That such taste hairs also perceive airborne substances is very likely since there is no distinct difference between smell and taste, but as yet electrophysiological tests are lacking.

11.6 Conclusion

We can safely assume that olfaction plays a crucial role in the sensory world of spiders although we know little about the nature of the chemical compounds involved, the receptors involved and virtually nothing about the production sites. Olfaction pertains to finding and assessing mates, detecting and luring prey and avoiding predators and extends to social interactions (Trabalon 2013). Accordingly, we first need to identify the nature of the chemical signals that are produced, copied and exploited and then move on to explore their flexible use depending on developmental state and environmental factors. Finally, we need to understand spider communication as a multimodal world that encompasses many sensory modalities: chemical, acoustic and visual.

Acknowledgements I cordially thank Ken Haynes, Michael Schmitt and Maxene Graze for critically reading the manuscript and many colleagues for providing colour photographs of some of the treated species.

References

Aisenberg A, Baruffaldi L, González M (2010) Behavioural evidence of male volatile pheromones in the sex-role reversed wolf spiders *Allocosa brasiliensis* and *Allocosa alticeps*. Naturwissenschaften 97:63–70

Allan RA, Elgar MA, Capon RJ (1996) Exploitation of an ant chemical alarm signal by the zodariid spider *Habronestes bradleyi* Walckenaer. Proc R Soc B 263:69–73

Barth FG (2002) A spider's world: senses and behavior. Springer, Berlin

Bell RD, Rypstra AL, Persons MH (2006) The effect of predator hunger on chemically mediated antipredator responses and survival in the wolf spider *Pardosa milvina* (Araneae: Lycosidae). Ethology 112:903–910

Benamú MA, Schneider MI, Sanchez NE (2010) Effects of the herbicide glyphosate on biological attributes of *Alpaida veniliae* (Araneae, Araneidae), in laboratory. Chemosphere 78:871–876

Bonte D (2013) Cost-benefit balance of dispersal and the evolution of conditional dispersal strategies in spiders. In: Nentwig W (ed) Spider ecophysiology. Springer, Heidelberg (this volume)

Bonte D, Van Dyck H, Bullock JM, Coulon A, Delgado M, Gibbs M, Lehouck V, Matthysen E, Mustin K, Saastamoinen M, Schtickzelle N, Stevens VM, Vandewoestijne S, Baguette M, Barton K, Benton TG, Chaput-Bardy A, Clobert J, Dytham C, Hovestadt T (2012) Costs of dispersal. Biol Rev 87:290–312

Chapman T, Davies SJ (2004) Functions and analysis of the seminal fluid proteins of male *Drosophila melanogaster* fruit flies. Peptides 25:1477–1490

Chinta SP, Goller S, Lux J, Funke S, Uhl G, Schulz S (2010) The sex pheromone of the wasp spider *Argiope bruennichi*. Angew Chem Int Ed. doi:10.1002/anie.200906311

Clark RJ, Harland DP, Jackson RR (2000) Speculative hunting by an araneophagic salticid spider. Behaviour 137:1601–1612

Cross FR, Jackson RR (2009) How cross-modality effects during intraspecific interactions of jumping spiders differ depending on whether a female-choice or mutual-choice mating system is adopted. Behav Processes 80:162–168

Cross FR, Jackson RR, Pollard SD (2009) How blood-derived odor influences mate-choice decisions by a mosquito-eating predator. Proc Natl Acad Sci USA 106:19416–19419

Dicke M, Grostal P (2001) Chemical detection of natural enemies by arthropods: an ecological perspective. Annu Rev Ecol Syst 32:1–23

Eberhard WG (1981) The single line web of *Phoroncidia studo* Levi (Araneae: Theridiidae): a prey attractant? J Arachnol 9:229–232

Eberhard WG, Guzman Gomez S, Catley KM (1993) Correlation between spermathecal morphology and mating systems in spiders. Biol J Linn Soc 50:197–209

Elias DO, Hebets EA, Hoy RR, Mason AC (2005) Seismic signals are crucial for male mating success in a visual specialist jumping spider (Araneae: Salticidae). Anim Behav 69:931–938

Foelix RF (2011) The biology of spiders. Oxford University Press, Oxford

Foelix RF, Chu-Wang I-W (1973) The morphology of spider sensilla. II. Chemoreceptors. Tissue Cell 5:461–478

Foellmer MW, Fairbairn DJ (2003) Spontaneous male death during copulation in an orb weaving spider. Proc R Soc Lond B 270:S183–S185

Francke W, Schulz S (1999) Pheromones. In: Barton D, Nakanishi K, Meth-Cohn O, Mori K (eds) Comprehensive natural products chemistry, vol 8. Elsevier, Amsterdam

Gaskett AC (2007) Spider sex pheromones: emission, reception, structures, and function. Biol Rev 82:27–48

Gemeno C, Yeargan KV, Haynes KF (2000) Aggressive chemical mimicry by the bolas spider *Mastophora hutchinsoni*: identification and quantification of a major prey's sex pheromone components in the spider's volatile emissions. J Chem Ecol 26:1235–1243

Griesinger LM, Evans SC, Rypstra AL (2011) Effects of a glyphosate-based herbicide on mate location in a wolf spider that inhabits agroecosystems. Chemosphere 84:1461–1466

Haynes KF, Yeargan KV, Gemeno C (2001) Detection of prey by a spider that aggressively mimics pheromone blends. J Insect Behav 14:535–543

Haynes KF, Gemeno C, Yeargan KV, Millar JG, Johnson KM (2002) Aggressive chemical mimicry of moth pheromones by a bolas spider: how does this specialist predator attract more than one species of prey? Chemoecology 12:99–105

Heiling AM, Cheng K, Herberstein ME (2004) Exploitation of floral signals by crab spiders (*Thomisus spectabilis*, Thomisidae). Behav Ecol 15:321–326

Herberstein ME, Wignall A (2011) Deceptive signals in spiders. In: Herberstein ME (ed) Spider behaviour. Flexibility and versatility. Cambridge University Press, Cambridge
Hosken DJ, Stockley P (2003) Benefits of polyandry: a life history perspective. Evol Biol 33:173–194
Hostettler S, Nentwig W (2006) Olfactory information saves venom during prey-capture of the hunting spider *Cupiennius salei* (Araneae: Ctenidae). Funct Ecol 20:369–375
Irwin RE, Adler LS, Brody AK (2004) The dual role of floral traits: pollinator attraction and plant defence. Ecology 85:1503–1511
Jackson RR, Clark RJ, Harland DP (2002) Behavioural and cognitive influences of kairomones on an araneophagic jumping spider. Behaviour 139:749–775
Jackson RR, Nelson XJ, Sune GO (2005) A spider that feeds indirectly on vertebrate blood by choosing female mosquitoes as prey. Proc Natl Acad Sci USA 102:15155–15160
Jerhod E, Stoltz JA, Andrade MCB, Schulz S (2010) Acylated serine derivative: a new class of arthropod pheromones of the Australian redback spider, *Latrodectus hasselti*. Angew Chem Int Ed 49:1–5
Johanson BG, Jones TM (2007) The role of chemical communication in mate choice. Biol Rev 82:265–289
Johnson A, Revis O, Chadwik Johnson J (2011) Chemical prey cues influence the urban microhabitat preferences of Western black widow spiders, *Latrodectus hesperus*. J Arachnol 39:449–453
Johnston RE (2003) Chemical communication in rodents: from pheromones to individual recognition. J Mammal 84:1141–1162
Junker RR, Blüthgen N (2010) Floral scents repel facultative flower visitors, but attract obligate ones. Ann Bot 105:777–782
Junker RR, Bretscher S, Dötterl S, Blüthgen N (2011) Phytochemical cues affect hunting-site choices of a nursery web spider (*Pisaura mirabilis*) but not a crab spider (*Misumena vatia*). J Arachnol 39:113–117
Jurenka R (2004) Insect pheromone biosynthesis. Top Curr Chem 239:97–132
Kasumovic MM, Andrade MCB (2004) Discrimination of airborne pheromones by mate-searching male western black widow spiders (*Latrodectus hesperus*): species- and population-specific responses. Can J Zool 82:1027–1034
Kevan PG, Greco CF (2001) Contrasting patch choice behaviour by immature ambush predators, a spider (*Misumena vatia*) and an insect (*Phymata americana*). Ecol Entomol 26:148–153
Koh TH, Seah WK, Yap L-MYL, Li D (2009) Pheromone-bases female mate choice and its effect on reproductive investment in a spitting spider. Behav Ecol Sociobiol 63:923–930
Lehmann LM, Walker SE, Persons MH (2004) The influence of predator sex on chemically mediated antipredator response in the wolf spider *Pardosa milvina* (Araneae: Lycosidae). Ethology 110:323–339
Levi HW (1993) The orb-weaver genus *Kaira* (Araneae, Araneidae). J Arachnol 21:209–225
Li D, Jackson RR (2005) Influence of diet-related chemical cues from predators on the hatching of egg-carrying spiders. J Chem Ecol 31:333–342
Lürling M, Scheffer M (2007) Info-disruption: pollution and the transfer of chemical information between organisms. Trends Ecol Evol 22:374–379
McGregor PK (2005) Communication. In: Bolhuis JJ, Giraldear L-A (eds) The behaviour of animals. Mechanisms, function, and evolution. Blackwell, Oxford
Michalkova V, Pekár S (2009) How glyphosate altered the behaviour of agrobiont spiders (Araneae: Lycosidae) and beetles (Coleoptera: Carabidae). Biol Control 51:444–449
Nelson XJ, Jackson RR (2011a) Flexibility in the foraging strategies of spiders. In: Herberstein ME (ed) Spider behaviour. Flexibility and versatility. Cambridge University Press, Cambridge
Nelson XJ, Jackson RR (2011b) Flexible use of anti-predator defences. In: Herberstein ME (ed) Spider behaviour. Flexibility and versatility. Cambridge University Press, Cambridge
Nentwig W (1987) The prey of spiders. In: Nentwig W (ed) Ecophysiology of spiders. Springer, Berlin

Nessler SH, Uhl G, Schneider JM (2007) Genital damage in the orb-weaving spider *Argiope bruennichi* (Araneae: Araneidae) maximises paternity success. Behav Ecol 18:174–181
Papke MD, Riechert SE, Schulz S (2001) An airborne female pheromone associated with male attraction and courtship in a desert spider. Anim Behav 61:877–886
Peckham GW, Peckham EG (1887) Some observations on the mental powers of spiders. J Morphol 1:383–419
Pekár S (2013) Side-effect of synthetic pesticides on spiders. In: Nentwig W (ed) Spider ecophysiology. Springer, Heidelberg (this volume)
Persons MH, Lynam EC (2004) *Pardosa milvina* (Araneae: Lycosidae) spiderling movement in the presence of conspecific and heterospecific silk and excreta. J Arachnol 32:341–344
Persons MH, Rypstra AL (2000) Preference for chemical cues associated with recent prey in the wolf spider *Hogna helluo* (Araneae: Lycosidae). Ethology 106:27–35
Persons MH, Rypstra AL (2001) Wolf spiders show graded antipredator behavior in the presence of chemical cues from different sized predators. J Chem Ecol 27:2493–2503
Persons MH, Walker SE, Rypsta AL, Marshall SD (2001) Wolf spider predator avoidance tactics and survival in the presence of diet-associated predator cues (Araneae: Lycosidae). Anim Behav 61:43–51
Persons MH, Walker SE, Rypsta AL (2002) Fitness costs and benefits of antipredator behavior mediated by chemotactic cues in the wolf spider *Pardosa milvina* (Araneae: Lycosidae). Behav Ecol 13:386–392
Punzo F, Kukoyi O (1997) The effects of prey chemical cues on patch residence time in the wolf spider *Trochosa parthenus* (Chamberlin) (Lycosidae) and the lynx spider *Oxyopes salticus* Hentz (Oxyopidae). Bull Brit Arachnol Soc 10:323–326
Schäfer MA, Misof B, Uhl G (2008) Effects of body size of both sexes and female mating history on male mating behavior and paternity success in a spider. Anim Behav 76:75–86
Schneider JM, Gilberg S, Fromhage L, Uhl G (2006) Sexual conflict over remating in a cannibalistic spider. Anim Behav 71:781–788
Schonewolf KW, Bell R, Rypstra AL, Persons MH (2006) Field evidence of an airborne enemy avoidance kairomone in wolf spiders. J Chem Ecol 32:1565–1576
Schulz S (2004) Semiochemistry of spiders. In: Cardé RT, Miller JG (eds) Advances in insect chemical ecology. Cambridge University Press, Cambridge
Schulz S, Toft S (1993) Identification of a sex pheromone from a spider. Science 260:1635–1637
Searcy LE, Rypstra AL, Persons MH (1999) Airborne chemical communication in the wolf spider *Pardosa milvina*. J Chem Ecol 25:2527–2533
Simmons LW (2001) Sperm competition and its evolutionary consequences in the insects. Princeton University Press, Princeton
Singer F, Riechert SE, Xu H, Morris AW, Becker E, Hale JA, Noureddine MA (2000) Analysis of courtship success in the funnel web spider *Agelenopsis aperta*. Behaviour 137:93–117
Stowe MK, Tumlinson JH, Heath RR (1987) Chemical mimicry: bolas spiders emit components of moth prey species sex pheromones. Science 236:964–967
Stowe MK, Turlings TCJ, Loughrin JH, Lewis WJ, Tumlinson JH (1995) The chemistry of eavesdropping, alarm, and deceit. Proc Natl Acad Sci USA 92:23–28
Tichy H, Gingl E, Ehn R, Papke M, Schulz S (2001) Female sex pheromone of a wandering spider (*Cupiennius salei*): identification and sensory reception. J Comp Physiol A 187:75–78
Townley MA, Tillinghast E (2013) Aggregate silk gland secretions of araneoid spiders. In: Nentwig W (ed) Spider ecophysiology. Springer, Heidelberg (this volume)
Trabalon M (2013) Chemical communication and contact cuticular compounds in spiders. In: Nentwig W (ed) Spider ecophysiology. Springer, Heidelberg (this volume)
Uhl G (1994) Genital morphology and sperm storage in *Pholcus phalangioides* (Fuesslin) (Pholcidae; Araneae). Acta Zool 75:13–25
Uhl G, Elias DO (2011) Communication. In: Herberstein ME (ed) Spider behaviour. Flexibility and versatility. Cambridge University Press, Cambridge

Uhl G, Nessler SH, Schneider JM (2010) Securing paternity in spiders? A review on occurrence and effects of mating plugs and male genital mutilation. Genetica 138:75–104
Vereecken NJ, McNeil JN (2010) Cheaters and liars: chemical mimicry at its best. Can J Zool 88:725–752
Watson PJ (1986) Transmission of a female sex pheromone thwarted by males in the spider *Linyphia litigiosa* (Linyphiidae). Science 233:219–220
Wrinn KM, Evans SC, Rypstra AL (2012) Predator cues and an herbicide affect activity and emigration in an agrobiont wolf spider. Chemosphere 87:390–396
Wyatt TD (2003) Pheromones and animal behaviour: communication by smell and taste. Cambridge University Press, Cambridge
Xiao Y, Zhang J, Li S (2009) A two-component female-produced pheromone of the spider *Pholcus beijingensis*. J Chem Ecol 35:769–778
Yeargan KV (1994) Biology of bolas spiders. Annu Rev Entomol 39:81–99
Yeargan KV, Quate LW (1996) Juvenile bolas spiders attract psychodid flies. Oecologia 106:266–271

Chapter 12
Karyotypes, Sex Chromosomes, and Meiotic Division in Spiders

Tereza Kořínková and Jiří Král

12.1 Introduction

Karyotypes of roughly 1,300 species of arachnids have been studied so far. For spiders, we were able to trace published chromosome data on nearly 700 species belonging to 64 families; about 120 of them were only determined to the genus level. However, most data concern the advanced and most diverse clade, the entelegyne araneomorphs. In contrast to their enormous species diversity, entelegynes exhibit conservative karyotypes. The cytogenetic knowledge on the other spiders, namely, mesothelids, mygalomorphs, and haplogynes, is still poor. In spite of this, obtained data suggest that the karyotype diversity of mygalomorphs and haplogynes is higher than that of entelegyne spiders.

12.2 Diploid Numbers and Chromosome Architecture

In keeping with the exceptional species diversity, spiders exhibit a great diversity in diploid chromosome numbers (2n). Reported 2n of spider males range from 7 (Suzuki 1954, this study) (Fig. 12.1a) to 110 (Král et al. 2011). The only mesothelid karyotyped so far (*Heptathela kimurai*) exhibits 2n♂ = 96 (Suzuki 1954). Diploid numbers of mygalomorph males vary from 14 (*Atypus affinis*, Atypidae; Řezáč et al. 2006) to 110 (*Poecilotheria formosa*, Theraphosidae; Král et al. 2011). Interestingly, most mygalomorphs karyotyped so far possess higher diploid numbers than araneomorph spiders. In males of haplogynes, 2n varies from 7 (*Ariadna lateralis*, Segestriidae; Suzuki 1954) to 38 (*Austrochilus* sp.,

T. Kořínková • J. Král (✉)
Laboratory of Arachnid Cytogenetics, Department of Genetics and Microbiology, Faculty of Science, Charles University, Prague, Czech Republic
e-mail: spider@natur.cuni.cz

W. Nentwig (ed.), *Spider Ecophysiology*,
DOI 10.1007/978-3-642-33989-9_12, © Springer-Verlag Berlin Heidelberg 2013

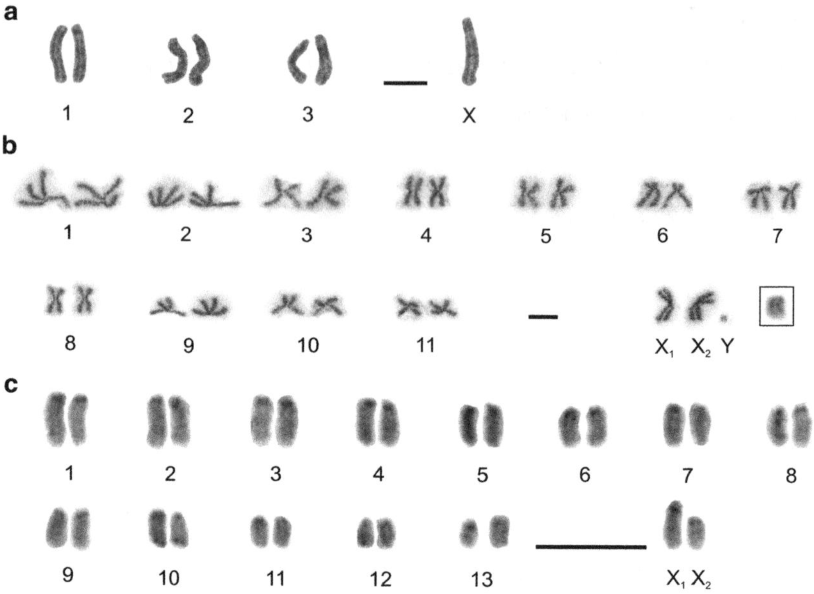

Fig. 12.1 Male karyotypes of araneomorph spiders. (**a**) *Dasumia carpathica* (Dysderidae, 2n♂ = 7, X0), spermatogonial metaphase composed of holokinetic chromosomes. (**b**) *Kukulcania* aff. *hibernalis* (Filistatidae, 2n♂ = 25, X_1X_2Y), metaphase II. Most chromosomes are metacentric except submetacentric pairs # 6 and 7. Note the metacentric morphology of the Y chromosome (*inset*). (**c**) *Pardosa wagleri* (Lycosidae, 2n♂ = 28, X_1X_20), spermatogonial metaphase, C-banding. All chromosomes are acrocentric; they contain a centromeric block of heterochromatin. Pair # 10 includes also telomeric heterochromatin. Note the unequal size of X chromosomes (X_1:X_2 = 1:0.73). Similarly to most other spiders, multiple X chromosomes and autosomes have the same amount and distribution of constitutive heterochromatin. Bar = 10 μm (**a**, **c** provided by M. Forman, **b** provided by I.M. Ávila Herrera)

Austrochilidae; Král et al. 2006). Entelegyne spiders exhibit a similar range of chromosome numbers from 2n♂ = 10 (*Uloborus danolius*, Uloboridae; Parida and Sharma 1987) to 2n♂ = 49 (*Araneus ventricosus*, Araneidae; Wang et al. 1993).

Concerning chromosome morphology, the karyotypes of most spiders are composed of monocentric chromosomes (i.e. standard chromosomes with a centromere). Mygalomorphs show various proportions of monoarmed (acrocentric and subtelocentric) and biarmed (metacentric and submetacentric) chromosomes (Fig. 12.2) (Král et al. 2011). Most haplogyne spiders exhibit a predominance of biarmed chromosomes (Fig. 12.1b) (Král et al. 2006). In entelegynes, on the other hand, chromosomes are prevailingly monoarmed (Fig. 12.1c) (Suzuki 1954; see also Appendix, this volume). Predominance of monoarmed chromosomes was reported also in the only mesothelid studied so far (Suzuki 1954). A very different structure of chromosomes has been discovered in representatives of the haplogyne families Dysderidae and Segestriidae, belonging to the superfamily Dysderoidea (Diaz and Saez 1966), an ancient lineage found already at Cretaceous strata (Penney and Selden 2011). Karyotypes of these spiders are composed of

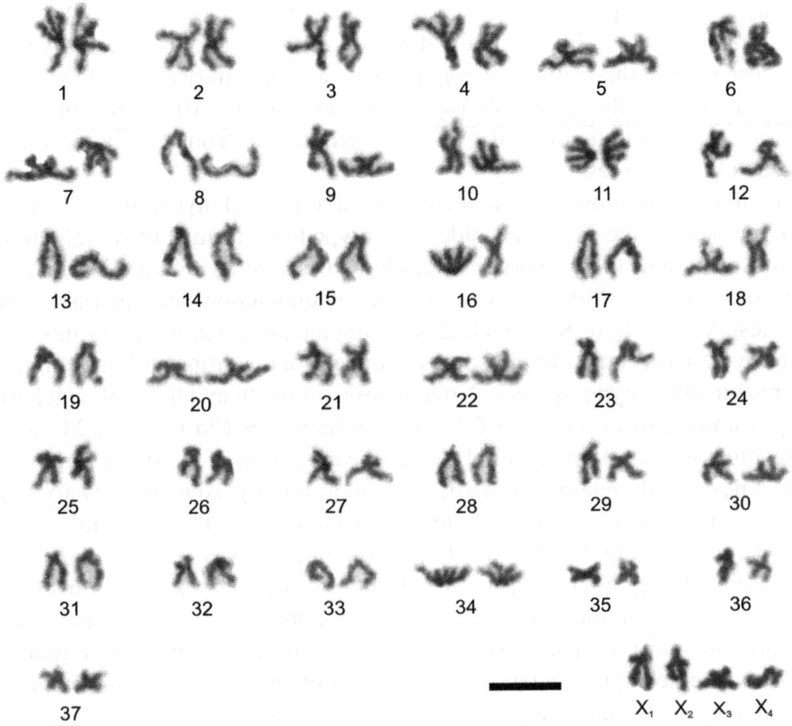

Fig. 12.2 Male karyotype of the mygalomorph *Avicularia minatrix* (Theraphosidae, 2n♂ = 78, $X_1X_2X_3X_40$), metaphase II. Autosomes are biarmed except two subtelocentric (# 27 and 32) and nine acrocentric (# 8, 13–15, 17, 19, 28, 31, 33) pairs. X chromosomes are positively heteropycnotic and exhibit a metacentric (X_2, X_3), submetacentric (X_1), and acrocentric (X_4) morphology. Bar = 10 μm (provided by L. Krkavcová)

holokinetic (= holocentric) chromosomes, which lack a primary constriction with the centromere (Fig. 12.1a). Instead, kinetochores occupy the major part of length of these chromosomes. As a result, products of breakages (fragments) or fusions (fused chromosomes) often segregate regularly during divisions and can thus be more easily fixed in a population (Řezáč et al. 2007). This is in contrast to monocentric chromosomes, where chromosome fusions and breakages might produce aberrant products (dicentric chromosomes or acentric fragments). Holokinetic chromosomes are with all probability derived from the monocentric ones and have originated independently in several lineages of protists, plants, and animals including arachnids (Viera et al. 2009). In spiders they occur in no other clade than Dysderoidea, thus suggesting to be a chromosomal synapomorphy of the superfamily (Král et al. 2006). Interestingly, some dysderoids exhibit the lowest chromosome numbers among spiders (Suzuki 1954, this study, Fig. 12.1a). According to Diaz et al. (2010), spider holokinetic chromosomes have originated by fusions of large numbers of chromosomes. Holokinetic chromosomes have also been reported in two entelegynes, namely, one species of the genus *Argiope* (Araneidae) (Amalin et al. 1992) and one species of the genus *Oxyopes* (Oxyopidae) (Barrion et al. 1989). However, these propositions are without much doubt a misinterpretation. Karyotypes of araneids (including the genus *Argiope*; Datta and Chatterjee 1988) and most oxyopids (including the genus *Oxyopes*; Stávale et al. 2011) are formed by acrocentrics, the morphology of which resembles in some respects that of holokinetic chromosomes.

Based on high chromosome numbers in mesothelids and mygalomorphs, Suzuki (1954) hypothesised the ancestral spider karyotype to be formed by a high number of chromosomes and the karyotype evolution to proceed via reduction of 2n by chromosome fusions. Most indices of this phenomenon are available from entelegynes. According to Král et al. (2006), the ancestral karyotype of this group should have 42 acrocentric chromosomes in males—a condition still quite frequent among recent entelegyne spiders. Nonetheless, more than 80 % of entelegyne karyotypes known so far consist of 21–30 chromosomes (2n♂ = 22, 24, and 28 being the most frequent). Interestingly, after the reduction of 2n, the chromosomes of entelegynes usually retain their acrocentric morphology. Acrocentric karyotypes with lower chromosome numbers could be derived from the ancestral ones by tandem fusions (Suzuki 1954). An alternative process was suggested by White (1973), namely, the cycles of centric fusions and subsequent pericentric inversions. According to our opinion, this scenario is supported by the fact that centric fusions are the most frequent source of chromosome polymorphism found in populations of entelegyne spiders. Karyotypes containing biarmed autosomes are rare in entelegynes. Remarkably, they are usually formed exclusively by biarmed chromosomes, and the chromosome number is half of that in related species possessing acrocentric chromosomes. Such karyotypes have probably originated by quick series of centric fusions of all ancestral acrocentric chromosomes (Rowell 1985). Karyotypes consisting of biarmed chromosomes have been found in most species of the family Dictynidae (Král 1995) and in several species of other families (e.g. *Larinioides patagiatus*, Araneidae: Hackman 1948; *Neoscona scylla*, Araneidae: Suzuki 1954; some populations of *Delena cancerides*, Sparassidae: Rowell 1985). Exceptional are entelegyne karyotypes composed of biarmed chromosomes with the same or very similar diploid number as in related forms with an acrocentric set. In this case, the karyotype was saturated by biarmed chromosomes presumably via pericentric inversions (Stávale et al. 2010).

Data on functional structures of spider chromosomes are limited. Centromeric and some telomeric regions are usually formed by constitutive heterochromatin (Fig. 12.1c). Nevertheless, the total amount of constitutive heterochromatin in spider chromosomes is only moderate or low and heterochromatin is rare outside these regions (Dolejš et al. 2011). Telomeres of eukaryotes contain specific DNA repeats that show a considerable evolutionary conservatism. Telomeric repeats of most arthropods are formed by the "insect" motif $(TTAGG)_n$. This sequence is absent in spiders and the composition of their telomeric DNA is unknown (Vítková et al. 2005). Spider karyotypes studied so far contain a low number of nucleolus organiser regions. These structures usually have a terminal position (Král et al. 2006; Rodríguez Gil et al. 2007; Dolejš et al. 2011). In contrast to other spiders, the nucleolus organiser regions of haplogynes are often placed also on sex chromosomes (Král et al. 2006).

12.3 Sex Chromosomes

A notable feature of spider karyotypes is the predominance of unusual multiple X chromosome systems. The vast majority of studied spiders exhibit the system ♂X_1X_2/♀$X_1X_1X_2X_2$, often assigned as X_1X_20, where 0 indicates the absence of the Y chromosome (Fig. 12.1c). Males and females thus differ by two chromosomes in diploid number. Systems with multiple X chromosomes are rare in other animals, and usually they occur as a derived type. In spiders, however, the X_1X_20 system is probably ancestral as inferred from its presence in the most primitive recent spiders, namely, the representative of the suborder Mesothelae (Suzuki 1954) and the basal mygalomorph families Atypidae and Dipluridae (see Řezáč et al. 2006 for a review). Based on their different sizes (Fig. 12.1c) and absence of chiasmata during their pairing at male meiosis (Fig. 12.3b, c), both X chromosomes are considered nonhomologous. There are two hypotheses on the origin of the X_1X_20 system. This sex chromosome determination has with all probability evolved from a single X chromosome either by a non-disjunction followed by differentiation of the newly formed X chromosome (White 1940; Postiglioni and Brum-Zorrilla 1981) or by fission. The weak point of the former hypothesis is that non-disjunction of X may break the balance between male and female sex factors. In the latter hypothesis, the question is how each of the fission products gains a functional centromere. Previously reported explanations are either the fission of an aberrant dicentric chromosome X (Pätau 1948; Suzuki 1954) or the fission of a metacentric, where one of the arms would retain the original centromere and the other would be translocated to a centric fragment (Bole-Gowda 1950; White 1973). Another possible scenario is a simple fission of the centromere of a biarmed X chromosome, which can generate two products with functional centromeres.

The X_1X_20 system has been found in 77 % of spiders karyotyped so far (Araújo et al. 2005) and largely prevails in the entelegyne lineage of araneomorphs, whereas in mygalomorphs and haplogynes it is infrequent. Despite the evolutionary stability of the X_1X_20 mode, secondary systems have originated from it in some spiders. In some entelegynes, the $X_1X_2X_30$ or the rare $X_1X_2X_3X_40$ system have supposedly arisen by non-disjunctions (Postiglioni and Brum-Zorrilla 1981; Datta and Chatterjee 1988). The $X_1X_2X_30$ system has been found in 13 entelegyne families so far (Kořínková and Král, unpublished) (Fig. 12.3e). In contrast to this, the $X_1X_2X_3X_40$ system has only been detected in a few entelegyne species (Datta and Chatterjee 1988). Král et al. (2011) found it also in two mygalomorphs and hypothesised that it evolved by duplication of the ancestral X_1X_20 system, namely, via nondisjunctions or polyploidisation. Furthermore, an X0 system was found in one family of mygalomorphs (Řezáč et al. 2006), six families of haplogynes (Král et al. 2006) (Fig. 12.1a), and 12 families of entelegynes (Cokendolpher 1989; Kořínková and Král, unpublished). The occurrence of X0 sex chromosome system in these spider lineages suggests its several independent origins. Concerning entelegynes, this system is hypothesised to have arisen by centric fusion (Bole-Gowda 1952), by tandem fusion (Bole-Gowda 1950), or by deletions of one

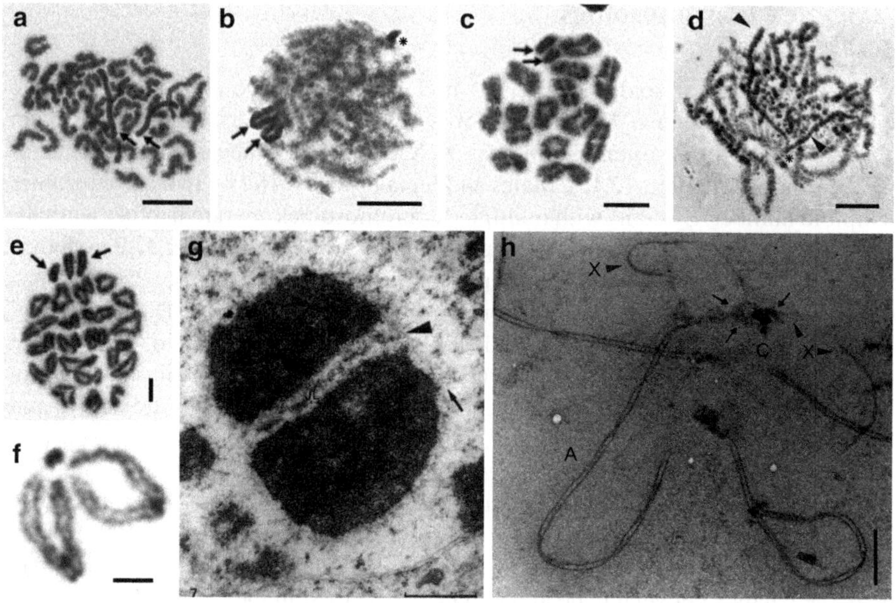

Fig. 12.3 Behaviour of sex chromosomes in the germline. (**a**, **b**) *Pax islamita* (Zodariidae, X_1X_20), (**a**) part of spermatogonial prometaphase. X_1 and X_2 chromosomes (*arrows*) show different condensation than autosomes being placed close each other. (**b**) Male pachytene. Ends of two loop X chromosomes (*arrow*) show end-to-end pairing. X chromosomes are positively heteropycnotic and placed on the periphery of the nucleus (*asterisk*, nucleolus). (**c**) *Pellenes tripunctatus* (Salticidae, 2n♂ = 28, X_1X_20), late diakinesis of male. X chromosomes (*arrows*) are positively heteropycnotic and pair in parallel on the periphery of the plate. (**d**) *Malthonica silvestris* (Agelenidae, X_1X_20), female pachytene. Bivalents X_1X_1 and X_2X_2 (*arrowheads*) are heterochromatinic and associated by centromeric regions (*asterisk*) at the nuclear periphery. Sex chromosome bivalents exhibit an imperfect chromomere pattern. (**e**) *Stegodyphus lineatus* (Eresidae, 2n♂ = 43, $X_1X_2X_30$), male diakinesis. X chromosomes (*arrows*) exhibit parallel pairing on the periphery of the plate. (**f**) *Loxosceles rufescens* (Sicariidae), male diakinesis, X_1X_2Y trivalent. (**g**) *Schizocosa malitiosa* (Lycosidae, X_1X_20), section through male pachytene nucleus showing paired X chromosomes, transmission electron microscopy. Pairing of sex chromosomes is ensured by the junction lamina (*arrowhead*). Note the high condensation of chromatin and perichromosome fibrillar zone (*arrow*); (**h**) *M. silvestris* (X_1X_20), spread synaptonemal complexes, transmission electron microscopy. Two X univalents (X) and a bivalent (A) are connected by attachment plaques (*arrows*). Attachment plaques are associated with a centrosome (C). Scale bar = 10 μm (**a**–**d**), 5 μm (**e**, **f**), 2 μm (**h**), 0.5 μm (**g**) (**d**, **h** modified from Král 2007; **f** from Král et al. 2006; **g** from Benavente and Wettstein 1977; **a**–**c**, **e** provided by M. Forman)

chromosome of the X_1X_20 system (Suzuki 1954). The centric fusion of the acrocentric X_1 and X_2 chromosomes of entelegynes led to a biarmed chromosome X, whereas the tandem fusion has produced an acrocentric X chromosome.

Another mechanism of sex chromosome evolution in spiders is via gonosome--autosome rearrangements. The resulting sex chromosome determinations (so-called neo-sex chromosome systems) always contain a Y chromosome. The $X_1X_2X_3Y$ system of the salticid genera *Habronattus* and *Evarcha* arose from the X_1X_20

system by a tandem fusion between one X and an autosome (Maddison 1982). The XY systems of *Atypus affinis* (Atypidae) (Řezáč et al. 2006) and *Leptoneta infuscata* (Leptonetidae) (Král et al. 2006) have presumably evolved by fusion of an autosome and the X chromosome of the X0 system. One of the most complicated sex chromosome systems described so far has been found in the Australian sparassid *Delena cancerides* ($X_1X_2X_30$). Here, hybridisations between populations with different autosome–autosome and autosome–sex chromosome fusions have produced systems, some of which comprise even the whole karyotype except one autosome pair (Sharp and Rowell 2007). In another entelegyne spider, *Malthonica ferruginea* (Agelenidae), Král (2007) has found an $X_1X_2X_3X_4X_5Y$ neo-sex chromosome system derived also from the $X_1X_2X_30$ determination.

Another system containing a Y chromosome is the X_1X_2Y sex chromosome determination that was described in the genus *Loxosceles* (Sicariidae) (Silva 1988; Silva et al. 2002). It was later found also in other haplogyne families, namely, Drymusidae, Filistatidae, Hypochilidae, and Pholcidae (Král et al. 2006). In most representatives studied so far, the X_1 and X_2 are large metacentric chromosomes of similar size, whereas the Y is a metacentric microchromosome. These chromosomes also exhibit a specific meiotic behaviour. From premeiotic interphase to metaphase of the first meiotic division (metaphase I), arms of the X chromosomes show a distal end-to-end pairing with both arms of the Y chromosome (Silva et al. 2002; Král et al. 2006) (Fig. 12.3f). The origin of the X_1X_2Y system is unknown. Silva (1988) suggested it has arisen by gonosome–autosome rearrangements. The specific structure and meiotic behaviour of the X_1X_2Y chromosomes indicate that spider families with this system form a monophyletic lineage (Král et al. 2006). Most of them belong to the scytodoid clade (Coddington and Levi 1991). The broad distribution of the X_1X_2Y system indicates its antiquity. However, it has undergone gradual transformation in some families, namely, into the XY system as in Diguetidae (probably by pericentric inversions of the X_1 and X_2 chromosome, followed by their fusion into one metacentric element) and further (by elimination of the Y chromosome) into the X0 system found in many pholcids. This indicated even some basal araneomorph families exhibiting only the X0 system (e.g. Scytodidae) might in fact belong to the X_1X_2Y clade (Král et al. 2006).

Sex chromosomes of spiders exhibit a specific behaviour in the male germline. Multiple X chromosomes as well as sex chromosomes of the X_1X_2Y and X0 systems are sometimes distinguishable already at spermatogonial mitosis. Sex chromosomes can associate and exhibit a different condensation compared to autosomes (Fig. 12.3a). Often the distinct condensation results in a different intensity of staining (negative and positive heteropycnosis, respectively) (Král et al. 2006, 2011; Král 2007). The distinct condensation and heteropycnosis of sex chromosomes reflect their inactivation by heterochromatinisation (Datta and Chatterjee 1988). At prophase I (and sometimes also metaphase I), sex chromosomes are heteropycnotic and pair without chiasmata, being placed at the periphery of the nucleus (Fig. 12.3b, c). The sex chromosome attachment is often initiated already at premeiotic interphase. The pairing either involves both ends of "loop-like" sex chromosomes (Fig. 12.3b, f), or the chromosomes show a parallel ("side-by-side") attachment (Fig. 12.3c, e). The former mode of pairing is typical for haplogynes (Král et al. 2006), and it was found also in some mygalomorphs and basal entelegynes. It is probably ancestral in spiders (Král et al. 2011). The latter type of pairing is typical for entelegynes (Král 2007) and some haplogynes with holokinetic chromosomes (Benavente and Wettstein 1980).

Ultrastructural studies revealed an amazing diversity of structures ensuring the parallel attachment of sex chromosomes in spiders. The X chromosomes of lycosids (X_1X_20) and sparassids ($X_1X_2X_30$) are associated by a laminar structure, which is presumably derived from the synaptonemal complex (structure ensuring meiotic pairing of homologous chromosomes) and named the junction lamina (Benavente and Wettstein 1977) (Fig. 12.3g). In contrast to this, in the $X_1X_2X_30$ system of *Tegenaria domestica* (Agelenidae), the pairing is mediated by chromatin projections (Benavente et al. 1982), and in the haplogyne spider *Segestria florentina* (Segestriidae) (X_1X_20) with holokinetic chromosomes, the pairing of the sex chromosomes is ensured by their direct attachment (Benavente and Wettstein 1980).

In the homogametic sex of all organisms studied so far, sex chromosomes behave in the same way as autosomes during meiosis. Pairing of homologous autosomes as well as sex chromosomes is initiated during a specific stage (zygotene) of prophase I. In contrast to this, homologous sex chromosomes of spider females are already paired at premeiotic interphase (Král et al. 2011) being inactivated by heterochromatinisation until prophase I (Král 2007; Král et al. 2011). Furthermore, the sex chromosome bivalents exhibit an end-to-end association during these stages (Fig. 12.3d). Associated ends are formed by a telomeric (mygalomorphs) or centromeric (entelegynes) block of heterochromatin, respectively. The unusual meiotic behaviour of female sex chromosomes probably acts against pairing and recombination between chromosomes belonging to different X chromosome bivalents. The suppression of recombination among the sex chromosome bivalents has accelerated the structural differentiation of particular sex chromosomes (Král 2007; Král et al. 2011) originated by non-disjunctions.

In addition to the sex chromosomes described above, karyotypes of entelegynes (Král 2007; Král et al. 2011) and mygalomorphs (Král et al. 2011) also include an unusual sex chromosome pair exhibiting a specific behaviour during male meiosis. Ultrastructural studies in some entelegynes revealed that this pair displays an end-to-end pairing with X univalents, mediated by attachment plaques. The chromosomes of the pair are morphologically undistinguishable, but probably differentiated on molecular basis. A complex formed by the pair and the other sex chromosomes is associated with the centrosome during prophase I (Fig. 12.3h). The most plausible explanation suggests that these chromosomes represent an ancestral sex chromosome pair (proto-X–proto-Y), which generated the multiple sex chromosomes by non-disjunctions (Král 2007). The sex chromosome pair has been found to be facultatively heterochromatinised in some haplogynes (Král et al. 2006) and mygalomorphs (Král et al. 2011). This inactivation probably prevents chromosomes of the pair from recombination of regions that are already differentiated (Král et al. 2011).

12.4 Modifications of Meiotic Division

Some spiders exhibit modifications of meiosis, mostly at prophase and metaphase I in males. A peculiar bipolarisation of the nucleus has been found at male diakinesis of Agelenidae (Revell 1947, this study) (Fig. 12.4a) and Lycosidae (Chemisquy et al. 2008). Bivalents form two unequal groups (one of which includes also the multiple X chromosomes) at nearly opposite sides of the nucleus, so that the situation resembles the arrangement of chromosomes at anaphase I. This process is preceded by the orientation of the bivalents to the centrosome at early prophase I. Consequently, with sister centrosomes moving apart to establish the spindle, some bivalents move too so that the nucleus becomes bipolar at diakinesis (Revell 1947). During metaphase I, all bivalents form a standard metaphase plate. The occurrence of the nucleus bipolarisation in two distantly related entelegyne families, Agelenidae and Lycosidae, suggests that this phenomenon is widespread in entelegyne spiders. The unusual arrangement of bivalents may be related to a unique ultrastructure of meiotic spindle, which has been found in lycosid males (Wise 1984). Each autosomal bivalent is from prometaphase I encased in a double membrane tube, which during late prometaphase I extends to the proximity of the spindle poles (Fig. 12.4c). Sex chromosomes lack the membrane tubes.

Male bivalents of some diplurid mygalomorphs are peculiar by late expression of chiasmata, which appear at the end of prophase I only (Král et al. 2011). The delayed manifestation of chiasmata (so-called cryptochiasmatic meiosis) represents an evolutionary step between chiasmatic and achiasmatic meiosis (i.e. the absence of recombinations and chiasmata at heterogametic sex) (White 1973).

Achiasmatic meiosis has been suggested in males of two families with holokinetic chromosomes, namely, Dysderidae and Segestriidae (Benavente and Wettstein 1980). However, our data suggest that the absence of chiasmata is only seeming due to the period of considerable despiralisation of chromatin at late prophase I (i.e. after pachytene) (the so-called diffuse stage) (Fig. 12.4b). The diffuse stage of these spiders is extremely long; the bivalents recondense not earlier than at diakinesis or even prometaphase I. In *Dysdera*, the expression of chiasmata is further complicated by their precocious completion (Diaz et al. 2010). A diffuse stage has also been found in males of the other haplogynes (presumably as a synapomorphy of this lineage, Král et al. 2006) as well as in some entelegynes (Mittal 1966). Furthermore, it seems to occur in females of all spiders (Král et al. 2011). A diffuse stage has so far been discovered in various plant and animal groups. It is characterised by a considerable despiralisation of chromatin, which probably reflects an enhanced transcriptional activity during this period. In contrast to autosomes, sex chromosomes are heterochromatinised and therefore probably inactive during the diffuse stage of spiders (Král et al. 2006) (Fig. 12.4b).

Another meiotic modification is the so-called inverted meiosis that was found in males of *Dysdera crocata* (X0) (Benavente and Wettstein 1980) possessing holokinetic chromosomes. In this spider, chromatids of the X chromosome segregate to the opposite poles already at the anaphase of the first meiotic division

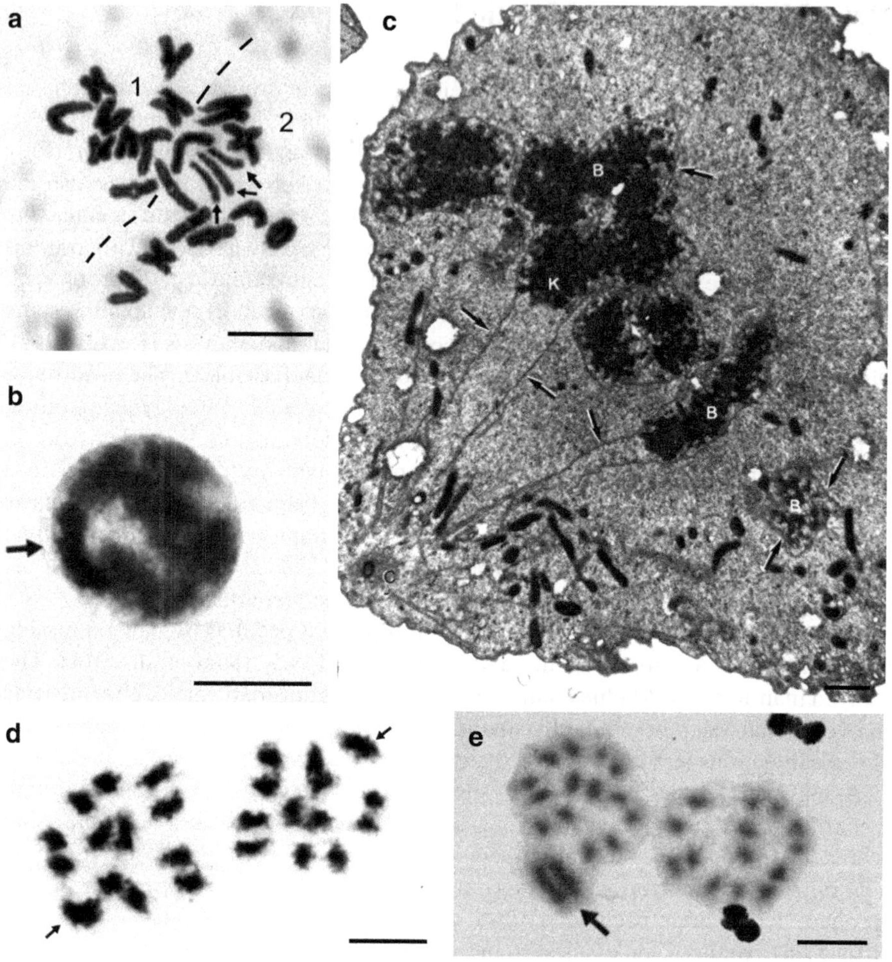

Fig. 12.4 Male modifications of meiotic division. (**a**) *Malthonica ferruginea* (Agelenidae), diakinesis. Bivalents form two groups (*1*, *2*). One group includes multiple X chromosomes (*arrows*). (**b**) *Dysdera ninnii* (Dysderidae), diffuse stage. Note the positively heteropycnotic X chromosome (*arrow*) on the periphery of the nucleus. (**c**) *Rabidosa rabida* (Lycosidae), section through a spermatocyte near metaphase I, transmission electron microscopy. Bivalents (*B*) are arranged at the equator of the cell. One kinetochore (*K*) was included in the plane of the section. Each bivalent and bundle of kinetochore microtubules is encased in a double membrane tube. These tubes (marked by *two counter arrows*) extend to within a few microns of the centrosome (*C*). (**d**) *Dysdera crocata*, anaphase I. The X chromosome (*arrows*) is formed by one chromatid only, i.e. exhibits inverted meiosis. (**e**) *Dysdera ninnii*, anaphase I. The X chromosome (*arrow*) is composed of two chromatids, i.e. exhibits standard meiosis. In contrast to autosomes, chromatids of X chromosome are closely attached. Furthermore, X chromosome displays positive heteropycnosis and delayed segregation to the pole. Bar = 10 μm (**a**, **b**, **d**, **e**), 1 μm (**c**) (**c** modified from Wise 1984; **d** modified from Král et al. 2006)

(Fig. 12.4d). Thus, this behaviour is opposite to the standard meiosis, which is characterised by separation of sister chromatids during the anaphase of the second meiotic division. Standard meiotic behaviour of the X chromosome has been found in other species of the genus *Dysdera* (this study, Fig. 12.4e) as well as in the family Segestriidae (Rodríguez Gil et al. 2002). The phenomenon of inverted meiosis is restricted to some organisms with holokinetic chromosomes, and it is possible due to the absence of the centromere connection at holokinetic chromosomes (Viera et al. 2009).

12.5 Conclusions

Spiders show a considerable diversity of diploid numbers, chromosome morphology, and sex chromosomes. Most spiders exhibit standard chromosomes except the superfamily Dysderoidea whose chromosomes are holokinetic. Unusual multiple sex chromosomes of spiders have received more attention than any other aspect of their cytogenetics. The so-called X_1X_20 sex chromosome system (♂X_1X_2/♀$X_1X_1X_2X_2$) is considered ancestral for spiders and occurs in most studied species. There are several hypotheses on its origin. The most plausible one, assuming duplication of a single X chromosome, is supported by recent findings of unusual sex chromosome behaviour at meiosis of spider females. This behaviour may prevent pairing and recombination of nonhomologous X chromosomes. Despite its evolutionary stability, the X_1X_20 system has been transformed in some lineages by X–X fusions, by X chromosome duplications or sex chromosome–autosome rearrangements. Finally, some spiders exhibit modifications of meiotic division, mostly in males. The most intriguing modification concerns the bipolarisation of prophase I nucleus in entelegyne males. Moreover, just before formation of metaphase plate, each bivalent is enclosed into a double membrane tube.

Acknowledgements We are indebted to our colleagues M. Forman, I.M. Ávila Herrera, and L. Krkavcová who kindly provided original data, images, and karyotypes of selected species for Figs. 12.1, 12.2, and 12.3. Furthermore, we are very grateful to D. Ubick (California Academy of Sciences, San Francisco) for species determination of *Kukulcania*. We apologise to the authors whose work was not cited due to the space constraints. This study was supported by the project of the Grant Agency of the Czech Academy of Sciences (project IAA601110808).

References

Amalin DM, Barrion AA, Rueda LM (1992) Morphology and cytology of *Argiope catenulata* (Doleschall) (Araneae: Araneidae). Asia Life Sci 1:35–44

Araújo D, Cella DM, Brescovit AD (2005) Cytogenetic analysis of the neotropical spider *Nephilengys cruentata* (Araneomorphae, Tetragnathidae): standard staining, NORs, C-bands and base-specific fluorochromes. Braz J Biol 65:193–202

Barrion AA, Amalin DM, Casal CV (1989) Morphology and cytology of the lynx spider *Oxyopes javanus*. Philipp J Sci 118:229–237

Benavente R, Wettstein R (1977) An ultrastructural cytogenetic study on the evolution of sex chromosomes during the spermatogenesis of *Lycosa malitiosa* (Arachnida). Chromosoma 64:255–277

Benavente R, Wettstein R (1980) Ultrastructural characterization of the sex chromosomes during spermatogenesis of spiders having holocentric chromosomes and a long diffuse stage. Chromosoma 77:69–81

Benavente R, Wettstein R, Papa M (1982) Ultrastructural analysis of the $X_1X_2X_30$ sex chromosome system during the spermatogenesis of *Tegenaria domestica* (Arachnida). J Cell Sci 58:411–422

Bole-Gowda BN (1950) The chromosome study in the spermatogenesis of two lynx-spiders (Oxyopidae). Proc Zool Soc Bengal 3:95–107

Bole-Gowda BN (1952) Studies on the chromosomes and sex-determining mechanism in four hunting spiders (Sparassidae). Proc Zool Soc Bengal 5:51–70

Chemisquy MA, Rodríguez Gil SG, Scioscia CL, Mola LM (2008) Cytogenetic studies of three Lycosidae species from Argentina (Arachnida, Araneae). Genet Mol Biol 31:857–867

Coddington JA, Levi HW (1991) Systematics and evolution of spiders (Araneae). Annu Rev Ecol Syst 22:565–592

Cokendolpher JC (1989) Karyotypes of three spider species (Araneae: Pholcidae, *Physocyclus*). J New York Entomol Soc 97:475–478

Datta SN, Chatterjee K (1988) Chromosomes and sex determination in 13 araneid spiders of North-Eastern India. Genetica 76:91–99

Diaz MO, Saez FA (1966) Karyotypes of South-American Araneida. Mem Inst Butantan 33:153–154

Diaz MO, Maynard R, Brum-Zorrilla N (2010) Diffuse centromere and chromosome polymorphism in haplogyne spiders of the families Dysderidae and Segestriidae. Cytogenet Genome Res 128:131–138

Dolejš P, Kořínková T, Musilová J, Opatová V, Kubcová L, Buchar J, Král J (2011) Karyotypes of central European spiders of the genera *Arctosa*, *Tricca*, and *Xerolycosa* (Araneae: Lycosidae). Eur J Entomol 108:1–16

Hackman W (1948) Chromosomenstudien an Araneen mit besonderer Berücksichtigung der Geschlechtschromosomen. Acta Zool Fenn 54:1–101

Král J (1995) The karyotype studies in central Europe's species of the superfamilies Amaurobioidea and Dictynoidea (Araneida). In: Růžička V (ed) Proc 15th Eur Coll Arachnol, Inst Entomol, České Budějovice

Král J (2007) Evolution of multiple sex chromosomes in the spider genus *Malthonica* (Araneae: Agelenidae) indicates unique structure of the spider sex chromosome systems. Chromosome Res 15:863–879

Král J, Musilová J, Šťáhlavský F, Řezáč M, Akan Z, Edwards RL, Coyle FA, Almerje CR (2006) Evolution of the karyotype and sex chromosome systems in basal clades of araneomorph spiders. Chromosome Res 14:859–880

Král J, Kořínková T, Forman M, Krkavcová L (2011) Insights into the meiotic behavior and evolution of multiple sex chromosome systems in spiders. Cytogenet Genome Res 133:43–66

Maddison WP (1982) XXXY sex chromosomes in males of the jumping spider genus *Pellenes* (Araneae: Salticidae). Chromosoma 85:23–37

Mittal OP (1966) Karyological studies on the Indian spiders. VI. Chromosome number and sex-determining mechanism in the family Araneidae. Res Bull Panjab Univ 17:335–351

Parida BB, Sharma NN (1987) Chromosome number, sex mechanism and genome size in 27 species of Indian spiders. Chromosome Inf Serv 43:11–13

Pätau K (1948) X-segregation and heterochromasy in the spider *Aranea reaumuri*. Heredity 2:77–100

Penney D, Selden PA (2011) Fossil spiders: the evolutionary history of a mega-diverse order. Siri Scientific Press, Manchester

Postiglioni A, Brum-Zorrilla N (1981) Karyological studies on Uruguayan spiders II. Sex chromosomes in spiders of the genus *Lycosa* (Araneae-Lycosidae). Genetica 56:47–53

Revell SH (1947) Controlled X-segregation at meiosis in *Tegenaria*. Heredity 1:337–347

Řezáč M, Král J, Musilová J, Pekár S (2006) Unusual karyotype diversity in the European spiders of the genus *Atypus* (Araneae: Atypidae). Hereditas 143:123–129

Řezáč M, Král J, Musilová J, Pekár S (2007) The spider genus *Dysdera* (Araneae, Dysderidae) in central Europe: revision and natural history. J Arachnol 35:432–462

Rodríguez Gil SG, Mola LM, Papeschi AG, Scioscia CL (2002) Cytogenetic heterogeneity in common haplogyne spiders from Argentina (Arachnida: Araneae). J Arachnol 30:47–56

Rodríguez Gil SG, Merani MS, Scioscia CL, Mola LM (2007) Cytogenetics in three species of *Polybetes* Simon 1879 from Argentina (Araneae, Sparassidae) I. Karyotype and chromosome banding pattern. J Arachnol 35:227–237

Rowell DM (1985) Complex sex-linked fusion heterozygosity in the Australian huntsman spider *Delena cancerides* (Araneae: Sparassidae). Chromosoma 93:169–176

Sharp HE, Rowell DM (2007) Unprecedented chromosomal diversity and behaviour modify linkage patterns and speciation potential: structural heterozygosity in an Australian spider. J Evol Biol 20:2427–2439

Silva D (1988) Estudio cariotípico de *Loxosceles laeta* (Araneae: Loxoscelidae). Rev Perú Entomol 31:9–12

Silva RW, Klisiowicz DR, Cella DM, Mangili OC, Sbalqueiro IJ (2002) Differential distribution of constitutive heterochromatin in two species of brown spider: *Loxosceles intermedia* and *L. laeta* (Araneae, Sicariidae), from the metropolitan region of Curitiba, PR (Brazil). Acta Biol Par (Curitiba) 31:123–136

Stávale LM, Schneider MC, Araújo D, Brescovit AD, Cella DM (2010) Chromosomes of Theridiidae spiders (Entelegynae): Interspecific karyotype diversity in *Argyrodes* and diploid number intraspecific variability in *Nesticodes rufipes*. Genet Mol Biol 33:1–6

Stávale LM, Schneider MC, Brescovit AD, Cella DM (2011) Chromosomal characteristics and karyotype evolution of Oxyopidae spiders (Araneae, Entelegynae). Genet Mol Res 10:752–763

Suzuki S (1954) Cytological studies in spiders. III. Studies on the chromosomes of fifty-seven species of spiders belonging to seventeen families, with general considerations on chromosomal evolution. J Sci Hiroshima Univ, Ser B 15:23–136

Viera A, Page J, Rufas JS (2009) Inverted meiosis: the true bugs as a model to study. In: Benavente R, Volf J-N (eds) Genome dynamics V. Meiosis. Karger, Basel

Vítková M, Král J, Traut W, Zrzavý J, Marec F (2005) The evolutionary origin of insect telomeric repeats, $(TTAGG)_n$. Chromosome Res 13:145–156

Wang Y, Song D, Wang X, Yang Z (1993) Preliminary studies on the chromosome of four species of spiders. Acta Arachnol Sinica 2:110–113

White MJD (1940) The origin and evolution of multiple sex chromosome mechanisms. J Genet 40:303–336

White MJD (1973) Animal cytology and evolution. Cambridge University Press, Cambridge

Wise D (1984) The ultrastructure of an intraspindle membrane system in meiosis of spider spermatocytes. Chromosoma 90:50–56

Chapter 13
Male Reproductive System of Spiders

Peter Michalik and Elisabeth Lipke

13.1 Introduction

Spider reproduction is characterized by several special traits (e.g. Eberhard 2004). For example, spider males evolved a unique way of transferring sperm into the female by using its pedipalps which are modified into secondary copulatory organs. These so-called palpal organs are not innervated and lack sense organs as well as muscles—possible consequences of this unique genital morphology were recently discussed in a review paper by Eberhard and Huber (2010). The complexity of the palpal organ can reach from simple (pyriform) form as in mygalomorph spiders to highly organized forms as in most entelegyne spiders (Kraus 1984; Coddington 1990). The copulatory organs are not directly associated with the primary genital system, which is located in the opisthosoma. The seminal fluid (spermatozoa and secretion; see below) is extruded from the gonopore and transferred into the so-called spermophor (tube-like invagination) of the palpal organ by using a sperm web as a carrier, which results in a temporary exposure of the sperm to the environment (Foelix 2011). In the majority of spiders, the spermophor is porous and surrounded by a glandular epithelium (e.g. Lopez 1987; Suhm et al. 1996). The mechanism of sperm uptake (sperm induction) and expulsion during copulation is still not clear. It was assumed that capillary forces or resorption of secretion released into the spermophor prior to sperm induction plays a role in this process (e.g. Lopez 1987; Suhm et al. 1996). Nevertheless, not all spider species possess glands associated to the spermophor. Thus, other mechanism of sperm induction and ejaculation must exist (Eberhard and Huber 2010) as suggested for mesothelid spiders where the flexible spermophor is compressed during sperm transfer by applying haemolymph pressure (Kraus 1984; Haupt 2003).

P. Michalik (✉) • E. Lipke
Department of General and Systematic Zoology, Zoological Institute and Museum, Ernst-Moritz-Arndt-University, Greifswald, Germany
e-mail: michalik@uni-greifswald.de

W. Nentwig (ed.), *Spider Ecophysiology*,
DOI 10.1007/978-3-642-33989-9_13,

In contrast to the numerous studies on the palpal organs of spiders, little is known about the primary male genital organs, although knowledge of the organization of sperm and presence of seminal secretion are of high relevance to understand pre- and postcopulatory processes (Herberstein et al. 2011). For example, spider sperm are always transferred encapsulated and inactive to the female (Alberti 1990, 2000) and the seminal fluid contains of one or more types of secretion (Michalik and Huber 2006; Michalik 2009). Even though the function and consequences of these traits are essential to understand spider reproduction no information is available until now. First hints of the influence of the seminal secretion are given by behavioral studies—females of the wolf spider *Schizocosa malitiosa* are reluctant to remate after the first copulation (Aisenberg and Costa 2005). In this chapter we give an overview of the present state of knowledge of the morphology of the primary genital system, spermatozoa and seminal secretion.

13.2 Male Reproductive System

13.2.1 General Remarks

The first description of the primary male genital system of spiders was given by Bertkau (1875). According to his description the genital system consists of paired tubular testes and thin deferent ducts. The deferent ducts are fused near the genital opening forming the ejaculatory duct, which opens into the gonopore in the epigastric furrow. Accessory glands as known from Amblypygi or Uropygi are lacking which might be the result of the evolution from indirect sperm transfer with spermatophores to direct sperm transfer using the palpal organs (Alberti 2005). Even though Bertkau (1875) studied only few species, he already noted differences in the gross morphology. Consecutive studies revealed a considerable diversity in the organization of the male genital system (Fig. 13.1). For example, the testes can be fused distally (Mesothelae) (Michalik 2007), proximally (Atypidae, Segestriidae, Dysderidae and Scytodidae) (Bertkau 1878; Michalik 2009) or completely (synapomorphy of Oonopidae) (Burger and Michalik 2010). The deferent ducts can be short or highly convoluted (Michalik 2009) and in some taxa with modified parts, as, e.g. so-called ampullae in the sicariid *Loxosceles intermedia* (Costa-Ayub and Faraco 2007) or vesicles in the dictynid *Nigma flavescens* (Michalik 2009). The ejaculatory duct is usually inconspicuous, but can be enlarged as the seminal vesicle in theridiids (Knoflach 1998; Michalik 2009) or widened towards the deferent ducts as in the philodromid *Philodromus dispar* (Crome 1951). The function of these different structures is not known, but it can be assumed that temporary sperm storage might occur in those areas.

The architecture of the spider male gonads reflects the general organization known for animals (e.g. White-Cooper et al. 2009). The spider testis consists of spatiotemporally arranged spermatogenic cells at different developmental stages

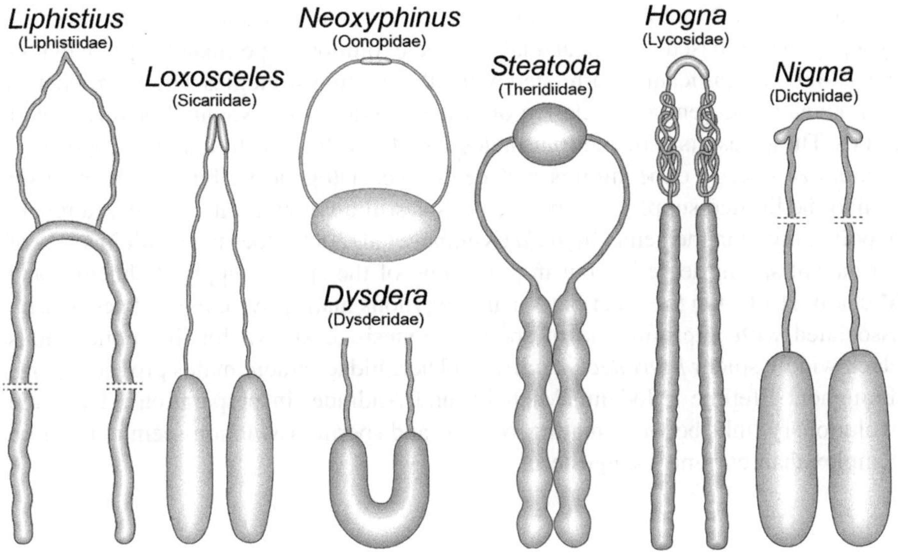

Fig. 13.1 Schematic drawings of the male reproductive system of selected taxa (modified according to Michalik 2009)

which are embedded in the somatic tissue (e.g. Alberti and Weinmann 1985). The only exception from this general pattern is reported for the telemid spider *Telema tenella*. Here, the subspherical testis shows no cyst-like organization, but possesses a small cavity filled with spermatozoa, which is directly connected to the lumen of the deferent duct (Lopez and Juberthie-Jupeau 1983).

Spermatogonia and early stages of spermatogenesis are located in the periphery, whereas late spermatids and spermatozoa are present in the central part of the testis (e.g. Michalik et al. 2006). All stages of spermatogenesis are present resulting in a heterogeneous appearance of the testis. The somatic (epithelial) cells bear microvilli at the apical pole and are attached to each other by septate junctions and zonulae adhaerentes. Furthermore, the somatic cells are characterized by a high number of Golgi, endoplasmic reticulum and vesicles indicating a high secretory activity (e.g. Alberti and Weinmann 1985; Alberti and Coyle 1991; see below).

13.2.2 Permanent Sperm Depletion

In general, spermatogenesis starts in subadult stages and continues throughout the male's life (e.g. Michalik and Uhl 2005). Exceptions to this organization are only described for some orbicularian spiders (see also Appendix, this volume). In the one-palped spider *Tidarren argo* (Theridiidae), male spermatogenesis is terminated before the final molt resulting in an extreme decreased volume of the testis (Michalik et al. 2010). The same phenomenon is reported for species of the genera *Nephila* and *Nephilengys* (Nephilidae; Michalik and Rittschof 2011; Schneider and

Michalik 2011). Moreover, the spermiogenesis is nearly synchronous. As revealed by a phylogeny-based statistical analysis, the evolution of permanent sperm depletion (PSD) is significantly correlated with other mating strategies which limit males to monogamy as genital mutilation or sexual cannibalism (Michalik and Rittschof 2011). These results indicate that males which are limited to a single copulation economize energy expenditures and cease spermatogenesis, but the mechanism behind the limited supply of germline cells is still unknown. Interestingly, a regain of polygamy as in the genus *Nephila* (Kuntner et al. 2009) does not result in a loss of permanent sperm depletion but in portioning of the sperm supply (Schneider and Michalik 2011). On the other hand, a monogynous mating system is not necessarily associated with a termination of spermatogenesis as shown for the monogynous black widow spider *Latrodectus hasselti* (Theridiidae) where males produce sperm throughout lifetime (Modanu, Michalik and Andrade, in preparation). Thus, the evolutionary links between mating systems and sperm production seem to be more complex than currently suggested.

13.3 Spermatozoa: Development and Evolution

First histological studies of spider spermatozoa and spermatogenesis reach back into the nineteenth century (Bertkau 1877), but detailed descriptions of sperm structures became possible only by using electron microscopy. Thus, the first ultrastructural study of spider spermatozoa conducted by Ōsaki (1969) on the mesothelid *Heptathela kimurai* was a turning point in spider spermatology. In the following decades nearly 60 species were studied revealing an astonishing structural diversity at different organizational levels (e.g. Alberti 1990, 2000; Michalik et al. 2003, 2004a, b, 2005, 2006; Michalik and Hormiga 2010).

13.3.1 Spermatogenesis

Spermatogenesis follows the general pattern known for animals and begins with the cleavage of spermatogonia in the periphery of the testis, and after two meiotic divisions, spermatids remain connected by cellular bridges. Early spermatids are mainly characterized by a spherical nucleus and a homogenous electron-lucent cytoplasm (e.g. Alberti et al. 1986; Michalik and Huber 2006). Further cell components are mitochondria, centrioles and a Golgi apparatus. The Golgi vesicles fuse to form an acrosomal vacuole (acrosomal vesicle) at the anterior pole of the nucleus. The nucleus is surrounded by a so-called manchette of microtubules. The function of the manchette is not known for spider sperm, but is likely important for the shaping and condensation of the nucleus (e.g. Hermo et al. 2010). The two centrioles (distal and proximal centriole) are located at the posterior part of the spermatid and oriented perpendicular to each other (e.g. Alberti et al. 1986; Alberti 1990).

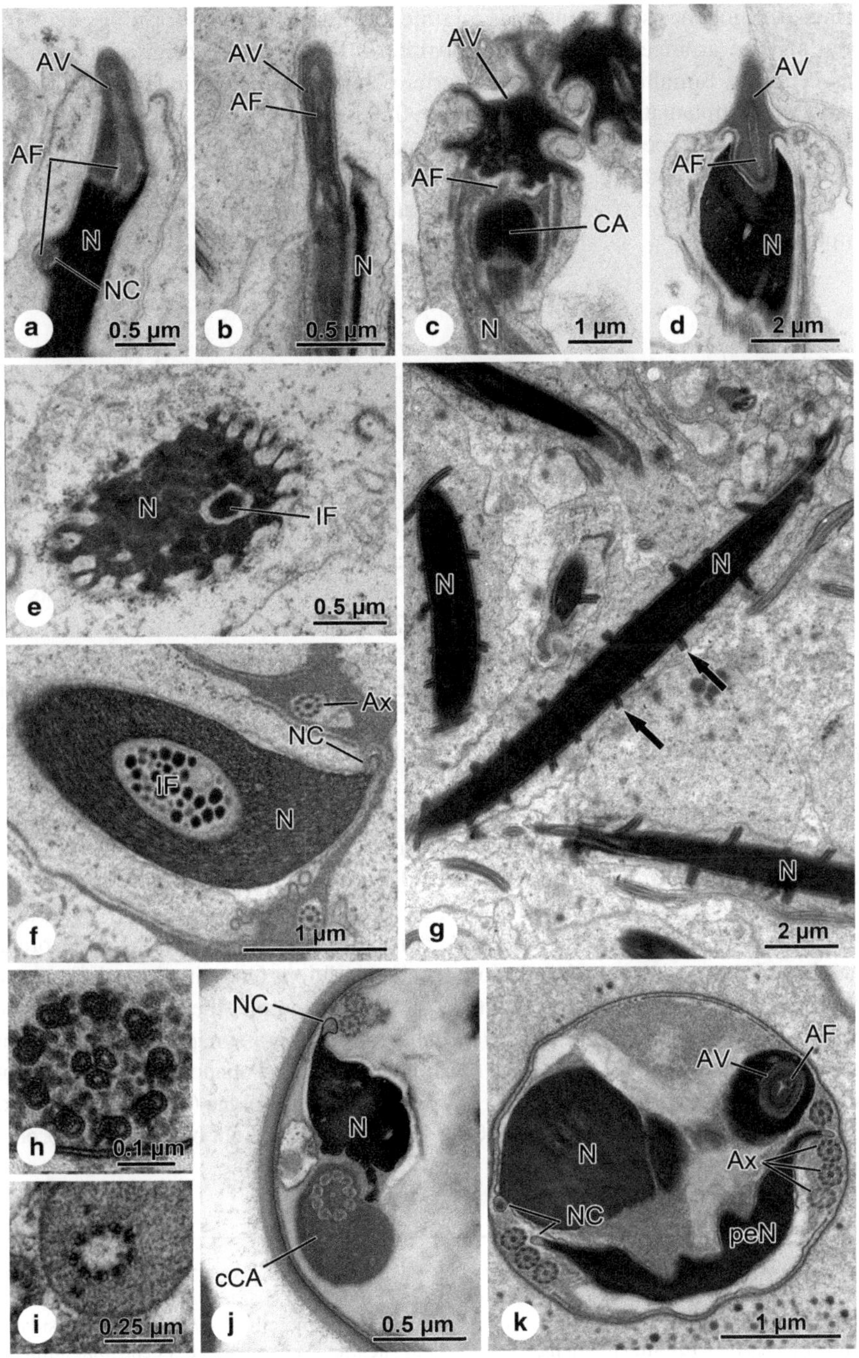

Fig. 13.2 Overview of the diversity of the different sperm cell components. TEM. (**a–d**) Longitudinal section of the acrosomal complex, (**a**) *Liphistius* (Liphistiidae), (**b**) *Pholcus* (Pholcidae), (**c**) *Tetragnatha* (Tetragnathidae) and (**d**) *Heteropoda* (Sparassidae). (**e–g**) Cross section of the

During spermiogenesis, the centrioles migrate towards the nucleus forming a flagellar tunnel and are arranged in a "tandem position". The axoneme originates from the distal centriole. Simultaneously, the posterior pole of the nucleus indents resulting in the so-called implantation fossa (Alberti 1990).

In mid- and late spermatids, the cell components such as the nucleus show a high diversity in shape and organization between taxa (Fig. 13.2e–g). In the following we will give a short overview of the structural diversity of the main cell components.

13.3.1.1 Acrosomal Complex

The spherical acrosomal vacuole of early spermatids elongates and is deeply indented from its base forming a subacrosomal space (Alberti 1990). Within the subacrosomal space, the acrosomal filament (perforatorium) originates and extends into a narrow indentation in the periphery of the nucleus, the so-called nuclear canal. As revealed by TRITC-phalloidin labeling, which reacts specifically to polymeric and oligomeric forms of actin, the acrosomal filament consists of F-actin fibrils (Costa-Ayub and Faraco 2007). At the end of spermiogenesis, the acrosomal filament extends until the end of the nuclear canal in Mesothelae and Hypochilidae (Ōsaki 1969; Alberti and Coyle 1991; Michalik 2007) or is shorter in length like in Mygalomorphae (Alberti et al. 1986; Tripepi et al. 1990) and haplogyne and entelegyne spiders (e.g. Michalik and Huber 2006; Michalik and Hormiga 2010).

The shape of the acrosomal vacuole can differ enormously between taxa. As summarized in Alberti (1990), the vacuole can be cone-shaped as, e.g. in Mesothelae, Mygalomorphae and Filistatidae (Alberti and Weinmann 1985; Alberti et al. 1986; Fig. 13.2a) or elongated and fingerlike as, e.g. in Pholcidae (Alberti and Weinmann 1985; Michalik et al. 2005; Fig. 13.2b). According to Alberti (1990), the widespread cone-shaped type represents the plesiomorphic condition. Organizations different to this simple type are known from few taxa only. For example, the corkscrew-shaped acrosomal vacuole of the genus *Tetragnatha* seems to be a synapomorphic trait for this group (Michalik et al. 2006; Fig. 13.2c), and the arrow-shaped indented acrosomal vacuole reported from Pisauridae, Lycosidae and Oxyopidae (Reger 1970; Ōsaki 1972; Wu et al. 1997; Michalik et al. 2012; see Fig. 13.2d) might be a synapomorphic trait for the RTA clade (Michalik and Ramirez, unpublished).

Fig. 13.2 (continued) nucleus, **(e)** *Chileotaxus* (Synotaxidae), **(f)** *Gelanor* (Mimetidae) and **(g)** *Pholcus*. **(h, i)** Axoneme, **(h)** *Pholcus* (Original: R. Dallai) and **(i)** *Linyphia* (Linyphiidae). **(j)** Chambered centriolar adjunct in *Heteropoda*. **(k)** Coiled sperm cell of *Schizocosa* (Lycosidae). Abbreviations: *AF* acrosomal filament, *AV* acrosomal vacuole, *Ax* axoneme, *CA* centriolar adjunct, *cCA* chambered centriolar adjunct, *IF* implantation fossa, *N* nucleus, *NC* nuclear canal, *peN* postcentriolar elongation of the nucleus

13.3.1.2 Nucleus

The spherical nucleus of early spermatids is characterized by a heterogeneous nucleoplasm. The differentiation of the nucleus begins with the condensation of the chromatin that is organized in a fibrillar pattern (e.g. Alberti and Weinmann 1985). Usually there are some electron-dense and electron-lucent areas present during the condensation process, but at the end of spermiogenesis, the chromatin appears highly condensed and homogenous (e.g. Alberti and Weinmann 1985; Michalik et al. 2004a). During the condensation process, the nucleus elongates in an asymmetric way and extends beyond the implantation fossa resulting in a postcentriolar nuclear elongation—a character which was considered as synapomorphy for Amblypygi and Araneae (see Alberti 2000). At the end of spermiogenesis, the length and shape of the pre- and postcentriolar part of the nucleus varies across spiders (Fig. 13.2e–g). For example, the postcentriolar elongation is usually short in basal taxa as in Mesothelae and Mygalomorphae (Alberti et al. 1986; Michalik 2007), but can be extremely elongated as in the genus *Tetragnatha* (Michalik et al. 2006). The implantation fossa can be small or deeply indented into the nucleus as in some pholcid spiders (Alberti and Weinmann 1985; Michalik et al. 2005), and it can be filled with granules or homogenous material (Alberti 1990; Fig. 13.2f). The evolution of this remarkable structural diversity is not understood at all, but it can be assumed that nuclear characters are highly homoplastic (Michalik and Ramirez, unpublished data). The functional implications of the different shapes of the nucleus are unknown since studies addressing the function of the sperm components are still lacking.

13.3.1.3 Axoneme and Associated Structures

The axoneme of spiders consists of a 9+3 microtubular pattern (e.g. Ōsaki 1969)—a character which was considered as synapomorphy for Tetrapulmonata (Weygoldt and Paulus 1979). As revealed by tannic acid analyses, the doublets consist of an A-subtubule of 13 protofilaments and a B-subtubule with 11 protofilaments (Dallai et al. 1995). The central tubules consist of 13 protofilaments and are connected by spokes to the doublets (Fig. 13.2h). A deviation from this pattern is only described for Linyphioidea (= Linyphiidae + Pimoidae; Hormiga 2003). Here, the axoneme lost the central tubules resulting in a 9 + 0 microtubular pattern (Michalik and Alberti 2005; Michalik and Hormiga 2010; Fig. 13.2i). Moreover, the axoneme of "linyphioids" is remarkably shorter compared to all other spider species observed so far (Michalik and Hormiga 2010). The evolution towards different axoneme architecture is not understood for spiders, but as hypothesized by Michalik and Alberti (2005), the aberrant organization might result in a different spermatozoal movement as shown for other chelicerates (Ishijima et al. 1988). The motility performance might have an effect on the paternity success and thus drive the evolution towards a different sperm phenotype—a hypothesis which needs still to be tested for spiders.

Associated structures of the axoneme are exclusively described from its proximal part. For example, in basal taxa and some haplogynes, the axoneme is often surrounded by glycogen (Ōsaki 1969; Alberti and Weinmann 1985; Michalik et al. 2004a), whereas in most araneomorph spiders the axonemal basis is often embedded or adjacent to electron-dense material, the so-called centriolar adjunct (Alberti 1990). The centriolar adjunct usually extends into the implantation fossa. Exceptions to this general organization are only described for members of the RTA clade. The axonemal basis is completely embedded in the centriolar adjunct which extends distally resulting in a flower-like appearance in cross sections (Alberti 1990). This so-called chambered centriolar adjunct was suggested to be a synapomorphic trait for the RTA clade (Michalik et al. 2013; Fig. 13.2j). Interestingly, Alberti (1990) mentioned adjacent thin electron-dense streaks around the axoneme for the oecobiid spider *Uroctea durandi*—a group which was suggested to be closely related to the RTA clade (Miller et al. 2010).

13.3.2 Coiling Process and Sperm Conjugates

Spermiogenesis in Araneae is completed after the coiling process (Alberti 1990). During this process the main cell components (acrosomal vacuole, nucleus and axoneme) coil within the sperm cell (Fig. 13.2k)—a phenomenon which is also reported for Pseudoscorpiones, Pedipalpi and Ricinulei (Alberti 2000). The flagellar tunnel disappears and the axoneme coils around the nucleus in the periphery of the cell. After the coiling process, the roundish sperm cells appear compact and the cell components are usually densely packed within the cytoplasm (e.g. Alberti and Weinmann 1985; Alberti and Coyle 1991). According to Bertkau (1877), the evolution of coiled sperm cells could be linked to the peculiar transfer mode in spiders. Nevertheless, the occurrence of coiled spermatozoa in other arachnid groups suggests a more complex evolutionary scenario.

A peculiar structure reported for some haplogyne taxa is the so-called vesicular area (Alberti and Weinmann 1985). Here, a large area of the sperm cell is "filled" with homogenous often electron-dense material (Michalik et al. 2004a; Michalik and Huber 2006). The origin of the vesicular area is still under debate. As suggested by Alberti and Weinmann (1985), intracellular vesicles fuse to form these vacuole-like areas (see also Michalik et al. 2004a; Michalik and Huber 2006). The cell components partly protrude into the vesicular area and get secondarily surrounded by a membrane. On the contrary, Costa-Ayub and Faraco (2007) suggested, based on their study on the synspermia formation of the brown spider *Loxosceles intermedia* (Sicariidae), that the main cell components are not coiled within the cell but instead the whole cell coils partially. Consequently, the homogenous areas (= vesicular area) between the membrane-surrounded cell components are of extracellular origin. A similar organization is reported for collembolans where an extracellular cavity is formed during the so-called winding process of the sperm cell

(Dallai et al. 2004). But in contrast to collembolan spermatozoa, no connection between the vesicular area and the extracellular space could be shown for spiders.

The final step in spermiogenesis is the formation of an extracellular secretion sheath around the coiled sperm. The sheath is usually produced by secretion of the epithelium of the deferent duct, but few studies reported a sheath formation from the distal testis lumen (Michalik and Huber 2006; Michalik et al. 2013). The secretion sheath can be single or multilayered and the thickness can vary remarkably. It was assumed that the sheath protects the spermatozoa during sperm induction and sperm transfer and/or in the female sperm storage organs (Bertkau 1877; Alberti 1990). In spiders three different transfer forms can be distinguished—coenospermia (several individual sperm cells are surrounded by a common secretion sheath), cleistospermia (each individual sperm cell is surrounded by a secretion sheath) and synspermia (a certain number of sperm cells fuse at the end of spermiogenesis and are surrounded by a common secretion sheath) (see Alberti 2000). A fourth type of transfer form was described for telemid spiders (Juberthie et al. 1981). This so-called spermatophore is produced in the distal deferent duct, but as suggested by Alberti (1990), it can represent a subtype of coenospermia or a cluster of cleistospermia. The evolution of the sperm transfer forms is still not clear in spiders (see Alberti and Coyle 1991). Coenospermia are secondary sperm conjugates (Higginson and Pitnick 2011) and typical for Mesothelae and Mygalomorphae, whereas cleistospermia are exclusive for Araneomorphae (Michalik et al. 2004b). Synspermia are primary sperm conjugates (Higginson and Pitnick 2011) and are only described from several haplogyne families (Alberti and Weinmann 1985; Michalik et al. 2004a; Costa-Ayub and Faraco 2007). Figure 13.3 shows a possible scenario of the evolution of transfer forms. It can be assumed that the different types of sperm transfer forms in spiders have multiple independent origins as coenospermia are also described for the haplogyne Filistatidae and the synspermia are reported from other not closely related haplogyne families. As summarized by Higginson and Pitnick (2011), the evolution of sperm conjugates could depend on several factors as sperm–egg interactions, sperm cooperation or sperm–female interactions as recently shown in a multimethodological approach for diving beetles (Higginson et al. 2012). For spiders no such study exists, but the highly diverse genital structures and sperm storage pattern suggest a correlation with sperm conjugation and sperm morphology as recently shown for flatworms (Schärer et al. 2011).

13.4 Seminal Secretions

Since accessory glands are absent in the spider male genital system, the secretion transferred to the female is produced by the somatic tissue of testes and deferent ducts. The structural diversity of the secretion within the seminal fluid is considerably high and presumably taxon specific (Michalik 2009). It was suggested that parts of the secretion droplets in testis and deferent ducts are likely used for the

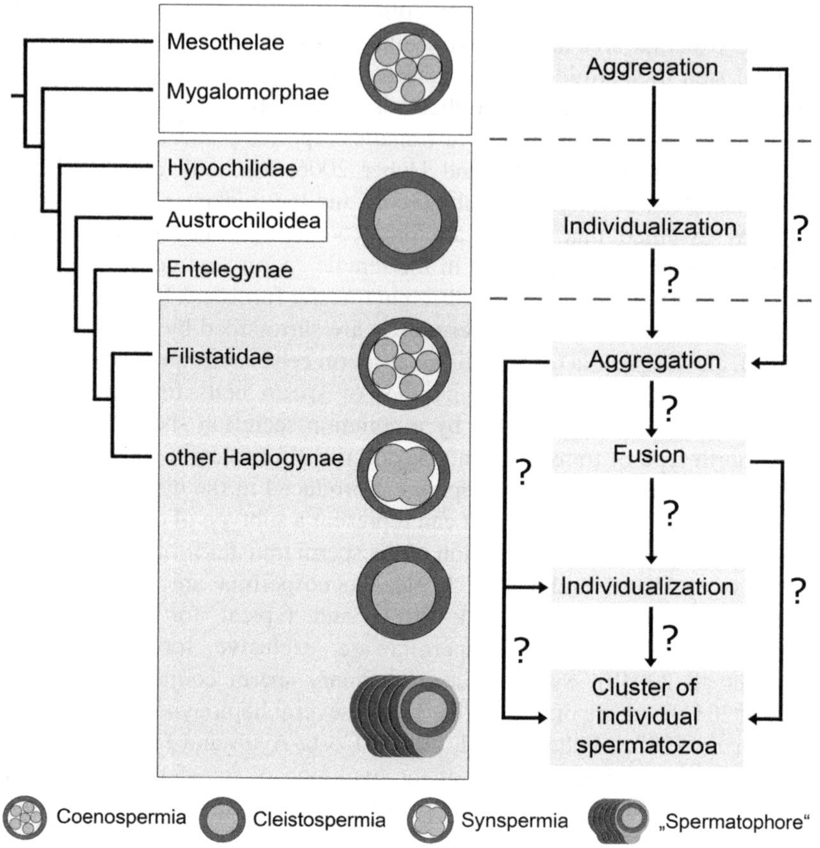

Fig. 13.3 Sperm transfer forms in spiders and their possible evolution

formation of the secretion sheath, but most secretions are transferred with the seminal fluid. Interestingly, not only one type of secretion can be present in the seminal fluid. For example, the seminal fluid of pholcid spiders can contain up to three different types of secretion droplets (Michalik and Uhl 2005). The function of the seminal secretion is unknown for spiders, but as shown indirectly by behavioral studies, sperm-associated substances seem to have an influence on the female reluctance in the wolf spider *Schizocosa malitiosa* (Aisenberg and Costa 2005). As shown for insects, male transferred substance can influence female physiology remarkably (e.g. Adams and Wolfner 2007). On the other hand, female secretion can influence sperm storage and sperm usage as shown for *Drosophila* flies (Wolfner 2011). A similar complex network of interactions can be assumed for spiders since females and males have complex genitalia with associated glandular epithelia which presumably contribute or change the composition of sex-related substances (Huber 2005; Herberstein et al. 2011; Fig. 13.4).

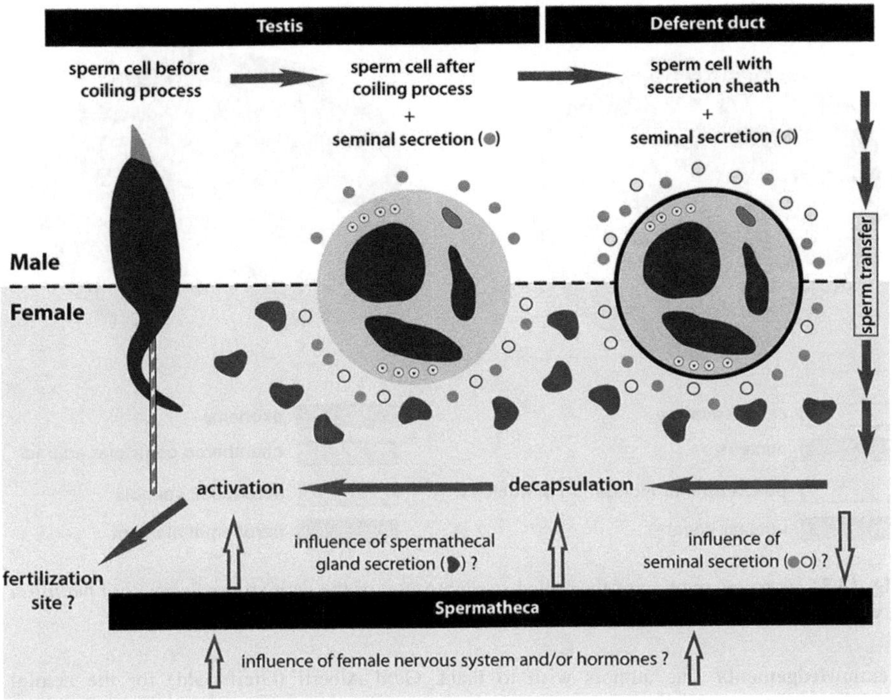

Fig. 13.4 Schematic illustration of the transformation of spermatozoa and the appearance and interactions of sex-related secretions (from Herberstein et al. 2011)

13.5 Conclusions and Outlook

The male reproductive system of spiders is very diverse across taxa regarding its gross morphology and sperm ultrastructure. Numerous studies showed that sperm characters have high phylogenetic potential. Nevertheless, the evolution of the different sperm cell components and sperm conjugates is not understood. Future analyses should focus on the evolutionary morphology of sperm structures including new methodology as 3D reconstruction based on transmission electron microscopy sections (Michalik et al. 2013; Fig. 13.5). This will not only help to understand the organization of the often complex spermatozoa but also increase the reproducibility across taxa.

Whereas most studies focused on the phylogenetic implications of sperm structures, the functional implications were never addressed. The main reason might be the difficult accessibility of motile sperm cells since all spermatozoa are encapsulated and activated in the female only (Herberstein et al. 2011). But in order to understand spider reproduction knowledge about processes involving sperm activation, influence of sex-related secretion and fertilization is urgently needed.

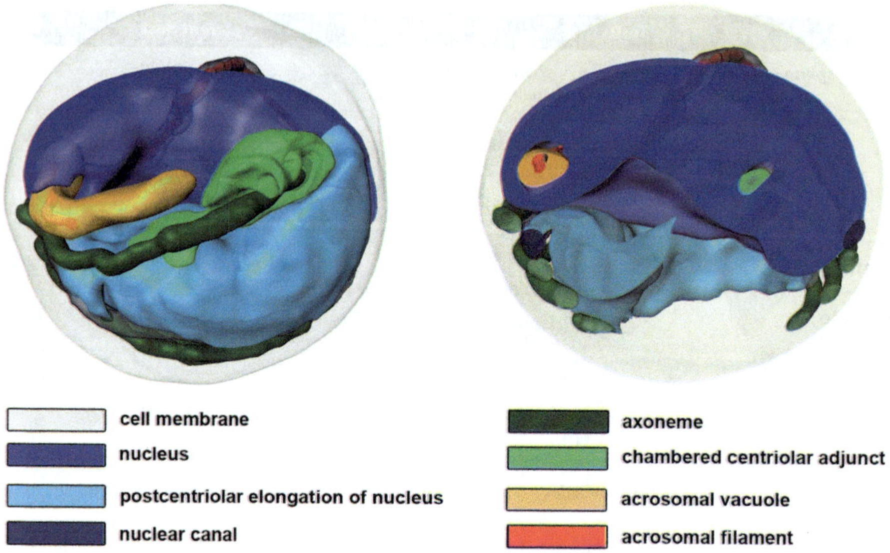

Fig. 13.5 3D reconstruction of the coiled spermatozoon of the wolf spider *Schizocosa malitiosa* (Lycosidae)

Acknowledgements The authors wish to thank Gerd Alberti (Greifswald) for the helpful comments on the manuscript.

References

Adams EM, Wolfner MF (2007) Seminal proteins but not sperm induce morphological changes in the *Drosophila melanogaster* female reproductive tract during sperm storage. J Insect Physiol 53:319–331. doi:10.1016/j.jinsphys.2006.12.003

Aisenberg A, Costa FG (2005) Females mated without sperm transfer maintain high sexual receptivity in the wolf spider *Schizocosa malitiosa*. Ethology 111:545–558

Alberti G (1990) Comparative spermatology of Araneae. Acta Zool Fenn 190:17–34

Alberti G (2000) Chelicerata. In: Adiyodi KG, Adiyodi RG (eds) Reproductive biology of invertebrates, vol 9B. Wiley, Chichester, pp 311–388

Alberti G (2005) Double spermatogenesis in Chelicerata. J Morphol 266:281–297

Alberti G, Afzelius BA, Lucas SM (1986) Ultrastructure of spermatozoa and spermatogenesis in bird spiders (Theraphosidae, Mygalomorphae, Araneae). J Submicrosc Cytol 18:739–753

Alberti G, Coyle FA (1991) Ultrastructure of the primary male genital system, spermatozoa, and spermiogenesis of *Hypochilus pococki* (Araneae, Hypochilidae). J Arachnol 19:136–149

Alberti G, Weinmann C (1985) Fine structure of spermatozoa of some labidognath spiders (Filistidae, Segestriidae, Dysderidae, Oonopidae, Scytodidae, Pholcidae; Araneae; Arachnida) with remarks on spermiogenesis. J Morphol 185:1–35

Bertkau P (1875) Über den Generationsapparat der Araneiden. Arch Naturgesch 41:235–262

Bertkau P (1877) Über die Übertragungsorgane und die Spermatozoen der Spinnen. Verh Nat Ver Preuss Rheinlande 34:28–32

Bertkau P (1878) Versuch einer natürlichen Anordnung der Spinnen nebst Bemerkungen zu einzelnen Gattungen. Arch Naturgesch 44:351–409

Burger M, Michalik P (2010) The male genital system of goblin spiders: evidence for the monophyly of Oonopidae (Arachnida: Araneae). Am Mus Novit 3675:1–13

Coddington JA (1990) Ontogeny and homology in the male palpus of orb-weaving spiders and their relatives, with comments on phylogeny (Araneoclada: Araneoidea, Deinopoidea). Smithson Contrib Zool 496:1–52

Costa-Ayub CLS, Faraco CD (2007) Ultrastructural aspects of spermiogenesis and synspermia in the brown spider *Loxosceles intermedia* (Araneae: Sicariidae). Arthropod Struct Dev 36: 41–51. doi:10.1016/j.asd.2006.07.004

Crome W (1951) Die grobe Morphologie des männlichen Genitalapparates einiger Araneen. D Zool Z 1:169–186

Dallai R, Afzelius BA, Witalinski W (1995) The axoneme of the spider spermatozoon. B Zool 62: 335–338

Dallai R, Fanciulli PP, Frati F, Paccagnini E, Lupetti P (2004) Sperm winding in Collembola. Pedobiologia 48:493–501

Eberhard WG (2004) Why study spider sex: special traits of spiders facilitate studies of sperm competition and cryptic female choice. J Arachnol 32:545–556

Eberhard WG, Huber BA (2010) Spider genitalia: precise maneuvers with a numb structure in a complex lock. In: Leonard JL, Córdoba-Aguilar A (eds) Evolution of primary sexual characters in animals. Oxford University Press, Oxford

Foelix RF (2011) Biology of spiders. Oxford University Press, Oxford

Haupt J (2003) The Mesothelae—a monograph of an exceptional group of spiders (Araneae: Mesothelae). Zoologica 154:1–102

Herberstein ME, Schneider JM, Uhl G, Michalik P (2011) Sperm dynamics in spiders. Behav Ecol 22(4):692–695. doi:10.1093/beheco/arr053

Hermo L, Pelletier RM, Cyr DG, Smith CE (2010) Surfing the wave, cycle, life history, and genes/proteins expressed by testicular germ cells. Part 2: Changes in spermatid organelles associated with development of spermatozoa. Microsc Res Tech 73:279–319. doi:10.1002/Jemt.20787

Higginson DM, Miller KB, Segraves KA, Pitnick S (2012) Female reproductive tract form drives the evolution of complex sperm morphology. Proc Natl Acad Sci USA 109:4538–4543. doi:10.1073/pnas.1111474109

Higginson DM, Pitnick S (2011) Evolution of intra-ejaculate sperm interactions: do sperm cooperate? Biol Rev 86:249–270. doi:10.1111/j.1469-185X.2010.00147.x

Hormiga G (2003) *Weintrauboa*, a new genus of pimoid spiders from Japan and adjacent islands, with comments on the monophyly and diagnosis of the family Pimoidae and the genus *Pimoa* (Araneoidea, Araneae). Zool J Linn Soc 139:261–281

Huber BA (2005) Sexual selection research on spiders: progress and biases. Biol Rev 80:363–385

Ishijima S, Sekiguchi K, Hiramoto Y (1988) Comparative study of the beat patterns of American and Asian Horseshoe Crab sperm: evidence for a role of the central pair complex in forming planar waveforms in flagella. Cell Motil Cytoskeleton 9:264–270

Juberthie C, Lopez A, Kovoor J (1981) Spermiogenesis and spermatophore in *Telema tenella* Simon (Araneae: Telemidae)—an ultrastructural study. Int J Inver Rep Dev 3:181–191

Knoflach B (1998) Mating in *Theridion varians* Hahn and related species (Araneae: Theridiidae). J Nat Hist 32:545–604

Kraus O (1984) Male spider genitalia: evolutionary changes in structure and function. Abh Naturwiss Ver Hamburg 27:373–382

Kuntner M, Coddington JA, Schneider JM (2009) Intersexual arms race? Genital coevolution in Nephilid spiders (Araneae, Nephilidae). Evolution 63:1451–1463

Lopez A (1987) Glandular aspects of sexual biology. In: Nentwig W (ed) Ecophysiology of spiders. Springer, Berlin

Lopez A, Juberthie-Jupeau L (1983) Structure et ultrastructure de la spermathéque chez *Telema tenella* Simon (Araneae, Telemidae). Mem Biospeol 10:413–418

Michalik P (2007) Spermatozoa and spermiogenesis of *Liphistius* cf. *phuketensis* (Mesothelae, Araneae, Arachnida) with notes on phylogenetic implications. Arthropod Struct Dev 36: 327–335. doi:10.1016/j.asd.2007.03.001

Michalik P (2009) The male genital system of spiders (Arachnida, Araneae) with notes on the fine structure of seminal secretions. Contrib Nat Hist 12:959–972

Michalik P, Aisenberg A, Postiglioni R, Lipke E (2013) Spermatozoa and spermiogenesis of the wolf spider *Schizocosa malitiosa* (Lycosidae, Araneae) and its functional and phylogenetic implications. Zoomorphology. doi:10.1007/s00435-012-0166-z

Michalik P, Alberti G (2005) On the occurrence of the 9+0 axonemal pattern in the spermatozoa of sheetweb spiders (Linyphiidae, Araneae). J Arachnol 33:569–572

Michalik P, Dallai R, Giusti F, Alberti G (2004a) The ultrastructure of the peculiar synspermia of some Dysderidae (Araneae, Arachnida). Tissue Cell 36:447–460. doi:10.1016/j.tice.2004.08.001

Michalik P, Dallai R, Giusti F, Mercati D, Alberti G (2005) Spermatozoa and spermiogenesis of *Holocnemus pluchei* (Scopoli, 1763) (Pholcidae, Araneae). Tissue Cell 37:489–497. doi:10.1016/j.tice.2005.09.003

Michalik P, Gray MR, Alberti G (2003) Ultrastructural observations of spermatozoa and spermiogenesis in *Wandella orana* Gray, 1994 (Araneae: Filistatidae) with notes on their phylogenetic implications. Tissue Cell 35:325–337. doi:10.1016/S0040-8166(03)00054-5

Michalik P, Haupt J, Alberti G (2004b) On the occurrence of coenospermia in mesothelid spiders (Araneae: Heptathelidae). Arthropod Struct Dev 33:173–181. doi:10.1016/j.asd.2004.02.001

Michalik P, Hormiga G (2010) Ultrastructure of the spermatozoa in the spider genus Pimoa: new evidence for the monophyly of Pimoidae plus Linyphiidae (Arachnida: Araneae). Am Mus Novit 3682:1–17

Michalik P, Huber BA (2006) Spermiogenesis in *Psilochorus simoni* (Berland, 1911) (Pholcidae, Araneae): evidence for considerable within-family variation in sperm structure and development. Zoology 14–25. doi:10.1016/j.zool.2005.09.010

Michalik P, Knoflach B, Thaler K, Alberti G (2010) Live for the moment—adaptations in the male genital system of a sexually cannibalistic spider (Theridiidae, Araneae). Tissue Cell 42:32–36. doi:10.1016/j.tice.2009.06.004

Michalik P, Rittschof CC (2011) A comparative analysis of the morphology and evolution of permanent sperm depletion in spiders. PLoS One 6(1). doi:10.1371/journal.pone.0016014

Michalik P, Sacher P, Alberti G (2006) Ultrastructural observations of spermatozoa of several tetragnathid spiders with phylogenetic implications (Araneae, Tetragnathidae). J Morphol 267 (2):129–151. doi:10.1002/jmor.10226

Michalik P, Uhl G (2005) The male genital system of the cellar spider *Pholcus phalangioides* (Fuesslin, 1775) (Pholcidae, Araneae): development of spermatozoa and seminal secretion. Front Zool 2:12. doi:10.1186/1742-9994-2-12

Miller JA, Carmichael A, Ramirez MJ, Spagna JC, Haddad CR, Rezac M, Johannesen J, Kral J, Wang XP, Griswold CE (2010) Phylogeny of entelegyne spiders: affinities of the family Penestomidae (NEW RANK), generic phylogeny of Eresidae, and asymmetric rates of change in spinning organ evolution (Araneae, Araneoidea, Entelegynae). Mol Phylogenet Evol 55: 786–804. doi:10.1016/j.ympev.2010.02.021

Ōsaki H (1969) Electron microscope study on the spermatozoon of the Liphistiid spider *Heptathela kimurai*. Acta Arachnol 22:1–12

Ōsaki H (1972) Electron microscope study on spermiogenesis in the spider, *Oxyopes sertatus*. Jpn J Zool 16:184–199

Reger JF (1970) Spermiogenesis in the spider *Pisaurina* sp.: a fine structure study. J Morphol 130:421–434

Schärer L, Littlewood DTJ, Waeschenbach A, Yoshida W, Vizoso DB (2011) Mating behavior and the evolution of sperm design. Proc Natl Acad Sci USA 108:1490–1495. doi:10.1073/pnas.1013892108

Schneider JM, Michalik P (2011) One-shot genitalia are not an evolutionary dead end—regained male polygamy in a sperm limited spider species. BMC Evol Biol 11. doi:10.1186/1471-2148-11-197

Suhm M, Thaler K, Alberti G (1996) Glands in the male palpal organ and the origin of the mating plug in *Amaurobius* species (Araneae: Amaurobiidae). Zool Anz 234:191–199

Tripepi S, Perotta E, Saita A (1990) Spermiogenesis in the spider *Hapalopus tripepii* (Dresco) (Araneae, Mygalomorphae): an ultrastructural analysis. Bull Br Arachnol Soc 8:171–176

Weygoldt P, Paulus HF (1979) Untersuchungen zur Morphologie, Taxonomie und Phylogenie der Chelicerata. II. Cladogramme und Entfaltung der Chelicerata. Z Zool Syst Evolutionsforsch 17:117–177

White-Cooper H, Doggett K, Ellis RE (2009) The evolution of spermatogenesis. In: Birkhead TR, Hosken D, Pitnick S (eds) Sperm biology: an evolutionary perspective. Academic, Oxford, pp 151–183

Wolfner MF (2011) Precious essences: female secretions promote sperm storage in *Drosophila*. PLoS Biol 9(11):e1001191. doi:10.1371/journal.pbio.1001191

Wu XM, Song DX, Chen DY (1997) Ultrastructure of spermatozoa and spermatogenesis in wolf spider *Pardosa astrigera* (Araneae: Lycosidae). Acta Zool Sinica 43:17–26

Part V
Venom

The venom of a spider usually contains many different components, attacking several targets in a potential prey item at the same time. This high redundancy may be seen as evolutionary strategy, insurance against resistance development or a comparable target response, or just as an effect of combinatorial chemistry. Despite the efficiency of their high-tox weapons, some spiders reduce their venom usage, other prefer to destroy target tissue with enzymes instead of blocking receptors, while another spider group switches to the application of sticky glue. Despite their long success story as venomous animals, spiders are not really dangerous to humans and lethal outcomes of spider bites are unusual.

Chapter 14
Main Components of Spider Venoms

Wolfgang Nentwig and Lucia Kuhn-Nentwig

14.1 Introduction

Venom glands are already present in the oldest spider group, the Mesothelae. The glands lie in the anterior portion of the cheliceral basal segment but are very small, and it is doubtful how much the venom contributes to the predatory success. In mygalomorph spiders, the well-developed venom glands are still in the basal segment of the chelicerae and produce powerful venom that is injected via the cheliceral fangs into a victim. In all other spiders (Araneomorphae), the venom glands have become much larger and reach into the prosoma where they can take up a considerable proportion of this body part. Only a few spiders have reduced their venom glands, either partially or completely (Uloboridae, Holarchaeidae and Symphytognathidae are usually mentioned) or modified them significantly (Scytodidae, see Suter and Stratton 2013).

As well as using venom, spiders may also use their chelicerae to overwhelm an item of prey. It is primarily a question of size whether a spider chews up small arthropods without applying venom or if it injects venom first. Very small and/or defenceless arthropods are picked up and crashed with the chelicerae, while larger, dangerous or well-defended items are carefully approached and only attacked with venom injection. Some spiders specialize on prey groups, such as noctuid moths (several genera of bola spiders among Araneidae), web spiders (Mimetidae), ants (*Zodarion* species in Zodariidae, aphantochiline thomisids, several genera among Theridiidae, Salticidae, Clubionidae and Gnaphosidae) or termites (Ammoxenidae). However, these more or less monophagous species amount only to roughly 2 % of all known spider species, while 98 % are polyphagous. From these considerations, it follows that the majority of spider venoms are not tailored to any given invertebrate or insect group but are rather unspecialized to be effective over a broad spectrum of prey types that spiders naturally encounter.

W. Nentwig (✉) • L. Kuhn-Nentwig
Institute of Ecology and Evolution, University of Bern, Bern, Switzerland
e-mail: wolfgang.nentwig@iee.unibe.ch

W. Nentwig (ed.), *Spider Ecophysiology*,
DOI 10.1007/978-3-642-33989-9_14, © Springer-Verlag Berlin Heidelberg 2013

Table 14.1 Main component categories and some characteristics from spider venoms

Venom compound category	Molecular size (kDa)	Function	Distribution	Molecular diversity
Low molecular mass compounds	<1	Various. Some are neurotransmitters, some support other compounds synergistically, often unknown	Probably in all spider venoms	Limited
Acylpolyamines	<1	Neurotoxins, act on ion channels	Main toxic category in Araneidae and Nephilidae, common in Agelenidae, also in a few further families	Limited
Linear peptides	1–9	Destroy membranes	Known form Lycosoidea and Zodariidae, probably also in a few more families	High
Cysteine-rich mini-proteins	2–15	Neurotoxins, act on ion channels	Known from most families	Very high
Large proteins	110–140	Destroy membranes	Theridiidae	Potentially high (unknown)
Enzymes	18–44	Destroy membranes and tissue	Probably in all spider venoms, main category in Sicariidae	Limited

While so far none of the venoms from specialized spider species have been investigated, we do have in the meantime a broad overview on the composition of the venom of a few well-investigated species. According to recent reviews (Vassilevski et al. 2009; Kuhn-Nentwig et al. 2011), six main groups of components can be distinguished for spider venoms: low molecular mass compounds, acylpolyamines, linear peptides, cysteine-rich mini-proteins, large proteins and enzymes (Table 14.1).

14.2 Low Molecular Mass Compounds

Most scientists investigating spider venom focus on substances other than low molecular mass compounds which are usually more a side result of research. From what is known at present however, it can be stated that spider venoms contain a vast variety of such compounds. They are comprised of organic acids, nucleotides and nucleosides, amino acids, amines, polyamines and other substances, many of them functioning as neurotransmitters.

The venom cation concentrations of *Cupiennius salei* (Ctenidae) are 9 mM Na^+, 215 mM K^+ and 1 mM Ca^{2+}. This is in contrast to the cation concentrations of its haemolymph (about 200 mM Na^+, 10 mM K^+ and 4 mM Ca^{2+}), meaning that the potassium concentration in the venom is increased more than 20-fold, compared to the haemolymph, and, respectively, the sodium concentration is decreased (Kuhn-Nentwig et al. 1994). At such concentrations, potassium is able to induce depolarization of excitable cell membranes, causing paralysis of a prey item. Potassium is also known as an effective synergist of neurotoxins in *Cupiennius salei* (Wullschleger et al. 2004, 2005). There are no comparable data for other spiders but we assume that such inverse ion concentrations are widespread among spiders.

Citric acid has been detected in 48 spider species from 16 families in concentrations of 16–147 mM (Kuhn-Nentwig et al. 2011). There are several reasons for its presence in spider venoms: it could prevent bacterial growth; it could be part of the venom glands' buffer system to compensate for highly cationic cytolytic peptides and acylpolyamines; it could function as an enhancer to enforce the effect of other substances or as a divalent metallic ion chelator; and finally it could also contribute to a partial inhibition of zinc ion metalloproteases and Ca^{2+}-dependent enzymes (i.e. phospholipases A_2) in venom glands. After injection into a prey item, the high citric acid concentration would be diluted, thus activating the enzymes (Odell et al. 1999). Further organic acids from spider venoms are lactic acid and phosphoric acid.

Sulphated and other nucleosides are known from the venoms of 30 species from 11 families (Schroeder et al. 2008; Kuhn-Nentwig et al. 2011). In the venom of several *Loxosceles* species (Sicariidae), these compounds are very common and constitute even up to about 50 % of the venom total dry mass in the venom of the agelenid *Tegenaria agrestis* (Taggi et al. 2004). These and related substances can induce paralytic and also lethal effects in insects.

All biogenic amino acids have been identified from spider venoms. Taurine has been detected in the venoms of *Cupiennius salei* and the black widow *Latrodectus tredecimguttatus* (Theridiidae). Glutamate and GABA, several biogenic amines such as histamine, tyramine, serotonin, octopamine, dopamine and 5-hydroxytryptamine (5-HT; serotonin), as well as polyamines (spermine, spermidine, putrescine and cadaverine), have also been identified from spider venoms. These substances influence the nervous system of insects or act directly as neurotransmitters. Acetylcholine, choline, noradrenaline and adrenaline are further neurotransmitters in the insect nervous system which were also identified from several spider venoms (Rash and Hodgson 2002; Schroeder et al. 2008).

Histamine is well known from bee and wasp venoms. In vertebrates it produces pain and is therefore seen as a defensive substance (Bettini 1978). This "pain theory" is not really convincing for spiders because only large spiders (Ctenidae, Sparassidae, Theraphosidae) suffer from vertebrate predators, while for smaller spiders, invertebrates are more important predators. In insects, histamine acts as a neurotransmitter, targeting ionotropic receptor channels, i.e. it enhances synergistically the effect of mini-proteins that affect ion channels (Wullschleger et al. 2005).

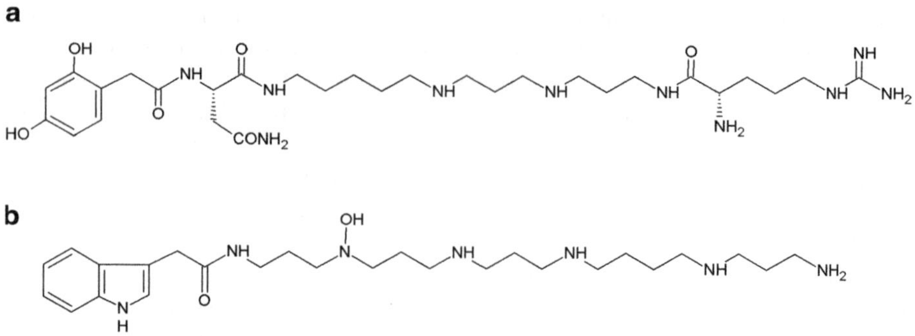

Fig. 14.1 Characteristic acylpolyamine toxins from spider venom. (**a**) Argiopine from *Argiope lobata* (Araneidae) as an example for an amino acid-containing acylpolyamines with phenolic end group and (**b**) AGEL 489 from *Agelenopsis aperta* (Agelenidae) as an example for a non-amino acid-containing acylpolyamines with an indolic end group. Adapted from Kuhn-Nentwig et al. (2011)

14.3 Acylpolyamines

Acylpolyamines are a class of neuroactive compounds, with a polyamine structure that frequently contains an aromatic acyl end moiety (indolic or phenolic) at its end (350–1,000 Da). A high structural diversity within these compounds is obtained by a combination of different acyl groups with polyamine chains, varying in length, number of amide bonds and functional groups. There are two groups of acylpolyamine toxins; those containing amino acids and those not containing them (Fig. 14.1). In a similar manner, they both provoke a reversible paralysis caused by blocking activated postsynaptic glutamate receptor channels (McCormick and Meinwald 1993; Schäfer et al. 1994; Itagaki and Nakajima 2000). So far 176 different acylpolyamines have been identified from 20 species in eight families (Kuhn-Nentwig et al. 2011), and such a high diversity of compounds is seen as fulfilling the principles of combinatorial chemistry to optimise binding to a high diversity of targets.

Amino acids containing acylpolyamines have so far only been found in the orb weaver families Araneidae and Nephilidae. A total of 82 different toxins have been described: 18 from six *Araneus* and *Argiope* species, 64 from three *Nephila* and two *Nephilengys* species (Itagaki et al. 1997; Palma et al. 1998; McCormick and Meinwald 1993). Araneidae and Nephilidae are regarded as two separate but closely related families that obviously developed amino acids containing acylpolyamines as one of their main toxin groups. There are no acylpolyamines records from other families in the superfamily Araneoidea, namely, none from Tetragnathidae and Linyphiidae (see also Appendix, this volume).

Non-amino acids containing acylpolyamine toxins have been identified from three mygalomorph families (Ctenizidae, Hexathelidae and Theraphosidae) and three other families (Agelenidae, Amaurobiidae and Pisauridae), but not from Araneoidea. In all families, only one to four different compounds per species were found, but in both so far investigated agelenids, a much higher diversity was found (*Agelenopsis aperta* 68 and *Hololena curta* 12 compounds) (Chesnov et al. 2001; Tzouros et al. 2005). In contrast to Araneidae and Nephilidae, these six families primarily rely on other venom components, mainly mini-proteins. We assume that non-amino acids containing acylpolyamine toxins may possibly be much more widespread than it is known so far.

14.4 Linear Peptides

Linear peptides of diverse length and cationic charge have been identified in the venom of some spider species. These peptides adopt an α-helical conformation and are able to destroy prokaryotic and/or eukaryotic cell membranes (Kuhn-Nentwig 2003, 2009). Therefore, they have also been named cytolytic, membranolytic or antimicrobial peptides, and because of their cationic charge, the term small cationic peptides is also used. A few small linear peptides isolated so far only from the ctenid *Phoneutria nigriventer* are exceptional by not adopting an α-helix.

Linear peptides usually do not contain cysteine residues and have a higher amount of lysines and arginines; thus, they have relatively high positive net charges between +3 and +10. They are disordered in aqueous solutions, but adopt an α-helical structure in the presence of negatively charged membranes. These basic peptides are attracted to the cell surfaces by electrostatic interactions between their positively charged side chains of lysine and arginine and negatively charged membrane phospholipid headgroups and phosphate groups containing lipopolysaccharides or teichoic acids of cellular membranes. Different membrane disrupting models have been discussed, but what they do all have in common is that the membrane is finally destroyed (Fig. 14.2).

The molecular mass range of these "short" peptides varies from 1,910 to 5,221 Da which corresponds to 18–48 amino acids. There are 72 records concerning eight families and 17 species: 36 records from *Cupiennius salei* (Ctenidae), 12 records from *Lachesana tarabaevi* (Zodariidae) and the remaining records from lycosids, oxyopids and a few other families. Besides these short peptides, 16 further "large" peptides have been reported for *L. tarabaevi*, composed of 69–75 amino acid residues (7,880–8,571 Da) and with a net charge of +14. These peptides exhibit two α-helical regions connected by a short sequence of four amino acids, and it is assumed that two short peptides form the large peptide in a "head-to-tail" orientation. Their insecticidal effect is superior to the effects of the small peptides (Kozlov et al. 2006; Vassilevski et al. 2008; Kuhn-Nentwig et al. 2011).

The cytolytic acting peptides from *Cupiennius salei* are grouped into several cupiennin families; in *Lachesana tarabaevi* the short peptides are called latarcins, the large cyto-insectotoxins. In both species, these peptides are very important in the prey-killing process, see also Kuhn-Nentwig and Nentwig (2013). Nevertheless, this action is supported by other venom components (low molecular mass compounds, neurotoxic mini-proteins and enzymes). If these results can be

a

GFGALFKKFLAKKVAKTVAKQAAQGAKYVVNKQME

Cupiennin 1a

[2K38]

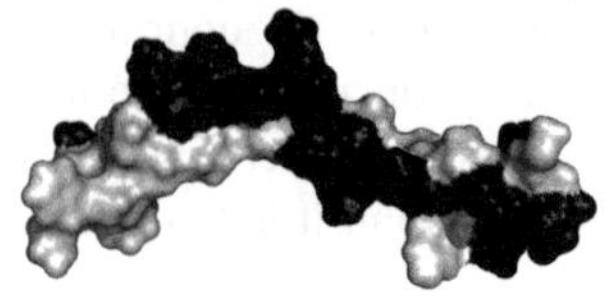

b

GLFGKLIKKFGRKAISYAVKKARGKH

Latarcin 2a

[2G9P]

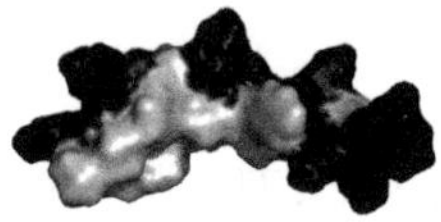

Fig. 14.2 Amino acid sequences and three-dimensional structures of (**a**) short cationic α-helical peptides from the ctenid *Cupiennius salei* and (**b**) the zodariid *Lachesana tarabaevi*. Positively charged amino acids in the amino acid sequences are given in *grey*. In the spatial structure of the peptides charged/polar residues are shown in *black* and hydrophobic amino acids in *grey*. The Protein Data Bank codes are given in *brackets*. Adapted from Kuhn-Nentwig et al. (2011)

generalized to Ctenidae, Lycosidae, Oxyopidae and related families (Lycosoidea), the development of cytolytic peptides could represent a new evolutionary step of this group and of Zodariidae.

14.5 Mini-proteins

Mini-proteins comprise peptides with molecular masses between 2,650 and 14,800 Da with most toxins between 3,000 and 9,000 Da. They contain 6–14 cysteines (Table 14.2) and exert a typical complex pattern of disulphide bridges. The most common pattern follows the "inhibitor cysteine knot" (ICK) motif, exhibiting a spatial structure consisting of a β-hairpin and a "knot" built by the C^3–C^6 bond penetrating the ring formed by the two other bonds and the involved amino acids (Norton and Pallaghy 1998). In proteins with six cysteines, the disulphide bridges are arranged as C1–C4, C2–C5 and C3–C6, and with eight cysteines, the arrangement is C1–C4, C2–C5, C3–C8 and C6–C7 in which the fourth disulphide bridge C6–C7 is introduced into the extended β-hairpin structure.

Nearly 1,000 different mini-proteins are known so far from the venom of 60 spider species belonging to 20 families (Kuhn-Nentwig et al. 2011). This characterizes mini-proteins as the most common venom compound in spiders, probably occurring in the venoms of most spider families, and there appear to be only a few families relying on other main components (see below). The increasing number of transcriptomic studies indicates that many different mini-proteins occur in one venom and more than one hundred different toxins or variations of a few different toxin types are not uncommon in a single species (e.g. Tang et al. 2010; Zhang et al. 2010).

Table 14.2 Frequency of the number of cysteines in mini-proteins with ICK motif for major spider taxa (data from Kuhn-Nentwig et al. 2011)

	Number of toxins recorded per given number of cysteines							
	6	7	8	9	10	11	12	14
Hexathelidae	61	1	18		8			
Theraphosidae	122	4	27		5	1		
Other mygalomorphs	2	1	8					
Haplogynae		1	14		12		7	
Agelenidae	28	2	25		2		7	
Amaurobiidae			4					
Sparassidae	7							
Ctenidae	14	7	25	7	14	2	8	
Lycosidae		9	111	8	18	1	5	6
Oxyopidae					3			
% of peptides	38.2	4.1	39.2	2.4	10.1	0.7	4.4	1.0

Mini-proteins act selectively against a specific target and form a tight and stable toxin-target complex. They mainly act on membrane proteins in electro-excitable cell membranes (neuronal and muscular cells), primarily modulating ion channels such as calcium (Ca^{2+}), sodium (Na^+) and potassium (K^+), but also on mechano-, chemo-, and thermosensitive receptors, and inhibit, activate or delay voltage-activated channels so that their normal ion regulation is affected (see Herzig and King 2013). Usually, these toxins are already very effective at nanomolar concentrations, which is at least one order of magnitude better than the average concentration needed for the unspecific membrane destruction by cytolytic peptides.

The fast transport of impulses along and between excitable cells is caused by the movement of sodium (Na^+) ions across the membranes of excitable cells via voltage-gated sodium (Na_v) channels. These ion channels enable the coordination of locomotion, and for this reason, they are usually the most abundant ion channels in nerve and muscle tissue. This makes sodium channels also the first target for paralytic neurotoxins. There is a high diversity in this family of sodium channels made of at least nine members: $Na_v1.1$ to $Na_v1.9$. Depending on the precise mode of action, three toxin types are distinguished: β-toxins shift the voltage dependence of Na_v channels activation, δ-toxins delay the inactivation of Na_v channels and μ-toxins inhibit Na_v channels. About 40 % of all functionally characterized mini-proteins are sodium channel inhibitors.

In most tissues of animals, especially in the nervous system and in muscles, calcium channels regulate the release of neurotransmitters. They can be inhibited by toxins causing a long-lasting specific blockade of these presynaptic voltage-gated calcium (Ca_v) channels, typically called omega-toxins (ω-toxins). There are six recognised types and many subtypes of voltage-gated calcium channels (L, N, P, Q, R and T) differing in their electrophysiological characteristics and reacting differently to different inhibitors or activators (Rash and Hodgson 2002). In spider venoms, roughly 35 % of all functionally characterized mini-proteins are calcium channel inhibitors.

Potassium channels comprise a highly diverse and large group of ion channels, regulating different cellular processes in the body. Voltage-gated potassium (K_v) channels consist of four main subunits and several accessory subunits to form an ion pore. Spider κ-toxins inhibit voltage-activated potassium channels and account for 25 % of all functionally characterized mini-proteins.

14.6 Large Proteins

The largest compounds in spider venom are proteins with a molecular mass between 110 and 140 kDa. They have so far been found exclusively in the genera *Latrodectus* (black widows), *Steatoda* and *Achaearanea* (Theridiidae). Investigations into the venom of *Latrodectus* species concern primarily the North American L. *mactans,* the Eurasian *L. tredecimguttatus* and the Australian L. *hasselti*. In all cases, the venom consists of seven major proteins selectively toxic to three groups: (1) α-latrotoxin (α-LTX) which is vertebrate selective and has a molecular mass of 130 kDa; (2) five latroinsectotoxins (LIT), namely, α-, β-, γ-, δ-, and έ-LIT, which are selective for insects and have a molecular mass between 110 and 140 kDa; and (3) the 120 kDa α-latrocrustotoxin (α-LTX) which is selective for crustaceans (Grishin 1998). These masses correspond to 1,000–1,200 amino acid residues with a rather high level of homology of over 30 % residue identity (Vassilevski et al. 2009).

α-LTX has a high affinity to form dimers that aggregate into tetramers and then insert into the lipid membrane of nerve cells, thus forming a central channel. This pore acts as a nonselective cation channel allowing a massive influx of extracellular Ca^{2+} into the nerve, which leads to vesicular exocytosis (Rohou et al. 2007). In other words, this represents an exhaustive neurotransmitter release from a variety of nerves that depletes the synaptic vesicles, blocks the signal transmission and causes muscular paralysis. α-LTX causes secretion of all known neurotransmitter types and the effects of latroinsectotoxins and latrocrustotoxin are rather similar. The highly toxic effect of the *Latrodectus* toxins is enhanced by an 8 kDa peptide that is not toxic and cannot form membrane pores by itself, but augments the affinity of α-LTX to membranes (Kiyatkin et al. 1995).

14.7 Enzymes

A variety of enzymes have been identified in spider venoms from 49 species, belonging to 14 families (Kuhn-Nentwig et al. 2011). Most of these enzymes can be separated into two groups, one that cleaves polymers in the extracellular matrix, and the other one targeting phospholipids and related compounds in membranes. The overall purpose of a co-injection of enzymes with toxins into a prey's tissue is obvious: by destroying the barrier of extracellular matrix and cell membranes, the toxins can reach their targets faster. Additionally, the proteolytic activity of some of these enzymes may facilitate the subsequent preoral digestion.

Hyaluronidases, historically the first enzymes found in spider venoms, cleave the mucopolysaccharide hyaluronan (hyaluronic acid), a major constituent of the extracellular matrix. This facilitates the spread of toxins and further venomous compounds, and therefore it has been frequently termed "spreading factor" (e.g. Bettini 1978). Hyaluronidases differ slightly between spider species since their apparent molecular mass varies between 32 and 44 kDa. A second major group of enzymes targeting the extracellular matrix is collagenases. These are matrix metalloproteases, cleaving peptide bonds between proline and other amino acid residues in collagen, a key compound in the animal extracellular matrix.

Cell membranes of living organisms consist of a lipid bilayer that is mainly composed of phospholipids. Inner and outer surfaces differ in chemical composition with phosphatidylcholine, sphingomyelin and a variety of glycolipids determining the outer surface. These molecules are the targets of various hydrolase enzymes breaking down the phosphodiester bond (phosphodiesterase), degrading sphingomyelin (sphingomyelin phosphodiesterase or sphingomyelinase D) or hydrolysing phospholipids (phospholipases). Since sphingomyelinase D also hydrolyses lysoglycerophospholipids or lysophosphatic acid, it is now usually called phospholipase D. Phospholipase D became a famous case among spider venom enzymes because sicariids are the spider family relying to the highest degree on the activity of these enzymes when subduing a prey (Binford et al. 2009; see also Binford 2013). *Loxosceles* venom is very potent, and due to its enzymatic nature, it is the only venom which causes necrotic effects in humans.

It is obviously very efficient to support the effect of neurotoxic components by enzymes which facilitate spreading. So far, enzymes targeting the extracellular matrix or membrane compounds have only been found in a few spider species, but we assume that it is much more common in spider venoms.

14.8 Evolutionary Strategies of Spider Venom

Spiders possess venom glands to produce venom that they use primarily to paralyse and /or kill their prey items. The main strategy to reach these goals seems to rely mainly on mini-proteins. Most spiders investigated so far possess a variety of mini-proteins in their venom glands. It is, however, difficult to decide whether mini-proteins or low molecular mass compounds represent the plesiomorphic repertoire of all spiders, since we have no information on the venom composition of the most plesiomorphic spider group (Mesothelae).

Unfortunately, we still have only very limited knowledge on spider venoms, derived from some 0.4 % of all known spider species, belonging to less than one third of the known families. This research bias was driven by a selection towards large and easily accessible spiders and species of medical importance, while other important and large groups were completely neglected. At this stage, generalisations are risky but some general trends are nevertheless observable.

Since the venoms from many mygalomorph species contain many mini-proteins, a variety of low molecular mass compounds and enzymes, we assume that early in the evolution of spider venoms, this turned out to be a very well-functioning and reliable mixture. Nevertheless, numerous modifications, changes and replacements have occurred, which we summarise here as three main lines of evolutionary change:

1. Mini-proteins are permanently modified. Structurally, mini-proteins can be considered superficially as similar peptides, most of them in the range of 3–8 kDa. However, the variation of only a few features such as molecular target (functional diversity), sequence variations, number of cysteines or size (structural diversity) yields a nearly endless number of different toxins. In many spider species, dozens of mini-proteins occur and the record so far consists of 166 mini-proteins known from one peptidomic study on the theraphosid *Haplopelma hainanum* (Tang et al. 2010). Quite often, the difference between the structures of two peptides consists only in an exchange of one amino acid, which can result in drastic changes of potency and/or selectivity.
2. Supporting mechanisms or synergisms between mini-proteins and other components, namely, low molecular mass compounds, linear peptides and enzymes, frequently increase their effect. This can be achieved either by directly attacking neuronal or muscular cells or by destroying their membranes or target tissue so that mini-proteins have easier access to their targets. The first may be achieved by a variety of neurotransmitters and further neuroactive compounds, the latter by linear peptides or enzymes. This enables the spider to inject less venom to get the same result, suggesting that energetic reasons are the main driver (Wullschleger et al. 2004, 2005).
3. If mini-proteins can be substituted by something better, it is obvious that they should be replaced, partially or completely. There are at least three examples of spider groups where mini-proteins have been more or less completely replaced by "better" compounds. The first is given by Araneidae and Nephilidae, relying mainly or exclusively on amino acids containing acylpolyamines. A second example concerns Theridiidae which replaced mini-proteins by large neurotoxic proteins. Thirdly, the venom of sicariids contains predominantly a highly effective phospholipase D. Also the increasing appearance of linear peptides in ctenids and some related families (Lycosoidea) can be interpreted as an approach of substituting venom components by such cytolytic peptides.

These changes and modifications are part of the evolutionary arms race to optimize the venom composition permanently, but they also pose the question as to which kind of venom may be "better". Since the amount of venom needed to subdue a given prey item is a crucial factor, comparative biotests may give an answer. Tests with 14 spider species from 12 families showed that the LD_{50} for the cockroach *Blatta orientalis* varies between 0.3 and 542 ng dry venom/mg insects, i.e. over more than three orders of magnitude. The best results, however, were achieved from venoms with very different venom strategies (the theraphosid *Avicularia metallica* with mini-proteins, the sicariid *Loxosceles deserta* with phospholipase D and the theridiid *Latrodectus hesperus* with large neurotoxic proteins) and varied in a much smaller range between 0.9 and 10 ng dry venom/mg insect (Friedel and Nentwig 1989; Nentwig et al. 1992). This indicates that different main classes of spider venom components can be more or less equally toxic and successful, but the variation within venom components or spider families is much larger.

14.9 Conclusions

Spider venoms contain a huge diversity of compounds that can be classified into six major categories: low molecular mass compounds, acylpolyamines, linear cationic peptides, cysteine-rich mini-proteins, large neurotoxic proteins and enzymes. The venoms from many mygalomorph species, containing several mini-proteins, a variety of low molecular mass compounds and enzymes, represent a very well-functioning and reliable mixture and may be seen as the basic form of spider venoms. Nevertheless, numerous modifications, changes and replacements have occurred. At least three spider groups developed very different venom compositions: Araneidae and Nephilidae rely mainly on amino acids containing acylpolyamines, Theridiidae have developed large neurotoxic proteins and Sicariidae venoms predominantly contain phospholipase D.

References

Bettini S (1978) Arthropod venoms. Handbook of experimental pharmacology. Springer, Berlin

Binford G (2013) The evolution of a toxic enzyme in Sicariid spiders. In: Nentwig W (ed) Spider ecophysiology. Springer, Heidelberg (this volume)

Binford GJ, Bodner MR, Cordes MH, Baldwin KL, Rynerson MR, Burns SN, Zobel-Thropp PA (2009) Molecular evolution, functional variation, and proposed nomenclature of the gene family that includes sphingomyelinase D in sicariid spider venoms. Mol Biol Evol 26:547–566

Chesnov S, Bigler L, Hesse M (2001) The acylpolyamines from the venom of the spider *Agelenopsis aperta*. Helv Chim Acta 84:2178–2197

Friedel T, Nentwig W (1989) Immobilizing and lethal effects of spider venoms on the cockroach and the common mealbeetle. Toxicon 27:305–316

Grishin EV (1998) Black widow spider toxins: the present and the future. Toxicon 36:1693–1701

Herzig V, King G (2013) The neurotoxic mode of action of venoms from the spider family Theraphosidae. In: Nentwig W (ed) Spider ecophysiology. Springer, Heidelberg (this volume)

Itagaki Y, Nakajima T (2000) Acylpolyamines: mass spectrometric analytical methods for Araneidae spider acylpolyamines. J Toxicol Toxin Rev 19:23–52

Itagaki Y, Fujita T, Naoki H, Yasuhara T, Andriantsiferana M, Nakajima T (1997) Detection of new spider toxins from a *Nephilengys borbonica* venom gland using on-line m-column HPLC continuous flow (FRIT) FAB LC/MS and MS/MS. Nat Toxins 5:1–13

Kiyatkin NI, Kulikovskaya IM, Grishin EV, Beadle DJ, King LA (1995) Functional characterization of black widow spider neurotoxins synthesised in insect cells. Eur J Biochem 230:854–859

Kozlov SA, Vassilevski AA, Feofanov AV, Surovoy AY, Karpunin DV, Grishin EV (2006) Latarcins, antimicrobial and cytolytic peptides from the venom of the spider *Lachesana tarabaevi* (Zodariidae) that exemplify biomolecular diversity. J Biol Chem 281:20983–20992

Kuhn-Nentwig L (2003) Antimicrobial and cytolytic peptides of venomous arthropods. Cell Mol Life Sci 60:2651–2668

Kuhn-Nentwig L (2009) Cytolytic and antimicrobial peptides in the venom of scorpions and spiders. In: de Lima ME (ed) Animal toxins: state of the art. Editora UFMG, Belo Horizonte

Kuhn-Nentwig L, Nentwig W (2013) The cytotoxic mode of action of the venom of *Cupiennius salei* (Ctenidae). In: Nentwig W (ed) Spider ecophysiology. Springer, Heidelberg (this volume)

Kuhn-Nentwig L, Schaller J, Nentwig W (1994) Purification of toxic peptides and the amino acid sequence of CSTX-1 from the multicomponent venom of *Cupiennius salei* (Araneae: Ctenidae). Toxicon 32:287–302

Kuhn-Nentwig L, Stöcklin R, Nentwig W (2011) Venom composition and strategies in spiders: is everything possible? Adv Insect Physiol 40:1–86

McCormick KD, Meinwald J (1993) Neurotoxic acylpolyamines from spider venoms. J Chem Ecol 19:2411–2451

Nentwig W, Friedel T, Manhart C (1992) Comparative investigations on the effect of the venom of 18 spider species onto the cockroach *Blatta orientalis* (Blattodea). Zool Jb Physiol 96:279–290

Norton RS, Pallaghy PK (1998) The cystine knot structure of ion channel toxins and related polypeptides. Toxicon 36:1573–1583

Odell GV, Fenton AW, Ownby CL, Doss MP, Schmidt JO (1999) The role of venom citrate. Toxicon 37:407–409

Palma MS, Itagaki Y, Fujita T, Naoki H, Nakajima T (1998) Structural characterization of a new acylpolyaminetoxin from the venom of Brazilian garden spider *Nephilengys cruentata*. Toxicon 36:485–493

Rash LD, Hodgson WC (2002) Pharmacology and biochemistry of spider venoms. Toxicon 40: 225–254

Rohou A, Nield J, Ushkaryov YA (2007) Insecticidal toxins from black widow spider venom. Toxicon 49:531–549

Schäfer A, Benz H, Fiedler W, Guggisberg A, Bienz S, Hesse M (1994) Polyamine toxins from spiders and wasps. In: Cordell GA, Brossi A (eds) The alkaloids, chemistry and pharmacology. Academic, San Diego

Schroeder FC, Taggi AE, Gronquist M, Malik RU, Grant JB, Eisner T, Meinwald J (2008) NMR-spectroscopic screening of spider venom reveals sulfated nucleosides as major components for the brown recluse and related species. Proc Natl Acad Sci USA 105:14283–14287

Suter RB, Stratton GE (2013) Predation by spitting spiders—elaborate venom gland, intricate delivery system. In: Nentwig W (ed) Spider ecophysiology. Springer, Heidelberg (this volume)

Taggi AE, Meinwald J, Schroeder FC (2004) A new approach to natural products discovery exemplified by the identification of sulfated nucleosides in spider venom. J Am Chem Soc 126: 10364–10369

Tang X, Zhang Y, Hu W, Xu D, Tao H, Yang X, Li Y, Jiang L, Liang S (2010) Molecular diversification of peptide toxins from the tarantula *Haplopelma hainanum* (*Ornithoctonus hainana*) venom based on transcriptomic, peptidomic, and genomic analyses. J Proteome Res 9:2550–2564

Tzouros M, Chesnov S, Bienz S, Hesse M, Bigler L (2005) New linear polyamine derivatives in spider venoms. Toxicon 46:350–354

Vassilevski AA, Kozlov SA, Samsonova OV, Egorova NS, Karpunin DV, Pluzhnikov KA, Feofanov AV, Grishin EV (2008) Cyto-insectotoxins, a novel class of cytolytic and insecticidal peptides from spider venom. Biochem J 411:687–696

Vassilevski AA, Kozlov SA, Grishin EV (2009) Molecular diversity of spider venom. Biochemistry (Moscow) 74:1505–1534

Wullschleger B, Kuhn-Nentwig L, Tromp J, Kämpfer U, Schaller J, Schürch S, Nentwig W (2004) CSTX-13, a highly synergistically acting two-chain neurotoxic enhancer in the venom of the spider *Cupiennius salei* (Ctenidae). Proc Natl Acad Sci USA 101:11251–11256

Wullschleger B, Nentwig W, Kuhn-Nentwig L (2005) Spider venom: enhancement of venom efficacy mediated by different synergistic strategies in *Cupiennius salei*. J Exp Biol 208:2115–2121

Zhang Y, Chen J, Tang X, Wang F, Jiang L, Xiong X, Wang M, Rong M, Liu Z, Liang S (2010) Transcriptome analysis of the venom glands of the Chinese wolf spider *Lycosa singoriensis*. Zoology (Jena) 113:10–18

Chapter 15
The Neurotoxic Mode of Action of Venoms from the Spider Family Theraphosidae

Volker Herzig and Glenn F. King

15.1 Introduction

The family Theraphosidae includes 939 species from 121 genera (Platnick 2012). Theraphosid spiders are typically large (body length up to 10 cm; leg span up to 30 cm), and this group includes the largest extant spider, *Theraphosa blondi*. They primarily hunt on or near the ground, although there are a few arboreal species. These spiders are commonly known as "tarantulas," which is somewhat misleading because the "real" tarantula (*Lycosa tarantula*, an Italian wolf spider) was described in 1758 by Linnaeus, the inventor of the binominal nomenclature for taxonomic classification. We are not aware of any mention of the term "tarantula" before 1758, and thus we suggest using the term "theraphosid" for all members of the family Theraphosidae, in line with the naming of their venom proteins as theraphotoxins (King et al. 2008b). "Bird-eating spiders" are another common misnomer used for theraphosids, since they mostly do not prey on birds. This term originates from a copper engraving done by the German artist Maria Sibylla Merian when she visited Surinam in 1699–1701, which shows a theraphosid spider eating a hummingbird.

Theraphosids belong to the infraorder of spiders known as mygalomorphs in which the fangs move in an up-and-down fashion and the venom glands are located in the basal segment of the chelicerae. This contrasts with modern araneomorph spiders in which the fangs move laterally against one another and the venom glands extend into the prosoma (Kuhn-Nentwig et al. 2011). Although theraphosids are found on all continents except Antarctica, they are most abundant in subtropical and tropical climates. They do not rely on silk for capturing prey. Instead, they wait in their burrow entrance (or in the close vicinity) for passing prey, which is quickly grasped with the fangs and incapacitated by venom injection. Theraphosids are generalist predators, and larger species will hunt not just invertebrates but also

V. Herzig • G.F. King (✉)
Institute for Molecular Bioscience, The University of Queensland, St. Lucia, QLD, Australia
e-mail: v.herzig@imb.uq.edu.au; glenn.king@imb.uq.edu.au

W. Nentwig (ed.), *Spider Ecophysiology*,
DOI 10.1007/978-3-642-33989-9_15, © Springer-Verlag Berlin Heidelberg 2013

small vertebrate prey including amphibians, reptiles, birds, and mammals. Thus, theraphosid venoms contain toxins active against a wide range of vertebrate and invertebrate species, which has made theraphosid venom peptides valuable as pharmacological tools and as leads for the development of new drugs (see Stoecklin, this volume; Saez et al. 2010) and bioinsecticides (Windley et al. 2012).

As a group, the venoms of theraphosids have been more widely studied than the venoms of any other family of spiders. Their large size, high venom yields, and long lifespan (>20 years) enable them to be kept for extended periods under laboratory conditions in order to collect venom for research purposes. Furthermore, many theraphosids are widely distributed in the pet trade, which facilitates sourcing of specimens for research.

15.2 Theraphosid Bites

Despite their size and their fearsome reputation, there has not been a single documented fatality caused by theraphosid bites (see also Nentwig and Kuhn-Nentwig, this volume). Studies from Brazil (Lucas et al. 1994) and Australia (Isbister et al. 2003) have shown that theraphosid bites only induce mild localized symptoms such as pain, bleeding from the bite site, and, in rare cases, mild systemic effects such as nausea and vomiting. Theraphosid bites are therefore unlikely to pose a serious health threat, despite some Australian species exhibiting lethal effects in canines (Isbister et al. 2003). There may be exceptions as some African and Asian theraphosid species, in particular the genus *Poecilotheria*, have been involved in more severe, systemic envenomations in keepers of pet spiders (Ahmed et al. 2009). Symptoms after *Poecilotheria* envenomation include severe pain, vomiting, muscle cramps throughout the entire body that can occur up to several weeks after the bite, breathing difficulties, increased heart rate, and even unconsciousness. The severity of these bites is supported by our own observation that spiders from the genus *Poecilotheria* have a very high venom yield compared with other theraphosids of similar size (Herzig 2010).

15.3 Theraphosid Venom Composition

Similar to venoms from other spiders, the composition of theraphosid venoms is extremely complex (Fig. 15.1) due to the presence of a variety of low molecular compounds, peptides, and enzymes (Escoubas and Rash 2004; Kuhn-Nentwig et al. 2011), as outlined in the following sections.

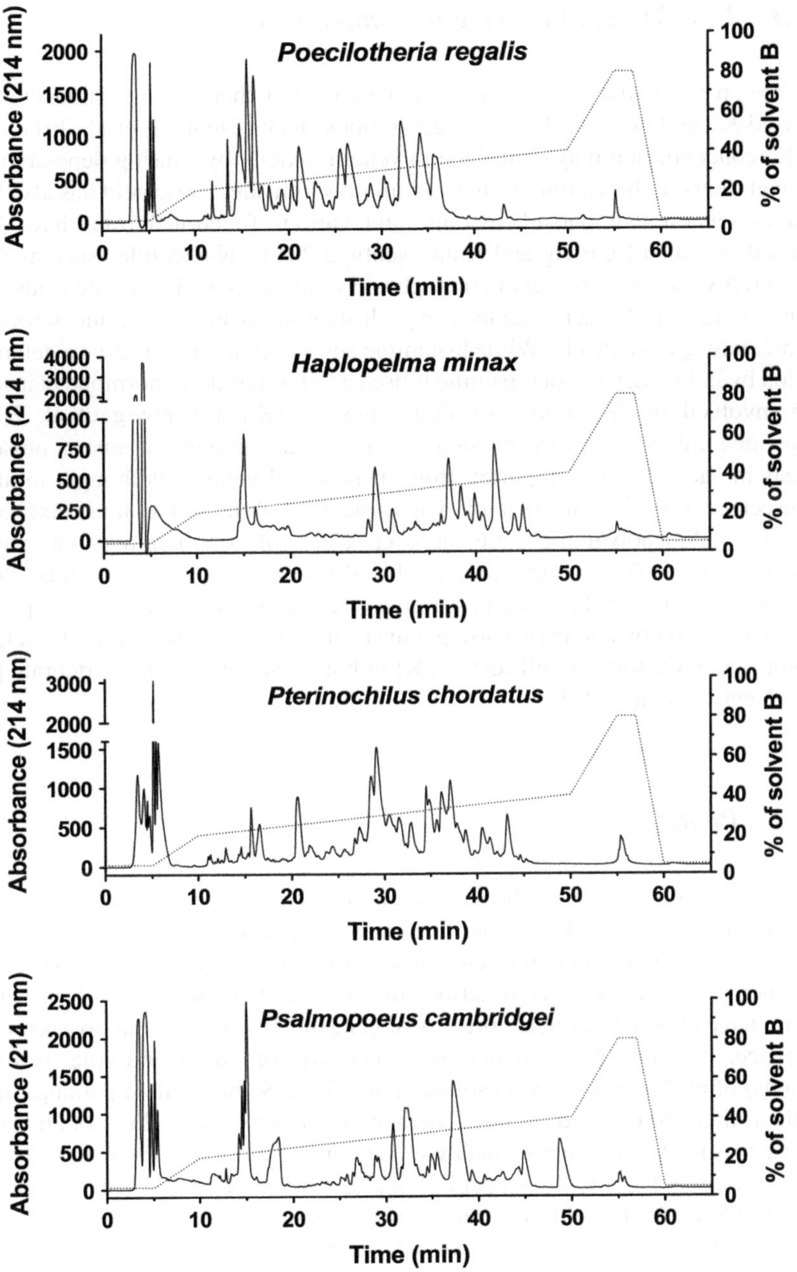

Fig. 15.1 Reverse-phase HPLC chromatograms showing the relative complexity of selected theraphosid venoms. Dried venom (1.0 mg) from four different female theraphosids was dissolved in solvent A [0.1 % formic acid (FA) in water], loaded onto a C18 analytical column (Phenomenex; 5 μm, 4.6 mm × 250 mm), and eluted with a gradient of solvent A and solvent B (90 % acetonitrile and 0.1 % FA in water) (*dashed line*, right ordinate axis). Peak elution was

15.3.1 Low Molecular Weight Compounds

It has been reported that, in comparison to hemolymph, theraphosid venom contains increased K^+ and decreased Na^+ concentrations (Kuhn-Nentwig et al. 2011). The high K^+ concentration may contribute to venom toxicity by causing depolarization of axonal fibers in the vicinity of the venom injection site. Citric acid has also been found in several theraphosid venoms, and various functional roles have been discussed (see also Nentwig and Kuhn-Nentwig 2013). Nucleotides such as ATP, ADP, and AMP are also present in theraphosid venoms, as well as amino acids (e.g., glutamic acid) and biogenic amines (e.g., histamine, octopamine, and serotonin) (Kuhn-Nentwig et al. 2011). While histamine and serotonin might serve a defensive function by inducing pain, octopamine is used as a fight or flight hormone in insects, being involved in the control of flight muscles (Kuhn-Nentwig et al. 2011). Octopamine might therefore be used to affect the activity of insect muscles, whereas the acetylcholine reported from theraphosid venoms is thought to affect the insect central nervous system, where acetylcholine is the main excitatory transmitter. Acylpolyamines are another class of small molecular weight compounds (<1,000 Da) found in theraphosid venoms (Escoubas and Rash 2004; Kuhn-Nentwig et al. 2011). Acylpolyamines are composed of a polyamine plus an aromatic end moiety, and they cause paralysis primarily by modulating the activity of ionotropic receptors (AChR and GluR) in both invertebrate and vertebrate prey (Kuhn-Nentwig et al. 2011).

15.3.2 Peptides

The main component of theraphosid venoms are disulfide-rich peptides, also called mini-proteins, most of which have masses ranging from 2.6 to 7.2 kDa (27–65 amino acid residues) although a few are larger (Escoubas and Rash 2004); almost 1,000 such peptides have been described to date (Herzig et al. 2011). These venom peptides have evolved through a process of gene duplication and massive sequence divergence, so that each theraphosid typically produces multiple isoforms (paralogs) of each peptide toxin (Sollod et al. 2005). Some of these paralogs differ by only a single amino acid residue. Theraphotoxins account for 25 % of all venom proteins in the ArachnoServer database (Herzig et al. 2011) even though they represent only ~2 % of all spiders. They are overrepresented due the factors outlined in Sect. 15.1, including their large size and ready availability through the pet trade. In the following sections, we will explain the reason for the extraordinary stability of most theraphosid venom peptides (Sect. 15.4) and provide some insights into their mode of action (Sect. 15.5).

Fig. 15.1 (continued) monitored using absorbance at 214 nm (*solid line*, left ordinate axis.) The venoms were sourced from Asian (*upper two panels*), African (*third panel*), and American (*bottom panel*) theraphosid species

15.3.3 Enzymes

Enzymes reported from theraphosid venoms include hyaluronidase, phosphodiesterase, and proteases (see also Nentwig and Kuhn-Nentwig 2013). No sequence information has been obtained for any of these proteins, and hence their evolutionary relationship to enzymes from other spiders and other venomous animals remains unclear (Kuhn-Nentwig et al. 2011). It has been speculated that these enzymes serve the purpose of destroying the extracellular matrix and cell membranes in order to facilitate the anatomical spread of smaller peptide toxins in envenomated prey or predators. A possible predigestive role for these enzymes has also been discussed (Kuhn-Nentwig et al. 2011), but most preoral digestion is likely to be carried out by enzymes in the digestive juices that are regurgitated onto the prey.

15.4 Three-Dimensional Structure of Theraphotoxins

3D structures have been determined for only 44 spider-venom peptides, and half of these are theraphotoxins. 20 of the 22 theraphotoxin structures that have been solved contain an inhibitor cystine knot (ICK) motif (Fig. 15.2a); one comprises a Kunitz-type fold (Fig. 15.2b); and one has an unusual structure and disulfide configuration (Fig. 15.2c). All of these theraphotoxins contain three disulfide bonds, and only the single theraphotoxin with a Kunitz-type fold is larger than 4.3 kDa. Thus, the structure of theraphotoxins in the mass range 4.3–13 kDa, as well as those with more than three disulfide bonds, remains largely unexplored. Thus, additional theraphotoxin 3D scaffolds are certain to be discovered in future.

The dominance of ICK motifs in the solved structures of theraphotoxins is reflective of the widespread dominance of this motif in spider venoms. ICK toxins, also known as knottins (Gracy and Chiche 2011), represent a basal recruitment into the venoms of spiders. Ancestrally recruited ICK genes have been extensively duplicated and diversified to the point where they dominate the peptidome of most spider venoms. There are two properties that account for the dominance of this structural motif: stability and sequence plasticity. The ICK motif is defined as an antiparallel β-sheet stabilized by a cystine knot. In spider toxins, the β-sheet typically comprises only two β-strands (Fig. 15.2a) although a third N-terminal strand is sometimes present. The cystine knot consists of a ring formed by two disulfide bridges and the intervening sections of peptide backbone, with a third disulfide bond piercing the ring to create a pseudo-knot. This "knot" provides these small peptides with exceptional chemical, thermal, and biological stability; they are resistant to extremes of pH, organic solvents, high temperatures, and proteases. Moreover, the marked insensitivity of the ICK scaffold to changes in intercystine residues has enabled spiders to develop diverse pharmacologies using subtle variations on the same disulfide framework (Saez et al. 2010; Gracy and Chiche 2011). The stability and pharmacological plasticity conferred by the ICK motif has

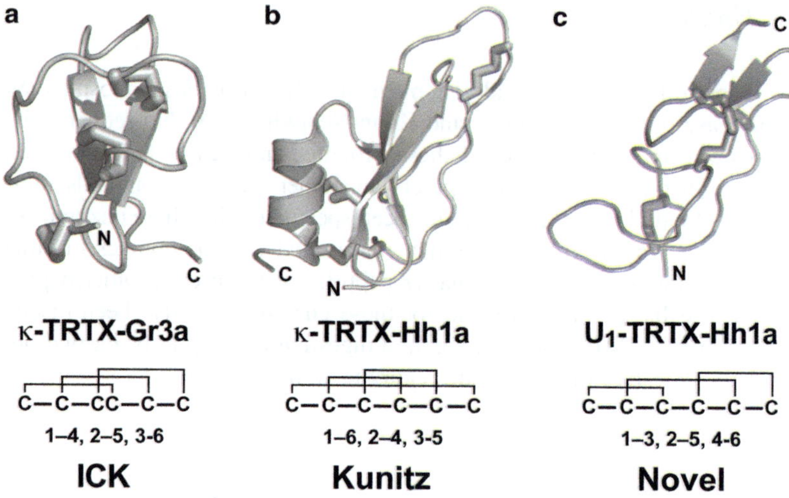

Fig. 15.2 Examples of the three structural folds reported thus far for theraphotoxins: (**a**) κ-TRTX-Gr3a from *Grammostola rosea* (VSTX1; PDB coordinate file 1S6X) contains an inhibitor cystine knot (ICK) motif; (**b**) κ-TRTX-Hh1a from *Haplopelma schmidti* (huwentoxin-XI; PDB coordinate file 2JOT) has a Kunitz-type fold; (**c**) U_1-TRTX-Hh1a also from *Haplopelma schmidti* (huwentoxin-II; PDB coordinate file 1I25) has a novel disulfide connectivity and 3D scaffold. The N- and C-termini are labeled on each structure, and disulfide bonds are shown as thick tubes. Disulfide connectivities are shown below each structure

resulted in many theraphotoxins becoming leads for the development of new therapeutics (Saez et al. 2010; Klint et al. 2012; Oldrati et al. 2013) and bioinsecticides (Windley et al. 2012).

15.5 Mechanism of Action of Theraphosid Peptide Toxins

Theraphosid toxins act on a range of molecular targets, with the majority having a neurotoxic mode of action (Fig. 15.3). The primary molecular targets are ion channels, including voltage-gated potassium, calcium and sodium channels, mechanosensitive channels, transient receptor potential channels, and acid-sensing ion channels. Some theraphosid toxins were identified as lectins or protease inhibitors. However, the molecular target has not been determined for ~60 % of all described theraphotoxins (Herzig et al. 2011), and therefore it is quite likely that the range of targets identified for these toxins will increase in future as a wider range of in vitro assays are developed. In Sects. 15.5.1–15.5.8, we provide an overview of the major molecular targets of theraphotoxins before we discuss in Sect. 15.6 how different venom components interact to ensure rapid and permanent immobilization of prey.

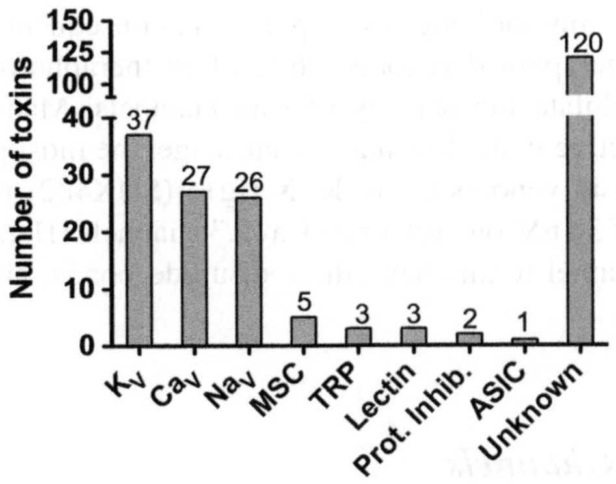

Fig. 15.3 Summary of molecular targets for the 200 theraphosid peptide toxins listed in the ArachnoServer database (www.arachnoserver.org; accessed February 21, 2012) (Herzig et al. 2011). The majority of toxins with known molecular target act on voltage-gated ion channels (K_V, Ca_V, and Na_V), followed by mechanosensitive ion channels (MSCs) and transient receptor potential (TRP) channels. Lectins and serine protease inhibitors are also found in theraphosid venoms as well as toxins targeting acid-sensing ion channels (ASICs). Note that some toxins have more than one molecular target and therefore the cumulative number of toxins in the histogram is >200

15.5.1 Potassium Channels

The functions of voltage-gated potassium (K_V) channels are almost as diverse as the myriad of different K_V channel subtypes. The involvement of K_V channels in neurotransmitter release by inducing the repolarization phase of the action potential in vertebrates and invertebrates makes them a primary target for spider toxins. Not surprisingly, more than one third of all theraphotoxins with a known target act on K_V channels, typically with half maximal inhibitory concentration (IC_{50}) values in the nanomolar range. These toxins range in size from 29 to 55 residues, with all of them containing three disulfide bonds (Herzig et al. 2011). The most potent of these toxins is κ-TRTX-Sc1a (stromatoxin-1) from *Stromatopelma calceatum* with an IC_{50} of 1.2 nM on vertebrate K_V4.2 channels (Herzig et al. 2011).

15.5.2 Calcium Channels

Voltage-gated calcium (Ca_V) channels are key signal transducers that convert cell membrane depolarization into an influx of extracellular calcium ions, thereby mediating a wide range of critical intracellular processes including muscle contraction and neurotransmitter release. Whereas vertebrates express ten different subtypes with some functional overlap, insects express only three subtypes with apparently little or no functional redundancy as knockout of the corresponding channel genes is embryonic lethal (King 2007). Thus, Ca_V channels are clearly excellent targets for immobilizing insect prey, and consequently, this pharmacology is common in theraphosid venoms; 26 % of all theraphotoxins with a known mode of action modulate the activity of Ca_V channels. Many theraphosid Ca_V channel toxins are active in the low nanomolar range; the most potent Ca_V channel toxin from theraphosid venoms is ω-TRTX-Hg1a (SNX482) from *Hysterocrates gigas* with an IC_{50} of 30 nM on vertebrate $Ca_V2.3$ channels (Herzig et al. 2011). All theraphosid Ca_V channel toxins have three disulfide bonds, and their size varies from 28–42 residues.

15.5.3 Sodium Channels

Voltage-gated sodium (Na_V) channels provide the current pathway for the rapid depolarization of excitable cells that is required to initiate action potentials. Nine different subtypes of mammalian Na_V channels have been described, whereas only one subtype is present in invertebrates with a high level of sequence conservation (87–98 % identity) across a wide range of insect orders (King et al. 2008a). The presence of a single Na_V channel subtype in insects makes it an ideal target for spider toxins that aim to incapacitate invertebrate prey as exemplified by the fact that many chemical insecticides target this channel. It is therefore no surprise that a quarter of all theraphotoxins with known mode of action affect Na_V channels. Theraphosid Na_V channel toxins vary in size from 29 to 39 residues, all with three disulfide bonds, and many of them are active in the low nanomolar range; the most potent is β/ω-TRTX-Tp2a (protoxin-2) from *Thrixopelma pruriens* with an IC_{50} of 0.3 nM on human $Na_V1.7$ channels (Klint et al. 2012). All Na_V channel toxins from spider venoms are allosteric modifiers or "gating modifiers"; they do not occlude the channel pore but instead bind to an allosteric site that induces a conformational change in the channel that alters the equilibrium between the open, closed, and inactivated states (Klint et al. 2012). These toxins can induce excitatory symptoms in prey, ultimately leading to paralysis (e.g., by inhibiting channel inactivation like the δ-theraphotoxins), or they can induce a flaccid paralysis (e.g., by causing a hyperpolarizing shift in the voltage dependence of channel activation like the β-theraphotoxins).

15.5.4 Mechanosensitive Channels

Mechanosensitive channels (MSCs) have been implicated in a variety of physiological processes, and they are thought to play a key role in modification of the electrical and contractile activity of muscle tissue. Five theraphotoxins, each containing 31–35 residues and three disulfide bonds, are currently known to act on MSCs (Bowman et al. 2007). The most potent of these mechanotoxins is M-TRTX-Gr1b (GsMTx4) from *Grammostola rosea* with a K_d of 240 nM against stretch-activated channels in rat astrocytes. However, it remains unclear whether the activity of these toxins on MSCs contributes directly to incapacitation of prey or predators, or whether other channels are the primary target of these toxins. The fact that one of these toxins, κ-TRTX-Ps1b (phrixotoxin-2) from *Paraphysa scrofa*, blocks $K_V4.2$ channels with an IC_{50} of 34 nM compared with its 176-fold higher K_d of 6 μM on MSCs in rat astrocytes supports this hypothesis.

15.5.5 Transient Receptor Potential Channels

Three toxins from the venom of the Trinidad chevron tarantula *Psalmopoeus cambridgei*, each containing 34–35 residues and three disulfide bonds, are weak agonists (ED_{50} of 320 nM–11.9 μM) of a subtype of transient receptor potential (TRP) channel known as TRPV1 (Siemens et al. 2006). TRPV1 is involved in body temperature regulation and the nociceptive response to scalding heat. TRPV1 is also the receptor for capsaicin, the active component of chilli peppers, and activation of the receptor in vertebrates produces a burning pain. Thus, the TRPV1 agonists from theraphosid venoms may play a defensive role against vertebrate predators. However, these toxins are also active at K_V channels, and it is possible that vertebrate TRPV1 channels are an incidental target of the toxins and that their primary target in terms of the ecology of theraphosids is an entirely different channel that they target with higher potency.

15.5.6 Acid-Sensing Ion Channels

π-TRTX-Pc1a (PcTx1), a 40-residue toxin with three disulfide bonds from the venom of *Psalmopoeus cambridgei*, is the most potent and selective blocker known of acid-sensing ion channel (ASIC) 1a, with an IC_{50} of 0.9 nM on rat ASIC1a (Saez et al. 2010). Block of ASIC1a is associated with an anti-nociceptive effect in mammals (Saez et al. 2010) which would counteract the putative defensive role of theraphosid toxins that activate TRPV1. Moreover, since ASICs are restricted to chordates (Saez et al. 2010), such a toxin would be ineffective against non-vertebrate prey and predators. Thus, it is possible that π-TRTX-Pc1a might have other yet to be identified molecular targets. Another possibility is that this toxin plays a role in deadening the sensory circuits of prey (discussed in Sect. 15.6).

15.5.7 Other Targets

Two isoforms of a 55-residue serine protease inhibitor have been isolated from the venom of the Chinese species *Haplopelma schmidti* (previously *H. huwenum*)

(Liang 2004). Although these toxins contain three disulfide bonds, they deviate from the typical ICK scaffold found in most theraphosid toxins and instead contain a Kunitz domain (Fig. 15.2b). Inhibiting serine proteases could help prevent inactivation of other venom toxins by proteases in the prey or predator. Three lectins (32–34 residues, 3 disulfide bonds) have also been isolated from the venom of the same theraphosid. However, their role in prey incapacitation remains unclear, as they have very low toxicity in both vertebrates and insects (Liang 2004).

15.5.8 Unknown Targets

The molecular target remains unknown for ~60 % of all reported theraphosid venom peptides. The main reason is that many toxin sequences have been derived from venom-gland transcriptomes and there have been no studies of their activity or molecular target. Many of these toxins are likely to act on one of the molecular targets mentioned in Sects. 15.5.1–15.5.7. However, the large variation in size (27–116 residues) and in the number of disulfide bonds (1–5) in these toxins suggests that some of them might act on new molecular targets that have not yet been reported for theraphotoxins. Interesting examples of theraphosid toxins with unknown molecular target include the peptides U_1-TRTX-Pc1a and U_2-TRTX-Pc1a, from the venom of *Psalmopoeus cambridgei*, which are reported to block development of the intra-erythrocyte stage of the malaria parasite (Saez et al. 2010).

15.6 Toxin Cabals Maximize the Ability of Venom to Incapacitate Prey

The concept of toxin cabals was first introduced by Olivera to explain how groups of venom peptides from marine cone snails could act synergistically to achieve the same physiological end point in targeted prey (Olivera and Cruz 2001). For example, the "lightning-strike cabal" causes immediate tetanic immobilization of fish prey; it requires at least one δ-toxin that inhibits inactivation of Na_V channels plus a second κ-toxin that blocks K_V channels. The simultaneous block of K_V channels and sustained opening of Na_V channels cause massive depolarization of axonal fibers near the venom injection site, which leads to tetanic paralysis. Both δ- and κ-theraphotoxins are present in the venom of the Chinese theraphosid *Chilobrachys guangxiensis* (previously *C. jingzhao*), indicating that at least some theraphosid venoms might contain lightning-strike cabals.

Some fish-hunting cone snails also contain a set of individually paralytic peptide toxins that together form a "motor cabal" that potently abolishes neuromuscular transmission (Olivera and Cruz 2001); this cabal sets in after the lightning-strike cabal. Numerous μ-, β-, and ω-theraphotoxins could be classified within the motor cabal. In contrast to the lightning-strike and motor cabals, some fish-hunting cone snails appear to employ a "nirvana cabal" to deaden the sensory circuitry and place prey in a relaxed and sedated state. This nirvana cabal is achieved by combining an antagonist of NMDA receptors (a specific type of ionotropic glutamate receptors) and a potent analgesic. Free glutamate present in theraphosid venoms will activate ionotropic glutamate receptors in the prey immediately after venom injection, which will allow acylpolyamines present in the venom to access these channels where they act as open-channel blockers. Theraphotoxins with analgesic effects have also been reported, such as β/ω-TRTX-Tp2a (protoxin II) and π-TRTX-Pc1a which target $Na_V1.7$ and ASIC1a channels, respectively (Saez et al. 2010). Thus, a nirvana-type cabal might be present in some theraphosid venoms where its role might be to make large and potentially dangerous prey easier to handle. The analogies between cone snail and theraphosid venoms both at the level of individual toxins and their groupings into cabals indicate that they have independently evolved similar strategies for immobilizing prey.

Other components in the venom serve to maximize the potency of the disulfide-rich peptide toxins that act on the neuronal circuitry of prey and predators. For example, hyaluronidase most likely facilitates the anatomical spread of these toxins in vertebrates, while serine protease inhibitors presumably serve to protect the disulfide-rich toxins from proteolytic degradation by prey/predator proteases. Thus, while the biochemical composition of a theraphosid venom might seem inexplicably and perhaps unnecessarily complex at first, a closer examination reveals that the venom is an extremely fine-tuned system of different components, sometimes with a seemingly contradictory mode of action, that act together in a synergistic, time-dependent manner to maximize the overall effect of the venom in the prey. Moreover, in addition to venom components that are responsible for incapacitating prey, other venom components that induce pain, such as histamine, serotonin, and TRPV1 agonists, are likely to have a mainly defensive purpose.

15.7 Conclusions

The majority of the 939 extant theraphosid spiders are large, generalist predators of both invertebrate and vertebrate prey. Their size and popularity in the pet trade facilitates obtaining large amounts of a diverse range of venoms for research, explaining why one quarter of all peptide toxins currently described in the ArachnoServer database originate from theraphosid venoms, despite accounting for only 2 % of the taxonomic diversity of spiders. Theraphosid venoms are composed of a complex cocktail of low molecular weight compounds, peptides, and enzymes. The main venom components are peptide toxins in the range of 2.6–7.2 kDa with a neurotoxic mode of action. Most theraphotoxins contain an inhibitor cystine knot motif, which endows them with exceptional chemical and thermal stability as well as resistance to proteases. By acting on ion channels in the nervous system, such as K_V, Ca_V, or Na_V channels, these peptides induce paralysis and/or death of prey. Other toxins in their venoms inhibit proteolytic degradation, while venom enzymes aid in the spread of toxins in prey and predators. Besides toxins that are involved in prey incapacitation, theraphosid venoms also contain components that induce pain in vertebrates, making them useful for predator deterrence.

Acknowledgements We acknowledge financial support from the Australian Research Council (Discovery Grant DP1095728) and thank Pierre-Alain Chaumeil from the Queensland Facility for Advanced Bioinformatics for extracting data from ArachnoServer.

References

Ahmed N, Pinkham M, Warrell DA (2009) Symptom in search of a toxin: muscle spasms following bites by Old World tarantula spiders (*Lampropelma nigerrimum*, *Pterinochilus murinus*, *Poecilotheria regalis*) with review. Q J Med 102:851–857

Bowman CL, Gottlieb PA, Suchyna TM, Murphy YK, Sachs F (2007) Mechanosensitive ion channels and the peptide inhibitor GsMTx-4: history, properties, mechanisms and pharmacology. Toxicon 49:249–270

Escoubas P, Rash L (2004) Tarantulas: eight-legged pharmacists and combinatorial chemists. Toxicon 43:555–574

Gracy J, Chiche L (2011) Structure and modeling of knottins, a promising molecular scaffold for drug discovery. Curr Pharm Des 17:4337–4350

Herzig V (2010) Update zum DeArGe-Spinnengiftprojekt. Arachne 15:26–31

Herzig V, Wood DLA, Newell F, Chaumeil PA, Kaas Q, Binford GJ, Nicholson GM, Gorse D, King GF (2011) ArachnoServer 2.0, an updated online resource for spider toxin sequences and structures. Nucleic Acids Res 39:D653–D657

Isbister GK, Seymour JE, Gray MR, Raven RJ (2003) Bites by spiders of the family Theraphosidae in humans and canines. Toxicon 41:519–524

King GF (2007) Modulation of insect Ca_v channels by peptidic spider toxins. Toxicon 49:513–530

King GF, Escoubas P, Nicholson GM (2008a) Peptide toxins that selectively target insect Na_v and Ca_v channels. Channels 2:100–116

King GF, Gentz MC, Escoubas P, Nicholson GM (2008b) A rational nomenclature for naming peptide toxins from spiders and other venomous animals. Toxicon 52:264–276

Klint JK, Senff S, Rupasinghe DB, Er SY, Herzig V, Nicholson GM, King GF (2012) Spider-venom peptides that target voltage-gated sodium channels: pharmacological tools and potential therapeutic leads. Toxicon 60:478–491. doi:10.1016/j.toxicon.2012.1004.1337

Kuhn-Nentwig L, Stoecklin R, Nentwig W (2011) Venom composition and strategies in spiders: is everything possible? Adv Insect Physiol 40:1–86

Liang S (2004) An overview of peptide toxins from the venom of the Chinese bird spider *Selenocosmia huwena* Wang [=*Ornithoctonus huwena* (Wang)]. Toxicon 43:575–585

Lucas SM, Da Silva Junior PI, Bertani R, Cardoso JL (1994) Mygalomorph spider bites: a report on 91 cases in the state of Sao Paulo, Brazil. Toxicon 32:1211–1215

Nentwig W, Kuhn-Nentwig L (2013) Main components of spider venoms. In: Nentwig W (ed) Spider ecophysiology. Springer, Heidelberg (this volume)

Oldrati V, Bianchi E, Stöcklin R (2013) Spider venom components as drug candidates. In: Nentwig W (ed) Spider ecophysiology. Springer, Heidelberg (this volume)

Olivera BM, Cruz LJ (2001) Conotoxins, in retrospect. Toxicon 39:7–14

Platnick N (2012) The world spider catalog, Version 13.0. American Museum of Natural History. http://research.amnh.org/entomology/spiders/catalog/index.html. Accessed 26 October 2012

Saez NJ, Senff S, Jensen JE, Er SY, Herzig V, Rash LD, King GF (2010) Spider-venom peptides as therapeutics. Toxins 2:2851–2871
Siemens J, Zhou S, Piskorowski R, Nikai T, Lumpkin EA, Basbaum AI, King D, Julius D (2006) Spider toxins activate the capsaicin receptor to produce inflammatory pain. Nature 444:208–212
Sollod BL, Wilson D, Zhaxybayeva O, Gogarten JP, Drinkwater R, King GF (2005) Were arachnids the first to use combinatorial peptide libraries? Peptides 26:131–139
Windley MJ, Herzig V, Dziemborowicz SA, Hardy MC, King GF, Nicholson GM (2012) Spider-venom peptides as bioinsecticides. Toxins 4:191–227

[illegible]

Chapter 16
The Cytotoxic Mode of Action of the Venom of *Cupiennius salei* (Ctenidae)

Lucia Kuhn-Nentwig and Wolfgang Nentwig

16.1 Introduction

The venom of the ctenid spider *Cupiennius salei* (Fig. 16.1) is rich in components which belong to different functional groups. Besides low molecular mass compounds, the venom contains several disulphide-rich peptides, also called mini-proteins, which act as neurotoxins on ion channels or as enhancers of neurotoxins. Likewise, a variety of small cytolytic peptides, which destroy membranes very efficiently, and enzymes are present in the venom. Neurotoxins with cytolytic activity, cytolytic α-helical small cationic peptides and enzymes most probably attacking connective tissue and phospholipid membranes cause the overall cytotoxic effect of this venom. Synergistic and enhancing interactions between components enable the spider to achieve a maximum of toxicity with a minimum of venom quantity.

16.2 Low Molecular Mass Compounds

The ion concentrations in the venom of *Cupiennius salei* are determined as Na^+ 8.9 mM, K^+ 215 mM and Ca^{2+} 0.94 mM (Kuhn-Nentwig et al. 1994). The high K^+ ion content synergistically increases the insecticidal activity of the main neurotoxins CsTx-1 and CsTx-9 (Wullschleger et al. 2005). These concentrations are the reverse of the concentrations found in the hemolymph of *Cupiennius salei* (Na^+ 223 mM, K^+ 6.79 mM and Ca^{2+} 4.0 mM) (Loewe et al. 1970).

The venom contains all 20 standard amino acids, most of which at concentrations below 25 pmol/μl; only glycine is more common (43.3 pmol/μl).

L. Kuhn-Nentwig (✉) • W. Nentwig
Institute of Ecology and Evolution, University of Bern, Bern, Switzerland
e-mail: lucia.kuhn@iee.unibe.ch

W. Nentwig (ed.), *Spider Ecophysiology*,
DOI 10.1007/978-3-642-33989-9_16,

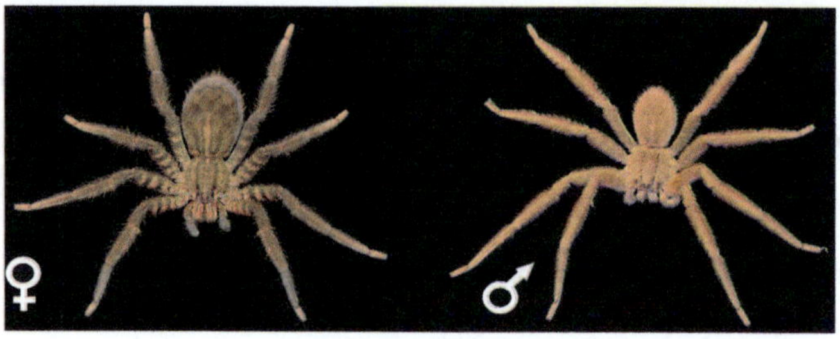

Fig. 16.1 Adult *Cupiennius salei* (Ctenidae). *Left* female spider, *right* male spider, dorsal view

Remarkable is the frequent occurrence of taurine (70.0 pmol/ml). Histamine was determined with a concentration of 5.7 nmol/μl, and the polyamines putrescine and cadaverine could be detected only in traces (3–18 pmol/μl) (Kuhn-Nentwig et al. 1994). Histamine is a known enhancer of the neurotoxic activity of the main neurotoxin CsTx-1 (Wullschleger et al. 2005).

16.3 Disulphide-Rich Peptides

Until now, 43 different cysteine-containing peptides have been identified from the cDNA library of *Cupiennius salei* venom glands (Kuhn-Nentwig, unpublished results), and for 17 of them, the amino acid sequence data have been published (Kuhn-Nentwig et al. 2004; Trachsel et al. 2012). Their molecular masses range between 3.5 and 9.9 kDa, and some of them exhibit a C-terminal amidation as posttranslational modification. Most of them contain an inhibitor cystine knot (ICK) motif where the disulphide bridge bonds are between C1–C4, C2–C5, C3–C8 and C6–C7. However, some of the peptides are characterised by the presence of only two cysteines (CsTx-16), and some contain up to 14 cysteines.

The expressed neurotoxins and neurotoxin-like structures can be divided into three groups: The first and most frequently expressed group (78.4 %) comprises the main neurotoxin CsTx-1, followed by CsTx-9, CsTx-10 and CsTx-11 and the enhancer peptides CsTx-8, CsTx-12 and CsTx-13 (Trachsel et al. 2012; Wullschleger et al. 2004) (Fig. 16.2). Interestingly, C-terminally truncated homologues of CsTx-1 (described as CsTx-2a, b), CsTx-9 (CsTx-7), CsTx-8 and CsTx-12 (CsTx-14) and CsTx-13 (CsTx-15) have been isolated in small quantities from the venom and seem to be rather posttranslational products than true translation products. All these peptides exhibit a high homology to peptides also identified in a cDNA library of *Lycosa singoriensis* (Zhang et al. 2010).

The second group is composed of neurotoxin-like structures (20 %) and we named some of them (5.4 %) "ancient" neurotoxins or neurotoxin-like structures because their structure is related to neurotoxins which have been published for the agelenids *Agelenopsis aperta* and *Agelena orientalis*, the ctenid *Phoneutria nigriventer* and the lycosid *Geolycosa* sp. In the *Cupiennius salei* venom, this group comprises

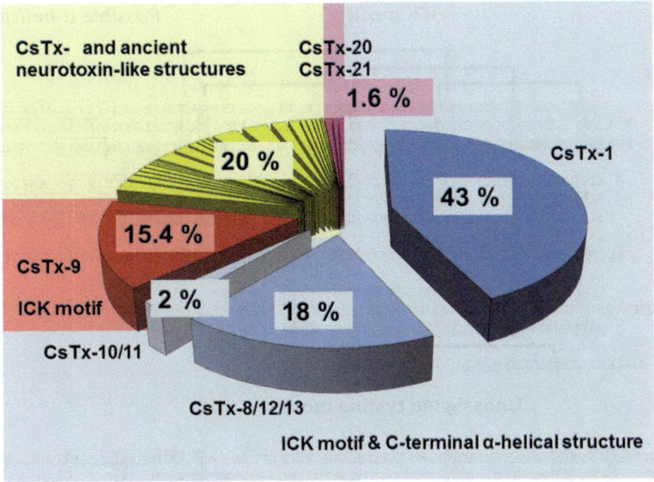

Fig. 16.2 Distribution of expressed neurotoxins and neurotoxin-like structures ($N = 625$ contigs; 90 % assemblage) in the venom of *Cupiennius salei*. Group 1 peptides (compare text) exhibit the ICK motif and possess C-terminally an α-helical structure (coloured in *blue*) but include also CsTx-9, exhibiting only the ICK motif (coloured in *red*). Group 2 peptides (CsTx peptides and ancient neurotoxin-like structures) are expressed only in low amounts (marked in *yellow*). Group 3 peptides (CsTx-20/21) exhibit no propeptide after the signal peptide (coloured in *pink*)

24 peptides; they account only for 20 % of the total expressed neurotoxin-like structures, and their concentrations in the venom are very low (Fig. 16.2). Therefore, we assume that a functional context can be excluded for some of them and evolutionary reasons could explain their existence.

The third peptide group differs from group 1 and 2 by the absence of an acidic propeptide between the signal peptide and the mature peptide (1.6 %). CsTx-20 and CsTx-21 homologues are more acidic peptides with isoelectric points (pI's) between 4.85 and 6.06, exhibiting 10 cysteines and molecular masses between 7.2 and 9.9 kDa (Kuhn-Nentwig, unpublished results; Trachsel et al. 2012) (Fig. 16.3).

16.3.1 Neurotoxins

The most abundant neurotoxin in the venom of *Cupiennius salei* is CsTx-1 with concentrations between 1.4 and 3.3 mM. CsTx-1 exhibits an ICK motif and is composed of 74 amino acid residues with a highly cationic amidated C-terminus. This peptide is the most insecticidal neurotoxin in the venom, and its insecticidal activity (LD_{50} 0.35 pmol/mg *Drosophila*) is three to four times increased through synergistic interactions with other neurotoxins, enhancers or cytolytic peptides

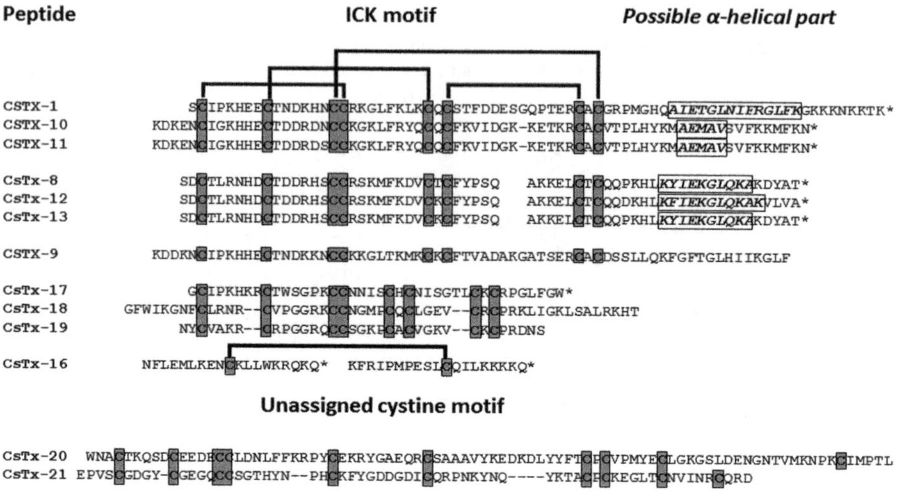

Fig. 16.3 Overview of cysteine-containing neurotoxins and neurotoxin-like structures identified in the venom of *Cupiennius salei*. Cysteine residues are in *bold type* and shaded in *grey*. The disulphide bridge pattern [ICK motif; single interchain disulphide bridge (CsTx-16) or unassigned disulphide bridge pattern (CsTx-20/21)] are indicated above the sequences. Possible α-helical parts in the C-terminal part of the peptides are boxed in *bold* and *italic*. C-terminal amidation is indicated by an *asterisk*

(Kuhn-Nentwig et al. 2004; Wullschleger et al. 2004, 2005). Furthermore, CsTx-1 blocks insect (cockroach) mid/low voltage-activated and high voltage-activated Ca_v channels and vertebrate L-type (GH3 cells) Ca_v channels (Kubista et al. 2007).

Reducing the C-terminal part of CsTx-1 results in considerable differences in the LD_{50} values obtained in bioassays on *Drosophila* flies. In the case of CsTx-2a (lacking the last 13 amino acid residues), the insecticidal activity is reduced to only 14 % and for CsTx-2b (lacking the last 14 amino acid residues) to less than 1 % (Fig. 16.3). It is obvious that the C-terminal part of CsTx-1 plays an important role in its toxicity even though a synthetically produced peptide containing only the C-terminal last 13 amino acids, which we named CT1-short, shows no insecticidal activity. Also, CT1-short has no effect on the insecticidal activity of CsTx-1, CsTx2a or CsTx2b when administered together (Kuhn-Nentwig et al. 2000).

Secondary structure prediction of the C-terminal part of CsTx-1 resulted in a putative α-helix. However, such a putative C-terminal α-helical structure is only possible in CsTx-1 and in CT1-long (this synthetically produced peptide corresponds to amino acid residues 45–74 of CsTx-1) and not in CsTx-2a, CsTx-2b and CT1-short, as verified by CD measurements of these peptides in the presence of membrane-mimicking trifluoroethanol. Investigations of CsTx-1 and CT1-long on prokaryotic and eukaryotic cell membrane systems exhibit an unspecific membranolytic activity of both peptides in the micromolar range on prokaryotic

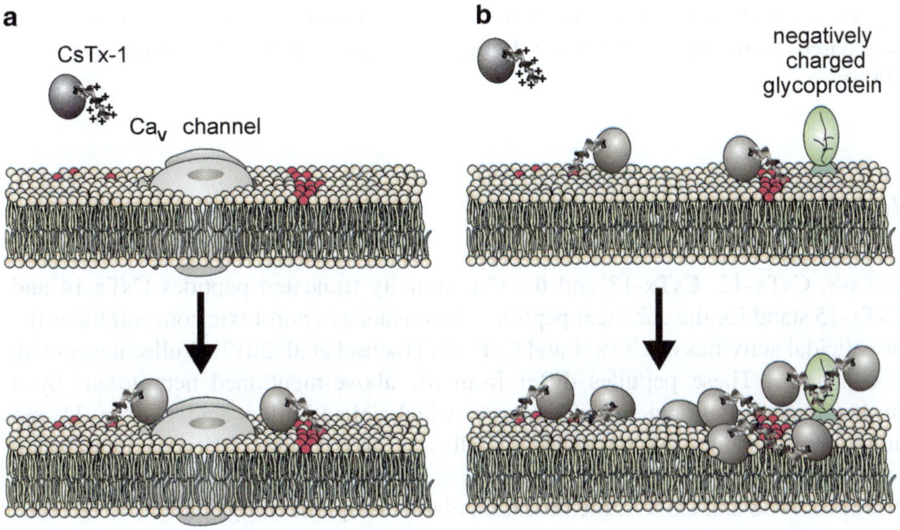

Fig. 16.4 (**a**) CsTx-1 inhibits a Ca^{2+} channel after attraction and anchoring with its highly cationic C-terminus to the cell membrane at negatively charged cell structures such as phospholipids with negatively charged head groups (*red*) or rafts of such negatively charged phospholipids. Ion-channel inhibition may take place by direct binding to the ion channel or binding to negatively charged lipid rafts, thus influencing the membrane architecture surrounding the ion channel. (**b**) In higher concentrations and in absence of the target ion channel, CsTx-1 acts membranolytic by disturbing the membrane architecture after binding to negatively charged phospholipids, negatively charged lipid rafts or negatively charged glycoproteins

and eukaryotic cells. This membranolytic activity exhibits different preferences of CsTx-1 and CT1-long depending on the tested membrane system (Kuhn-Nentwig et al. 2012).

Synergistic interactions between neurotoxins and low molecular mass compounds or between neurotoxins and enhancer peptides or cytolytic peptides are well documented (Adams 2004; Wullschleger et al. 2004, 2005), but a synergistic interaction within one peptide points to a further possibility to increase the toxicity of venom compounds. CsTx-1 exhibits two structurally different domains: In the N-terminal position, the ICK motif is responsible for the Ca^{2+} channel inhibition, and in the C-terminal position, an α-helix motif exhibits cytolytic activity (Kuhn-Nentwig et al. 2012) (Fig. 16.4).

Besides CsTx-1, so far two further neurotoxins in the venom of *Cupiennius salei*, CsTx-10 and CsTx-11, and the enhancer peptides, CsTx-8, CsTx-12 and CsTx-13, possess a comparable C-terminal extension in which secondary structure predictions identified a putative C-terminal α-helical structure. Also from another spider, the miturgid *Cheiracanthium punctorium*, a two-domain modular toxin CpTx-1a has been reported (Vassilevski et al. 2010), which could point to a common strategy of higher entelegyne spiders for the enhancement of the toxicity of one peptide. This strategy is also known for some scorpion venom-derived peptides exhibiting a putative α-helical N-terminus and a CSαβ motif fold originated from three disulphide bridges located C-terminally (Kuhn-Nentwig 2009).

16.3.2 Enhancer Peptides

CsTx-8, CsTx-12, CsTx-13 and the C-terminally truncated peptides CsTx-14 and CsTx-15 stand for the enhancer peptides that enhance in non-toxic concentrations the insecticidal activities of CsTx-1 and CsTx-9 (Trachsel et al. 2012; Wullschleger et al. 2004, 2005). These peptides differ from the above-mentioned neurotoxins by a further posttranslational modification in which, after amino acid residue 34, six amino acid residues are posttranslationally removed (Kuhn-Nentwig, unpublished results). This results in two peptide chains A and B, connected by the two disulphide bridges C3–C8 and C6–C7. Identification of the disulphide bridge pattern of CsTx-13 by nanoelectrospray tandem MS revealed the ICK motif as described above for CsTx-1 (Wullschleger et al. 2004). The C-terminus of chain B is also posttranslationally amidated in CsTx-8, CsTx-12 and CsTx-13. Interestingly, peptide chains B of CsTx-8, CsTx-12 and CsTx-13 exhibit, after secondary structure predictions, an α-helix motif which may act membranolytically as described for CsTx-1. The insecticidal activity of CsTx-13 (LD_{50} value 16.3 pmol/mg *Drosophila* fly) seems to be the weakest when compared with CsTx-9 (LD_{50} value 3.12 pmol/mg fly) and CsTx-2a (LD_{50} value 2.58 pmol/mg fly).

16.4 Cytolytic Peptides

16.4.1 Overview

The venom of *Cupiennius salei* contains many membranolytically acting peptides (cupiennins) with molecular masses between 1.5 and 4.2 kDa. They exert a strong cytolytic activity towards prokaryotic and eukaryotic cells (Kuhn-Nentwig et al. 2002b). The cupiennin 1 (a–d) and cupiennin 2 (a–e) families are characterized by 35 amino acid residues, pI's between 10.2 and 10.5 and net charges from +6 to +7. These highly cationic peptides exhibit hydrophobic N-termini composed of six amino acid residues which are followed by six repeats (cupiennin 1 family) or five repeats (cupiennin 2) of four amino acids, which form the central part of the peptide chain, with lysine always in first position. The C-termini are more polar, and more than 40 % of all amino acid residues are hydrophobic. Due to well-defined hydrophobic and hydrophilic areas within the α-helix, the peptides reach an amphiphilic conformation which is essential for their cytolytic activity (Powers and Hancock 2003).

Furthermore, small cationic peptides (SCP), as the cupiennin 3 (SCP 3a–d; 27 amino acid residues) and cupiennin 4 (SCP 4a, b; 27 amino acid residues) families, are characterised by the absence of two to three repeats in the central part of the peptides when compared with the cupiennin 1 family. Nevertheless, these peptides exhibit pI's between 10.4 and 11.2, have net charges from +5 to +7 and are also C-terminally amidated such as the cupiennin 1 and 2 families. Besides these, several N-terminally or C-terminally truncated forms of the cupiennin 1 (SCP 1a–h) and 4 families (SCP 4c–g) have been identified (Trachsel et al. 2012). Currently, it is not clear if some of them are posttranslational products of known cupiennins or products of simple, binary or complex precursors as described for the latarcins, membranolytic peptides from the venom of the zodariid spider *Lachesana tarabaevi* (Kozlov et al. 2006). Interestingly, the SCP families 6 and 7 have introduced besides lysine also arginine into the peptide chain to obtain a cationic character as it is known from other spider venom-derived cytolytic peptides as oxyopinins and latarcins (Kozlov et al. 2006).

16.4.2 Membranolytic Activity of the Cupiennin 1 Family

The cupiennin 1 family is the best investigated membranolytically acting peptide family from *Cupiennius salei*. CD measurements of cupiennin 1a in water exhibit a random coiled structure. In the presence of membrane-mimicking trifluoroethanol (50 %) or negatively charged phospholipid vesicles, the formation of an α-helix occurs. It is supposed that the peptides may be attracted to the cell surfaces by electrostatic interactions between their positively charged side chains of lysine and negatively charged membrane phospholipid head groups and other negatively charged components of bacteria, protozoa and eukaryotic cells (Kuhn-Nentwig et al. 2002b; Pukala et al. 2007a).

Determination of the solution structure of cupiennin 1a by nuclear magnetic resonance spectroscopy exhibits a helix-hinge-helix structure, a structural motif, which has frequently been identified in cationic cytolytic peptides. Well-defined helices are located between residues Gly3-Ala21 and Tyr28-Lys32, and the hinge region is supposed to be initiated by Gly25 (Pukala et al. 2007a). Analysing the role of the N- and C-terminal segments of cupiennin 1d shows that the cytolytic activity depends on the hydrophobic N-terminus and is modulated by the polar C-terminus (Kuhn-Nentwig et al. 2002a). With a length of ~30 Å of the N-terminal helix of cupiennin 1a, the peptide is able to span the bilayer of bacterial cell membranes and phosphatidylcholine bilayers, resulting in pore formation and membrane destruction (Cornell and Separovic 1983; Pukala et al. 2007a).

16.4.3 Biological Activity of the Cupiennin 1 Family

The cupiennin 1a family acts cytolytically on a variety of different bacteria cell membrane types in the submicromolar range (minimal inhibitory concentrations from 0.08 to 5 μM). Additionally, the eukaryotic pathogens trypanosomes and plasmodia, causing sleep sickness and malaria, are destroyed in submicromolar concentrations (IC_{50} values 0.029 to 0.658 μM). Eukaryotic cells, which dispose of negatively charged cell membrane structures, such as erythrocytes, rat skeletal myoblasts or different human leukemic and tumour cells are destroyed likewise in the sub- and micromolar range. In the case of human erythrocytes, it could be demonstrated that binding of cationic peptides is mediated by attraction to negatively charged sialic acids on the outer leaflet of these cells. A stereospecific mode of action of cupiennin 1a could be excluded (Kuhn-Nentwig et al. 2011).

Besides the direct effects of cupiennin 1a on membrane systems, this peptide also inhibits the formation of nitric oxides by neuronal nitric oxide synthase. The mechanism involves a complexation with calmodulin. Calmodulin is the regulatory protein for a variety of kinase phosphorylating enzymes and the eukaryotic cytoskeleton and it is essential for operations of neuronal nitric oxide synthase (Pukala et al. 2007b). Likewise, the production of superoxide by the NADPH oxidase in phorbol myristate acetate-stimulated granulocytes is additionally inhibited by cupiennin 1a (Kuhn-Nentwig et al. 2011). It is supposed that cupiennin 1a will interfere with many cellular functions and that it simultaneously destroys membrane parts of the neuronal tissues and muscle cells leading to a collapse of the cellular and neuronal functions.

16.5 Enzymes

Besides low molecular mass compounds and peptides with molecular masses up to 10 kDa, several proteins with molecular masses between 25 and 97 kDa have been identified in the *Cupiennius salei* venom. One of these proteins exhibits hyaluronidase activity and cleaves hyaluronan into fragments of varied molecular size (Kuhn-Nentwig et al. 1994).

Up to now the presence of hyaluronan (hyaluronic acid), a large linear polymer of repeating disaccharides of glucuronic acid and GlcNAc in arthropods, is still controversially discussed. On the one side, it is stated that hyaluronan, common in vertebrates, has not been found in arthropods and in *Drosophila* only chondroitin sulphate and heparan sulphate have been identified (Takeo et al. 2004; Toyoda et al. 2000). Additionally, no hyaluronan synthase genes were found searching the genomic sequencing project for *Drosophila* (DeAngelis 2002).

On the other hand, histochemical investigations of the mesenteric connective tissue of cockroaches and locusts indicate the presence of hyaluronan in various instars (Ashhurst and Costin 1971; Francois 1978; Treherne et al. 1982), but also chondroitin sulphate and heparan sulphate have been identified in internal organs of cockroaches (dos Santos et al. 2006). From *Hippasa partita* (Lycosidae), a highly substrate-specific hyaluronidase is known, which only cleaves hyaluronan, but not chondroitin and heparan sulphate (Nagaraju et al. 2007).

It has intensively been discussed that hyaluronidase acts as spreading factor, facilitating the access of neurotoxic and cytolytic venom components to their targets (Kuhn-Nentwig et al. 2011). This assumption is convincing for large mygalomorph spiders which may have small vertebrates as prey and which may need to defend themselves against vertebrate predators. In contrast to this, most araneomorph spiders do not target vertebrates. Nevertheless, hyaluronidase activity has been identified in their venoms, but its function as spreading factor still needs further clarification in terms of substrate specificity of the hyaluronidase and possible substrate availability within various prey items, e.g. such as basement membranes surrounding nerve and muscle tissues or connective tissues.

Preliminary results from the cDNA library of *Cupiennius salei* venom glands also show that it contains, besides several other enzymes, a hyaluronidase sequence with a high similarity to the hyaluronidase BmHYA1 [UniProtKB/TrEMBL: D1MBU1], identified from the venom of the Chinese red scorpion *Buthus martensii* Karsch (Feng et al. 2010). Additionally, a phospholipase C sequence with a high similarity to phospholipase C-like protein [UniProtKB/TrEMBL: C5J8D0] from the scorpion *Opisthacanthus cayaporum* venom glands (Silva et al. 2009) and putative phospholipase C [UniProtKB/TrEMBL: B7Q2N6 and B7P6Q6] similar to the tick *Ixodes scapularis* have been identified with the Blast algorithm (Kuhn-Nentwig and Piquemal, unpublished results).

16.6 Conclusions

The venom of *Cupiennius salei* is characterized by (1) a high diversity of cytolytic compounds (linear cytolytic peptides and a cysteine-rich peptide exhibiting two domains: the ICK motif and an α-helical cytolytically acting domain), (2) the neurotoxic activity of ion-channel inhibitors, (3) a highly active hyaluronidase and (4) synergistic interactions between many of these components (Fig. 16.5). The combined effects of synergistic and enhancing interactions between various components enable *Cupiennius salei* to inject a maximum of toxicity with a minimum of venom quantity, thus optimizing its venom investment.

Acknowledgements We thank Johann Schaller and Stefan Schürch for long-term cooperation in spider venom peptide amino acid sequence analyses and MS measurements and Reto Stöcklin (Atheris Laboratories).

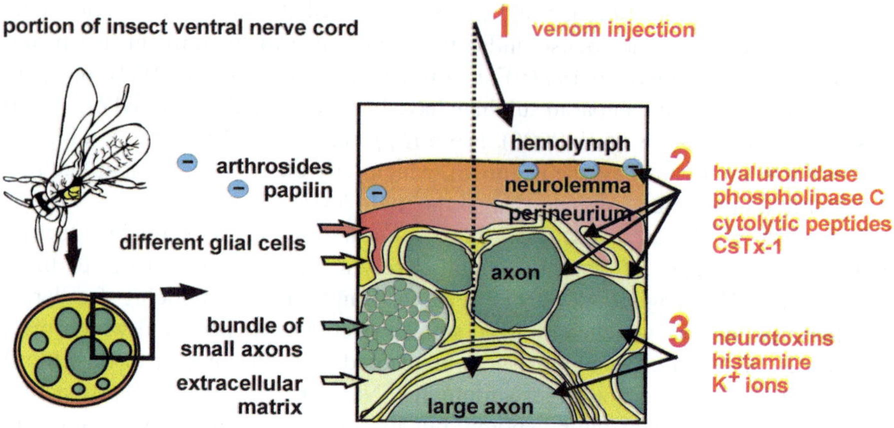

Fig. 16.5 The brain and the nerve system of insects are protected from direct contact with the hemolymph by the hemolymph-brain barrier (Treherne 1985). The rather permeable acellular neurolemma [containing glycoaminoglycans, negatively charged papilins and arthrosides (Kramerova et al. 2000; Sickmann et al. 1992)] is followed by the main barrier: the perineurium. Further, glial cells and axons are embedded in different glycosaminoglycans containing extracellular matrix (Francois 1978; Treherne et al. 1982). The envenomation process is shown here in three steps: *1*. *Cupiennius salei* venom can get directly or via hemolymph in contact with the neuronal tissue. *2*. Hyaluronidase and phospholipase C may act as spreading factor. The cytolytic peptides can be electrostatically attracted to negatively charged compounds and adopt an α-helical conformation which leads to cell membrane destruction. *3*. As a result, neurotoxins and other direct-acting substances have a better access to their neuronal targets cells

References

Adams ME (2004) Agatoxins: ion channel specific toxins from the american funnel web spider, *Agelenopsis aperta*. Toxicon 43:509–525

Ashhurst DE, Costin NM (1971) Insect mucosubstances. II. The mucosubstances of the central nervous system. Histochem J 3:297–310

Cornell BA, Separovic F (1983) Membrane thickness and acyl chain length. Biochim Biophys Acta 733:189–193

DeAngelis PL (2002) Evolution of glycosaminoglycans and their glycosyltransferases: implications for the extracellular matrices of animals and the capsules of pathogenic bacteria. Anat Rec 268:317–326

dos Santos AV, Onofre GR, Oliveira DM, Machado EA, Allodi S, Silva LC (2006) Heparan sulfate is the main sulfated glycosaminoglycan species in internal organs of the male cockroach, *Periplaneta americana*. Micron 37:41–46

Feng L, Gao R, Meng J, Gopalakrishnakone P (2010) Cloning and molecular characterization of BmHYA1, a novel hyaluronidase from the venom of Chinese red scorpion *Buthus martensii* Karsch. Toxicon 56:474–479

Francois J (1978) The ultrastructure and histochemistry of the mesenteric connective tissue of the cockroach *Periplaneta americana* L. (Insecta, Dictyoptera). Cell Tissue Res 189:91–107

Kozlov SA, Vassilevski AA, Feofanov AV, Surovoy AY, Karpunin DV, Grishin EV (2006) Latarcins, antimicrobial and cytolytic peptides from the venom of the spider *Lachesana tarabaevi* (Zodariidae) that exemplify biomolecular diversity. J Biol Chem 281:20983–20992

Kramerova IA, Kawaguchi N, Fessler LI, Nelson RE, Chen YL, Kramerov AA, Kusche-Gullberg M, Kramer JM, Ackley BD, Sieron AL, Prockop DJ, Fessler JH (2000) Papilin in development; a pericellular protein with a homology to the adamts metalloproteinases. Development 127: 5475–5485

Kubista H, Mafra RA, Chong Y, Nicholson GM, Beirao PS, Cruz JS, Boehm S, Nentwig W, Kuhn-Nentwig L (2007) CSTX-1, a toxin from the venom of the hunting spider *Cupiennius salei*, is a selective blocker of L-type calcium channels in mammalian neurons. Neuropharmacology 52:1650–1662

Kuhn-Nentwig L (2009) In: Lima ME (ed) Animal toxins: state of the art. Editora UFMG, Belo Horizonte, pp 249–268

Kuhn-Nentwig L, Schaller J, Nentwig W (1994) Purification of toxic peptides and the amino acid sequence of CSTX-1 from the multicomponent venom of *Cupiennius salei* (Araneae: Ctenidae). Toxicon 32:287–302

Kuhn-Nentwig L, Schaller J, Kämpfer U, Imboden H, Malli H, Nentwig W (2000) A lysine rich C-terminal tail is directly involved in the toxicity of CSTX-1, a neurotoxic peptide from the venom of the spider *Cupiennius salei*. Arch Insect Biochem Physiol 44:101–111

Kuhn-Nentwig L, Dathe M, Walz A, Schaller J, Nentwig W (2002a) Cupiennin 1d*: the cytolytic activity depends on the hydrophobic N-terminus and is modulated by the polar C-terminus. FEBS Lett 527:193–198

Kuhn-Nentwig L, Müller J, Schaller J, Walz A, Dathe M, Nentwig W (2002b) Cupiennin 1, a new family of highly basic antimicrobial peptides in the venom of the spider *Cupiennius salei* (Ctenidae). J Biol Chem 277:11208–11216

Kuhn-Nentwig L, Schaller J, Nentwig W (2004) Biochemistry, toxicology and ecology of the venom of the spider *Cupiennius salei* (Ctenidae). Toxicon 43:543–553

Kuhn-Nentwig L, Willems J, Seebeck T, Shalaby T, Kaiser M, Nentwig W (2011) Cupiennin 1a exhibits a remarkably broad, non-stereospecific cytolytic activity on bacteria, protozoan parasites, insects, and human cancer cells. Amino Acids 40:69–76

Kuhn-Nentwig L, Fedorova IM, Lüscher BP, Kopp LS, Trachsel C, Schaller J, Vu XL, Seebeck T, Streitberger K, Nentwig W, Sigel E, Magazanik LG (2012) A venom-derived neurotoxin, CsTx-1, from the spider *Cupiennius salei* exhibits cytolytic activities. J Biol Chem 287: 25640–25649

Loewe R, Linzen B, von Stackelberg W (1970) Die gelösten Stoffe in der Hämolymphe einer Spinne, *Cupiennius salei* Keyserling. Z Vergl Physiol 66:27–34

Nagaraju S, Devaraja S, Kemparaju K (2007) Purification and properties of hyaluronidase from *Hippasa partita* (funnel web spider) venom gland extract. Toxicon 50:383–393

Powers JP, Hancock RE (2003) The relationship between peptide structure and antibacterial activity. Peptides 24:1681–1691

Pukala TL, Boland MP, Gehman JD, Kuhn-Nentwig L, Separovic F, Bowie JH (2007a) Solution structure and interaction of cupiennin 1a, a spider venom peptide, with phospholipid bilayers. Biochemistry 46:3576–3585

Pukala TL, Doyle JR, Llewellyn LE, Kuhn-Nentwig L, Apponyi MA, Separovic F, Bowie JH (2007b) Cupiennin1a, an antimicrobial peptide from the venom of the neotropical wandering spider *Cupiennius salei*, also inhibits the formation of nitric oxide by neuronal nitric oxide synthase. FEBS Lett 274:1778–1784

Sickmann T, Weske B, Dennis RD, Mohr C, Wiegandt H (1992) Chemical distribution of glycosphingolipids in third-instar larval organs of the blowfly, *Calliphora vicina* (Insecta: Diptera). J Biochem 111:662–669

Silva EC, Camargos TS, Maranhao AQ, Silva-Pereira I, Silva LP, Possani LD, Schwartz EF (2009) Cloning and characterization of cDNA sequences encoding for new venom peptides of the Brazilian scorpion *Opisthacanthus cayaporum*. Toxicon 54:252–261

Takeo S, Fujise M, Akiyama T, Habuchi H, Itano N, Matsuo T, Aigaki T, Kimata K, Nakato H (2004) In vivo hyaluronan synthesis upon expression of the mammalian hyaluronan synthase gene in *Drosophila*. J Biol Chem 279:18920–18925

Toyoda H, Kinoshita-Toyoda A, Fox B, Selleck SB (2000) Structural analysis of glycosaminoglycans in animals bearing mutations in sugarless, sulfateless, and tout-velu. *Drosophila* homologues of vertebrate genes encoding glycosaminoglycan biosynthetic enzymes. J Biol Chem 275:21856–21861

Trachsel C, Siegemund D, Kämpfer U, Kopp LS, Bühr C, Grossmann J, Lüthi C, Cunningham M, Nentwig W, Kuhn-Nentwig L, Schürch S, Schaller J (2012) Multicomponent venom of the spider *Cupiennius salei*: a bioanalytical investigation applying different strategies. FEBS J 279:2683–2694

Treherne JE (1985) Blood-brain barrier. In: Kerkut GA, Gilbert LC (eds) Comprehensive insect physiology biochemistry and pharmacology. Pergamon, Oxford, pp 115–137

Treherne JE, Schofield PK, Lane NJ (1982) Physiological and ultrastructural evidence for an extracellular anion matrix in the central nervous system of an insect (*Periplaneta americana*). Brain Res 247:255–267

Vassilevski AA, Fedorova IM, Maleeva EE, Korolkova YV, Efimova SS, Samsonova OV, Schagina LV, Feofanov AV, Magazanik LG, Grishin EV (2010) Novel class of spider toxin: active principle from the yellow sac spider *Cheiracanthium punctorium* venom is a unique two-domain polypeptide. J Biol Chem 285:32293–32302

Wullschleger B, Kuhn-Nentwig L, Tromp J, Kämpfer U, Schaller J, Schürch S, Nentwig W (2004) CSTX-13, a highly synergistically acting two-chain neurotoxic enhancer in the venom of the spider *Cupiennius salei* (Ctenidae). Proc Natl Acad Sci USA 101:11251–11256

Wullschleger B, Nentwig W, Kuhn-Nentwig L (2005) Spider venom: enhancement of venom efficacy mediated by different synergistic strategies in *Cupiennius salei*. J Exp Biol 208: 2115–2121

Zhang Y, Chen J, Tang X, Wang F, Jiang L, Xiong X, Wang M, Rong M, Liu Z, Liang S (2010) Transcriptome analysis of the venom glands of the Chinese wolf spider *Lycosa singoriensis*. Zoology (Jena) 113:10–18

Chapter 17
The Evolution of a Toxic Enzyme in Sicariid Spiders

Greta Binford

17.1 Introduction

Venoms of predatory animals are diverse biochemical systems at the interface of basic sciences and medical issues. They draw attention because some venoms cause serious damage to humans. As described in other chapters in this volume (see Herzig and King 2013; Kuhn-Nentwig and Nentwig 2013; Nentwig and Kuhn-Nentwig 2013), venoms have more recently grabbed attention of researchers because they contain a wealth of chemical diversity with pharmacological potential (see Oldrati et al. 2013). Among venomous lineages, spiders are predicted to contain the largest wealth of chemical diversity (Escoubas and King 2009). Over 43,000 described species of spiders have evolved from a common ancestor that lived at least 300 million years ago. Venoms originated very early in the spider lineage and have been evolving as predatory tools in the context of diversifying tactics of prey capture. The end products are chemical cocktails in a single spider that include hundreds of toxins, most with remarkable target specificity. Venoms of only a tiny subset of spider species have been explored, and nothing is known about venoms from many major spider lineages (see Nentwig and Kuhn-Nentwig 2013).

The store of chemicals in venoms is particularly rich in peptides with neurotoxic or antibiotic properties that are exciting for drug discovery. Spiders, like all venomous predatory lineages, have independently evolved small venom peptides that are rich in cysteines that form disulfide bonds and have particularly diverse activities and target specificities (see Herzig and King 2013). In addition to these peptide toxins, venoms also have a diverse set of proteins (polypeptides >10 kDa) with toxic activity. In spiders, these are even less well explored than peptide neurotoxins. Two of the most well-characterized large protein toxins in spider venoms are found in the medically relevant lineages *Latrodectus*, the genus that

G. Binford (✉)
Lewis & Clark College, Portland, USA
e-mail: binford@lclark.edu

W. Nentwig (ed.), *Spider Ecophysiology*,
DOI 10.1007/978-3-642-33989-9_17, © Springer-Verlag Berlin Heidelberg 2013

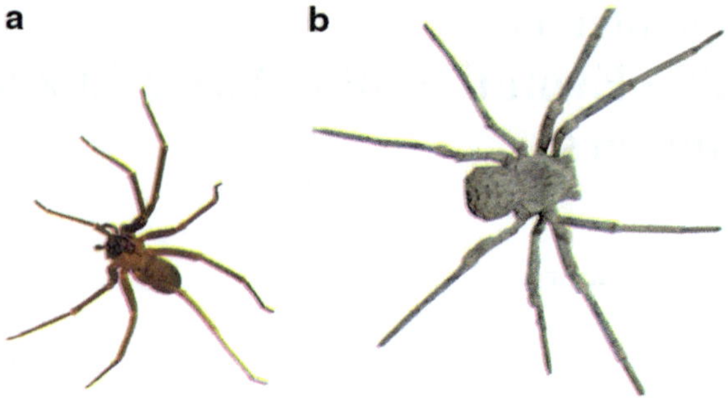

Fig. 17.1 (**a**) *Loxosceles arizonica;* (**b**) *Sicarius terrosus* (Sicariidae)

includes the black widow, and Sicariidae, the family that includes the brown recluse and six-eyed sand spiders. *Latrodectus* venoms have 130 kDa neurotoxins called latrotoxins that cause cramping and pain in humans. Venoms of sicariid spiders have 32–34 kDa toxic enzymes with sphingomyelinase D activity (SMase D) that are in a gene family called *SicTox*. These cause dermonecrosis in mammals and are present throughout Sicariidae. Both of these protein toxins cause medical problems in humans that have been well described elsewhere (Ushkaryov et al. 2004; Swanson and Vetter 2005, see Nentwig and Kuhn-Nentwig 2013), but notably, these are both toxins that are unique to a relatively small lineage of spiders. They thus represent opportunities to investigate the circumstances of evolutionary origin of novel toxic activity in venoms.

This chapter is a review of what we know about the diversity of SMase D and homologous proteins in the *SicTox* gene family that are expressed in sicariid venoms. Extensive analyses of venom composition and evolutionary dynamics of dermonecrotic enzymes in the *SicTox* gene family make them a great model to illustrate what we can learn about the distribution and evolution of a family of toxic proteins. I have three specific goals for this chapter: (1) to fully describe what we know about the diversity of a family of toxic proteins in spider venoms, (2) to discuss the evolutionary dynamics of recruitment and diversification of *SicTox* proteins for venom function, and (3) to illustrate the value of framing analyses of venom toxins in the context of the phylogenetic diversity and biology of the spiders.

17.2 Taxonomic Diversity and Biology of Sicariidae

It is helpful to frame description of the *SicTox* venom toxins in a context of diversity within the Sicariidae family, their phylogenetic position within spiders, and their feeding biology. The sicariid lineage is likely over 100 million years old (Binford et al. 2008) and is comprised of *Loxosceles* (including the brown recluse) and their sister genus *Sicarius* (6-eyed sand spiders) (Fig. 17.1). *Loxosceles* includes roughly 100 described species native to the Americas, West Indies, Mediterranean Europe, and Africa (Gertsch and Ennik 1983). *L. laeta* and *L. rufescens* have been introduced to regions outside of their native range, and *Loxosceles* are now on every continent except Antarctica. Relationships among *Loxosceles* and *Sicarius* are consistent with the most recent common ancestors of both genera being present on Western Gondwana before Africa and South America separated. Southern African *Loxosceles* are quite divergent from *Loxosceles* in Northwestern Africa and the Americas (Duncan et al. 2010). There is substantial diversity in South America and evidence of a single colonization of *Loxosceles* on North America across a proto-Caribbean land bridge (Binford et al. 2008, Fig. 17.2). In North America they have subsequently diversified into at least 50 species. The genus *Sicarius* includes 13 described South and Central American species and six from Southern Africa (Platnick 2012). These continental groups of *Sicarius* are reciprocally monophyletic, consistent with a common ancestor that predates the separation of Africa and South America.

For prey capture, *Loxosceles* spiders build small irregular webs in crevices, under rocks and debris, and in caves. While their webs help retain prey, *Loxosceles* also capture prey while wandering at night, making them facultative web foragers. *Sicarius* are less well-studied spiders that bury themselves in fine sand and have specialized hairs to which particles stick, presumably to aid in crypsis (Duncan et al. 2007). These spiders do not make webs, but ambush cursorial ground-dwelling prey by jumping from their sand cover. Within genera, species are largely homogeneous in behavior and habitat (pers obs.) outside of cave versus ground dwelling. Evidence to date suggests sicariids are generalist foragers of arthropods with diets that include particularly well-defended prey such as scorpions, ants, and other spiders.

Sicariids are haplogynes in a larger superfamily of Scytodoidea and are supported as sister to a clade that includes the families Drymusidae, Periegopidae, and Scytodidae (see also Appendix, this volume). Members of each of these lineages build webs, and *Scytodes* have the derived behavior of spitting glue on prey, as described by Suter and Stratton (2013).

17.3 Medical Risks Associated with Sicariid Bites

Across their native range, *Loxosceles* spiders inflict bites that cause dermonecrosis and sometimes systemic effects in humans (reviews: Swanson and Vetter 2005; Chaim et al. 2011; see also Nentwig and Kuhn-Nentwig 2013). Risks of *Loxosceles* bites in the United States are minor relative to risks in Latin America. Regions with the highest recorded rates of envenomation resulting in medical treatment (rarely death) are Mexico and dry areas of Peru, Chile, Brazil, and Argentina (da Silva et al. 2004). Medically significant *Loxosceles* bites also occur in Southern Africa (Newlands and Atkinson 1990). Risks associated with *Sicarius* bites are less well understood, but appear to differ between Africa and South America. There have been confirmed bites from African *Sicarius* that caused severe dermonecrosis and one death on record (van Aswegen et al. 1997). However, there are no records of medically relevant bites from *Sicarius* from South or Central America, and we have encountered no cultural fear of these animals when collecting in America.

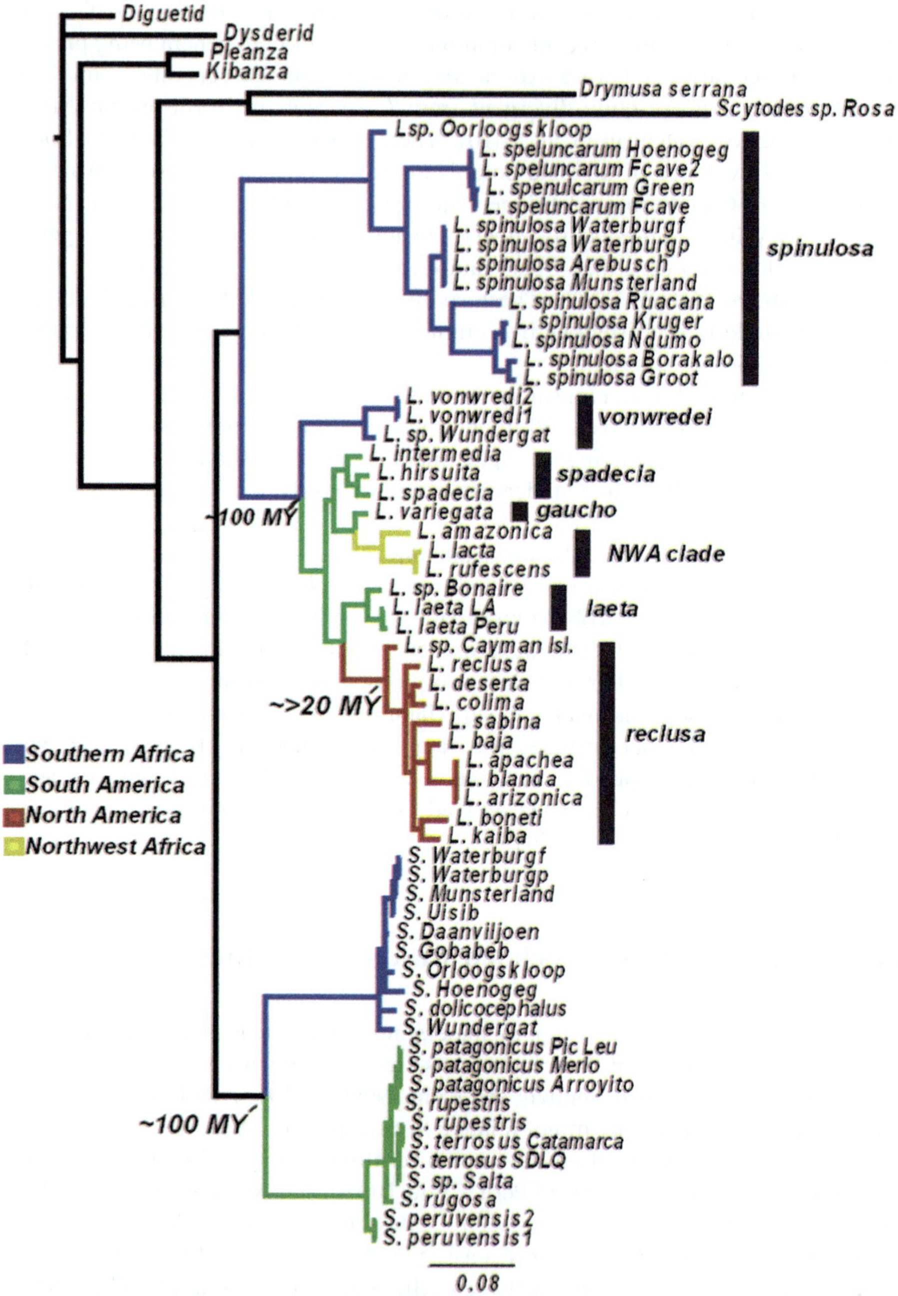

Fig. 17.2 Phylogenetic relationships of sicariids based on Binford et al. (2008). Lineages are colored based on native geographic region, and species groups of *Loxosceles* are labeled. Generic abbreviations of *L.* and *S.* refer to *Loxosceles* and *Sicarius*, respectively. NWA refers to the North West African clade

17.4 Venom Components

Despite the medical importance of sicariids, we have only scratched the surface of understanding the full set of components in their venoms. Nonetheless, venoms in this lineage are among the better understood of any spiders. The best-known sicariid toxins are the unusual SMase D enzymes that are key factors in the formation of lesions resulting from sicariid bites (Kurpiewski et al. 1981; Tambourgi et al. 1998). However, sicariid venoms also include a suite of other enzymes and low molecular weight components that are similar in sequence motif to other well-known toxins. I will first review in detail what we know about SMase D diversity and evolution, then discuss differences among sicariid taxa and what we can infer about evolutionary processes influencing venoms through comparative analyses. Then I will briefly describe other known components in sicariid venom.

17.4.1 SicTox Proteins

The mammalian dermonecrotic toxins that typify people's first impressions of venoms of the brown recluse are enzymes in a gene family we call *SicTox* (Binford et al. 2009). *SicTox* mRNA transcripts are among the most abundant in venom gland tissue of all sicariid species that have been studied in detail (Fernandes-Pedrosa et al. 2008; Gremski et al. 2010; pers. obs). Individual sicariid species have up to, and minimally, 12 distinct genes in this gene family (paralogs) that are expressed in venoms (Binford et al. 2009). The *SicTox* molecules are classic $(\alpha/\beta)_8$ TIM-barrel proteins (the most common enzyme structure known) and depend upon binding of a Mg^{2+} ion for catalysis. The first activity described for these enzymes was cleavage of the cell membrane phospholipid sphingomyelin at the D site (SMase D), leaving the cleavage products of choline and ceramide 1 phosphate. The link between the cleavage of cell membrane phospholipids and dermonecrosis involves a complex cascade of physiological events that ultimately activates an immune response involving the complement system; however, details of this mechanism are still being worked out (review in Chaim et al. 2011).

The activity of *SicTox* proteins, in isolation of the other chemicals in sicariid venom, is sufficient to cause dermonecrotic lesions in mammals (Kurpiewski et al. 1981). Recombinant *SicTox* proteins can hydrolyze lysophospholipids in addition to sphingomyelin (van Meeteren et al. 2004; Lee and Lynch 2005; Chaves-Moreira et al. 2011). The breadth of activities of these proteins that extends beyond SMase D has inspired reference in the literature to this gene family as phospholipase D (PLD). However, since these proteins are not homologous to other, more common PLD families that are expressed for nonvenomous function, referring to them as "PLDs," mask important differences in evolutionary history and protein structure (Binford et al. 2009).

17.4.2 Recruitment of SMase D Toxins in Spider Venom

The evolutionary history of SMase D as a venom toxin is particularly interesting because this enzymatic activity is relatively rare suggesting its presence in sicariid venom represents an independent recruitment event of a venom toxin (Binford and Wells 2003). Venom toxins with phospholipase activity are relatively common including phospholipase A (PLA) toxins that are widespread in snakes (Fry et al. 2009) and found in spiders (Foradori et al. 2005); however, snake venom PLAs are not homologous to *SicTox* genes. Outside of spiders, proteins homologous to SMase D have been isolated from mRNA transcripts in tick salivary glands (Alarcon-Chaidez et al. 2009) and venom of a buthid scorpion (Borchani et al. 2011). They are otherwise unknown in arachnids, most notably in other haplogynes that are close relatives of sicariids (Binford and Wells 2003). This disparate distribution suggests that the toxin has undergone independent evolutionary recruitment for toxic function minimally in sicariids, ticks (Fry et al. 2009), and at least one scorpion lineage. Interestingly, homologous toxins with SMase D activity are expressed in pathogenic *Corynebacterium* but not any other bacteria to date. All of these proteins are remote homologs of the ubiquitous protein domain family glycerophosphoryl diester phosphodiesterase (GDPD). The presence of homologs in bacteria could be the result of an independent recruitment from ancestral GDPDs (Fry et al. 2009) or of lateral transfer from metazoans to bacteria (Cordes and Binford 2006). Sequences of homologous genes available in databases are too divergent (~21 % identity between spiders and bacteria and between venom-expressed *SicTox* and GPDPs) to allow confident alignment necessary for the rigorous phylogenetic analyses required to understand patterns of relationships that are necessary to distinguish among these explanations.

17.4.3 Diversification of SMase D Toxins Postrecruitment

After the recruitment of *SicTox* genes for venom function, there is evidence of diversification within this gene family involving gene duplications and evolution of functional specificity. The gene family has duplication rates that are higher than general background levels in model organisms (Binford et al. 2009). The duplications lead to multiple copies within genomes. Work in our lab and others has isolated mRNAs of venom-expressed genes from a combined 21 sicariid species (Fernandes-Pedrosa et al. 2008; Binford et al. 2009; Gremski et al. 2010). The sampling represents the full phylogenetic breadth of sicariid taxa and reveals patterns of differences in the presence of and/or expression levels of different *SicTox* gene lineages among the different sicariid taxa. All sicariid taxa express *SicTox* genes in their venoms; however, different species groups in different regions of the world (Fig. 17.2) differ in the particular paralogs that are either present in the genome or are most highly expressed. Phylogenetic analyses support two distinct clades of *SicTox* genes. The α-clade of *SicTox* includes genes isolated only from *Loxosceles* from the New World and Northwest Africa (NWA clade Fig. 17.2). The β-clade includes genes expressed in venom glands of Southern African *Loxosceles* (the basal lineage of the genus) and both African and South American *Sicarius* (Binford et al. 2009). β-clade transcripts have also been isolated from New World *Loxosceles*, but are expressed at much lower levels than α-clade paralogs (Ramos-Cerrillo et al. 2004; Chaves-Moreira et al. 2011).

There is evidence of evolution of functional specificity of different venom-expressed copies of *SicTox* genes post-duplication. The evidence lies in differences in SMase D activities in crude venoms across sicariid species and differences inactivity of expressed α- and β-clade proteins. In fact, to some degree, differences in the patterns of expression of α- and β-clade *SicTox* genes among sicariid lineages correlate with differences in SMase D activity levels in venoms. Venoms across all New World and Northwest African *Loxosceles* and African *Sicarius* have high levels of SMase D activity, and Southern African *Loxosceles* have slightly reduced levels of SMase D activity. However, South American *Sicarius* are reduced by at least two orders of magnitude relative to African venoms (Binford et al. 2009). South American *Sicarius* express the same β-clade *SicTox* proteins as are found in S. African *Sicarius*. We do not know what has changed in South American *Sicarius* venoms to cause the apparent difference in activity. New World *Sicarius* express a diverse set of β-clade *SicTox*, so we suspect that there is a difference in functional specificity of the *SicTox* genes and the efficacy for hydrolyzing sphingomyelin is reduced. Alternatively, other components in the venoms may be blocking the activity or our ability to detect it with our assay.

So far, 14α-clade and 4β-clade clones have been expressed and their protein products assayed for SMase D activity, and some clones have been assayed for activity across a broader range of phospholipid substrates. At least one α-clade toxin from *L. reclusa* has activity that includes hydrolysis of five other phospholipids beyond sphingomyelin (Lee and Lynch 2005). β-clade clones from North and South American *Loxosceles* show highly reduced/no SMase D activity (reviewed in Binford et al. 2009) but are less well characterized with respect to substrate specificity (Ramos-Cerrillo et al. 2004; de Giuseppe et al. 2011). Moreover, residues in the binding pocket of the enzyme tend to show conservation in α-clade but differences between α- and β-clade and more variability among β-clade toxins (Binford et al. 2009; de Giuseppe et al. 2011). Understanding the mechanism underlying evolution of substrate specificity is an ongoing effort.

17.4.4 Functional Role of SMase D in Prey Capture

Differences in functional specificity of *SicTox* proteins in sicariid venoms have likely been honed by natural selection to help with prey capture. While most focus on the *SicTox* gene family has centered on activity in mammals, their natural prey consists largely, if not exclusively, of arthropod prey (Fischer et al. 2006;

Zobel-Thropp et al. 2012). We have recently determined that purified α-clade protein from *L. arizonica* has SMase D activity and is a potent insecticidal neurotoxin (Zobel-Thropp et al. 2012). The mechanism of neurotoxicity is unknown; however, it likely involves phospholipid head groups interacting with ion channels as the result of cleavage by SMase D (Xu et al. 2008). Interestingly, the lower SMase D activity in American *Sicarius* relative to all *Loxosceles* and African *Sicarius* correlates with a small but significant decrease in venom potency of whole crude venom on insects (Zobel-Thropp et al. 2010).

While our understanding of the specific role of these toxins in prey capture is still developing, proteins in the *SicTox* gene family are highly expressed in sicariid venoms and may have both neurotoxic and cytotoxic effects on prey.

17.4.5 Other Venom Components

While *SicTox* genes are prominently expressed in sicariid venoms, there is also a large set of other components that are acting in concert to immobilize prey. One- and two-dimensional electrophoretic separations (Fig. 17.3) indicate that there are components ranging in size between 2 and 97 kDa and that the patterns of size groupings vary to some degree among species (Machado et al. 2005; Binford et al. 2009; Zobel-Thropp et al. 2010). Transcriptome and/or proteomic analyses of venoms from South American *L. laeta* (Fernandes-Pedrosa et al. 2008) and *L. intermedia* (dos Santos et al. 2009) combined with our own preliminary data for African and North American *Loxosceles* (unpublished) have identified other components including a broad range of proteases (Veiga et al. 2001; Young and Pincus 2001; da Silveira et al. 2007), enzymes that target connective tissues (e.g., hyaluronidase; Young and Pincus 2001), and small peptide neurotoxins (de Castro et al. 2004). I will detail a few of these below with the goal of illustrating our developing understanding of the complexity of the pool of components in these venoms.

Two interesting protein toxins that have been isolated from *Loxosceles* venom gland tissues but appear to be expressed at lower levels than *SicTox* genes are astacins and TCTPs. Astacins are the only other venom-expressed gene family to be formally described across multiple species of sicariids (Trevisan-Silva et al. 2010). They are a family of digestive, extracellular, or cell-surface-bound proteases involved in peptide processing. TCTPs are homologs of the translationally controlled tumor protein family. As the name implies, TCTPs were first discovered from human breast cancer tissues. These proteins are also known as histamine-releasing factors (HRF) because of their effects on humans. The role these 22 kDa proteins play in invertebrate prey capture is unclear, but they represent up to 4 % of transcripts sequenced through high-throughput sequencing (Sade et al. 2012). TCTPs influence inflammatory activity in vertebrates; however, it is not known how these toxins affect invertebrates and the role they may play in prey capture.

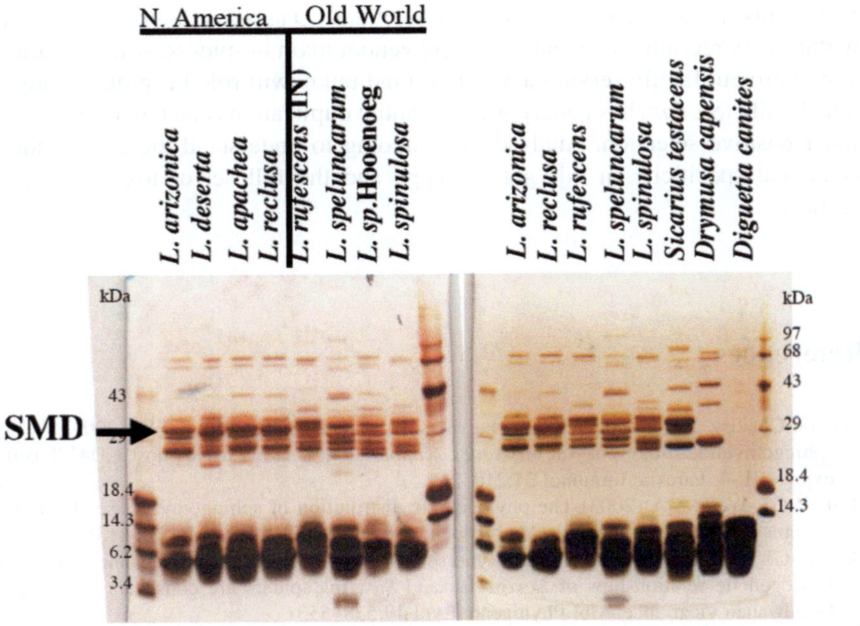

Fig. 17.3 Comparative 1-dimensional electrophoretic separations of proteins in a broad range of sicariid venoms (Binford and Wells 2003). The band that includes proteins in the size of SMase D (SMD) is noted with an area (32–35 kDa). The dense-staining region between ~4–14 kDa includes peptides in the size range of known neurotoxins from other spider species

In addition to apparent neurotoxic activity of *SicTox* genes, sicariid venoms also contain a diverse set of peptide neurotoxins that have characteristic peptide toxin motifs such as being rich in cysteines. Sicaritoxins (LiTx1-4) are the first formally described sicariid peptide neurotoxins. These have been isolated from *L. intermedia* (de Castro et al. 2004) and *L. laeta* (Fernandes-Pedrosa et al. 2008), and our comparative analyses indicate they are widespread in sicariids. Most isolates have principle structural motifs (PSM) (Kozlov and Grishin 2005) that are common in spider venom peptides that are tightly folded with inhibitory cysteine knot (ICK) scaffolds, typical of ion channel-blocking toxins (Escoubas 2006). Despite the recognizable motifs, the new toxins are clearly divergent from toxins in databases and will provide a wealth of opportunity for discovery.

17.5 Conclusions

Sicariid spiders, including *Loxosceles* and *Sicarius*, have an unusual venom toxin with sphingomyelinase D activity. This toxin is a sufficient causative agent for the formation of dermonecrotic lesions when sicariid spiders, including the brown recluse, bite people. It also serves as a potent insecticidal neurotoxin, rapidly and irreversibly immobilizing prey. Sphingomyelinase D appears to represent a distinct evolutionary recruitment event of a novel venom toxin in spiders. After recruitment from a protein family serving a different (and unknown) role in spider's body, the gene family (*SicTox*) has undergone substantial duplications and bursts of evolution under positive selection. Analyses are ongoing to understand the full breadth of functional specificity in this gene lineage and the full set of toxins in sicariid venoms.

References

Alarcon-Chaidez FJ, Boppana VD, Hagymasi AT, Adler AJ, Wikel SK (2009) A novel sphingomyelinase-like enzyme in *Ixodes scapularis* tick saliva drives host $CD4^+$ T cells to express IL-4. Parasite Immunol 31:210–219

Binford GJ, Wells MA (2003) The phylogenetic distribution of sphingomyelinase D activity in venoms of Haplogyne spiders. Comp Biochem Physiol B Biochem Mol Biol 135:25–33

Binford GJ, Callahan MS, Bodner MR, Rynerson MR, Nuñez PB, Ellison CE, Duncam RP (2008) Phylogenetic relationships of *Loxosceles* and *Sicarius* spiders are consistent with Western Gondwanan vicariance. Mol Phylogenet Evol 49:538–553

Binford GJ, Bodner MR, Cordes MHJ, Baldwin KL, Rynerson MR, Burns SN, Zobel-Thropp PA (2009) Molecular evolution, functional variation, and proposed nomenclature of the gene family that includes sphingomyelinase D in Sicariid spider venoms. Mol Biol Evol 26:547–566

Borchani L, Sassi A, Yekhlef RB, Safra I, El Ayeb M (2011) Heminecrolysin, a potential immunogen for monospecific antivenom production against *Hemiscorpius lepturus* scorpion. Toxicon 58:681–688

Chaim OM, da Silveira RB, Trevisan-Silva D, Ferrer VP, Sade YB, Boia-Ferreira M, Gremski LH, Gremski W, Senff-Ribeiro A, Takahashi HK, Toledo MS, Nader HB, Veiga SS (2011) Phospholipase-D activity and inflammatory response induced by brown spider dermonecrotic toxin: endothelial cell membrane phospholipids as targets for toxicity. Biochim Biophys Acta 1811:84–96

Chaves-Moreira D, Souza FN, Fogaca RTH, Mangili OC, Gremski W, Senff-Ribeiro A, Chaim OM, Veiga SS (2011) The relationship between calcium and the metabolism of plasma membrane phospholipids in hemolysis induced by brown spider venom phospholipase-D toxin. J Cell Biochem 112:2529–2540

Cordes MHJ, Binford GJ (2006) Lateral gene transfer of a dermonecrotic toxin between spiders and bacterial. Bioinformatics 22:264–268

da Silva PH, da Silveira RB, Appel MH, Mangili OC, Gremski W, Veiga SS (2004) Brown spiders and loxoscelism. Toxicon 44:693–709

da Silveira RB, Willie AC, Chaim OM, Appel MH, Silva DT, Franco CR, Toma L, Mangili OC, Gremski W, DietrichCP NHB, Veiga SS (2007) Identification, cloning, expression and functional characterization of an astacin-like metalloprotease toxin from *Loxosceles intermedia* (brown spider) venom. Biochem J 406:355–363

de Castro CS, Silvestre FG, Araújo SC, Yazbeck GM, Mangili OC, Cruz I, Chávex-Olórtegui C, Kalapothakis E (2004) Identification and molecular cloning of insecticidal toxins from the venom of the brown spider *Loxosceles intermedia*. Toxicon 44:273–280

de Giuseppe PO, Ullah A, Silva DT, Gremski LH, Wille AC, Chaves Moreira D, Ribeiro AS, Chaim OM, Murakami MT, Veiga SS, Arni RK (2011) Structure of a novel class II phospholipase D: catalytic cleft is modified by a disulphide bridge. Biochem Biophys Res Commun 409:622–627 dos Santos LD, Dias NB, Roberto J, Pinto AS, Palma MS (2009) Brown recluse spider venom: proteomic analysis and proposal of a putative mechanism of action. Protein Pept Lett 16:933–943

Duncan RP, Autumn K, Binford GJ (2007) Convergent setal morphology in sand-covering spiders suggests a design principle for particle capture. Proc R Soc B 274:3049–3056

Duncan RP, Rynerson MR, Ribera C, Binford GJ (2010) Diversity of *Loxosceles* spiders in Northwestern Africa and molecular support for cryptic species in the *Loxosceles rufescens* lineage. Mol Phylogenet Evol 55:234–248

Escoubas P (2006) Molecular diversification in spider venoms: a web of combinatorial peptide libraries. Mol Divers 10:545–554

Escoubas P, King GF (2009) Venomics as a drug discovery platform. Expert Rev Proteomics 6:221–224

Fernandes-Pedrosa MF, Junqueira de Azevedo ILM, Gonçalves de Andrade RM, Kobashi LS, Almeida DD, Ho PL, Tambourgi DV (2008) Transcriptome analysis of *Loxosceles laeta* (Araneae, Sicariidae) spider venomous gland using expressed sequence tags. BMC Genomics 9:279–290

Fischer ML, Vasconcellos-Neto J, Neto LGS (2006) The prey and predators of Loxosceles intermedia Mello-Leitao 1934 (Araneae, Sicariidae). J Arachnol 34:485–488

Foradori MJ, Smith SC, Smith E, Wells RE (2005) Survey for potentially necrotizing spider venoms, with special emphasis on *Cheiracanthium mildei*. Comp Biochem Physiol C141:32–39

Fry BG, Roelants K, Champagne DE, Scheib H, Tyndal JDA, King GF, Nevalainen TJ, Norman JA, Lewis RJ, Norton RS, Renjifo C, Rodríguez de la Vega RC (2009) The toxicogenomic multiverse: convergent recruitment of proteins into animal venoms. Annu Rev Genomics Hum Genet 10:483–511

Gertsch WJ, Ennik F (1983) The spider genus *Loxosceles* in North America, Central America, and the West Indies (Araneae, Loxoscelidae). Bull Am Mus Nat Hist 175:264–360

Gremski LH, da Silveira RB, Chaim OM, Probst CM, Ferrer VP, Nowatzki J, Weinschutz HC, Madeira HM, Gremski W, Nader HB, Senff-Ribeiro A, Veiga SS (2010) A novel expression profile of the *Loxosceles intermedia* spider venomous gland revealed by transcriptome analysis. Mol Biosyst 6:2403–2416

Herzig V, King G (2013) The neurotoxic mode of action of venoms from the spider family Theraphosidae. In: Nentwig W (ed) Spider ecophysiology. Springer, Heidelberg (this volume)

Kozlov S, Grishin E (2005) Classification of spider neurotoxins using structural motifs by primary structure features. Single residue distribution analysis and pattern analysis techniques. Toxicon 46:672–686

Kuhn-Nentwig L, Nentwig W (2013) The cytotoxic mode of action of the venom of *Cupiennius salei* (Ctenidae). In: Nentwig W (ed) Spider ecophysiology. Springer, Heidelberg (this volume)

Kurpiewski G, Forrester LJ, Barrett JT, Campbell BJ (1981) Platelet aggregation and sphingomyelinase D activity of a purified toxin from the venom of *Loxosceles reclusa*. Biochim Biophys Acta 678:467–476

Lee S, Lynch KR (2005) Brown recluse spider (*Loxosceles reclusa*) venom phospholipase D (PLD) generates lysophosphatidic acid (LPA). Biochem J 391:317–323

Machado LF, Laugesen S, Botelho ED, Ricart CAO, Fontes W, Barbaro KC, Roepstorff P, Sousa MV (2005) Proteome analysis of brown spider venom: identification of loxnecrogin isoforms in *Loxosceles gaucho* venom. Proteomics 5:2167–2176

Nentwig W, Kuhn-Nentwig L (2013) Main components of spider venoms. In: Nentwig W (ed) Spider ecophysiology. Springer, Heidelberg (this volume)

Nentwig W, Kuhn-Nentwig L (2013) Spider venoms potentially lethal to humans. In: Nentwig W (ed) Spider ecophysiology. Springer, Heidelberg (this volume)

Newlands G, Atkinson P (1990) Behavioural and epidemiological considerations pertaining to necrotic araneism in southern Africa. S Afr Med J 77:92–95

Oldrati V, Bianchi E, Stöcklin R (2013) Spider venom components as drug candidates. In: Nentwig W (ed) Spider ecophysiology. Springer, Heidelberg (this volume)

Platnick I (2012) The World Spider Catalog, version 12.1, online at http://research.amnh.org/iz/spiders/catalog/COUNTS.html

Ramos-Cerrillo B, Olvera A, Odell GV, Zamudio F, Paniagua-Solis J, Alagon A, Stock RP (2004) Genetic and enzymatic characterization of sphingomyelinase D isoforms from the North American fiddleback spiders *Loxosceles boneti* and *Loxosceles reclusa*. Toxicon 44:507–514

Sade YB, Bóia-Ferreira M, Gremski LH, da Silveira RB, Gremski W, Senff-Ribeiro A, Chaim OM, Viego SS (2012) Molecular cloning, heterologous expression and functional characterization of a novel translationally-controlled tumor protein (TCTP) family member from *Loxosceles intermedia* (brown spider) venom. Int J Biochem Cell Biol 44:170–177

Suter RB, Stratton GE (2013) Predation by spitting spiders—elaborate venom gland, intricate delivery system. In: Nentwig W (ed) Spider ecophysiology. Springer, Heidelberg (this volume)

Swanson DL, Vetter RS (2005) Bites of brown recluse spiders and suspected necrotic arachnidism. N Engl J Med 352:700–707

Tambourgi DV, Magnoli FC, van den Berg CW, Morgan BP, de Araujo PS, Alves EW, da Silva WD (1998) Sphingomyelinases in the venom of the spider *Loxosceles intermedia* are responsible for both dermonecrosis and complement dependent hemolysis. Biochem Biophys Res Commun 251:366–373

Trevisan-Silva D, Gremski LH, Chaim OC, da Silveira RB, Meissner GO, Mangili OC, Barbaro KC, Gremski W, Veiga SS, Senff-Ribeiro A (2010) Astacin-like metalloproteases are a gene family of toxins present in the venom of different species of the brown spider (genus *Loxosceles*). Biochimie 92:21–32

Ushkaryov YA, Volynski KE, Ashton AC (2004) The multiple actions of black widow spider toxins and their selective use in neurosecretion studies. Toxicon 43:527–542

Van Aswegen G, van Rooyen JM, van der Nest DG, Veldman FJ, de Villiers TH, Oberholzer G (1997) Venom of a six-eyed crab spider, *Sicarius testaceus* (Purcell, 1908), causes necrotic and haemorrhagic lesions in the rabbit. Toxicon 35:1149–1152

van Meeteren LA, Frederiks F, Giepmans BNG, Pedrosa MFF, Billington SJ, Jost BH, Tambourgi DV, Moolenaar WH (2004) Spider and bacterial sphingomyelinases D target cellular lysophosphatidic acid receptors by hydrolyzing lysophosphatidylcholine. J Biol Chem 279:10833–10836

Veiga SS, Zanetti VC, Braz A, Mangili OC, Gremski W (2001) Extracellular matrix molecules as targets for brown spider venom toxins. Braz J Med Biol Res 34:843–850

Xu Y, Ramu Y, Lu Z (2008) Removal of phospho-head groups of membrane lipids Immobilizes voltage sensors of K^+ channels. Nature 451:826–829

Young AR, Pincus SJ (2001) Comparison of enzymatic activity from three species of necrotizing arachnids in Australia: *Loxosceles rufescens, Badumna insignis* and *Lampona cylindrata*. Toxicon 39:391–400

Zobel-Thropp PA, Bodner MR, Binford GJ (2010) Comparative analyses of venoms from American and African *Sicarius* spiders that differ in sphingomyelinase D activity. Toxicon 55:1274–1282

Zobel-Thropp PA, Kerins AE, Binford GJ (2012) Sphingomyelinase D in sicariid spider venom is a potent insecticidal toxin. Toxicon 60:265–271

Chapter 18
Predation by Spitting Spiders: Elaborate Venom Gland, Intricate Delivery System

Robert B. Suter and Gail E. Stratton

18.1 Introduction

Spitting spiders (Araneae: Scytodidae) subdue prey by entangling them at a distance with a mixture of silk, glue, and perhaps venom (Fig. 18.1; Gilbert and Rayor 1985; Suter and Stratton 2009; Foelix 2011; see also a detailed behavioral description in the appendix of Japyassú and Machado 2010). All of the components of this mixture originate in the venom glands, a pair of relatively elaborate structures (Fig. 18.2) consisting of five histologically distinct regions, the anterior three producing venom and the posterior two, comprising the largest lobe of the gland, producing silk and glue (Kovoor and Zylberberg 1972).

In a sense, the venom gland and its products anchor a suite of tightly linked adaptations that constitute the predatory system of these remarkable spiders. These include not only the morphological delivery apparatus (venom ducts, modified fangs, and both cheliceral and prosoma musculature) but also behaviors such as the collection of chemical, tactile, and proprioceptive sensory data that indicate something about the location and identity of a potential prey item and the integration of this information in triggering a predatory attack.

R.B. Suter (✉)
Department of Biology, Vassar College, Poughkeepsie, NY, USA
e-mail: suter@vassar.edu

G.E. Stratton
Department of Biology, University of Mississippi, University, MS, USA
e-mail: byges@olemiss.edu

W. Nentwig (ed.), *Spider Ecophysiology*,
DOI 10.1007/978-3-642-33989-9_18,

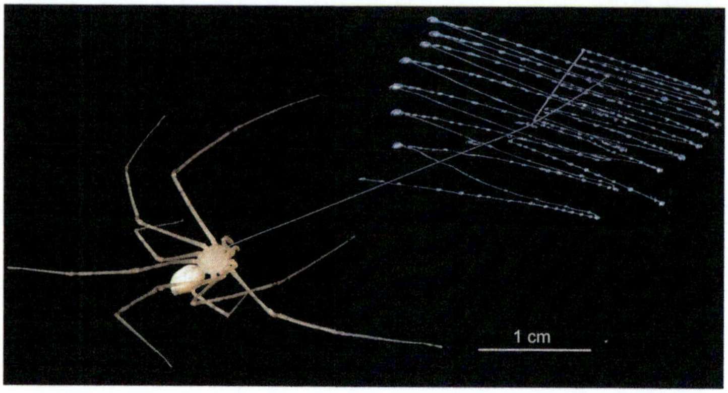

Fig. 18.1 A female *Scytodes thoracica* and her spit as it was deposited on a microscope slide. In the spit, two separate zigzag patterns can be seen, the left one having issued from the spider's left fang and the right one from the right fang. The spit was deposited from bottom to top as the chelicerae were extended. These spiders accomplish a full spitting episode in 18–32 ms. The single line joining the spider and the spit was added digitally. (Photo illustration by RB Suter)

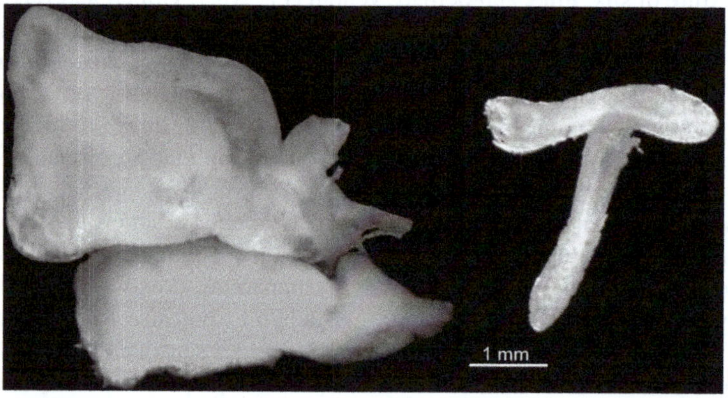

Fig. 18.2 Venom glands from *Scytodes* (*left*) and *Schizocosa* (*right*), a wolf spider. Lateral (*upper*) and ventral (*lower*) views of the venom glands of *S. thoracica* as compared to the venom glands of *Schizocosa duplex*. The glands, which were removed from spiders of about the same size, demonstrate both the more elaborate structure of the ones from *S. thoracica* and their larger relative size. (Photo illustration by RB Suter; Suter and Stratton 2005)

18.2 Venom Gland Contents

The mix of materials ejected from the spider's fangs during expectoration includes a gluey substance and a fibrous material (Fig. 18.3). It is not clear that the spit from *Scytodes* contains venom. Venom may be excreted without silk and glue (e.g., during envenomation after a prey item has been immobilized by spit; Clements and Li 2005; Japyassú and Machado 2010), so it may also be possible for the spider to eject silk and glue without venom, and because spit-immobilized

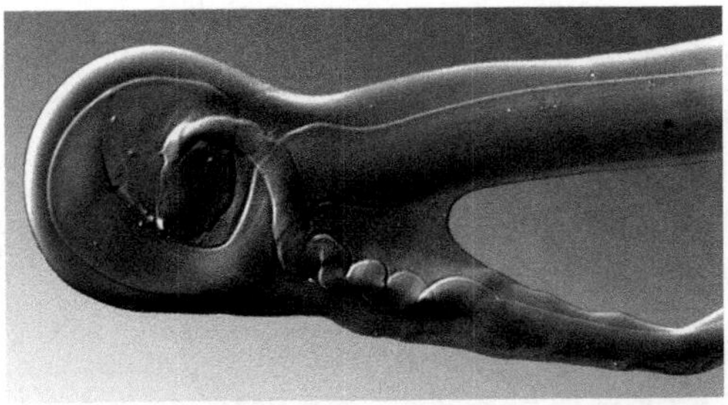

Fig. 18.3 The transition between the laterad sweep of silk deposition (*lower*) and the mediad sweep (*upper*) during a single oscillation cycle. Both silk and the surrounding glue are visible. This DICM image corresponds to one of the leftmost transitions shown in Fig. 18.1. (Photo by RB Suter)

prey are apparently unharmed by the spit itself (Clements and Li 2005), it may be that the spit contains no venom. Alternatively, the toxic components of the venom may be present but unable to penetrate the cuticle even at joints and other locations where the cuticle is thin and unsclerotized. The venom glands of spitting spiders appear not to produce sphingomyelinase D, the enzyme present in brown recluse (*Loxosceles* sp.) venoms that is largely responsible for the dermonecrotic lesions that form at the sites of recluse bites in humans (Binford and Wells 2003; Binford 2013). Neither the fibrous material, which we refer to as silk, nor the glue produced in the spitting spider's venom glands has been chemically characterized, although some histochemical data are available (Kovoor and Zylberberg 1972; Kovoor 1987) for *Scytodes velutina* (sub *S. delicatula*).

We know surprisingly little about the mechanical attributes of the silk. Within the first 0.3 s after expectoration, the silk contracts to about 60 % of its initial length, and it does so with sufficient force, about 0.3 mN in the aggregate, to compress the legs of thin-legged prey close to their bodies (Suter and Stratton 2009). During many predatory attacks, spitting spiders spit multiple times in close succession (Gilbert and Rayor 1985; Japyassú and Machado 2010), multiplying this compression and further restraining the prey. The silk has sufficient tensile strength to ensnare a prey item and to hold it relatively immobile while the attacking spider bites it and injects venom, usually on a protruding leg or antenna. The silk's tensile strength, as well as its stress vs. strain characteristics and their derivatives (e.g., Young's modulus, elasticity), has yet to be measured.

Inside the venom gland, much of the volume is occupied by the silk's immediate precursor. Transmission electron microscopy (Kovoor and Zylberberg 1972) reveals a substance that resembles what is seen in spiders' opisthosomal silk glands—a mass of closely packed fibrils in water solution constituting, together, a liquid crystalline material (Kerkam et al. 1991; Vollrath and Knight 2001). The other substance that occupies much of the venom gland's volume is the viscous gluey substance responsible for the spit's stickiness. By analogy, and perhaps by homology, this material may be related to the sticky material produced by the aggregate glands in the opisthosoma of araneoid spiders, although one should be skeptical about the possible homology since spitting spiders are only distantly related to the araneoids (Coddington 2005). The material from the aggregate glands is a solution of glycoproteins,

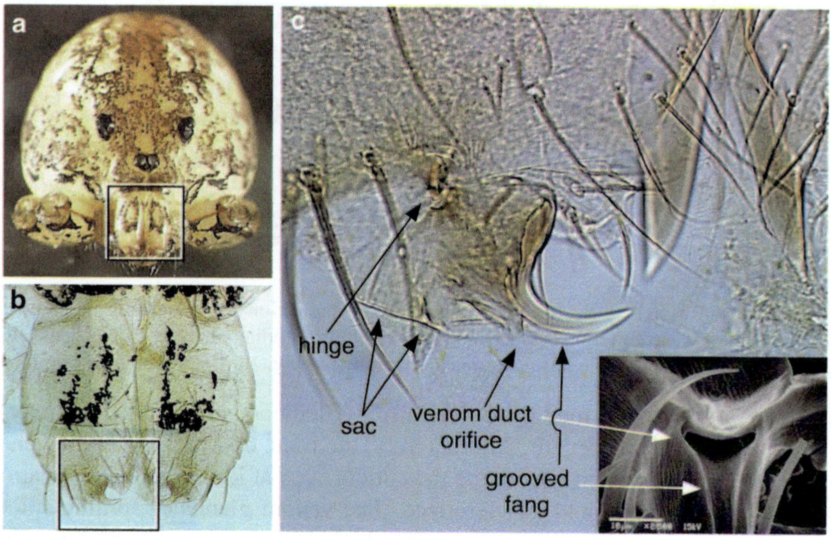

Fig. 18.4 Anterior view of the prosoma of *S. thoracica* (**a**), an enlarged view of the chelicerae (**b**), and two representations (**c**) of a fang and its associated hinge, sac, and venom duct orifice (Photo illustration by RB Suter; Suter and Stratton 2009)

neurotransmitter-like compounds, free amino acids, and other constituents, both organic and inorganic, which are low in concentration but may be functionally important (Choresh et al. 2009; Opell and Hendricks 2010; Sahni et al. 2010).

Work in other laboratories is under way to characterize the various components of the venom (sensu stricto) and of the glue and silk of *Scytodes*. We have hypothesized (Suter and Stratton 2009) that the silk is homologous to silk produced by the spider's opisthosomal silk glands because the genes coding for those fibroin proteins were already present in the venom gland cells of the spiders that, in evolution, gave rise to modern Scytodidae. Recent work by J. Garb and S. Correa (personal communication), comparing venom gland cDNAs to cDNAs derived from opisthosomal silk gland contents, does not support this hypothesis.

18.3 Organized Spit Pattern

When confronted with a prey item or, in the laboratory, when stimulated on the sternum with a hair, the spitting spider ejects a stream of spit through the opening of the venom duct located on each fang. In most spiders, this opening is near the distal end of the fang and on the fang's convex surface (see Fig. 2.6 in Foelix 2011). In spitting spiders, in contrast, the opening is at the proximal margin of the fang (Fig. 18.4), bordered on the fang side by a groove in the fang itself and on the cheliceral side by the somewhat sclerotized edge of the spider's cuticle.

If collected on a microscope slide and then visualized via differential interference contrast microscopy (DICM), the ejected spit can be seen as forming a highly organized pattern (Fig. 18.1), a zigzag that progresses from bottom to top. This - large-scale pattern forms because, during a spitting episode, the chelicerae are extended upward (providing the progression from bottom to top) at the same time that the fangs rapidly oscillate from side to side (providing the zigzags) (Suter and Stratton 2009).

There are also interesting but as yet unexplained smaller scale attributes of the overall pattern. For example, the mediad excursions of the zigzag are typically embellished with conspicuous and roughly evenly spaced droplets of fluid, whereas the laterad sweeps are thin and free of droplets (Fig. 18.1). Those familiar with the fine structure of the sticky spiral silk in the webs of orb-weaving araneids will recognize the similarity, in both structure and function, between those sticky droplet-covered silk lines and these in the spit of *Scytodes*. A notable difference, however, is that the components of the sticky spiral (silk and glue) are produced in different glands (the flagelliform and aggregate glands, respectively; Vollrath and Knight 2001; see also Townley and Tillinghast 2013) and combined from separate spinnerets as the silk is drawn out; in spitting spiders, all components are produced in a single gland and are extruded together through the same orifice.

18.4 Biomechanics of Spitting

The organized spit pattern (Fig. 18.1) is formed in 18–32 ms. During this brief interval, about 30 cm of glue-laden silk is ejected from each fang at rates as high as 28 m/s, a remarkable achievement considering (a) that the ejected material has both solid and liquid components and (b) that it is ejected through a very narrow opening (in adults of *Scytodes thoracica*, the crescent-shaped orifice is about 14 μm wide). Just as remarkable, because of what it implies about the underlying mechanism, is the fact that during this same brief interval, the fangs generate the zigzag pattern by oscillating at 300–1,700 Hz (Suter and Stratton 2009). These two processes, concerning fluid dynamics and fang dynamics, respectively, are considered below.

18.4.1 Fluid Dynamics

When spiders lay down silk from their opisthosomal silk glands and spinnerets, the silk is initially extruded through a valve and spigot as a concentrated, ion-modulated, water solution of silk proteins. This non-Newtonian liquid (Vollrath and Knight 2001) flows out in response, presumably, to the difference in hydrostatic pressure between the pressurized opisthosoma and the air outside the spider (Vollrath and Knight 2001; Foelix 2011). The proto-silk's irreversible hardening into a solid silk strand is achieved when the proto-silk is pulled and stretched by the spider's motion or weight or, during ballooning, by air currents (Eberhard 1987; Suter 1999; Vollrath and Knight 2001; Foelix 2011).

In contrast, when a spitting spider ejects its combination of silk and glue through an opening at the base of its fang, the only propellant is the hydrostatic pressure difference—no external object or force pulls on the proto-silk, stressing it and causing solidification. Moreover, the ejection is very rapid despite occurring through a narrow opening. The only way this can work is if the proto-silk material, in combination with the glue that is ejected with it, experiences shear thinning as it moves through the narrow opening. Shear thinning also makes opisthosomal silk extrusion possible, and it is from studies of that process that we know of its importance in spider silk spinning. Shear thinning refers to the reduction in viscosity that occurs when a non-Newtonian fluid experiences shear forces as microscopic streams of the fluid move past each other at different velocities. See Vollrath and Knight (2001) for a full and lucid account of the role of shear thinning in silk production.

Even with shear thinning of the ejecting fluid, the rate of ejection is still impressive. A possible explanation is that the hydrostatic pressure inside the spitting spider's prosoma is temporarily elevated during spitting, accelerating the ejected material to a higher velocity than would be possible with the spiders resting hydrostatic pressure. This may be the case: Millot (1949, in Foelix 2011) stated that an abrupt contraction of the muscles along the posterior margin of the spider's prosoma compresses its contents and raises the hydrostatic pressure. The elevated hydrostatic pressure might also provide enough energy to deform the proto-silk sufficiently during ejection to cause it to solidify, the process accomplished by pulling and stretching in the formation of opisthosomal silk (above). We do not yet know whether the presumed parallels between the spitting spider's prosomal silk formation and opisthosomal silk formation in these and other spiders are valid.

18.4.2 Fang Dynamics

Arthropods and many other animals ordinarily move individual body parts by contracting muscles, sometimes augmenting those forces by using energy stored in tendons or other elastic structures (Sensenig and Shultz 2003; Vogel 2009). Spiders use muscles in the same way, although the extension of some leg joints is accomplished with hydraulics rather than with direct muscle action (Anderson and Prestwich 1975; see also Kropf 2013). The motions of a spider's fangs are usually controlled by the extensor and flexor muscles located in its chelicerae (Foelix 2011). For spitting spiders, though, that may not be the whole story.

18.4.2.1 A Role for Muscles

The contraction and relaxation of muscles takes time, and that limits the number of times per second that a muscle can contract. As a consequence, in biological systems, anatomical structures rarely oscillate above 500 Hz, and when they do, asynchronous muscle contraction is usually involved (Syme and Josephson 2002). In asynchronous muscle contraction, a single muscle twitch drives multiple oscillations of a structure. This is only possible when the structure (e.g., a fly's wing) is mechanically coupled to an elastic element (the thoracic cuticle) that can alternately store and release mechanical energy (Josephson et al. 2000).

The fang oscillation frequencies typically achieved by spitting spiders far exceed 500 Hz (278–1,781 Hz; Suter and Stratton 2009), and the fang is mechanically coupled to such elastic structures as the cuticle of the chelicera and the fang-chelicera hinge. Could asynchronous muscle contraction account for the high-frequency oscillations of the fangs? Because asynchronous muscle contraction is only known in insects (Josephson et al. 2000), this hypothetical fang-driving mechanism seems unlikely in the spitting spider. On the other hand, high-frequency oscillations of spider appendages are rare outside of the Scytodidae, so asynchronous contraction may have been overlooked because it was not sought.

18.4.2.2 A Role for Hydrodynamic Forcing

Many high-frequency biological oscillations are *controlled* by muscles but are not directly *driven* by them—that is, these oscillations do not require that muscles contract at frequencies well above 500 Hz. The oscillations in a song bird's syrinx, for example, are caused by a fluid (air) flowing past flaps of tissue (the tympaniform membranes) that are both elastic and under tension. Each flap oscillates at a frequency that is primarily dependent on its length, its elasticity, and the tension it is under (Gardner et al. 2001). Muscles influence the flap's tension and length and certainly provide the pressure differential that causes the air to move past the flap, but they do not directly drive the oscillation.

The oscillations of a spitting spider's fang could be caused by something akin to the hydrodynamic forcing that causes a tympaniform membrane to vibrate. A closer analog is the whipping back and forth of a squirting hose when nobody is holding its nozzle. The necessary conditions are a fluid flowing through or past the fang, a mass (the fang) that is set in motion in reaction to the fluid flow, and elastic constraints on the motion of the fang. All of these conditions are satisfied when *Scytodes* ejects its mixture of glue and silk. Moreover, the ejection is rapid, causing a greater reaction force than it would if the ejection were slower, and the fangs are small (Suter and Stratton 2005, 2011), raising their oscillation frequency relative to what it would be if the fangs were more massive (Suter and Stratton 2009).

18.4.2.3 Corroboration?

Our working hypothesis has been that hydrodynamic forcing provides the impetus for the oscillations of the spitting spider's fangs. This appeals both because the alternative, asynchronous muscle contraction is not known in arachnids and because the high-velocity fluid ejection can so tidily account for the motion of

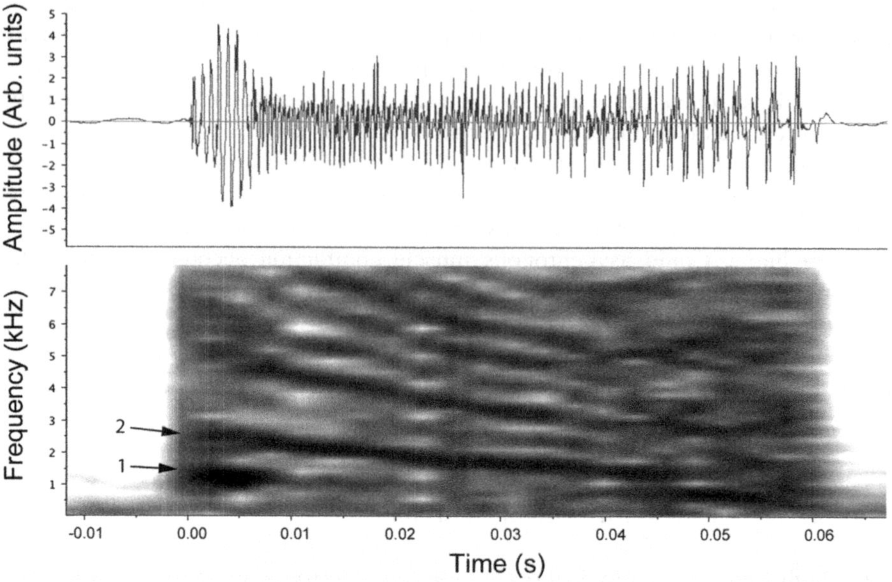

Fig. 18.5 Waveform (*upper*) and sound spectrogram (*lower*) of a single spitting episode involving both fangs ejecting a mixture of silk and glue. In the spectrogram, the dark lines with the lowest frequencies (1 and 2) would normally be considered the first and second harmonics of a sound. In this case, however, they represent the times when both fangs are oscillating in phase with each other (1; before 0.01 s and beginning again just before 0.05 s) and when the two fangs are oscillating out of synchrony (2; between 0.007 and 0.045 s). These changes from synchrony to asynchrony and back are also visible in the waveform. Note the declining frequency of the fang oscillations over time in both representations

the diminutive fangs. However, in the absence of a scaled-up working physical model that mimics the motions of the fangs during expectoration, we have looked for other kinds of corroborating support for the hypothesis.

One piece of corroboration comes by way of laser Doppler vibrometry. *Scytodes*' spitting is energetic enough to vibrate the spider's prosoma, and these vibrations are easily detected by a laser Doppler vibrometer. The resulting sound spectrograph of a single spitting event (Fig. 18.5) reveals that the fangs' oscillation frequencies decrease with time and that the fangs sometimes oscillate in synchrony but often do not. If the fangs' oscillations were driven directly by muscular contractions, one might expect the nervous system either to alternate the left and right fangs or to synchronize their motions, but one would not expect the two fangs to slide in and out of synchrony. On the other hand, because minor differences in fang mass and fang-associated elastic elements would cause differences in hydrodynamically forced oscillations, that kind of sliding in and out of synchrony is just what one would expect if the fangs were propelled by fluid flow (i.e., lacking the controls of a bilaterally organized nervous system).

Our search for corroboration also produced evidence that does not support the hydrodynamic forcing hypothesis. First, there is a tight mathematical relationship between oscillation frequency and the mass of the fang because it is the fang that

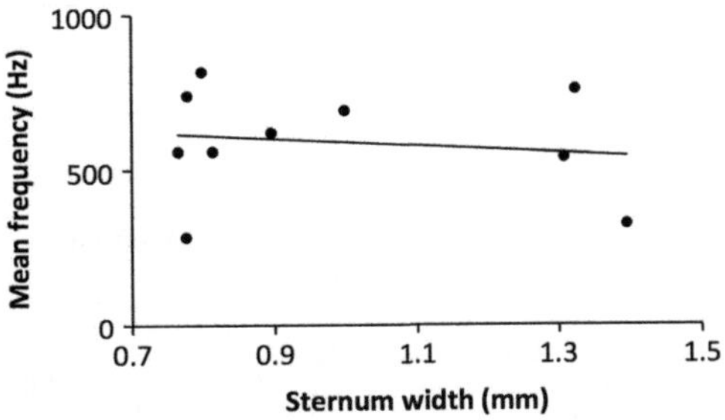

Fig. 18.6 If the fang oscillations of *S. thoracica* were propelled solely by hydrodynamic forcing, one would expect the oscillation frequency to fall with increasing fang mass. Fang size varies directly with spider size (Suter and Stratton 2011), so the expectation would become that oscillation frequency would vary with a reliable measure of spider size. There appears to be no such relationship ($N = 10$ spiders, two replications for each; $r^2 = 0.036$, $P = 0.652$) even over a range of sizes that correspond to a sixfold range in spider mass

must be accelerated at each extreme of its motion—when all else is equal, a more massive fang must oscillate at a lower frequency (Suter and Stratton 2011). But the fangs of larger spitting spiders do not oscillate more slowly than the fangs of much smaller spiders of the same species (Fig. 18.6).

Second, when we studied high-speed video showing the dynamics of fang motion in great temporal detail (at 10,000 fps), the oscillations were periodic but not sinusoidal (Fig. 18.7). Under hydrodynamic forcing, at least in the absence of a valve that changes fluid output at some constant phase of a cycle, one would expect motion that is both continuous and roughly sinusoidal. Instead, the motion of the fang tips is discontinuous and approximately triangular.

These new pieces of evidence point in opposite directions and thus do not help either to falsify or to confirm our hypothesis that the rapid oscillations of the spitting spider's fangs are driven by hydrodynamic forcing.

18.5 Conclusions

Spitting spiders of all sizes attack other arthropods by ejecting a mixture of silk and glue at them, immobilizing them long enough to allow safe envenomation. The spitting requires considerable metabolic expenditure both in the form of maintenance of impressively large venom glands and in the form of the biosynthesis of the proto-silk and the glue that are the primary constituents of the spit. Sensory input allowing identification and localization of the potential prey and neural coordination of the spit ejection itself also must occur. It may be that an important part of expectoration, the oscillations of the fangs (contributing to the zigzag pattern of spit that effectively covers a prey item), once triggered, does not require further neural input, but that possibility remains to be confirmed.

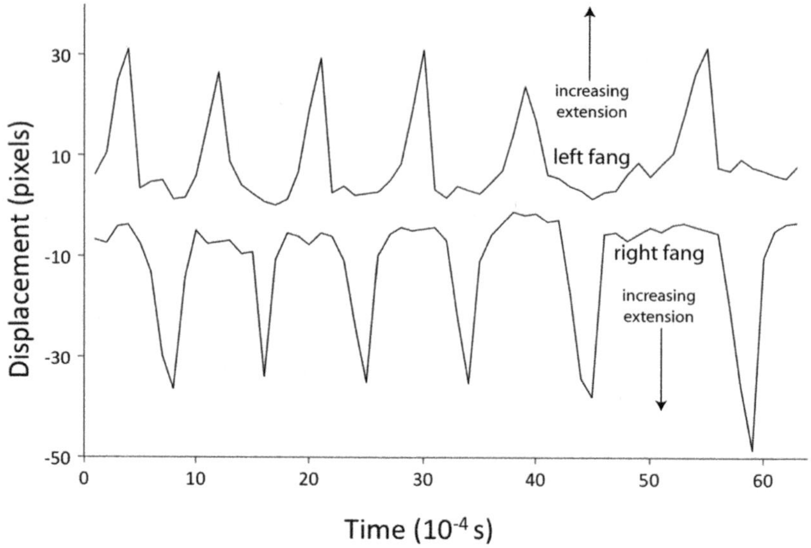

Fig. 18.7 Movements of the tips of the fangs during a portion of a single spitting episode. In this instance, the left and right fangs did not extend simultaneously, but they did maintain a relatively constant phase relationship with each other. Neither fang oscillated in a sinusoidal manner. The fang positions were digitized by applying imaging software (ImageJ) to video frames captured at 10,000 fps

The chemical, physical, and microfluidic properties of the proto-silk and glue mixture are also crucial for the proper functioning of the spitting system; at the moment, these properties are poorly understood except insofar as they may be similar to the analogous properties of the proto-silk and glue produced in the opisthosomal silk glands of spiders.

The evolution of the integrated spitting system of *Scytodes* remains veiled. Its component parts (hypertrophy and differentiation of the venom glands; modification of the fluid dynamic characteristics of the venom gland contents; migration of the venom duct opening from the distal to the proximal surface of the fang; elaboration and refinement of the neural support networks) now constitute an intricate, functioning system. It is not easy to discern the small steps that led from a pholcid-like spider to the current scytodid plan.

References

Anderson JF, Prestwich KN (1975) The fluid pressure pumps of spiders (Chelicerata, Araneae). Z Morph Tiere 81:257–277

Binford G (2013) The evolution of a toxic enzyme in sicariid spiders. In: Nentwig W (ed) Spider ecophysiology. Springer, Heidelberg (this volume)

Binford GJ, Wells MA (2003) The phylogenetic distribution of sphingomyelinase D activity in venoms of haplogyne spiders. Comp Biochem Physiol B 135:25–33

Choresh O, Bayarmagnai B, Lewis RV (2009) Spider web glue: two proteins expressed from opposite strands of the same DNA sequence. Biomacromolecules 10:2852–2856

Clements R, Li D (2005) Regulation and non-toxicity of the spit from the pale spitting spider *Scytodes pallida* (Araneae: Scydodidae). Ethology 111:311–321

Coddington JA (2005) Phylogeny and classification of spiders. In: Ubick D, Paquin P, Cushing PE, Roth V (eds) Spiders of North America: an identification manual. American Arachnological Society

Eberhard WG (1987) How spiders initiate airborne lines. J Arachnol 15:1–9

Foelix RF (2011) Biology of spiders. Oxford University Press, Oxford

Gardner T, Cecchi G, Magnasco M, Laje R, Mindlin GB (2001) Simple motor gestures for birdsongs. Phys Rev Lett 87:208101–1

Gilbert C, Rayor LS (1985) Predatory behavior of spitting spiders (Araneae: Scytodidae) and the evolution of prey wrapping. J Arachnol 13:231–241

Japyassú HF, Machado FA (2010) Coding behavioural data for cladistic analysis: using dynamic homology without parsimony. Cladistics 26:625–642

Josephson RK, Malamud JG, Stokes DR (2000) Asynchronous muscle: a primer. J Exp Biol 203:2713–2722

Kerkam K, Viney C, Kaplan D, Lombardi S (1991) Liquid crystallinity of natural silk secretions. Nature 349:596–598

Kovoor J (1987) Comparative structure and histochemistry of silk-producing organs in arachnids. In: Nentwig W (ed) Ecophysiology of spiders. Springer, Berlin

Kovoor J, Zylberberg L (1972) Histologie et infrastructure de la glande chélicérienne de *Scytodes delicatula* Sim. (Araneidae, Scytodidae). Ann Sci Nat Zool 14:333–388

Kropf C (2013) Hydraulic system of locomotion. In: Nentwig W (ed) Spider ecophysiology. Springer, Heidelberg (this volume)

Opell BD, Hendricks ML (2010) The role of granules within viscous capture threads of orb-weaving spiders. J Exp Biol 213:339–346

Sahni V, Blackledge TA, Dhinojwala A (2010) Viscoelastic solids explain spider web stickiness. Nat Commun 1:19. doi:10.1038/ncomms1019

Sensenig AT, Shultz JW (2003) Mechanics of cuticular elastic energy storage in leg joints lacking extensor muscles in arachnids. J Exp Biol 206:771–784

Suter RB (1999) An aerial lottery: the physics of ballooning in a chaotic atmosphere. J Arachnol 27:281–293

Suter RB, Stratton GE (2005) *Scytodes* vs. *Schizocosa*: predatory techniques and their morphological correlates. J Arachnol 33:7–15

Suter RB, Stratton GE (2009) Spitting performance parameters and their biomechanical implications in *Scytodes* (Araneae, Scytodidae). J Insect Sci 9:62, http://www.insectscience.org/9.62/

Suter RB, Stratton GE (2011) Does allometric growth explain the diminutive size of the fangs of *Scytodes* (Araneae, Scytodidae)? J Arachnol 39:174–177

Syme DA, Josephson RK (2002) How to build fast muscles: synchronous and asynchronous designs. Integr Comp Biol 42:762–770

Townley MA, Tillinghast E (2013) Aggregate silk gland secretions of araneoid spiders. In: Nentwig W (ed) Spider ecophysiology. Springer, Heidelberg (this volume)

Vogel S (2009) Glimpses of creatures in their physical worlds. Princeton University Press, Princeton

Vollrath F, Knight DP (2001) Liquid crystalline spinning of spider silk. Nature 410:541–548

Chapter 19
Spider Venoms Potentially Lethal to Humans

Wolfgang Nentwig and Lucia Kuhn-Nentwig

19.1 Introduction

Spiders have one pair of venom glands, and only a few families have reduced them completely (Uloboridae, Holarchaeidae) or modified them to another function (Symphytognathidae or Scytodidae, see Suter and Stratton 2013). All other 42,000 known spider species (99%) utilize their venom to inject it into prey items, which subsequently become paralysed or are killed. Spider venom is a complex mixture of hundreds of components, many of them interacting with cell membranes or receptors located mainly in the nervous or muscular system (Herzig and King 2013). Spider venom, as it is today, has a 300-million-yearlong history of evolution and adaptation and can be considered as an optimized tool to subdue prey.

In Mesothelae, the oldest spider group with less than 100 species, the venom glands lie in the anterior part of the cheliceral basal segment. They are very small and do not support the predation process very effectively. In Mygalomorphae, the venom glands are well developed and fill the basal cheliceral segment more or less completely. Many of these 3,000 species are medium- to large-/very large-sized spiders, and they have created the image of being dangerous beasts, attacking and killing a variety of animals, including humans. Although this picture is completely wrong, it is persistent and contributes considerably to human arachnophobia. The third group of spiders, Araneomorphae or "modern spiders", comprises 93% of all spider species. The venom glands are enlarged and extend to the prosoma; the openings of the venom ducts are moved from the convex to the concave side of the cheliceral fangs and enlarged as well. These changes save the chelicerae from the necessity of being large, and hence, on the average, araneomorph spiders are much smaller than mygalomorphs. Nevertheless, they possess relatively large venom glands, situated mainly in the prosoma, and may also have rather potent venom.

W. Nentwig (✉) • L. Kuhn-Nentwig
Institute of Ecology and Evolution, University of Bern, Bern, Switzerland
e-mail: wolfgang.nentwig@iee.unibe.ch

W. Nentwig (ed.), *Spider Ecophysiology*,
DOI 10.1007/978-3-642-33989-9_19,

This evolutionary development leads to the strange situation that while most large mygalomorphs are rather harmless to humans, among the araneomorph spiders, there are some groups possessing venoms which may affect humans considerably.

Most spiders never attack or bite humans. Moreover, several conditions must be fulfilled to call a spider dangerous or harmful for humans. (1) Spiders must be able to survive in the human environment or somehow coexist with humans. (2) After contact with the human skin, spiders must be willing to bite, i.e. they need a minimum of aggressive or self-defence behaviour. (3) Only spiders with more than 8–10 mm body length can penetrate the relatively thick human skin with their chelicerae. Since most spiders are smaller than 8–10 mm, do not live in man-made habitats and do not intend to attack and bite, most species are absolutely harmless to humans.

Most spider bites are unintentional, because a spider is accidentally squeezed and forced to bite. Most bites are harmless and can best be compared to a mosquito or wasp sting. For some spider genera, however, severe responses of human victims have been reported, leading to serious medical problems and sometimes also to death. This group comprises five genera with a few dozen species "of medical importance for humans" which are the topic of this chapter.

The most dangerous venomous animals are snakes, causing at least 20 fatalities per million humans annually (Table 19.1). Among arthropods, scorpions are the most dangerous (0.1–1.4 deaths per million humans), but also bees and wasps (up to 0.2 deaths) are dangerous. Spiders are considered to be less dangerous, but there is only poor documentation of the frequency of spider bites. Russell (1991) assumes <200 fatalities, globally per year, due to spider bites and Langley (2005) gives an annual average of six fatal issues for the USA. These figures refer to 0.04 and 0.02 deaths per million people per year, but both reports are based on statistics where no spider species verification was required.

The problem with spiders is that people often tend to name any sting or bite as a "spider bite", irrespectively of the presence of a spider or not. Physicians and other health professionals tend to identify all kinds of creatures as spiders and even confirm "spider bites" without having seen any animal. Also the presence of two fang marks is no reliable indication of a spider bite. For a more detailed insight into these problems, see Isbister and Gray (2002a, b), Isbister and White (2004) and Vetter and Isbister (2008). Following these considerations, three prerequisites must be fulfilled to accept a spider bite as such: (1) evidence of a bite through symptoms such as pain or discomfort, (2) collection of the spider during or immediately after the bite and (3) identification of the spider by an expert.

19.2 Spiders Lethal to Humans

19.2.1 Australian Funnel-Web Spiders (Atrax and Hadronyche, Hexathelidae)

The only group of mygalomorph spiders of medical importance comprises about 40 species of *Atrax* and *Hadronyche* (Australian funnel-web spiders, Hexathelidae),

Table 19.1 Numbers of bites and stings and documented deaths per year for the two most venomous animal groups (snakes and scorpions), for bees and wasps and for spiders

Venomous animal group	Region	Bites or stings per year	Deaths per year	Human population (millions)	Death rate per million	Reference
Snakes	World	5,400,000	125,345	5,840	21.5	Chippaux (1998)
Scorpions	World	115,000 (recorded)	180	2,264	0.08	Chippaux and Goyffon (2008)
–	World	1,190,000 (recorded + estimated)	3,271	2,264	1.4	Chippaux and Goyffon (2008)
Bees and wasps	USA	NIA	48	265	0.18	Langley (2005)
–	Australia	NIA	0.35	18	0.02	McGain et al. (2000)
Spiders	World	NIA	<200	5,300	0.038 (1)	Russell (1991)
–	USA	NIA	6	265	0.023 (1)	Langley (2005)
–	World	NIA	0 to <5	6,200	0 to <0.001 (2)	This study

The human population refers to the considered region and time when data were compiled

(1) Data do not refer to verified spider bites only and thus are much too high and not reliable

(2) According to this study, no proven record of fatal issue due to spider bites has been recorded for the last 20 years. In some cases, however, documentation is poor; therefore, we conclude that at most, "a few" fatal issues may have happened which is here translated as "0 to <5". On the basis of the medium year 2000 of the considered period, this refers to 0 to <0.001 deaths per million

NIA no information available

most commonly *A. robustus*, *H. infensa*, *H. versuta*, *H. formidabilis*, *H. venenata*, *H. cerberea* and a few still undescribed *Hadronyche* species (Isbister et al. 2005). These are large to very large spiders with typical orthognath chelicerae, dark brown to black and 15–45 mm body length. Most species are terrestrial, some arboreal in hollow tree trunks. They build silken tubular retreats with irregular silk lines radiating from the entrance ("funnel"). Spiders may be concentrated in suitable habitats, thus forming "colonies". They usually sit in the entrance of their burrow and grasp prey items that touch the silken lines. Adult males usually leave their burrow to search for females.

Atrax and *Hadronyche* species are restricted to the south-eastern coast of Australia where they can be found in moist to dry woodland and semiopen habitats. They prefer sheltered microhabitats, and so urban areas offer ideal possibilities. Spiders are frequently found around stone walls, garden rocks, dead wood and logs or heaps of building materials. This high spider abundance in the densely populated parts of Australia causes regular bites by Australian funnel-web spiders, but the frequency of bites is unknown. White et al. (1995) give an estimate of 30–40 annual bites, of which only one tenth needs medical treatment since most bites are "dry" (no venom injected) or effects are so minor that no health institution is consulted. Most bites occurred at the extremities, usually on the fingers.

In the most comprehensive analysis of these species, Isbister et al. (2005) gathered information on 138 cases where the collected spider could be attributed to *Atrax* and *Hadronyche* species. In 88% of cases adult males caused a bite, obviously in their mate-searching phase. Severe symptoms occurred in 31% of cases and were only caused by male spiders, 69% were considered as minor or moderate and nearly half of the severe cases happened to children. Half of the severe cases were due to bites of *Atrax robustus*, and one seventh of all cases could be attributed to *Hadronyche formidabilis* and *H. cerberea*, respectively. Before the introduction of antivenom in 1981, 13 fatalities occurred, half of them in children. After 1981 no further death was recorded. In more than 90% of antivenom application, a complete positive response could be achieved, and so the antivenom therapy is considered to be safe (Isbister et al. 2005).

A spider bite is usually very painful for about 30 min. Most cases are restricted to local symptoms (skin redness, sometimes piloerection, sweating and muscle fasciculation). Systemic symptoms usually appear 10 min to 1 h after the bite and are usually considered to be severe. They start with tongue spasms, followed by nausea and vomiting, abdominal pain, sweating and dyspnoea. Among the severe cases, about 75% showed sweating and hypertension, about 50% increased salivation, agitation, vomiting, tachycardia, fasciculation and pulmonary oedema. Recommended treatment of such cases is the use of funnel-web spider antivenom (Isbister et al. 2005).

The effective components of the venom of Australian funnel-web spiders are the δ-atracotoxins, 42 amino acid residue containing polypeptides with four disulphide bridges following the inhibitor cystine-knot motif. These neurotoxins bind with high affinity to mammalian sodium channels and cause a prolongation of the action potential duration, whereas binding to most invertebrate sodium channels is only with low affinity. The alerted neuronal excitability explains the intense muscle fasciculation, which is seen clinically during systemic envenomation, whereas effects on the autonomic nervous system such as vomiting or salivation are probably due to an accompanying excessive neurotransmitter release (Nicholson et al. 2004).

19.2.2 Recluse Spiders (Loxosceles, Sicariidae)

The worldwide occurring genus *Loxosceles* contains about 100 species, 85 of which are native to the Americas, and some have been globally distributed as alien species (e.g. the European *L. rufescens*). Recluse spiders are small- to medium-sized spiders of 5–12 mm body length, often yellowish or reddish to brown, and they have only six eyes. Spiders hide in crevices under rocks or under bark, where they spin a retreat and add a few silk lines to the surrounding surface, which alert them when a prey entangles. Most species occur primarily in arid to semiarid areas, and some are therefore predestined to a synanthropic way of living, including inside buildings. Males live up to two years, females up to three years. Only a few species are of medical importance, e.g. *L. reclusa* in North America, *L. laeta*, *L. gaucho* and *L. intermedia* in South America, *L. parrami* in South Africa and *L. rufescens* in the Mediterranean area. Recluse spiders can survive several months without food, are relatively tolerant to each other, and can reach high densities inside buildings such as up to 1,250 in one barn or 2,055 in one home. Interestingly, in this house, no family member had been bitten for over six years. In houses, spiders typically can be found under the folded flap of card boxes, inside cupboards, behind pictures and furniture, in shoes and in clothes left on the floor (White et al. 1995; Vetter 2008). This also describes the possibilities of getting unintentionally in contact with a recluse spider, which may then in turn feel threatened and may bite to defend. Typically, this results in bites in the extremities or, when a person is asleep or is getting dressed, bites to other exposed body parts. *Loxosceles* bites and the associated syndrome are often called loxoscelism.

Vetter (2008) distinguishes four categories of *Loxosceles* bites: (1) unremarkable bites with very little damage and self-healing; (2) mild reaction with skin redness, itching, and slight lesion but typically self-healing; (3) necrotic skin lesion; and (4) systemic or viscerocutaneous, affecting the vascular system, very rare, and potentially fatal. In contrast to the public perception, most recluse bites belong to categories (1) and (2), thus typically heal without treatment. In severe cases, *Loxosceles* venom causes vascular constriction at the bite site. After 3 h leukocytes infiltrate the tissue, dermal oedema occurs or arises after 6 h and itching, inflammation and ischaemia develop. Now the affected tissue causes pain, and a characteristic blister appears with eschar formation, which falls off later. This exposes soft tissue, which may take months to heal. In category (3), two third of all cases heal without complication. However, the most severe cases may result in 40 cm large necrotic lesions, healing only after several months and leaving an ugly scar. In very rare cases of systemic reactions, *Loxosceles* venom may cause hemolysis, intravascular coagulation, sepsis, renal failure and possibly death (White et al. 1995; Vetter 2008).

Phospholipase D (also known as sphingomyelinase D) is the major active compound in the venom of *Loxosceles* spiders. It is a rare enzyme in organisms, mainly known from sicariid spiders and microorganisms (see also Binford 2013). Depending on the species, phospholipases D are 31–34 kDa enzymes which hydrolyse sphingomyelin and lysoglycerophospholipids at the outer surface of cell membranes, thus destroying the membranes. Obviously, *Loxosceles* rely to the highest degree among spiders on the activity of such enzymes when subduing a prey. Humans are extraordinary sensitive to phospholipase D, as rabbits and guinea pigs also are, but unlike rats and mice (Vetter 2008). Damage is greater in obese victims because the enzyme readily destroys poorly vascularised adipose tissue. Terms such as necrotic skin lesion, dermonecrosis, skin ulcerations or even necrotic araneism have been used to describe the clinical picture of severe recluse bites.

For most non-necrotic forms of *Loxosceles* bites, no specific therapy is needed besides rest and some cooling. Antivenom therapy is frequently used against loxoscelism, but its efficacy is controversial. Since a recluse bite causes no pain, which is quite unusual for spiders, patients often search medical help only after the first day and if the wound worsens. Therefore, no reliable data on the frequency of bites is available. Analyses of such patients with confirmed *Loxosceles* bites indicated an average of 40–60% of cases with necrosis and 2–16% of severe or systemic cases (Vetter 2008). For South America, White et al. (1995) reviewed several studies prior to 1988 with a total of 25 fatalities, collected over more than 30 years. While loxoscelism is the most frequently diagnosed cause of spider bites in the USA and in South America, *Loxosceles* species do not seem to be of major medical concern in other continents.

In addition to such figures, several studies showed that *Loxosceles* bites are overdiagnosed by physicians. Specifically in the USA, thousands of recluse bites were diagnosed from regions where *Loxosceles* did not occur or where they were extremely rare. Also misdiagnosis is frequent, as re-examinations showed that up to 80% of diagnosed recluse bites were caused by other agents, usually other arthropods, Lyme borreliosis, *Streptococcus* or *Staphylococcus aureus* infections or other pathogens (Vetter 2008).

19.2.3 Black Widow Spiders (Latrodectus, Theridiidae)

The worldwide-distributed genus *Latrodectus* contains 31 species, usually called (black, brown, grey, etc.) widow spiders. Widows are small- to medium-sized spiders of 8–18 mm body length, often black with characteristic red marks but also brown or grey. They build irregular cobwebs with sticky threads and a tubular retreat. Typically, widow spiders inhabit arid to semiarid habitats but are capable of invading a variety of different habitats, including rural and urban areas. Though widows are not aggressive, contact with humans can lead to bites, which may cause serious symptoms so that widows are regarded as the spiders with the highest medical importance. Most relevant species are the South American black widow

Fig. 19.1 *Above*: *Latrodectus geometricus* and *L. tredecimguttatus* (Theridiidae), © Barbara Thaler-Knoflach at www.araneae.unibe.ch. *Below*: *Phoneutria* sp. (Ctenidae), © Matjaz Kuntner

L. curacaviensis (Central and South America), the brown widow *L. geometricus* (Africa but introduced as alien species to all other continents) (Fig. 19.1), the redback *L. hasseltii* (Asia to Australia), the North American black widow *L. mactans* (North America but introduced as alien species elsewhere), the African black widow *L. indistinctus* (southern Africa) and the European black widow *L. tredecimguttatus* (Europe to China) (Fig. 19.1). Due to their affinity for human buildings and goods such as containers, widows are predisposed for easy spread by cargo transport. Widow bites and the associated syndrome are often referred to as latrodectism. Since the bites of different *Latrodectus* species, when analysed under comparable medical standards, cause very similar symptoms, latrodectism is considered as the same clinical syndrome worldwide (Isbister and White 2004).

Bites usually occur on the extremities (70–80%), less frequently on the trunk (20%) and rarely on the head or neck. In many cases, a bite causes significant effects, with severe and long-lasting pain in two-thirds of cases, and prevented patients from sleeping in one-third of cases. As in most other comparably small spiders, puncture marks or bleeding is rarely observed. Pain increases in more than half of the cases within the first hour and mostly radiates into the limbs or abdominal pain develops. According to a meta-analysis of Isbister and White (2004), the typical symptoms include sweating in about 70% of cases and systemic effects in 20–30% of cases (nausea and vomiting in less than 20%, raised temperature and neuromuscular effects in about 10%, hypertension in less than 10% of cases). Pain usually lasts 1–2 days and the other symptoms 1–4 days.

Nevertheless, many widow bites do not need any treatment because no symptoms appear or local pain disappears after some hours (White et al. 1995). The treatment of the more serious widow bites may be performed symptomatically (i.e. pain relief medicine, antispasmodic and relaxant drugs), but also antivenom is frequently applied if available and/or if the case seems to justify it. There is increasing discussion on the effectiveness of antivenom therapy, and there is considerable reluctance to administer it due to the frequency of allergic reactions. Nevertheless, given the potentially fatal consequences, Isbister and White (2004) conclude that antivenom may be justified in up to two third of cases.

The most active components in the venoms of *Latrodectus* are latrotoxins, of which α-latrotoxin is active against vertebrates. This is a 131 kDa protein which composes two tetramers in the neuronal membrane, thus forming a pore which allows mono- and bivalent cations, neurotransmitters, ATP and water to influx into the cell. This permanent firing blocks synapses and provokes cramps, it may cause cardiac disturbance and renal damage, and finally, it may even lead to a fatal outcome due to heart failure, myocardial ischaemia or renal failure.

The frequency of fatal issues with humans is highly debated since all latrodectism studies are considerably biased. They usually start with patients asking for medical help or often include only severe or hospitalised cases and therefore have a strong bias. Nevertheless, it is believed that 15% of *Latrodectus* bites do not lead to envenomation (Peterson 2006) and that many bites cause only minor effects, thus requiring no treatment or hospitalisation (White et al. 1995; Isbister and Gray 2003). Furthermore, it has often been mentioned that the frequency of widow bites is seriously overestimated since many diagnoses of latrodectism are wrong. There are no positive tests available, no single symptom supports such a diagnosis and only a spider collected while biting and subsequent identification by an expert can guarantee that it is a "verified *Latrodectus* bite". The fatalities attributed to latrodectism as reviewed in White et al. (1995) must be seen in this light. For Australia, no fatal case has been reported since 1956 when antivenom therapy became available, and also for the USA, no fatal case is listed in their annual reports among ten thousands of widow bites from the last years (American Association of Poison Control Centers 2012). Nevertheless, there are a few publications of fatal incidents of *Latrodectus* bites (e.g. Hoxha 2006; Gaisford and Kautz 2011), but all refer to non-verified spider bites.

19.2.4 Armed Spiders (*Phoneutria*, Ctenidae)

Eight species have been described for the Central and South American genus *Phoneutria*, all large spiders with body lengths between 20 and 45 mm (Fig. 19.1). Body colouration is usually grey to brown with reddish chelicerae and a black-striped pattern on the underside of the first two pairs of legs. *Phoneutria* species do not build webs but search at night for prey. These tropical species live primarily in the rainforest but invade rural and urban areas and are able to habitate human buildings. When disturbed, these spiders show a typical defensive behaviour in which the body is in an erect position, the first two pairs of legs raised up, thus showing the conspicuous striped pattern on their underside, while the spider sways from side to side. Common species are *P. nigriventer*, *P. keyserlingi*, *P. fera* and *P. reidyi*.

Especially in the large urban areas of the Brazilian east coast, spiders frequently encounter humans. Though the spiders are nocturnal, most bites occur during the day and in houses, mainly in the mating season (January to April) when the spiders are more active. *Phoneutria* spiders are frequently encountered in shoes, among rubbish and construction material and in banana bunches, which explains the high frequency of bites to the hands and feet. *Phoneutria* bites are very common and accompanied by strong pain. Almost all patients complain about burning pain, often spreading over the affected limb. Further local symptoms are swelling and a dilatation of blood vessels (hyperaemia). Children and young people are more sensitive to *Phoneutria* venom than older people. Mild envenomation is observed in 90% of cases and includes accelerated cardiac rate as a systemic symptom. Moderate cases (further 9%) are characterised by nausea, vomiting and sweating. Severe cases (less than 1%) show a reduced cardiac rate, hypotension, cardiac arrhythmia and acute pulmonary oedema. Symptomatic treatment includes analgesics to relieve pain as a standard treatment in >75% of cases. In some severe cases, antivenom is given, but nearly 20% receive no treatment at all (White et al. 1995; Bucaretchi et al. 2000).

During the last 100 years, 10 fatal cases have been reported from Brazil for *Phoneutria*. Most cases, however, are poorly documented, and only two meet today's standards, and the last fatal case occurred in 1985. Death is considered to be extremely rare and seems to be restricted to small children (Bucaretchi et al. 2000).

The venom of *Phoneutria* spiders contains several components, which may explain its toxic effects. Histamine causes pain, and a variety of ion-channel active peptides (in the size range of 3.5–8.6 kDa) inhibit or delay inactivation of Ca^{2+}, K^{+} and Na^{+} channels. This leads to a depolarization of muscle fibres and nerve terminals at the neuromuscular junction. Also several tachykinin peptides with molecular masses between 0.9 and 1.7 kDa could be identified from the venom of *P. nigriventer*. They are characterized by vasodilatory and neurohormonal activities and probably provoke the observed release of acetylcholine and catecholamines, which is responsible for some systemic effects (White et al. 1995; Kuhn-Nentwig et al. 2011).

19.2.5 Other Spider Species Which Are Considered to Be Dangerous

Large mygalomorph spiders ("tarantulas") are feared because of their size and the overall Hollywood image of being very dangerous. The reality, however, is different. Bites of Australian mygalomorph spiders other than the hexathelids mentioned in Sect. 19.2.1 (such as Theraphosidae, mouse spider *Missulena* and trap-door spiders of the families Idiopidae and Nemesiidae) cause local pain, skin redness and sometimes bleeding due to the size of the wound, but no severe effects (Isbister and Gray 2002b). A comparable result was attained in a major Brazilian study with 91 identified cases. Theraphosid bites were considered to be rare (<1% of spider bites), and envenoming is described as mild with main symptoms of local pain and minor skin redness (Lucas et al. 1994).

Other large spiders, such as huntsmen (usually sparassids of the genera *Heteropoda* and *Neosparassus*), are feared because of their size and leg span, because of their ability to climb walls and ceilings and because of their synanthropic way of life. These spiders flee from humans already when they are a long distance away, but when people try to catch them, they may get bitten. Clinical effects are described as an immediate and transient pain, disappearing already after a few minutes. Systemic effects are rare and minor; thus, sparassids are regarded as very harmless spiders (Isbister and White 2004).

Yellow sac spiders of the genus *Cheiracanthium* comprise about 180 species, but only a few species exert a synanthropic way of living and encounter humans more frequently, mostly *C. inclusum* and *C. mildei* in North America and *C. punctorium* in Europe to Central Asia. These medium-sized spiders (body length mostly below 10 mm) have relatively long chelicerae and can bite humans, outdoors or at home or also at night when sleeping while unintentionally squeezing the spider. The bite provokes pain and discomfort in all cases, lasts typically for less than 2 h and leads to local redness (85% of cases), swelling (30%) and itchiness (30%). Systemic effects were observed in 15% (headache, vomiting). Complete recovery occurred fast, and no long-lasting effects were observed, so that *Cheiracanthium* species can be categorized as rather harmless (Vetter et al. 2006).

Despite frequent citation in the medical literature and in newspapers, hobo spiders (this term refers only to the alien introduction of the European *Tegenaria agrestis* (Agelenidae) into Pacific United States) do not cause painful bites or necrosis to humans (Vetter and Isbister 2008). Also bites by other agelenids (*Agelena*, *Agelenopsis*, *Hololena* species) are very rare events and usually harmless (Vetter 2012). Wolf spiders (Lycosidae) are distributed worldwide, and some genera comprise species of medium body size (e.g. *Hogna* and *Lycosa*), but nevertheless, their bites cause only medium pain and generally mild, transient symptoms. Details can be found in reviews by Isbister and White (2004) and Vetter and Isbister (2008).

19.3 Necrosis After a Spider Bite?

The term necrotic arachnidism refers to the fact that the bite of several spider species may induce skin necrosis. Especially in the older literature, often uncritically referred to, there is no clear differentiation between a direct venom effect and potential secondary infections. Moreover, cases of unverified spider bites are included in such statistics. The non-scientific classification of spider bites ("putative, presumptive, probable, documented") as listed in the medical education literature (Sams et al. 2001) unfortunately suggests that it could be possible to identify spider bites and even the spider species according to symptoms. There is no way of correct spider identification without a specimen; thus, it is indispensable of keeping the gold standard of verified bites as described above. Such nebulous definitions of spider bites are probably the basis for frequently encountering necrotic arachnidism as being attributed to many spider species still today in the literature.

According to current knowledge, necrosis following a spider's bite is only caused by *Loxosceles* species (and the small genus *Sicarius*, endemic to South African deserts, both in the same family, Sicariidae). The venom contains phospholipase D (also known as sphingomyelinase D), an enzyme which destroys membranes, thus causing necrosis (see Binford, 2013). Careful analysis of other spider venoms proved the unique case of sicariids. Especially *Cheiracanthium* venom, for a long time also considered to cause necrosis, definitely does not cause necrosis (Vetter et al. 2006; Vetter and Isbister 2008). Necrotic arachnidism induced by sicariids must clearly be distinguished from secondary infections which may be caused by a variety of different agents, usually initiated by a physical injury or forms of skin blistering and ulceration (White et al. 1995).

19.4 Conclusions

Spiders from four families (*Atrax* and *Hadronyche* (Hexathelidae), *Loxosceles* (Sicariidae), *Latrodectus* (Theridiidae) and *Phoneutria* (Ctenidae), with several species each) may cause severe symptoms when biting humans. This is mainly due to a combination of factors: (1) synanthropic way of living (all taxa), (2) high densities in urban areas (*Latrodectus*, *Loxosceles*, *Phoneutria*), (3) large body size (*Atrax, Hadronyche, Phoneutria*), (4) ion-channel targeting neuropeptides or other venom compounds which are very potent and/or to which vertebrates/humans are very sensitive (*Atrax, Hadronyche, Latrodectus, Phoneutria*), (5) a venom enzyme causing necrosis (*Loxosceles*), and (6) aggressive behaviour towards humans (*Phoneutria*). Several decades ago, fatal incidents were reported for all four groups, but medical documentation as well as medical standards in many countries were poor. In the last 2–3 decades, no fatalities were reported for *Atrax, Hadronyche, Latrodectus* and *Phoneutria.* The situation with *Loxosceles* is less clear, but on a global scale, it is obvious that fatalities due to spider bites are now close to zero, which refers to an annual mortality of <0.001 per million humans. Thus, spiders are by far less dangerous than bees and wasps (Table 19.1).

References

American Association of Poison Control Centers (2012) National Poison Data System. Available at www.aapc.org

Binford G (2013) The evolution of a toxic enzyme in sicariid spiders. In: Nentwig W (ed) Spider ecophysiology. Springer, Heidelberg (this volume)

Bucaretchi F, Deus Reinaldo CR, Hyslop S, Madureira PR, De Capitani EM, Vieira RJ (2000) A clinico-epidemiological study of bites by spiders of the genus *Phoneutria*. Rev Inst Med Trop Sao Paulo 42:17–21

Chippaux J-P (1998) Snake-bites: appraisal of the global situation. Bull World Health Organ 76:515–524

Chippaux J-P, Goyffon M (2008) Epidemiology of scorpionism: a global appraisal. Acta Trop 107:71–79

Gaisford K, Kautz DD (2011) Black widow spider: a case study. Dimens Crit Care Nurs 30:79–86

Herzig V, King G (2013) The neurotoxic mode of action of venoms from the spider family Theraphosidae. In: Nentwig W (ed) Spider ecophysiology. Springer, Heidelberg (this volume)

Hoxha R (2006) Two Albanians die from black widow spider bites. BMJ 333:278

Isbister GK, Gray MR (2002a) A prospective study of definite spider bites, with expert spider identification. Q J Med 95:723–731

Isbister GK, Gray MR (2002b) Bites by Australian mygalomorph spiders (Araneae, Mygalomorphae), including funnel-web spiders (Atracinae) and mouse spiders (Actinopodidae: *Missulena* spp.). Toxicon 43:133–140

Isbister GK, Gray MR (2003) Latrodectism: a prospective cohort study of bites by formally identified redback spiders. Med J Aust 179:88–91

Isbister GK, White J (2004) Clinical consequences of spider bites: recent advances in our understanding. Toxicon 43:477–492

Isbister GK, Gray MR, Balit CR, Raven RJ, Stokes BJ, Porges K, Tankel AS, Turner E, White J, Fisher MMD (2005) Funnel-web spider bites: a systematic review of recorded clinical cases. Med J Aust 182:407–411

Kuhn-Nentwig L, Stöcklin R, Nentwig W (2011) Venom composition and strategies in spiders: is everything possible? Adv Insect Physiol 60:1–86

Langley RL (2005) Animal-related fatalities in the United States – an update. Wilderness Environ Med 16:67–74

Lucas SM, Da Silva PJ, Bertani R, Costa Cardoso JL (1994) Mygalomorph spider bites: a report on 91 cases in the State of Sao Paulo, Brazil. Toxicon 32:1211–1215

McGain F, Harrison J, Winkel KD (2000) Wasp sting mortality in Australia. Med J Aust 173:198–200

Nicholson GM, Little MJ, Birinyi-Strachan LC (2004) Structure and function of δ-atracotoxins: lethal neurotoxins targeting the voltage-gated sodium channel. Toxicon 43:587–599

Peterson ME (2006) Black widow envenomation. Clin Tech Small Anim Pract 21:187–190

Russell FE (1991) Venomous arthropods. Vet Hum Toxicol 33:505–508

Sams HH, Dunnick CA, Smith ML, King LE (2001) Necrotic arachnidism. J Am Acad Dermatol 44:561–573

Suter RB, Stratton GE (2013) Predation by spitting spiders—elaborate venom gland, intricate delivery system. In: Nentwig W (ed) Spider ecophysiology. Springer, Heidelberg (this volume)

Vetter RS (2008) Spiders of the genus *Loxosceles* (Araneae, Sicariidae): a review of biological, medical and psychological aspects regarding envenomation. J Arachnol 36:150–163

Vetter RS (2012) Envenomation by spiders of the genus *Hololena* (Araneae: Agelenidae). Toxicon 60:312–314

Vetter RS, Isbister GK (2008) Medical aspects of spider bites. Annu Rev Entomol 53:409–429

Vetter RS, Isbister GK, Bush SP, Boutin LJ (2006) Verified bites by yellow sac spiders (genus *Cheiracanthium*) in the United States and in Australia: where is the necrosis? Am J Trop Med Hyg 74:1043–1048

White J, Cardoso JL, Fan HW (1995) Clinical toxicology of spider bites. In: Meier J, White J (eds) Clinical toxicology of animal venom and poisons. CRC, Boca Raton

Part VI
Silk

Spiders possess several silk gland types producing a variety of silk qualities, they fabricate either cribellate or non-cribellate silk, with or without sticky droplets. Silk is involved in all parts of a spider's life, from the retreat to draglines, sperm webs and cocoons, but its overall importance got silk as inimitable material for a variety of web types to catch prey. So, silk is the material which accounts for the overwhelming predacious success of many spider families. During 300 million years, the material properties of silk became unique, making it one of the most fascinating biomaterials.

Chapter 20
Spider Silk: Molecular Structure and Function in Webs

Todd A. Blackledge

20.1 Introduction

The evolutionary and ecological success of spiders is founded in part on numerous innovations in the chemistry, production, and utilization of their silk toolkits (Bond and Opell 1998; Blackledge et al. 2009). Opisthosomal silk glands in fossil *Attercopus* indicate that spider evolved silk at least 386 mya (Selden et al. 1991). The lack of spinnerets in *Attercopus* and other early "protospiders" suggests that the physical control over thread placement provided to spinning glands placed on muscled spinnerets played an important role in early silk and web evolution in Araneae (Selden et al. 1991). However, the diversification of the genes encoding spider silk proteins and the development of distinct silk gland morphologies were likely far more significant (Craig 2003; Garb et al. 2010). Silk production is largely uninvestigated outside of model species of orb-weaving spiders so that the tempo of evolution and the degree to which silk genes and spinning physiology might coevolve are unclear. Most work relating silk production and functions therefore focuses on orb-weaving spiders.

20.2 What Is Spider Silk?

Most spider silks consist of semicrystalline fibroins (fibrous proteins) where the amino acid chains are spatially constrained into rigid nanocrystal structures in some regions and amorphously free in other regions (Eisoldt et al. 2011). This gives silk a combination of high strength, due to the confinement, and high extensibility because the protein chains in the amorphous region are kinetically free and can

T.A. Blackledge (✉)
Department of Biology and Integrated Bioscience Program, The University of Akron, Akron, OH 44325-3908, USA
e-mail: blackledge@uakron.edu

W. Nentwig (ed.), *Spider Ecophysiology*,
DOI 10.1007/978-3-642-33989-9_20, © Springer-Verlag Berlin Heidelberg 2013

move when threads are stretched. Spider silks therefore require the continuous application of large forces to stretch a great distance before breaking. Although spider silks are impressively strong and extensible compared to most natural and synthetic materials (Gosline et al. 1999), it is this work of extension, termed "toughness," that is truly remarkable—some spider silks requiring five times more energy to break than an equivalent volume of Kevlar (Agnarsson et al. 2010).

The nano-structured, composite nature of spider silk is determined by an interaction between the amino acid sequence of the proteins and how fibers are assembled from liquid feedstocks. Both play important roles in controlling the material properties of spider silks. More important, amino acid sequence and spinning physiology provide potentially independent mechanisms that could be acted upon by natural selection in the evolution of the function of spider silks (Blackledge et al. 2011; Blackledge 2012). The end result is that most spiders spin "toolkits" of several diverse types of silks (Blackledge and Hayashi 2006), and individual silk types can vary significantly in their performance among species (Swanson et al. 2006, 2007; Sensenig et al. 2010).

20.3 Spider Silk Protein (Spidroin) Structure

A single spider silk protein is termed a spidroin, for "spider fibroin," and is typically quite large, 200–350 kDa (Ayoub et al. 2007; Garb et al. 2010). Spidroins consist of three regions—an initial N-terminus of ~100–400 amino acids, a long internal region of highly repetitive amino acid sequences, and a final C-terminus (Fig. 20.1) (Ayoub et al. 2007). The N- and C-termini are remarkably conserved across different types of silks (Garb et al. 2010), even among silk genes that diverged >240 MYA. This pattern suggests that the terminal regions are critical for shared properties among silk types, such as how diverse spidroins are stored as liquid dopes and then subsequently transformed into solid fibers (Eisoldt et al. 2011). The internal regions, in contrast, are highly divergent among silk types—to the extent that they cannot be homologized (Gatesy et al. 2001). However, the amino acid sequences of the internal regions are remarkably similar for a single type of spidroin compared across species and have a strongly hierarchical organization. The internal region of a spidroin can be divided into repetitive modules that are each 40–200 amino acids long and are repeated 20–100 times in tandem (Ayoub et al. 2007; Garb et al. 2010). Each repetitive module is nearly identical in amino acid sequence to the others, likely due to homogenization from concerted evolution (Hayashi and Lewis 2000). A repetitive module consists mostly of short, highly stereotyped amino acid sequences called "functional motifs" that are approximately five to ten amino acids in length with spacer regions in between them.

Different functional motifs are predicted to form specific secondary structures. The identity and frequency of these secondary structures help explain variation in the material properties of spider silk. For instance, the crystalline regions in spider silk are formed from repeats of polyalanine and glycine-alanine. Both amino acids

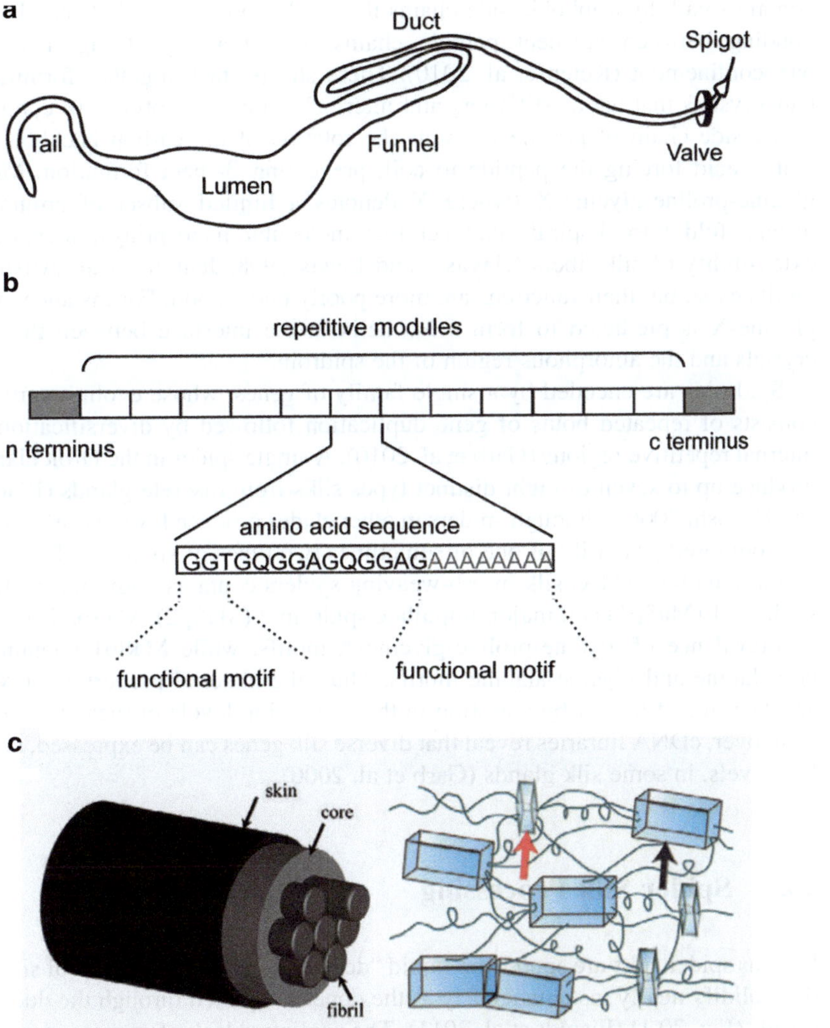

Fig. 20.1 Silk production and structure. (**a**) Spidroins are synthesized in the tail of the silk gland and are stored in the lumen as a liquid dope. As the dope passes through the spinning duct, a combination of shear forces, water uptake, and ion exchange produces a phase shift that causes nanocrystals to form and interlock the spidroins, thereby solidifying the fiber. The still wet silk fiber passes through a muscled valve that acts as a friction brake and controls the alignment of the spidroins along the fiber axis. (**b**) Spidroins consist of repeated modules of amino acids. Each module is composed largely of a few common functional motifs that form specific secondary structures. The N- and C-terminal regions are more heterogeneous, and their amino acid sequences are conserved among divergent silk types. (**c**) Hypothesized structure of major ampullate silk in an orb spider. A skin of lipids and glycoproteins surround a core of spidroins. The core likely consists of multiple fibrils. Individual spidroins are organized into highly crystalline domains embedded and an amorphous matrix. Two levels of crystalline domains are shown here. Adapted from Blackledge (2012) with panel C from Eisoldt et al. (2011)

contain small, hydrophobic side chains that fold peptides into β-sheets. Hydrogen bonding between adjacent peptide chains is particularly strong due to their nanoconfinement (Keten et al. 2010). These sheets stack together forming small nanocrystals that act as stiffeners and interlock multiple peptides. In contrast, the cyclic side chain of proline restrains the rotation of the carbon backbone of the amino acid forcing the peptide to coil, preventing β-sheet formation. Motifs of glycine-proline-glycine-X (where X denotes a limited subset of amino acids) instead fold into β-spirals that act like molecular nanosprings, increasing the extensibility of silk fibers (Hayashi and Lewis 1998; Jenkins et al. 2010). Other motifs exist, but their functions are more poorly understood. For instance, glycine-glycine-X is predicted to form a 3_{10} helix at the interface between the β-sheet crystals and the amorphous region of the spidroin.

Spidroins are encoded by a single family of genes, whose evolutionary history consists of repeated bouts of gene duplication followed by diversification of the internal repetitive regions (Garb et al. 2010). A single spider in the Orbiculariae can produce up to seven or eight distinct types silks from discrete glands (Blackledge and Hayashi 2006), although spiders in other clades produce fewer types. Most silks are composed primarily of one or sometimes two types of spidroins. For instance, major ampullate (MA) silk in orb-weaving spiders contains both major ampullate spidroin 1 (MaSp1) and major ampullate spidroin 2 (MaSp2). MaSp2 is distinct in its prevalence of glycine-proline-glycine-X motifs, while MaSp1 contains more polyalanine and glycine-alanine motifs. Thus, the material properties of MA silk are determined in part by variation in the expression levels of these two proteins. Moreover, cDNA libraries reveal that diverse silk genes can be expressed, at least at low levels, in some silk glands (Garb et al. 2006).

20.4 Spider Silk Processing

Fibrous spider silks are spun from liquid "dopes" stored in the lumens of silk glands that solidify nearly instantaneously as the dopes are pulled through the ducts of the glands (Fig. 20.1) (Eisoldt et al. 2011). The continued lack of success in accurately reproducing the remarkable material properties of native spider silk in fibers spun from recombinant or reconstituted silk proteins suggests that how a spider "spins" silk may be as critical as the composition of the dope. The process of fiber assembly is well-characterized for the MA silk glands of orb-weaving spiders (Eisoldt et al. 2011). Spidroins are secreted in the tail of the gland and stored as micelles in the lumen. The spidroins reach a remarkably high concentration, up to 50 % wt/vol, as the relatively hydrophilic terminal regions isolate the hydrophobic repetitive regions in the interior of the micelle. Fibers begin to solidify as the dope passes into the "S"-shaped spinning duct. Initially shear forces align the spidroins, while ion uptake and a slight drop in pH cause the terminal regions of spidroins to dimerize, thereby cross-linking molecules. This process exposes the repetitive regions of the spidroins, leading to the rapid formation of β-sheet crystals.

The still wet fiber is then subjected to further shear force that continue to align the crystals with the fiber axis as the fiber passes through a muscled valve that is present in the MA spinning duct, prior to exiting the spigot.

Orb spiders exert significant control over the shear forces applied to silk threads such that they can influence the relative alignment of the amorphous fraction of the fibroins. More force results in greater pre-straining of the silk making it measurably stiffer and less extensible. In fact, the material properties of MA silk produced by a single spider can vary by as much as 50 % under different spinning conditions due at least in part to this final draw down phase (Guinea et al. 2005).

Nutrition also influences the amino acid content of silk, which may be a response to the costs of synthesizing different amino acids (Blamires et al. 2012). The chemical composition of viscid glues appears particularly plastic (Higgins et al. 2001). However, the response is not universal (Blamires et al. 2010). Spiders change the performance of their silk in response to diet simply by altering structural properties of silk. The theridiid *Parasteatoda tepidariorum* spins thicker threads after encountering larger prey (Boutry and Blackledge 2008), and the thickness of threads closely scales with changes in body mass for many spiders (Blackledge and Zevenbergen 2007). *Argiope trifasciata* (Araneidae) even alters the properties of its dragline silk when shifting from horizontal to vertical surfaces (Garrido et al. 2002).

The final functional properties of a silk thread result from an interaction between the protein composition of the silk and spinning conditions. Amino acid sequence, determined by gene expression, sets the boundaries for a range of potential performance properties, while the spinning conditions—acidification, ion exchange, water resorption, and drawdown of the wet fiber—determine precisely where in that "performance space" a particular silk thread occupies. It is also useful to distinguish between the material properties of silk threads versus actual mechanical performance (Fig. 20.2). The former measures the intrinsic performance of materials, regardless of dimensionality, while the mechanical performance of any object is determined as much by its size and shape as by the material from which it is built. Thus, a thin cotton thread and thick cotton rope have identical material properties but radically different mechanical performance (Fig. 20.2b).

20.5 Silk Diversity

Eight to nine types of silk occur in Orbiculariae (orb-weaving spiders; see also Appendix, this volume), differentiated by their glandular origins, with fewer types occurring in other clades. The material properties and chemical compositions of each type of silk are distinct (Blackledge and Hayashi 2006) as are their functions (Fig. 20.3). Up to six types of silk play an integral role in the function of most webs. Adhesive silks—viscid aggregate glue and dry cribellate fibrils—are covered in Townley and Tillinghast (2013) and Opell (2013). The other silks are briefly outlined below.

Major ampullate (MA) silk forms the backbones of most araneomorph prey capture webs and is used extensively in other structures such as draglines,

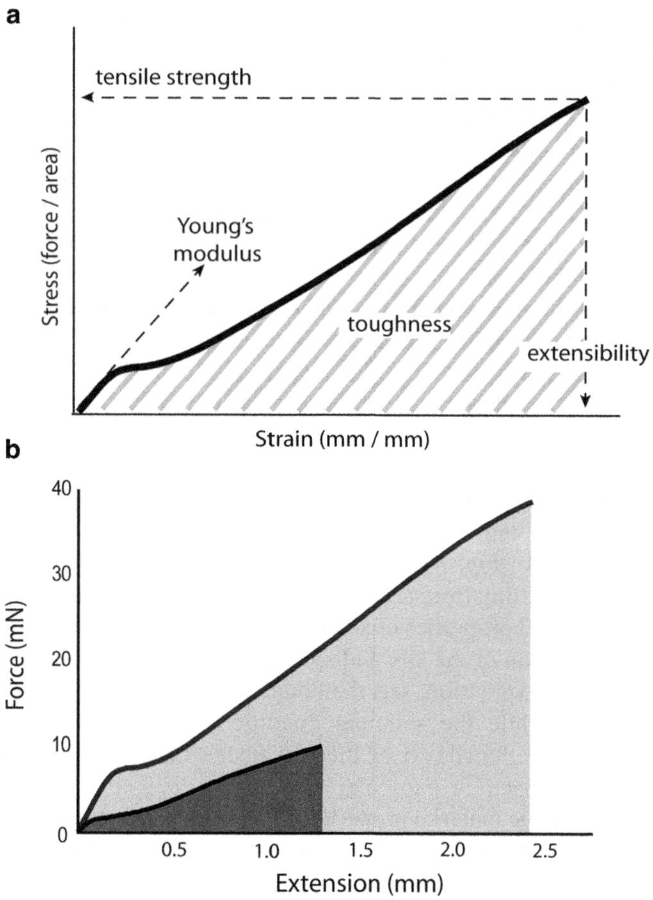

Fig. 20.2 Performance of silk threads. (**a**) Several material properties are typically quantified for silk threads. Young's modulus measures the stiffness in the initial "elastic region" where nearly all the energy of deformation is stored internally so that silk recovers completely when relaxed. Toughness measures the total energy per volume of silk necessary to extend a thread to failure. (**b**) Material properties are independent of the dimensions of the silk. Thus, two silk threads with identical material properties could perform very differently. Here, a longer thicker thread in *gray* requires more force to break at a greater extension compared to a shorter, thinner silk in *black*. The area under the force-extension curve represents the work that thread can perform—resisting moving prey, etc.

ballooning threads, and retreats. MA silk is also the best characterized silk, with material properties, chemical composition, and structural data for a small, but phylogenetically diverse set of species. MA silk is typically stronger compared to other silks and is also relatively stiff and less extensible. The tensile properties of MA silk also vary phylogenetically—orbicularian spiders produce measurably stronger and tougher MA silk threads compared to their sister RTA clade and basal taxa (Swanson et al. 2006). These differences correlate with the origin of MaSp2, which is currently known only from Orbiculariae.

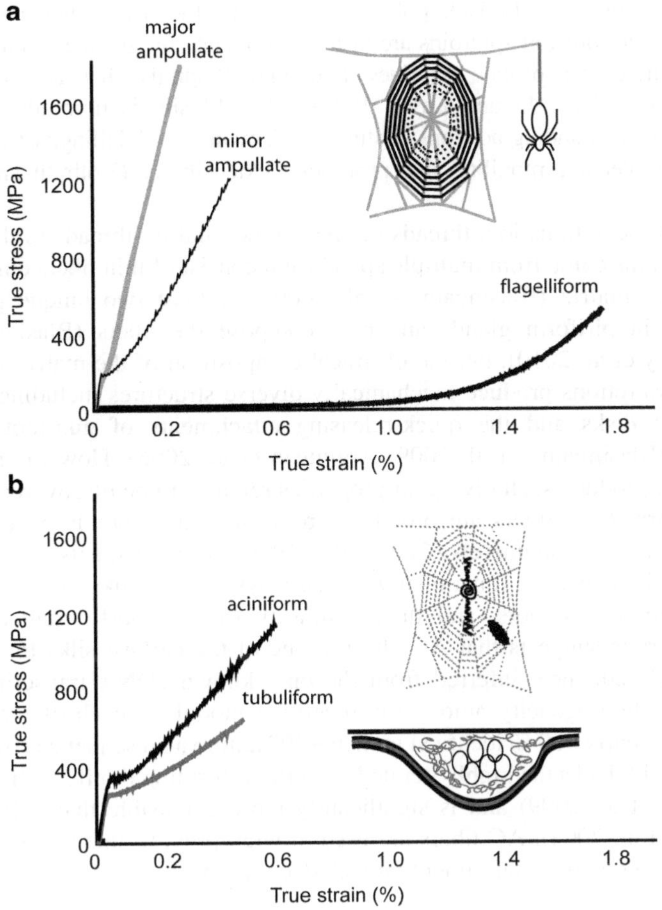

Fig. 20.3 The spider silk "toolkit." Araneid spiders produce five distinct fibrous silks as well as piriform silk that form attachment disks and adhesive silks that coat capture threads (see Townley and Tillinghast 2013; Opell 2013). (**a**) Orb webs are composed primarily of major ampullate silk backbones and glue-coated flagelliform silk capture spirals. These two silks differ greatly in their ultimate strength, extensibility, and stiffness. Minor ampullate silk may be used to spin the temporary or auxiliary spiral, but evidence is unclear. (**b**) Aciniform fibers are sometimes added to webs to produce stabilimenta and are also used to wrap prey. Adult female spiders also possess tubuliform glands that are used, along with several other silk types, to construct egg sacs. From Blackledge et al. (2011)

Minor ampullate (MiA) silk is typically thinner than MA and also weaker but more extensible. Its composition is similar to MaSp1 and lacks the glycine-proline-glycine-X motif found in orb spider MA silk. MiA silk is used for aerial bridging and possibly the temporary spirals of orb webs. MiA threads also sometimes supplement draglines and web elements spun from MA.

Flagelliform (Fl) silk forms the axial fibers of capture threads in araneoid spiders (see Townley and Tillinghast 2013). It is notably more extensible and orders of magnitude more compliant than all other spider silks (Swanson et al. 2007). The repetitive modules in Fl lack polyalanine repeats, thereby reducing cross-linking among the spidroins. Fl spidroins are instead dominated by glycine-proline-glycine-X motifs, where the proline residues fold into β-spirals that act as molecular "nanosprings" (Hayashi and Lewis 1998). Also, Fl silk is normally hydrated by water in the surrounding aqueous coating (see Townley and Tillinghast 2013), which increases molecular mobility also promoting extensibility (Vollrath and Edmonds 1989).

Piriform secretions join threads to one another and fix threads to the substrate. The secretions come from multiple spigots and consist of thin fibers embedded in a cement-like matrix (Blasingame et al. 2009). At least two unique proteins are expressed in piriform glands and may compose the fibers (Blasingame et al. 2009; Perry et al. 2010), but the chemical composition of the matrix is unknown. Piriform secretions produce mechanically diverse structures including permanent attachment disks and the quick-releasing attachments of gumfoot threads in cobwebs (Blasingame et al. 2009; Argintean et al. 2006). How the same set of glands can produce such divergent properties remains to be discovered.

Aciniform (Ac) fibers are produced from numerous spigots on the posterior median and lateral spinnerets (Coddington 1989), forming sheets or bands of silk, which are likely used in functionally similar ways to early silks. Ac glands occur in all araneomorph spiders, and their similarity to the poorly studied glands in mygalomorphs suggests that Ac silk was one of the earliest silks to evolve. The amino acid sequence, inferred from the only known cDNA transcript, shows a remarkable homogeneity among the repetitive modules—each of the 14 known modules is nearly 100 % identical in their ~200 amino acid sequence (Hayashi et al. 2004). Ac silk lacks the highly iterated crystalline-forming motifs found in MA silk (Rousseau et al. 2009) and is significantly more extensible than other dry silks (Hayashi et al. 2004). AC fibers are used to wrap prey, to line the interior of egg sacs, and to construct stabilimentum web decorations.

20.6 Function of Silk in Prey Capture Webs

Prey capture webs range in complexity from simple, silk-lined burrows to irregular sheets of silk spread across a substrate to aerial webs suspended on discrete frameworks of MA silk (Blackledge et al. 2009). While some webs primarily extend a spider's sensory environment, such as the trip lines extending from many liphistiid and mygalomorph burrows, most prey capture webs also interact mechanically with prey—slowing or even adhering to the prey. The function of prey capture webs is determined by an interaction between the material and structural properties of silk threads and how those threads are arranged in webs (i.e., architecture). Orb webs offer the best understood example (Blackledge et al. 2011; Harmer et al. 2011; Blackledge 2012). The capture of flying insects by orb-weaving spiders is determined by how effectively their orb webs first intercept insects, then stops their flight, and finally retains the insects long enough for capture. In addition, orb webs must quickly transfer information about the location of trapped prey to the spider and resist perturbations in the environment such as wind.

Prey interception is influenced mostly by the location of a web and its architecture. Thus, it is largely independent of the material properties and structures of silk. Eberhard (1986) modeled the spacing between rows of the capture spiral to maximize the number of insects intercepted by orb webs and found the optimal solution was just larger than an insect's wingspan. For a given amount of silk, a narrower mesh reduced orb size too much and decreases the total prey contacting the web, while greater spacing allowed too many insects to fly through the plane of the web without contacting threads. Web visibility presents an important exception to the independence of prey interception from silk properties because thicker threads are structurally stronger but also more visible.

The importance of amino acid sequence for silk function is dramatically illustrated by how orb webs stop and retain prey. Orb webs consist largely of two fibrous silk types and multiple adhesive silks (see Townley and Tillinghast 2013; Opell 2013). MA silk comprises the supporting radii and frames of orbs, as well as the backbones of most other types of prey capture webs, while Fl silk forms the core of the capture spiral. Both types of silk are remarkably tough—requiring great energy to rupture. However, MA silk is approximately 1,000-fold stiffer than Fl, with a breaking strength of 1–2 GPa and extensibility of 30–60 %. Fl, in contrast, is weaker and breaks at an engineering stress of ~250 MPa after stretching three to seven times its original length (Fig. 20.3a). These differences in material properties are determined primarily by the preponderance of crystal-forming polyalanine and glycine-alanine amino acid motifs in MA versus glycine-proline-glycine motifs in Fl.

Orb webs must dissipate tremendous kinetic energy though the silk to stop prey in midflight. Energy is dissipated primarily by the high hysteresis (or damping capacity) of MA silk in the radii. As the MA silk extends, the kinetic energy is transferred to heat when spidroins in the amorphous fraction of the silk rearrange as threads extend under impact (Fig. 20.4a). This prevents insects from ricocheting out of the web. Silk threads are so thin that air interacts with them as a relatively viscous fluid, and some models of orb web function therefore suggest that aerial damping also dissipates substantial prey energy (Lin et al. 1995). However, recent empirical data show that aerial damping plays only a minor role, at least for medium to large orb webs (Fig. 20.4b) (Sensenig et al. 2012). While abundant, the Fl silk in the capture spiral is so compliant that it does very little work of stopping prey (Sensenig et al. 2012). Energy dissipation is concentrated in a remarkably small portion of the orb's surface—~25–30 % of the orb's surface accounts for 90 % of all the work of stopping prey (Sensenig et al. 2012) due to the nonlinear tensile behavior of the MA silk (Cranford et al. 2012). Because MA silk becomes more compliant after it yields, stress is concentrated in the local area of deformation until the silk stiffens significantly when approaching failure. As a result, even a moderately damaged orb web retains the ability to effectively stop insects elsewhere on the web's surface (Sensenig et al. 2012; Cranford et al. 2012).

Capture thread adhesion is investigated extensively, and many of the details are covered by Townley and Tillinghast (2013) and Opell (2013). However, the role of

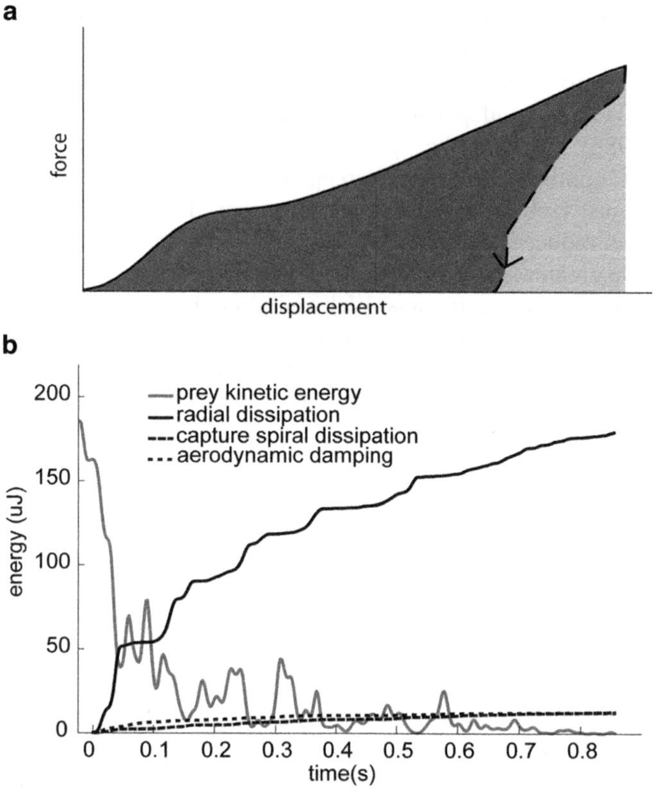

Fig. 20.4 Dissipation of prey energy. (**a**) Orb webs depend upon the high-energy damping of their major ampullate (MA) silk to stop prey. Approximately 60–70 % of the total work need to extend MA silk is transferred to heat. The 30–40 % that is stored (*light gray*) drives web oscillations. (**b**) The kinetic energy of prey can be dissipated through three routes—aerodynamic damping, damping in capture spiral silk, and damping in MA radial silk. At least for medium to large orb webs, most prey energy is dissipated through stretching of the radii in the local area of impact. Here, a 98 mg projectile strikes the web of *Araneus trifolium* (Araneidae) at time 0. Forward momentum is halted within the first 0.1 s, with energy stored in the silk being returned to the prey during each oscillation of the web (e.g., at 0.2 and 0.3 s). Although capture silk extends greatly during prey capture, the silk is so compliant that it does very little work. From Sensenig et al. (2012)

silk in prey retention at the whole web level is more poorly understood. Prey retention is particularly important for webs because many insects escape before spiders have adequate time to locate and subdue the prey. Simple architectural features of orb webs facilitate prey retention. Narrow spacing between rows of capture spiral increases the total number of threads adhering to any prey. The vertical orientation of most araneoid webs facilitates interception for vertical webs but, along with web asymmetry, also increases retention because insects that free themselves from one capture thread typically tumble down onto more adhesive threads (Opell et al. 2006). Townley and Tillinghast (2013) discuss how the mechanical performance of the axial Fl silk itself plays a critical role in promoting adhesion.

Silk also needs to effectively transmit information about the location of trapped prey to spiders. The MA silk in the radii of orb webs is particularly efficient at propagating longitudinal, rather than lateral or transverse, vibrations so that information is efficiently funneled to the hub of the web (Masters 1984). The degree to which this selective vibration transmission might be influenced by the crystallinity of the silk, which is itself oriented longitudinally and which varies substantially depending upon the ratio of MaSp1 versus MaSp2, is unknown. However, the tensioning of the radii by the spider during web building does influence vibration transmission. The tightening of radii during hub construction by hungry *Octonoba* (Uloboridae) tunes orb webs so that they respond more strongly to the vibrations of smaller trapped insects, presumably facilitating an increase in diet breadth (Watanabe 2000).

20.7 Evolution of Silk in Orb Spiders

Body size shifts orders of magnitude among different araneoid clades, so it is not surprising that it is a critical factor in the evolution of orb web architecture and silk properties (Sensenig et al. 2012). The stopping potential of orb webs, estimated as the maximum absorbable energy per cm^2 of web surface, increases in lineages of large-bodied spiders due to concerted changes in web architecture and silk properties. Larger spiders spin orb webs with disproportionately tightly packed threads and produce silk that is tougher than smaller species (Craig 1987b; Sensenig et al. 2010). The adhesive strength of viscid threads also coevolves with their tensile strength (Townley and Tillinghast 2013).

On the other hand, significant trade-offs exist in how silk functions during the interception, stopping, and retention of prey by orb webs. For instance, spiders can allocate a given amount of silk to an orb with a larger capture area that samples a larger portion of the air column, thereby increasing the total number of prey intercepted. But, this necessitates producing thinner or more widely spaced threads, thereby reducing stopping potential compared to a more compact web of the same volume of silk. Evolutionary changes in material properties of silks may mitigate some of these trade-offs. Spiders can use stronger, tougher silk in larger, widely spaced orb webs to maintain stopping potential (Craig 1987a; Sensenig et al. 2010).

20.8 Looking Beyond Orb Webs

The technological limitations of measuring silk properties for the very thin fibers produced in small-bodied lineages of araneid spiders, such as the Symphytognathidae, mean that substantial gaps in our knowledge extend to whole clades. More important, orbs represent only a minority of the prey capture webs spun by spiders (Shear 1986; Eberhard 1990). The functional properties of silks outside of araneoid orb webs are largely unexplored, with the exception of investigations into the adhesiveness of cribellate threads (see Opell 2013) and a few studies of the molecular structure (Dicko et al. 2008) and material properties of (Swanson et al. 2006; Boutry and Blackledge 2010) major ampullate silk.

The gumfoot trap threads of cobwebs are a notable exception that illustrates just how different the mechanics of prey capture can be from the orb web "model" because cobwebs capitalize on energy storage (Argintean et al. 2006). Gumfoot threads are held under tension by the cobweb and release from the ground when prey contact their gluey bases, pulling small prey into the air and resisting moment of larger prey (see also Townley and Tillinghast 2013). More discoveries clearly await intrepid investigators.

20.9 Conclusions

Integrating the molecular structure and properties of spider silk threads with the function and ecology of webs is a relatively new frontier (Harmer et al. 2011). Spiders inspire many biomimetic innovations such as sensors (Barth 2012), water-operated motors (Agnarsson et al. 2009), adhesives (Seidl and Vidoni 2013), and robotic legs (Spagna et al. 2007; Zentner 2013). Yet, these efforts pale in dimension next to the quest to synthesize spider silk in the laboratory for use by industry, the military, and medicine (Hinman et al. 2000; Altman et al. 2003; Scheibel 2004; Vollrath and Porter 2006). Web ecology and spider evolution bring new perspectives to this effort and yield discoveries such as using web-spinning behaviors to predict ultrahigh performance silks (Agnarsson et al. 2010) or new design principles for glues (Sahni et al. 2012). A holistic understanding of silk functions in webs is equally important, though, for understanding the evolution of spiders.

References

Agnarsson I, Dhinojwala A, Sahni V, Blackledge TA (2009) Spider silk as a novel humidity-driven biomimetic muscle. J Exp Biol 212:1990–1994

Agnarsson I, Kuntner M, Blackledge TA (2010) Bioprospecting finds the toughest biological material: extraordinary silk from a giant riverine orb spider. PLoS One 5(9):e11234. doi:11210.11371/journal.pone.0011234. doi:10.1371/journal.pone.0011234

Altman GH, Diaz F, Jakuba C, Calabro T, Horan RL, Chen JS, Lu H, Richmond J, Kaplan DL (2003) Silk-based biomaterials. Biomaterials 24:401–416

Argintean S, Chen J, Kim M, Moore AMF (2006) Resilient silk captures prey in black widow cobwebs. Appl Phys A Mater Sci Process 82:235–241

Ayoub NA, Garb JE, Tinghitella RM, Collin MA, Hayashi CY (2007) Blueprint for a high-performance biomaterial: full-length spider dragline silk genes. PLoS One 2(5):e514. doi:510.1371/journal.pone.0000514

Barth FG (2012) Learning from animal sensors: the clever "design" of spider mechanoreceptors. In: Lakhtakia A, MartinPalma RJ (eds) Bioinspiration, biomimetics, and bioreplication 2012, vol 8339. Proc SPIE. doi:833904 10.1117/12.902456

Blackledge TA (2012) Spider silk: a brief review and prospectus on research linking biomechanics and ecology in draglines and orb webs. J Arachnol 40:1–12

Blackledge TA, Hayashi CY (2006) Silken toolkits: biomechanics of silk fibers spun by the orb web spider *Argiope argentata*. J Exp Biol 209:2452–2461

Blackledge TA, Zevenbergen JM (2007) Condition dependent spider web architecture in the western black widow *Latrodectus hesperus*. Anim Behav 73:855–864

Blackledge TA, Scharff N, Coddington JA, Szüts T, Wenzel JW, Hayashi CY, Agnarsson I (2009) Reconstructing web evolution and spider diversification in the molecular era. Proc Natl Acad Sci USA 106:5229–5234

Blackledge TA, Kuntner M, Agnarsson I (2011) The form and function of spider orb webs: evolution from silk to ecosystems. Adv Insect Physiol 41:175–262

Blamires SJ, Chao IC, Tso IM (2010) Prey type, vibrations and handling interactively influence spider silk expression. J Exp Biol 213:3906–3910. doi:10.1242/jeb.046730

Blamires SJ, Wu CL, Tso IM (2012) Variation in protein intake induces variation in spider silk expression. PLoS One 7(2):e31626. doi:10.1371/journal.pone.0031626

Blasingame E, Tuton-Blasingame T, Larkin L, Falick AM, Zhao L, Fong J, Vaidyanathan V, Visperas A, Geurts P, Hu XY, La Mattina C, Vierra C (2009) Pyriform spidroin 1, a novel member of the silk gene family that anchors dragline silk fibers in attachment discs of the black widow spider, *Latrodectus hesperus*. J Biol Chem 284:29097–29108. doi:10.1074/jbc.M109.021378

Bond JE, Opell BD (1998) Testing adaptive radiation and key innovation hypotheses in spiders. Evolution 52:403–414

Boutry C, Blackledge TA (2008) The common house spider alters the material and mechanical properties of cobweb silk in response to different prey. J Exp Zool 309A:542–552

Boutry C, Blackledge TA (2010) Evolution of supercontraction in spider silk: structure-function relationship from tarantulas to orb-weavers. J Exp Biol 213:3505–3514

Coddington JA (1989) Spinneret silk spigot morphology: evidence for the monophyly of orb-weaving spiders, Cyrtophorinae (Araneidae), and the group Theridiidae plus Nesticidae. J Arachnol 17:71–96

Craig CL (1987a) The ecological and evolutionary interdependence between web architecture and web silk spun by orb web weaving spiders. Biol J Linn Soc 30:135–162

Craig CL (1987b) The significance of spider size to the diversification of spider-web architectures and spider reproductive modes. Am Nat 129:47–68

Craig CL (2003) Spider webs and silk: tracing evolution from molecules to genes to phenotypes. Oxford University Press, New York

Cranford SW, Tarakanova A, Pugno NM, Buehler MJ (2012) Nonlinear material behaviour of spider silk yields robust webs. Nature 482(7383):72–76

Dicko C, Porter D, Bond J, Kenney JM, Vollrath F (2008) Structural disorder in silk proteins reveals the emergence of elastomericity. Biomacromolecules 9:216–221. doi:10.1021/Bm701069y

Eberhard WG (1986) Effects of orb-web geometry on prey interception and retention. In: Shear WA (ed) Spiders, webs, behavior and evolution. Stanford University Press, Stanford

Eberhard WG (1990) Function and phylogeny of spider webs. Annu Rev Ecol Syst 21:341–372

Eisoldt L, Smith A, Scheibel T (2011) Decoding the secrets of spider silk. Mater Today 14 (3):80–86

Garb JE, DiMauro T, Vo V, Hayashi CY (2006) Silk genes support the single origin of orb webs. Science 312(5781):1762

Garb JE, Ayoub NA, Hayashi CY (2010) Untangling spider silk evolution with spidroin terminal domains. BMC Evol Biol 10:243. doi:10.1186/1471-2148-10-243

Garrido MA, Elices M, Viney C, Pérez-Rigueiro J (2002) Active control of spider silk strength: comparison of drag line spun on vertical and horizontal surfaces. Polymer 43:1537–1540

Gatesy J, Hayashi C, Motriuk D, Woods J, Lewis R (2001) Extreme diversity, conservation, and convergence of spider silk fibroin sequences. Science 291(5513):2603–2605

Gosline JM, Guerette PA, Ortlepp CS, Savage KN (1999) The mechanical design of spider silks: from fibroin sequence to mechanical function. J Exp Biol 202:3295–3303
Guinea GV, Elices M, Pérez-Rigueiro J, Plaza GR (2005) Stretching of supercontracted fibers: a link between spinning and the variability of spider silk. J Exp Biol 208:25–30
Harmer AMT, Blackledge TA, Madin JS, Herberstein ME (2011) High-performance spider webs: integrating biomechanics, ecology and behaviour. J R Soc Interface 8:457–471. doi:10.1098/rsif.2010.0454
Hayashi CY, Lewis RV (1998) Evidence from flagelliform silk cDNA for the structural basis of elasticity and modular nature of spider silks. J Mol Biol 275:773–784
Hayashi CY, Lewis RV (2000) Molecular architecture and evolution of a modular spider silk protein gene. Science 287(5457):1477–1479
Hayashi CY, Blackledge TA, Lewis RV (2004) Molecular and mechanical characterization of aciniform silk: uniformity of iterated sequence modules in a novel member of the spider silk fibroin gene family. Mol Biol Evol 21:1950–1959
Higgins LE, Townley MA, Tillinghast EK, Rankin MA (2001) Variation in the chemical composition of orb webs built by the spider *Nephila clavipes* (Araneae, Tetragnathidae). J Arachnol 29:82–94
Hinman MB, Jones JA, Lewis RV (2000) Synthetic spider silk: a modular fiber. Trends Biotechnol 18:374–379
Jenkins JE, Creager MS, Butler EB, Lewis RV, Yarger JL, Holland GP (2010) Solid-state NMR evidence for elastin-like beta-turn structure in spider dragline silk. Chem Commun 46:6714–6716. doi:10.1039/c0cc00829j
Keten S, Xu ZP, Ihle B, Buehler MJ (2010) Nanoconfinement controls stiffness, strength and mechanical toughness of beta-sheet crystals in silk. Nat Mater 9:359–367. doi:10.1038/nmat2704
Lin LH, Edmonds DT, Vollrath F (1995) Structural engineering of an orb-spider's web. Nature (London) 373(6510):146–148
Masters WM (1984) Vibrations in the orbwebs of *Nuctenea sclopetaria* (Araneidae): 1. Transmission through the web. Behav Ecol Sociobiol 15:207–216
Opell BD (2013) Cribellar thread. In: Nentwig W (ed) Spider ecophysiology. Springer, Heidelberg (this volume)
Opell BD, Bond JE, Warner DA (2006) The effects of capture spiral composition and orb-web orientation on prey interception. Zoology 109:339–345
Perry DJ, Bittencourt D, Siltberg-Liberles J, Rech EL, Lewis RV (2010) Piriform spider silk sequences reveal unique repetitive elements. Biomacromolecules 11:3000–3006. doi:10.1021/bm1007585
Rousseau ME, Lefevre T, Pezolet M (2009) Conformation and orientation of proteins in various types of silk fibers produced by *Nephila clavipes* Spiders. Biomacromolecules 10:2945–2953. doi:10.1021/bm9007919
Sahni V, Labhasetwar DV, Dhinojwala A (2012) Spider silk inspired functional microthreads. Langmuir 28:2206–2210
Scheibel T (2004) Spider silks: recombinant synthesis, assembly, spinning, and engineering of synthetic proteins. Microb Cell Fact 3:14. doi:10.1186/1475-2859-3-14
Seidl T, Vidoni R (2013) Adhesion to flat surfaces: from spiders to stickers. In: Nentwig W (ed) Spider ecophysiology. Springer, Heidelberg (this volume)
Selden PA, Shear WA, Bonamo PM (1991) A spider and other arachnids from the Devonian of New York [USA], and reinterpretations of Devonian Araneae. Palaeontology (Durham) 34:241–282
Sensenig A, Agnarsson I, Blackledge TA (2010) Behavioural and biomaterial coevolution in spider orb webs. J Evol Biol 23:1839–1856. doi:10.1111/j.1420-9101.2010.02048.x
Sensenig A, Lorentz KA, Kelly SP, Blackledge TA (2012) Spider orb webs rely on radial threads to absorb prey energy. J R Soc Interface 9:1880–1891
Shear WA (ed) (1986) Spiders, webs, behavior, and evolution. Stanford University Press, Stanford

Spagna JC, Goldman DI, Lin P-C, Koditschek DE, Full RJ (2007) Distributed mechanical feedback in arthropods and robots simplifies control of rapid running on challenging terrain. Bioinspir Biomim 2:9–18

Swanson BO, Blackledge TA, Summers AP, Hayashi CY (2006) Spider dragline silk: correlated and mosaic evolution in high performance biological materials. Evolution 60:2539–2551

Swanson BO, Blackledge TA, Hayashi CY (2007) Spider capture silk: performance implications of variation in an exceptional biomaterial. J Exp Zool A 307A:654–666

Townley MA, Tillinghast E (2013) Aggregate silk gland secretions of araneoid spiders. In: Nentwig W (ed) Spider ecophysiology. Springer, Heidelberg (this volume)

Vollrath F, Edmonds DT (1989) Modulation of the mechanical properties of spider silk by coating with water. Nature 340(6231):305–307

Vollrath F, Porter D (2006) Spider silk as a model biomaterial. Appl Phys A 82:205–212

Watanabe T (2000) Web tuning of an orb-web spider, *Octonoba sybotides*, regulates prey-catching behaviour. Proc R Soc Lond B Biol Sci 267(1443):565–569

Zentner L (2013) Modelling and application of the hydraulic spider leg mechanism. In: Nentwig W (ed) Spider ecophysiology. Springer, Heidelberg (this volume)

Chapter 21
Aggregate Silk Gland Secretions of Araneoid Spiders

Mark A. Townley and Edward K. Tillinghast

21.1 Introduction

Most spiders have multiple types of spinneret-associated silk glands used for different purposes. Spiders of the ecribellate superfamily Araneoidea, which include builders of sticky-droplet orb webs, cobwebs, and stereotyped aerial sheet webs (Blackledge et al. 2009; see also Appendix, this volume), typically have major ampullate (MaA), piriform (Pi), minor ampullate (MiA), aciniform (Ac), cylindrical (tubuliform), flagelliform (Fl), and aggregate (Ag) types of silk glands (Kovoor 1987; Blackledge et al. 2011). Of these, the Ag least conform to expectations for a silk gland, especially in terms of the form and composition of their products (Vollrath 1999). Rather than the largely proteinaceous fibers produced by typical silk glands that are essentially dry soon after being drawn, Ag are known for their sticky aqueous secretions with a substantial nonprotein component (in addition to water). This gland type only occurs in araneoid spiders (Kovoor 1977a), and indeed, it provides one of the autapomorphies delimiting this large clade (Schütt 2003) containing about a quarter of the 42,751 extant described spider species (Platnick 2012). In their best known contexts, as produced by orb web and cobweb builders, Ag secretions stick to prey, thereby helping to detain them and aiding in their capture. Additional roles in certain families will be mentioned below.

Araneoids typically have two pairs of Ag, though there are various examples, distributed across several families, of taxa in which some or all of these have been secondarily lost (e.g., Peters 1993; Schütt 1995; Agnarsson 2004; Miller 2007; Rix and Harvey 2010). In addition, Ag degenerate in many adult males (e.g., araneids, tetragnathids, most theridiids, some linyphiids) (e.g., Sekiguchi 1955; Agnarsson 2004;

M.A. Townley (✉)
Research Computing and Instrumentation, University of New Hampshire, Durham, NH, USA
e-mail: mtownley@cisunix.unh.edu

E.K. Tillinghast
Department of Biological Sciences, University of New Hampshire, Durham, NH, USA

W. Nentwig (ed.), *Spider Ecophysiology*,
DOI 10.1007/978-3-642-33989-9_21,

Miller and Hormiga 2004) and, rarely, adult females as well (Platnick et al. 1991), but they are retained, at least in part, in other adult males (e.g., most spiders in "clawless female clade" of Lopardo and Hormiga 2008, many linyphiids including all examined erigonines) (e.g., Miller and Hormiga 2004; Rix and Harvey 2010).

Droplets on silk fibers preserved in amber demonstrate that Ag had evolved by the Lower Cretaceous (~140 Ma) (Brasier et al. 2009; Selden and Penney 2010) and fossil araneoids, including a species of the nephilid *Nephila* (Selden et al. 2011), push this origin back to at least the Middle Jurassic (~165 Ma) (Selden and Penney 2010). A single origin for cribellate (deinopoid) and ecribellate (araneoid) orb webs, supported by molecular, morphological, and behavioral data, suggests that Ag secretions first came into use among cribellate orb web builders (Blackledge et al. 2009). A major conceptual difficulty in this scenario, however, has been envisioning the transition from prey capture using dry cribellar silk to capture using aqueous Ag droplets (e.g., Opell and Schwend 2009; Blackledge et al. 2011). Opell et al. (2011a) have proposed a very plausible hypothesis to explain how this shift may have occurred and have demonstrated that hybrid cribellar-Ag threads could have made an effective transitional form of sticky spiral.

21.2 Composition, Occurrence, and Function of Ag Secretions

First, two caveats: (1) What is currently known about specific Ag secretion components largely comes from published studies on about a dozen species from two families of orb web builders (Araneidae, Nephilidae) and one family of cobweb builders (Theridiidae); this from out of 11,675 described species in 16 araneoid families (Platnick 2012). (2) For none of the studied species is there as yet a reasonably complete description of all the secretion constituents, though *Nephila clavipes* comes closest.

Ag secretions are complex. In addition to water, a variety of small polar aliphatic compounds, inorganic salts, proteins, peptides, lipids, and aromatic toxins have been identified from one or more species, as detailed below.

21.2.1 Orb Web Builders

In ecribellate orb webs, Ag secretions from both pairs of Ag are used to coat a pair of Fl fibers as they are drawn from their spigots, thus producing the web's prey-catching sticky spiral (Peters 1987). Histochemical studies showing no differences between the two pairs of Ag (Kovoor 1987) suggest that their secretions are essentially the same (but see Sect. 21.3). Both pairs of Ag have ducts that, among silk gland types, are uniquely endowed with numerous nodules (Moon and Kim 2005) (Fig. 21.1). Their ultrastructure suggests a function transporting water and ions (phosphate especially), and possibly including some organic compounds,

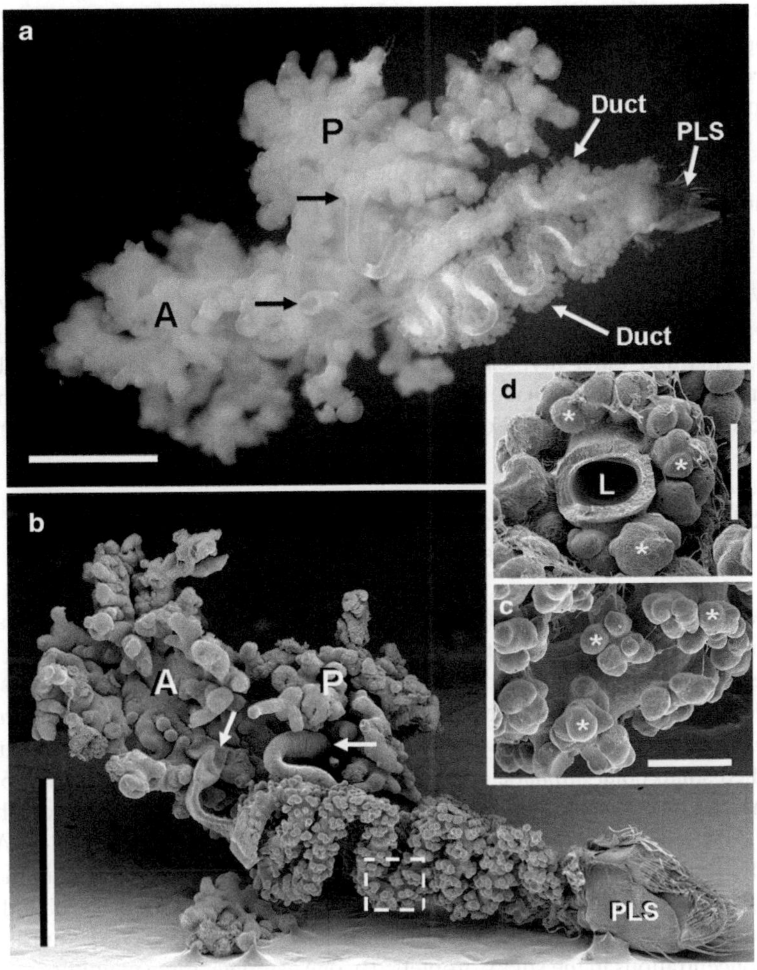

Fig. 21.1 The two aggregate silk glands from right side of opisthosoma of late penultimate instar male *Araneus cavaticus* (Araneidae). At most times the glands are more distended with luminal contents, making removal difficult, but at this time, not long before these glands atrophy in males, it is easier to separate the numerous outpocketings of the glands from the hepatopancreas in which they are embedded. *Unlabeled arrows* in (**a**) and (**b**) indicate where the duct connects to the body of the gland. (**a**) By light microscopy, the structured cuticular lining of the ducts makes their serpentine course apparent, especially evident from this angle in the duct of the anterior aggregate gland (*A*). Note the numerous nodules over much of the ducts' length. (**b**) Same aggregate glands as in (**a**) by scanning electron microscopy (SEM). The duct of the posterior aggregate gland (*P*) is largely obscured by that of the anterior gland (*A*). Boxed region of duct is shown at higher magnification in (**c**). (**d**) SEM of fractured duct showing radial distribution of nodules. *Asterisks* in (**c**) and (**d**) indicate a few of the many nodules present in these micrographs. *L* lumen, *PLS* posterior lateral spinneret. Scale bars: (**a**, **b**) 1 mm; (**c**, **d**) 100 μm

into the lumen of the duct (Kovoor and Zylberberg 1979), with energy provided by glycogen reserves (Kovoor and Zylberberg 1979; Tillinghast and Chase 1985). The Ag secretions are applied to the Fl fibers as a continuous liquid cylinder, but then, driven by surface tension, it breaks up into a series of connected sticky droplets such that, between droplets, the Fl fibers are still enveloped in Ag secretion (e.g., Edmonds and Vollrath 1992; Kane et al. 2010).

Though ecribellate orb web builders often eat and recycle their webs on a daily basis (Carico 1986), sticky spirals are typically very stable and have been stored for up to 10 months without significant changes in stickiness (Blackwall 1835; Edmonds and Vollrath 1992; Opell and Schwend 2008). In contrast, webs of *Cyrtarachne*, a moth-specialist araneid, lose their stickiness within 11 h (Cartan and Miyashita 2000). This and other atypical characteristics of sticky threads in the subfamily Cyrtarachninae indicate that their Ag secretion compositions differ in important respects from those of typical orb web builders (Cartan and Miyashita 2000).

21.2.1.1 Low-Molecular-Mass Components Including Lipids and Small Peptides

The water-soluble portion of a typical araneid or nephilid orb web is substantial, accounting for about one- to two-thirds of its desiccated mass, depending on the species and the individual web (see Townley et al. 1991). The bulk of this water-soluble material comes from the Ag. As determined for the araneid *Araneus cavaticus* in the lab, about 60% of the water-soluble mass can be attributed to a group of identified small polar aliphatic compounds (Table 21.1) (Townley et al. 2006), often net charged at the weakly acidic pH (4–5) of the Ag secretion (Schildknecht et al. 1972; Anderson and Tillinghast 1980; Tillinghast and Christenson 1984). Another roughly 10–20% of the water-soluble mass comes from inorganic ions including $H_2PO_4^-$, K^+, NO_3^-, Na^+, Cl^-, and Ca^{2+} (see Townley et al. 2006). The abundance of these organic and inorganic low-molecular-mass components (LMM) translates into high concentrations in the sticky droplets: By one estimate, three major organic LMM were each present at 1–2 M in webs of *Araneus diadematus* (Vollrath et al. 1990).

The organic LMM presented in Table 21.1 are those known to occur in relatively large quantity in the Ag secretions of one or more species. Several of these are related or identical to known or suspected neurotransmitters/neuromodulators (GABA, acetylcholine, β-alanine, taurine, glycine), consistent with hypotheses proposing a neural structure ancestry for spinneret-associated silk glands (Palmer 1990; Vollrath et al. 1990). With the exception of β-alaninamide, which has only been detected in webs of *Zygiella atrica*, each of these LMM has been detected in multiple species, though often in very different relative amounts and with omissions (in addition to β-alaninamide) in some species (Vollrath et al. 1990; Townley et al. 1991, 2006, 2012; Higgins et al. 2001). Thus, while selection has retained various members of a small group of organic LMM in different species, differences among species can still be dramatic, even between congeneric species. For example, in webs of *N. clavipes*, GABamide is one of three major organic LMM and glycine betaine is present in

Table 21.1 Organic low-molecular-mass solutes (LMM) in orb web sticky droplets[a]

Trivial name(s) {ionized form}	IUPAC name {ionized form}	Structure
Putrescine	Butane-1,4-diamine {butane-1,4-diaminium}	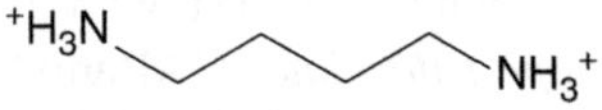
N-Acetylputrescine	*N*-(4-aminobutyl)acetamide {4-acetamidobutan-1-aminium}	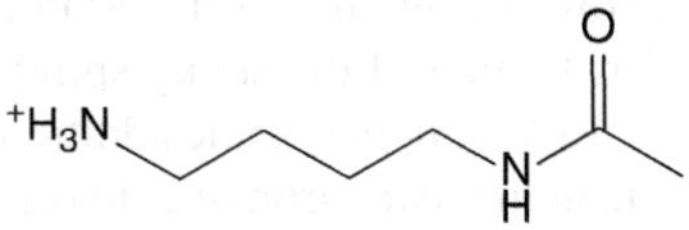
GABamide[b]; γ-aminobutyramide	4-Aminobutanamide {4-amino-4-oxobutan-1-aminium}	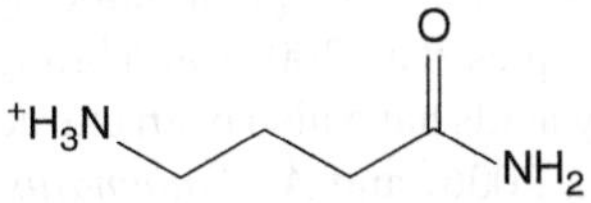
β-Alaninamide; β-aminopropionamide	3-Aminopropanamide {3-amino-3-oxopropan-1-aminium}	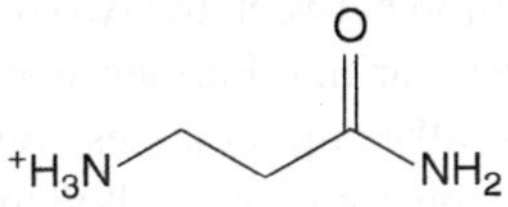
Choline	2-Hydroxy-*N*,*N*,*N*-trimethylethanaminium	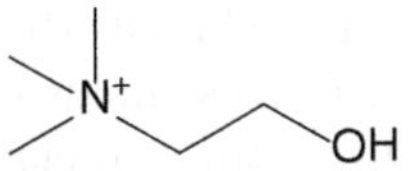
Betaine; glycine betaine	1-Carboxy-*N*,*N*,*N*-trimethylmethanaminium {2-(trimethylammonio) acetate}	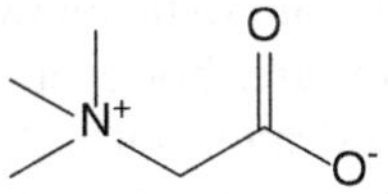
Glycine[c]	2-Aminoacetic acid {2-ammonioacetate}	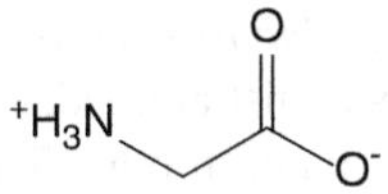
Isethionic acid {isethionate}	2-Hydroxyethanesulfonic acid {2-hydroxyethanesulfonate}	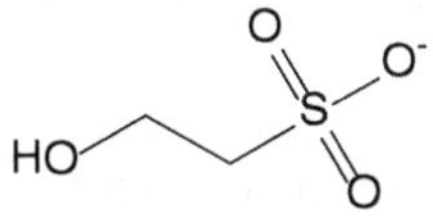
N-Acetyltaurine[d]	2-Acetamidoethanesulfonic acid {2-acetamidoethanesulfonate}	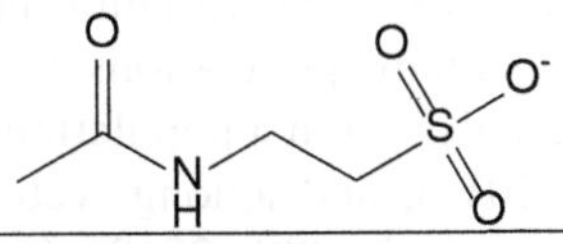

[a]Identifications from Fischer and Brander (1960), Vollrath et al. (1990), Townley et al. (1991), Higgins et al. (2001), and Townley et al. (2012). These are relatively abundant LMM in typical webs of one or more species. See Sect. 21.2.1.1 for other organic LMM reportedly present in reduced concentrations

[b]GABA (4-aminobutanoate) has also been reported as a somewhat atypical component in *Nephila clavipes* (Higgins et al. 2001) but is routinely present in aggregate gland secretions of the cobweb builder *Latrodectus* (see Sect. 21.2.2.1)

[c]Other free amino acids, notably proline and alanine, are also abundant in some species and are typically, though not invariably, in lower concentration than glycine (Townley et al. 2006; Townley and Tillinghast, unpublished)

[d]Taurine (2-aminoethanesulfonate) has also been observed in some species, most often as a minor LMM (<1 mol%), though larger amounts have also been seen (Higgins et al. 2001)

lesser quantity (Higgins et al. 2001), while in webs of *Nephila pilipes*, GABamide has not been detected and glycine betaine is a major organic LMM (Higgins et al., unpublished). In webs of the araneid *Argiope aurantia*, GABamide is easily the most abundant LMM with *N*-acetylputrescine absent or trivial, but in webs of *Argiope trifasciata*, GABamide is typically largely replaced by *N*-acetylputrescine (Townley et al. 1991, 2006). How these differences in composition affect the functioning of the sticky spiral is unknown.

In *N. clavipes*, some additional low-abundance organic LMM have been identified, including two aromatic toxins, [1-(3-diazenylphenyl)ethanol]iron (Marques et al. 2004) and 1-(2-guanidinoethyl)-1,2,3,4-tetrahydro-6-(hydroxymethyl)-β-carboline (Marques et al. 2005), and bradykinin and related peptides (Volsi et al. 2006). Various fatty acids have also been extracted from orb webs of *N. clavipes* (Schulz 2001; Salles et al. 2006) and *A. diadematus* (Vollrath et al. 1990), with differences in fatty acid composition between the two *N. clavipes* studies likely due to the use of solvents with different polarities. Further, when *N. clavipes* webs were extracted with more nonpolar solvents, other lipid classes were found to be much more abundant than fatty acids, most notably long-chain alkyl methyl ethers (1-methoxyalkanes), which accounted for 50–80% of extracted lipids (Schulz 2001). However, as Schulz (1999, 2001) points out, the source of these lipids is unknown, though dragline silk (i.e., MaA silk, used also in non-sticky orb web elements) contains relatively little lipid material, and 1-methoxyalkanes are so far known in nature only from spiders with Ag (three araneoid families). They are, however, not present in all such spiders as their absence in *A. diadematus* demonstrates (Schulz 1999). Consistent with an Ag secretion origin for at least some web lipids are lipid-like staining reactions in sticky droplets from webs of *Argiope lobata* (Peters 1995) and *N. clavipes* (Salles et al. 2006). *Kaira alba*, a male-moth-specialist araneid, apparently attracts its prey by emitting the sex pheromone of the female moths. This allomone is reportedly produced by the spider's morphologically modified Ag and volatilizes from the web (Lopez 1999, see also Uhl 2013).

Within a species, organic LMM composition is often consistent qualitatively and with respect to some quantitative relationships among LMM. Nevertheless, considerable intraspecific variation in LMM composition has been observed at multiple levels: among populations (Townley et al. 1991; Higgins et al. 2001), among individuals within a population (Vollrath et al. 1990; Higgins et al. 2001; Townley et al. 2012), and among webs from the same individuals (Higgins et al. 2001; Townley et al. 2006, 2012). Variation is likely even within a single orb web given differences in other sticky droplet parameters (volume and spacing) in different parts of the same web (Eberhard 1988; Edmonds and Vollrath 1992; Kane et al. 2010). Furthermore, stickiness of outer sticky spiral threads can be greater than that of inner threads without a difference in droplet volume per length of sticky spiral, likewise suggesting compositional differences between outer and inner threads, which may include LMM differences (Opell et al. 2009). Principal sources of intraspecific variation are not well established, but various interrelated possibilities have been indicated or suggested, including (1) both quantitative and qualitative aspects of diet, (2) the extent of web ingestion and recycling, (3) developmental changes including molt-intermolt and egg-laying cycles, (4) genetic differences, (5) changes in physical environment parameters (e.g., temperature, water vapor pressure, web site), (6) available silk gland supplies (taking in such related factors as web dimensions and the time between bouts of web building), and (7) gender (Eberhard 1988; Higgins et al. 2001; Crews and Opell 2006; Townley et al. 2006, 2012; Opell et al. 2009).

The spider's ability to synthesize different organic LMM varies from almost no (e.g., choline) to high (e.g., GABamide) synthetic capability (Higgins and Rankin 1999; Townley et al. 2006), with some synthesis possibly attributable to gut bacteria rather than the spider's tissues (Krejčík et al. 2010). Apparently as a result, there is a tendency for percentages of less readily synthesized organic LMM to decrease in webs of unfed spiders that are not allowed to recycle previous webs (Townley et al. 2006). The incorporation of substantial amounts of nutritionally essential LMM in the Ag secretion presumably makes web recycling especially important for maintaining adequate levels of these LMM. The same can be said for Ag secretion proteins, which contain especially high percentages of essential amino acids (39%) as compared with fibroins of more typical silk glands like the MaA (7%) (Tillinghast and Townley 1994). Indeed, the abundance of nutritional essentials in the sticky droplets, coupled with the observation that web recycling is largely restricted to araneoid orb web builders, has led to the suggestion that selection for web recycling is driven primarily by the advantages derived from retrieving components, not from non-sticky web elements like MaA silk fibers, but from Ag secretions (Tillinghast and Townley 1994; Benjamin and Zschokke 2003; Townley et al. 2006; Blackledge et al. 2009, 2011; see also Barrantes and Eberhard 2010).

No doubt a variety of functions can be ascribed to the LMM, but our understanding of them is still very limited. One role is water adsorption and retention (see also Sect. 21.2.1.2), some water being essential for maintaining extensibility within the Fl fibers and sticky droplet glue (Richter 1956; Fischer and Brander 1960; Schildknecht et al. 1972; Vollrath and Edmonds 1989; Bonthrone et al. 1992; Gosline et al. 2002; Guinea et al. 2010; Sahni et al. 2011a), both of which contribute to sticky spiral stickiness (Sect. 21.4). High collective LMM concentration helps resist drying through colligative water vapor pressure lowering (Vollrath et al. 1990). Additionally, some of the LMM are hygroscopic, taking up water from the air—choline especially over a range of relative humidities (Vollrath et al. 1990; Townley et al. 1991; Edmonds and Vollrath 1992). Lipid LMM at the surface of the sticky spiral might also aid water retention (Peters 1995; Schulz 1997). On the other hand, too much water will reduce stickiness by excessively increasing droplet extensibility and lubrication with surfaces and by reducing droplet viscosity and elasticity (Opell et al. 2011b; Sahni et al. 2011a). Thus, it might be expected that LMM composition and/or concentration differ among spiders adapted to different relative humidity environments (Opell et al. 2011b; Sahni et al. 2011a). One or more organic LMM, accompanied by water molecules, may make more direct contributions to sticky spiral extensibility by penetrating the Fl fibers' proteins and increasing protein mobility (Gosline et al. 1995, 2002), though Guinea et al. (2010) argue against such contributions from organic components. Speculatively, direct interactions between organic LMM and Ag glycoprotein glue may also enhance stickiness (Townley et al. 1991; Higgins et al. 2001; Blackledge et al. 2011). Antimicrobial properties have been suggested for various LMM including inorganic salts (Schildknecht et al. 1972), [1-(3-diazenylphenyl)ethanol]iron (Marques et al. 2004), and some lipid LMM (Schulz 1997, 2001; Salles et al. 2006). Some fatty acids may also repel ants (Salles et al. 2006). A role in subduing prey has been indicated for several of the low-abundance LMM found in *N. clavipes* webs that have toxic effects on honeybees (Marques et al. 2004, 2005; Salles et al. 2006; Volsi et al. 2006). Amphiphilic lipids may play significant roles (1) in increasing the wettability of prey epicuticles, thus increasing prey surface area in contact with Ag secretion, and (2) in making prey cuticle more permeable to other toxic Ag secretion components (Salles et al. 2006).

21.2.1.2 Proteins

Sticky droplets in orb webs contain abundant glycoprotein (Tillinghast et al. 1993), and it appears the stickiness of the droplets can be attributed primarily to this glycoprotein component rather than to the relatively minor capillary forces generated by the aqueous coating (Sahni et al. 2010, 2011a, b; Opell et al. 2011c). In *N. clavipes*, two genes, *aggregate spider glue 1* (*asg1*) and *asg2*, have been identified that are highly expressed in Ag but not in other silk glands or tissues (Choresh et al. 2009). The proteins they encode, ASG1 (45.2 kDa) and ASG2 (71.5 kDa), contain many sites with high potential for glycosylation. They appear to constitute at least a large part of the glycoprotein component of Ag secretion, and homologues of both proteins occur in the webs of other orb web builders (Choresh et al. 2009). A repetitive domain occurs in each protein. Remarkably, these two domains result from transcription of opposite strands of the same DNA segment, with 100% complementarity over a 353-bp span of the mRNA-derived sequences (Choresh et al. 2009). A high occurrence of proline in the N-terminal portion of ASG2, and sequence similarities to elastin and Fl fibroin repetitive domains, points to an elastic nature for ASG2, consistent with demonstrations of high extensibility and elasticity in the droplet glycoprotein (Richter 1956; Vollrath and Tillinghast 1991; Opell and Hendricks 2007, 2010; Opell et al. 2011b; Sahni et al. 2010, 2011a, b). Sequence similarities between ASG1 and chitin-binding proteins suggest a prey adhesion capacity in addition to that conferred by glycosylation, and a high percentage of charged amino acids in ASG1 may contribute to water retention (Choresh et al. 2009). Potential glycosylation sites identified for both ASG proteins are dominated by *O*-glycosylations, which agrees with biochemical analyses of a glycoprotein preparation from *A. aurantia* (Tillinghast et al. 1993). In the latter study, the glycoprotein preparation, containing *N*-acetylgalactosamine as the dominant sugar, also contained phosphorylated serine and threonine residues. As some serine and threonine residues have a high potential for both glycosylation and phosphorylation [e.g., of 45 potential *O*-glycosylation sites in ASG1 indicated in fig. 1 of Choresh et al. (2009), 38 are also potential phosphorylation sites as predicted by NetPhos 2.0 (Blom et al. 1999)], the extent and variability of each modification remain to be seen.

Proteinases (calpain- and jararhagin-like) have been detected as minor, but potentially significant, components in the Ag secretion of *N. clavipes* (Salles et al. 2006). Though without toxicity on their own when topically applied to honeybees, a synergistic toxic effect when mixed with web lipids was observed, a consequence perhaps of the lipids making insect cuticle permeable to the proteinases.

21.2.2 Cobweb Builders

Unlike the arrangement in orb web builders, there is apparently a division of labor between the two pairs of Ag among cobweb builders (theridioids: Theridiidae, Nesticidae) (Kovoor 1977b, 1987). One pair, more anterior than the other, are "typical" Ag similar to those of orb web builders. They have nodulated ducts and are presumably the source of the "web Ag secretion" applied to fibers within the web (e.g., gumfoot lines) for passively intercepting and sticking to prey (Kovoor 1977b). In contrast, the posterior "atypical" pair of Ag (also called lobed Ag) have very short non-nodulated ducts and are thought to be the source of the "combed Ag secretion" that is actively applied to prey (Kovoor 1977b) with the aid of setal combs on fourth leg tarsi during a sticky silk wrap attack (e.g., Agnarsson 2004; Barrantes and Eberhard 2007). The atypical Ag bear morphological and histochemical resemblances, presumably homoplastic, to the modified venom glands of the spitting spider *Scytodes*, the "spit" of which is likewise projected by the spider to hinder prey escape (Kovoor 1977b, 1987) (see Suter and Stratton 2013). Like the spit of *Scytodes*, combed Ag secretion is also used defensively against predators (Vetter 1980). The spigots serving both pairs of Ag have especially wide openings in theridioid spiders (e.g., Coddington 1989; Agnarsson 2004).

Gumfoot lines are the quintessential form of sticky thread within theridioid webs. They are dry over much of their length, with web Ag secretion applied over a short segment closest to the thread's attachment site to a substrate. However, web form among "cobweb builders" is actually highly variable, as is the occurrence and distribution of Ag secretion, both inter- and intraspecifically, and, if secretion is present at all in a given web, it may occur in situations other than on such gumfoot lines (Eberhard et al. 2008a). Available evidence indicates that the axial fibers in gumfoot lines are not Fl silk, as in orb web sticky spirals, but ampullate gland silk, with MaA silk alone or in combination with MiA silk proposed (Benjamin and Zschokke 2002; Blackledge et al. 2005a, b; Eberhard et al. 2008b). Fl silk instead reportedly accompanies combed Ag secretion during sticky silk wrap attacks, perhaps initiated by attaching to already drawn MaA silk (Eberhard 2010).

21.2.2.1 Low-Molecular-Mass Components, Including Lipids

As in many orb web builders, GABamide is a major organic LMM in web Ag secretion (from gumfoot lines) of the theridiid *Latrodectus* (Tillinghast and Christenson 1984), as well as in combed Ag secretion from these spiders (obtained by prodding the spider with a micropipette) (Kelly 1989; Tillinghast and Townley, unpublished). In both secretions, collected in the lab from *L. mactans* and *L. hesperus* indiscriminately (four collections of web Ag secretion, five of combed Ag secretion), we have also observed GABA (4-aminobutanoate) itself in amounts that sometimes rival or exceed those of GABamide, especially in combed Ag secretion. One or the other of these two compounds is typically the most abundant organic LMM in both secretions, though an as-yet-unknown LMM in combed Ag secretion (minor in web Ag secretion) can also take the prize. Other organic LMM in both secretions include several also known from orb web builders (Table 21.1), specifically choline (also observed by Kelly (1989)), glycine betaine, and isethionate. Differences between combed and web Ag secretions have also been seen (taking into account only identified organic LMM): (1) The lowest molar percentage of glycine betaine among combed Ag secretion samples (11.3 mol%) was 2.4 times greater than the highest percentage among web Ag secretion samples (4.8 mol%). (2) Free proline accounted for 7.3–15.7 mol% of organic LMM in combed Ag secretion but went undetected in web Ag secretion (in agreement with Kelly 1989). (3) Free glycine accounted for 6.5–38.8 mol% in combed Ag secretion but was detected in only one web Ag secretion collection (1.5 mol%). Thus, while the two secretions show some organic LMM similarities, they are not identical. Moreover, these results indicate that during gumfoot line formation, the secretion from the typical Ag is not significantly contaminated by atypical Ag secretion. Whether the converse applies as well, that combed Ag secretion is essentially free of contributions from the typical Ag, is unknown. Nor do we know if any contamination of combed Ag secretion differs depending on whether the secretion is used offensively or defensively.

Lipids are present on webs of *Latrodectus*, with 1-methoxyalkanes prominent, but just as with lipids extracted from orb webs, it has not been established that any of these derive from the Ag (Schulz 1999).

21.2.2.2 Peptides/Proteins

In *L. hesperus*, two water-soluble peptides, spider coating peptide 1 (SCP-1, 36 amino acids) and SCP-2 (14 amino acids), have been described that are products of the Ag (without a distinction made between typical and atypical Ag) but not of several other silk gland types (Hu et al. 2007). Curiously, these peptides were found not only in water extracts of gumfoot lines but in water extracts of egg sac fibers and web tangle connection joints. It could be that Ag secretion was present on the egg sacs and/or connection joints or that the genes encoding these peptides are also expressed in silk gland types that were not assayed during the study but contribute to egg sacs (Ac, Pi) and/or connection joints (Pi) (Hu et al. 2007). The former possibility gains plausibility from observations of Ag secretion occurring near egg sacs in some *Latrodectus* (Eberhard et al. 2008a) and of secretion sometimes found in *L. hesperus* webs other than on gumfoot lines (Eberhard et al. 2008a; Barrantes and Eberhard 2010). The functions of these peptides are not known. At least SCP-1 can bind metal ions, and by virtue of this property, two potential roles have been proposed, providing antimicrobial activity and aiding the gumfoot's glycoprotein glue in achieving its proper conformation (Hu et al. 2007).

A preliminary study of combed Ag secretion from *Latrodectus* concluded that at least two phosphoglycoproteins are present, at least one of which contains *N*-acetylgalactosamine (Kelly 1989). This is consistent with histochemical demonstrations of glycoprotein in the atypical as well as typical Ag (Kovoor 1977b, 1987) and suggests similarities to ASG1 and ASG2 of orb web builders (Sect. 21.2.1.2). Though these glycoproteins constituted the majority of the protein, the secretion also contained a group of proteins with lower apparent molecular masses (ranging between ~8–100 kDa) (Kelly 1989), at least some of which may correspond to pure protein grains seen by histochemistry only in the atypical Ag (Kovoor 1977b, 1987). Variability in the occurrence of some of these proteins may have been due to population or even species differences (*L. mactans* was given as the study species, but some *L. hesperus* were likely included). Additional histological and histochemical differences in the secretory products of typical versus atypical Ag have been described in *Latrodectus*, *Steatoda*, and *Argyrodes* (Kovoor 1977b; Kovoor and Lopez 1983).

21.2.3 Sheet Web Builders

Like orb web builders, the two pairs of Ag in linyphiid sheet web builders have histochemical characteristics similar to one another, indicating that their secretions are essentially the same (Kovoor 1987; Peters and Kovoor 1991). Unlike the Ag of orb web builders (Fig. 21.1) and the typical Ag pair of theridiids, the ducts of *Linyphia*'s Ag do not exhibit nodules. Nevertheless, a two-layered epithelium is common to all three (Kovoor and Zylberberg 1979; Peters and Kovoor 1991), in contrast to the simpler epithelium within the non-nodulated ducts of theridiid atypical Ag (Kovoor 1977b). Thus, Ag of *Linyphia* seem to occupy a middle ground in terms of luminal content processing within the ducts.

Linyphiid Ag secretion is apparently applied to Fl fibers, as in orb webs (Peters and Kovoor 1991; Benjamin et al. 2002). For the few species examined, the distribution of such coated threads in sheet and tangle portions of the web shows inter- and intraspecific variability, including an absence of Ag secretion in some webs (consistent for some species, variable in others) and the apparent use of Fl/Ag from only one side of the opisthosoma at times (Peters 1987; Peters and Kovoor 1991; Schütt 1995; Benjamin et al. 2002).

In *Linyphia* and *Microlinyphia*, Benjamin et al. (2002) observed that Ag secretion in the web dried shortly after being deposited and proposed that in these spiders the function of the secretion is not to adhere to and detain prey, but to cement intersecting fibers within the web, thus increasing web stability and perhaps facilitating the spider's movement within the web. They noted that this indicated compositional differences between the Ag secretions of linyphiids and orb web builders. Unfortunately, we are not aware of any studies on the composition of linyphiid Ag secretions. However, extracts of webs from ten species of linyphiids have revealed lipid components which, as in *Nephila* (Sect. 21.2.1.1) and *Latrodectus* (Sect. 21.2.2.1), include a variety of 1-methoxyalkanes (5–40% of extracted lipids) (Schulz 1997, 1999, 2004). But again, which, if any, of the extracted lipids have an Ag origin remains to be determined.

21.3 Organization Within Sticky Droplets

It is well known that sticky droplets in orb webs are not homogeneous (e.g., Richter 1956). When segments of sticky spiral laid on a glass slide are examined, a discrete material at the core of each droplet, surrounding the Fl fibers, is apparent (Fig. 21.2a). This material, accounting for about 15% of droplet volume (Opell and Hendricks 2010), has been referred to as a nodule (e.g., Vollrath and Tillinghast 1991) or granule (e.g., Opell and Hendricks 2010). To avoid confusion with the aforementioned nodules protruding from Ag ducts, we will use granule. Such granules have been observed in sticky droplets of various orb web builders (araneids, tetragnathids, and a theridiosomatid) (e.g., Opell and Hendricks 2010). In contrast, they have not been seen in theridiid gumfoot lines (Sahni et al. 2011a) (Fig. 21.2b). In orb web droplets examined by light/electron microscopy and optical surface profiling, a surface layer (perhaps lipidic), conical endcaps, and inhomogeneities in the region between granule and surface layer have also been indicated (Peters 1995; Cartan and Miyashita 2000; Salles et al. 2006; Opell and Hendricks 2009; Kane et al. 2010).

In large part because granule position appeared to coincide with lectin staining for N-acetylgalactosamine, the principal sugar in glycoprotein glue (Sect. 21.2.1.2), granules were taken to be packets of glycoprotein glue (Vollrath and Tillinghast 1991). However, after finding no positive relationship between granule size and sticky spiral stickiness (Opell and Hendricks 2009, 2010), Opell and Hendricks (2010) proposed a model in which the granule (which may itself contain glycoprotein) is an anchor for the actual glycoprotein glue, which forms a more voluminous, transparent layer around the granule and is in turn surrounded by an aqueous cover (Opell et al. 2011b). This model also seems more consistent with the ultrastructure of sticky droplets presented by Peters (1995) and with granules in collapsed sticky spiral segments not sticking to one another (Vollrath and Tillinghast 1991).

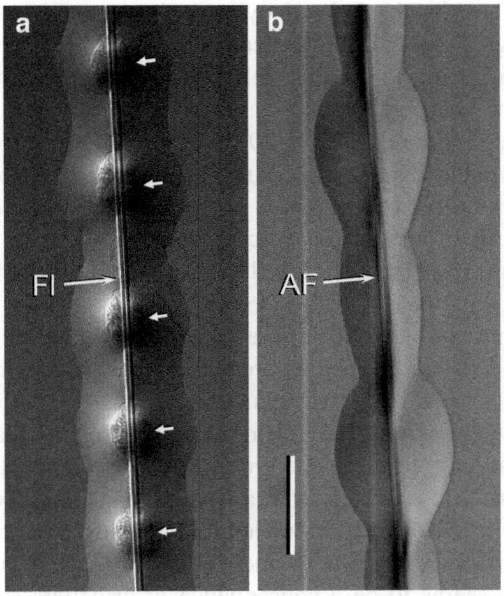

Fig. 21.2 (**a**) Span of sticky spiral from orb web of adult female *Argiope aurantia* (Araneidae), laid on a glass slide. Note granule (*unlabeled arrows*) at the center of each of five large droplets. *Fl*, pair of flagelliform silk gland fibers. (**b**) Span within gumfoot portion of gumfoot line from cobweb of adult female *Latrodectus hesperus* (Theridiidae), laid on a glass slide. Note absence of granules. *AF* axial fibers (see Sect. 21.2.2). Differential interference contrast optics. Both images at same magnification. Scale bar: 100 μm

Opell and Hendricks (2010) raised the possibility that one pair of Ag is the source of granule components, and the other is the source of the surrounding glycoprotein glue. In this context, it would be especially interesting to determine if ASG1 and ASG2 (Sect. 21.2.1.2) differ in their distribution within droplets and in their expression in the two pairs of Ag.

21.4 Physical Behavior of Ag Secretions

Sticky droplets of orb webs behave as viscoelastic materials (Sahni et al. 2010, 2011a, b), that is, with properties characteristic of both a liquid (viscosity is applicable, and there is a rate-dependent resistance to distortion) and solid (elasticity is applicable) (Vogel 2003). When a droplet is stretched quickly, as might occur when an insect first collides with the web, the droplet's liquid-like behavior is to the fore, with high viscous resistance generating a large adhesive force, good at retaining the insect. At low extension rates, such as might occur as the insect attempts to free itself, the viscous force component is less in evidence, and the droplet's solid-like behavior, apparently a manifestation of the glycoprotein glue's elasticity, is the primary provider of the smaller remaining adhesive force. Under these conditions, the droplet acts more like a rubber band, good at denying the insect purchase while expending its energy. When single droplets are stretched and given time ($\geq$15 min) to relax in this extended state, tension due to the glue's elasticity does not relax to zero but plateaus, still well above the contribution from capillary forces, at least at low to moderate relative humidities (Sahni et al. 2010, 2011a, b). Thus, orb web droplets behave as viscoelastic solids, indicating cross-linking within the glycoprotein glue. That this cross-linking is primarily noncovalent is suggested by the humidity dependence of the plateaued adhesive force within stretched droplets.

In contrast, tension within a stretched droplet on a gumfoot line plateaus nearly at zero, consistent with the behavior of a viscoelastic liquid and a virtual absence of cross-linking (Sahni et al. 2011a). These and other differences noted between orb web and gumfoot droplets (see below) are likely due in part to the absence of (glue-anchoring) granules in gumfoot droplets (Sahni et al. 2011a) (Fig. 21.2).

One might casually assume that, all other things being equal, a longer span of sticky spiral separating from a smooth surface would result in a greater total adhesive force than that resulting from a shorter span. However, in cribellate orb webs (see also Opell 2013), different length spans do not produce significantly different adhesive forces (Hawthorn and Opell 2003; Opell and Schwend 2009; Opell 2013). This seems to be a consequence of cribellate sticky spirals having relatively stiff axial (pseudoflagelliform) fibers and cribellar puffs. As a cribellar thread and surface pull away from each other, adhesive force is first generated along narrow bands near the edges of contact. The stiffness of the thread components results in only a small angle forming between the thread and surface before the two begin to separate. Consequently, once a thread pulls free at the edges, the intervening span peels off with little adhesive effect as the angle it makes to the surface continually increases (Opell and Schwend 2009).

In ecribellate orb webs, on the other hand, longer spans of sticky spiral do produce more total adhesive force than shorter spans, up to a point, thanks to the lower stiffness and greater extensibility of their components (Opell and Hendricks 2007). When an araneoid sticky spiral span and surface pull away from each other, with adhesive force generated at the edges of contact, a relatively large angle can form before edge droplets release from the surface. This results in significant recruitment of adhesive force, via the axial Fl fibers, from droplets further from the edges (Opell and Hendricks 2007). Each such interior droplet contributes only 0.70 times as much adhesive force as a neighbor that is closer to the edge of contact, so there is little additional adhesive force gained once a span of 20 droplets is reached (Opell and Hendricks 2009). This adhesive recruitment from more interior droplets is referred to as a suspension bridge mechanism (SBM), with the contacted surface, extended droplets and Fl fibers analogous to the suspended roadway, vertical cables, and upper horizontal cable, respectively, of a suspension bridge (Opell and Hendricks 2007). The total adhesive force generated by a span of sticky spiral is therefore made up of more than just the adhesive force between the glycoprotein glue and a surface. It also includes the forces needed to stretch the elastic droplets and axial Fl fibers (Sahni et al. 2010, 2011b; Opell et al. 2011b).

Indeed, estimates of the contribution from Fl fiber extension range from about one-third (Opell et al. 2008) to nearly half (Sahni et al. 2011b) of total sticky spiral stickiness. This SBM is not in operation in theridiid gumfoot lines because their sticky droplets readily detach from the axial fibers and their attachment disks readily detach from the substrate (Sahni et al. 2011a).

Another assumption that initially seems reasonable is that a stickier sticky spiral is necessarily a better sticky spiral. That this is not the case appears to reflect a selective pressure for a sticky spiral span to detach before actually breaking, thus maintaining web integrity and allowing the span to be reused (Agnarsson and Blackledge 2009). This check on stickiness operates at two levels. At a higher level, the adhesive force produced by a sticky spiral span does not exceed the force required to break its Fl fibers. This has been observed across 17 species of araneoid orb web builders for which these two forces covered a wide range of values and were strongly correlated (Agnarsson and Blackledge 2009). At a finer level, the stickiness of a droplet's glycoprotein glue does not exceed the force required to break the glycoprotein strand and thereby split the droplet. This was indicated by observing differences in stickiness when droplets contacted materials with different surface energies (Opell et al. 2011c) and is consistent with microscopic observations of droplets pulled from surfaces (Opell and Hendricks 2010) and with demonstrations that a sticky spiral span can generate the same adhesive force through many cycles of contact to and separation from a surface (Sahni et al. 2010). Because gumfoot lines function like spring-loaded traps that detach easily from the substrate (e.g., Argintean et al. 2006), considerations of web integrity and reuse and consequent limits on stickiness that are valid for orb webs do not seem to apply to gumfoot lines in cobwebs.

21.5 Conclusions

Information on Ag secretion composition is most plentiful for some of the more conspicuous orb web builders, with sequencing of two major glue glycoproteins especially significant (Choresh et al. 2009). But even among these spiders unidentified components remain, as do questions regarding the various components. Little is known with certainty, for example, about roles played by the various LMM, their synthesis and assemblage, and how inter- and intraspecific differences in their relative and absolute concentrations affect function. Among cobweb builders, the scant data available are almost entirely restricted to *Latrodectus,* and our ignorance is compounded by the division of the two pairs of Ag into morphologically, compositionally, and functionally distinct "typical" and "atypical" Ag types. With sheet web builders, where Ag secretions may serve a structural role within the web rather than sticking to prey (Benjamin et al. 2002), almost nothing is known in terms of composition. The discovery of additional Ag secretion constituents can be expected as the range of examined taxa expands.

The last few years have seen a surge in interest in the mechanical properties, adhesiveness, and organization of Ag sticky droplets and sticky threads. Among the results, (1) differences in the mechanical behavior and organization of sticky droplets from orb webs versus cobwebs have been revealed (Sahni et al. 2011a), (2) the importance of droplet and Fl fiber extensibility and elasticity to orb web stickiness have been demonstrated (Opell and Hendricks 2007, 2009; Opell et al. 2008; Sahni et al. 2010, 2011a), and (3) limits imposed by natural selection on orb web sticky spiral stickiness have been indicated (Agnarsson and Blackledge 2009; Opell et al. 2011b).

References

Agnarsson I (2004) Morphological phylogeny of cobweb spiders and their relatives (Araneae, Araneoidea, Theridiidae). Zool J Linn Soc 141:447–626

Agnarsson I, Blackledge TA (2009) Can a spider web be too sticky? Tensile mechanics constrains the evolution of capture spiral stickiness in orb-weaving spiders. J Zool 278:134–140

Anderson CM, Tillinghast EK (1980) GABA and taurine derivatives on the adhesive spiral of the orb web of *Argiope* spiders, and their possible behavioural significance. Physiol Entomol 5:101–106

Argintean S, Chen J, Kim M, Moore AMF (2006) Resilient silk captures prey in black widow cobwebs. Appl Phys A 82:235–241

Barrantes G, Eberhard WG (2007) The evolution of prey-wrapping behavior in spiders. J Nat Hist 41:1631–1658

Barrantes G, Eberhard WG (2010) Ontogeny repeats phylogeny in *Steatoda* and *Latrodectus* spiders. J Arachnol 38:485–494

Benjamin SP, Zschokke S (2002) Untangling the tangle-web: web construction behavior of the comb-footed spider *Steatoda triangulosa* and comments on phylogenetic implications (Araneae: Theridiidae). J Insect Behav 15:791–809

Benjamin SP, Zschokke S (2003) Webs of theridiid spiders: construction, structure and evolution. Biol J Linn Soc 78:293–305

Benjamin SP, Düggelin M, Zschokke S (2002) Fine structure of sheet-webs of *Linyphia triangularis* (Clerck) and *Microlinyphia pusilla* (Sundevall), with remarks on the presence of viscid silk. Acta Zool 83:49–59

Blackledge TA, Summers AP, Hayashi CY (2005a) Gumfooted lines in black widow cobwebs and the mechanical properties of spider capture silk. Zoology 108:41–46

Blackledge TA, Swindeman JE, Hayashi CY (2005b) Quasistatic and continuous dynamic characterization of the mechanical properties of silk from the cobweb of the black widow spider *Latrodectus hesperus*. J Exp Biol 208:1937–1949

Blackledge TA, Scharff N, Coddington JA, Szüts T, Wenzel JW, Hayashi CY, Agnarsson I (2009) Reconstructing web evolution and spider diversification in the molecular era. Proc Natl Acad Sci USA 106:5229–5234

Blackledge TA, Kuntner M, Agnarsson I (2011) The form and function of spider orb webs: evolution from silk to ecosystems. Adv Insect Physiol 41:175–262

Blackwall J (1835) Notice of several recent discoveries in the structure and economy of spiders. Trans Linn Soc Lond 16:471–485

Blom N, Gammeltoft S, Brunak S (1999) Sequence and structure-based prediction of eukaryotic protein phosphorylation sites. J Mol Biol 294:1351–1362

Bonthrone KM, Vollrath F, Hunter BK, Sanders JKM (1992) The elasticity of spiders' webs is due to water-induced mobility at a molecular level. Proc R Soc Lond B 248:141–144

Brasier M, Cotton L, Yenney I (2009) First report of amber with spider webs and microbial inclusions from the earliest Cretaceous (*c.* 140 Ma) of Hastings, Sussex. J Geol Soc Lond 166:989–997

Carico JE (1986) Web removal patterns in orb-weaving spiders. In: Shear WA (ed) Spiders: webs, behavior, and evolution. Stanford University Press, Stanford

Cartan CK, Miyashita T (2000) Extraordinary web and silk properties of *Cyrtarachne* (Araneae, Araneidae): a possible link between orb-webs and bolas. Biol J Linn Soc 71:219–235

Choresh O, Bayarmagnai B, Lewis RV (2009) Spider web glue: two proteins expressed from opposite strands of the same DNA sequence. Biomacromolecules 10:2852–2856

Coddington JA (1989) Spinneret silk spigot morphology: evidence for the monophyly of orbweaving spiders, Cyrtophorinae (Araneidae), and the group Theridiidae plus Nesticidae. J Arachnol 17:71–95

Crews SC, Opell BD (2006) The features of capture threads and orb-webs produced by unfed *Cyclosa turbinata* (Araneae: Araneidae). J Arachnol 34:427–434

Eberhard WG (1988) Behavioral flexibility in orb web construction: effects of supplies in different silk glands and spider size and weight. J Arachnol 16:295–302

Eberhard WG (2010) Possible functional significance of spigot placement on the spinnerets of spiders. J Arachnol 38:407–414

Eberhard WG, Agnarsson I, Levi HW (2008a) Web forms and the phylogeny of theridiid spiders (Araneae: Theridiidae): chaos from order. Syst Biodivers 6:415–475

Eberhard WG, Barrantes G, Madrigal-Brenes R (2008b) Vestiges of an orb-weaving ancestor? The "biogenetic law" and ontogenetic changes in the webs and building behavior of the black widow spider *Latrodectus geometricus* (Araneae Theridiidae). Ethol Ecol Evol 20:211–244

Edmonds DT, Vollrath F (1992) The contribution of atmospheric water vapour to the formation and efficiency of a spider's capture web. Proc R Soc Lond B 248:145–148

Fischer FG, Brander J (1960) Eine Analyse der Gespinste der Kreuzspinne. Hoppe-Seyler's Z Physiol Chem 320:92–102

Gosline J, Nichols C, Guerette P, Cheng A, Katz S (1995) The macromolecular design of spiders' silks. In: Sarikaya M, Aksay IA (eds) Biomimetics: design and processing of materials. AIP Press, Woodbury

Gosline J, Lillie M, Carrington E, Guerette P, Ortlepp C, Savage K (2002) Elastic proteins: biological roles and mechanical properties. Philos Trans R Soc Lond B 357:121–132

Guinea GV, Cerdeira M, Plaza GR, Elices M, Pérez-Rigueiro J (2010) Recovery in viscid line fibers. Biomacromolecules 11:1174–1179

Hawthorn AC, Opell BD (2003) van der Waals and hygroscopic forces of adhesion generated by spider capture threads. J Exp Biol 206:3905–3911

Higgins L, Rankin MA (1999) Nutritional requirements for web synthesis in the tetragnathid spider *Nephila clavipes*. Physiol Entomol 24:263–270

Higgins LE, Townley MA, Tillinghast EK, Rankin MA (2001) Variation in the chemical composition of orb webs built by the spider *Nephila clavipes* (Araneae, Tetragnathidae). J Arachnol 29:82–94

Hu X, Yuan J, Wang X, Vasanthavada K, Falick AM, Jones PR, La Mattina C, Vierra CA (2007) Analysis of aqueous glue coating proteins on the silk fibers of the cob weaver, *Latrodectus hesperus*. Biochemistry 46:3294–3303

Kane DM, Joyce AM, Staib GR, Herberstein ME (2010) Optical surface profiling of orb-web spider capture silks. Bioinspir Biomim. doi:10.1088/1748-3182/5/3/036004

Kelly SL (1989) The chemical composition of the defensive secretion of the spider *Latrodectus mactans* (Fabricius). MS thesis, University of New Hampshire, Durham

Kovoor J (1977a) La soie et les glandes séricigènes des arachnides. Ann Biol 16:97–171

Kovoor J (1977b) Données histochimiques sur les glandes séricigènes de la veuve noire *Latrodectus mactans* Fabr. (Araneae, Theridiidae). Ann Sci Nat Zool Biol Anim 12e Sér 19: 63–87

Kovoor J (1987) Comparative structure and histochemistry of silk-producing organs in arachnids. In: Nentwig W (ed) Ecophysiology of spiders. Springer, Berlin

Kovoor J, Lopez A (1983) Composition et histologie de l'appareil séricigène des *Argyrodes*, relations avec le comportement de ces araignées (Theridiidae). Rev Arachnol 5:29–43

Kovoor J, Zylberberg L (1979) Ultrastructure du canal des glandes agrégées et flagelliformes d'*Araneus diadematus* Clerck (Araneae, Araneidae). Zoomorphologie 92:217–239

Krejčík Z, Hollemeyer K, Smits THM, Cook AM (2010) Isethionate formation from taurine in *Chromohalobacter salexigens*: purification of sulfoacetaldehyde reductase. Microbiology 156:1547–1555

Lopardo L, Hormiga G (2008) Phylogenetic placement of the Tasmanian spider *Acrobleps hygrophilus* (Araneae, Anapidae) with comments on the evolution of the capture web in Araneoidea. Cladistics 24:1–33

Lopez A (1999) Un prédateur de papillons nocturnes extraordinaire: l'araignée *Kaira alba* (Hentz) et ses glandes à soie. Alexanor 21:195–201

Marques MR, Mendes MA, Tormena CF, Souza BM, Ribeiro SP, Rittner R, Palma MS (2004) Structure determination of an organometallic 1-(diazenylaryl)ethanol: a novel toxin subclass from the web of the spider *Nephila clavipes*. Chem Biodivers 1:830–838

Marques MR, Mendes MA, Tormena CF, Souza BM, Cesar LMM, Rittner R, Palma MS (2005) Structure determination of a tetrahydro-β-carboline of arthropod origin: a novel alkaloid-toxin subclass from the web of spider *Nephila clavipes*. Chem Biodivers 2:525–534

Miller JA (2007) Synaphridae of Madagascar (Araneae: Araneoidea): a new family record for the Afrotropical region. Proc Calif Acad Sci 4th Ser 58:21–48

Miller JA, Hormiga G (2004) Clade stability and the addition of data: a case study from erigonine spiders (Araneae: Linyphiidae, Erigoninae). Cladistics 20:385–442

Moon MJ, Kim TH (2005) Microstructural analysis of the capture thread spinning apparatus in orb web spiders. Entomol Res 35:133–140

Opell BD (2013) Cribellar thread. In: Nentwig W (ed) Spider ecophysiology. Springer, Heidelberg (this volume)

Opell BD, Hendricks ML (2007) Adhesive recruitment by the viscous capture threads of araneoid orb-weaving spiders. J Exp Biol 210:553–560

Opell BD, Hendricks ML (2009) The adhesive delivery system of viscous capture threads spun by orb-weaving spiders. J Exp Biol 212:3026–3034

Opell BD, Hendricks ML (2010) The role of granules within viscous capture threads of orb-weaving spiders. J Exp Biol 213:339–346

Opell BD, Schwend HS (2008) Persistent stickiness of viscous capture threads produced by araneoid orb-weaving spiders. J Exp Zool 309A:11–16

Opell BD, Schwend HS (2009) Adhesive efficiency of spider prey capture threads. Zoology 112: 16–26

Opell BD, Markley BJ, Hannum CD, Hendricks ML (2008) The contribution of axial fiber extensibility to the adhesion of viscous capture threads spun by orb-weaving spiders. J Exp Biol 211:2243–2251

Opell BD, Lipkey GK, Hendricks ML, Vito ST (2009) Daily and seasonal changes in the stickiness of viscous capture threads in *Argiope aurantia* and *Argiope trifasciata* orb-webs. J Exp Zool 311A:217–225

Opell BD, Tran AM, Karinshak SE (2011a) Adhesive compatibility of cribellar and viscous prey capture threads and its implication for the evolution of orb-weaving spiders. J Exp Zool 315A:376–384

Opell BD, Karinshak SE, Sigler MA (2011b) Humidity affects the extensibility of an orb-weaving spider's viscous thread droplets. J Exp Biol 214:2988–2993

Opell BD, Schwend HS, Vito ST (2011c) Constraints on the adhesion of viscous threads spun by orb-weaving spiders: the tensile strength of glycoprotein glue exceeds its adhesion. J Exp Biol 214:2237–2241

Palmer JM (1990) Comparative morphology of the external silk production apparatus of "primitive" spiders. PhD thesis, Harvard University, Cambridge

Peters HM (1987) Fine structure and function of capture threads. In: Nentwig W (ed) Ecophysiology of spiders. Springer, Berlin

Peters HM (1993) Functional organization of the spinning apparatus of *Cyrtophora citricola* with regard to the evolution of the web (Araneae, Araneidae). Zoomorphology 113:153–163

Peters HM (1995) Ultrastructure of orb spiders' gluey capture threads. Naturwissenschaften 82:380–382

Peters HM, Kovoor J (1991) The silk-producing system of *Linyphia triangularis* (Araneae, Linyphiidae) and some comparisons with Araneidae: structure, histochemistry and function. Zoomorphology 111:1–17

Platnick NI (2012) The world spider catalog, version 12.5. American Museum of Natural History, online at http://research.amnh.org/iz/spiders/catalog. doi:10.5531/db.iz.0001. Accessed 25 June 2012

Platnick NI, Coddington JA, Forster RR, Griswold CE (1991) Spinneret morphology and the phylogeny of haplogyne spiders (Araneae, Araneomorphae). Am Mus Novit 3016:1–73

Richter G (1956) Untersuchungen über Struktur und Funktion der Klebefäden in den Fanggeweben ecribellater Radnetzspinnen. Naturwissenschaften 43:23

Rix MG, Harvey MS (2010) The spider family Micropholcommatidae (Arachnida, Araneae, Araneoidea): a relimitation and revision at the generic level. ZooKeys 36:1–321

Sahni V, Blackledge TA, Dhinojwala A (2010) Viscoelastic solids explain spider web stickiness. Nat Commun 1:19. doi:10.1038/ncomms1019

Sahni V, Blackledge TA, Dhinojwala A (2011a) Changes in the adhesive properties of spider aggregate glue during the evolution of cobwebs. Sci Rep 1:41. doi:10.1038/srep00041

Sahni V, Blackledge TA, Dhinojwala A (2011b) A review on spider silk adhesion. J Adhes 87:595–614

Salles HC, Volsi ECFR, Marques MR, Souza BM, dos Santos LD, Tormena CF, Mendes MA, Palma MS (2006) The venomous secrets of the web droplets from the viscid spiral of the orb-weaver spider *Nephila clavipes* (Araneae, Tetragnathidae). Chem Biodivers 3:727–741

Schildknecht H, Kunzelmann P, Krauß D, Kuhn C (1972) Über die Chemie der Spinnwebe, I: Arthropodenabwehrstoffe, LVII. Naturwissenschaften 59:98–99

Schulz S (1997) The chemistry of spider toxins and spider silk. Angew Chem Int Ed Engl 36:314–326

Schulz S (1999) Structural diversity of surface lipids from spiders. In: Diederichsen U, Lindhorst TK, Westermann B, Wessjohann LA (eds) Bioorganic chemistry: highlights and new aspects. Wiley-VCH, Weinheim

Schulz S (2001) Composition of the silk lipids of the spider *Nephila clavipes*. Lipids 36:637–647

Schulz S (2004) Semiochemistry of spiders. In: Cardé RT, Millar JG (eds) Advances in insect chemical ecology. Cambridge University Press, Cambridge

Schütt K (1995) *Drapetisca socialis* (Araneae: Linyphiidae): web reduction—ethological and morphological adaptations. Eur J Entomol 92:553–563

Schütt K (2003) Phylogeny of Symphytognathidae *s.l.* (Araneae, Araneoidea). Zool Scr 32: 129–151

Sekiguchi K (1955) Differences in the spinning organs between male and female adult spiders. Sci Rep Tokyo Kyoiku Daigaku Sect B 8:23–32

Selden PA, Penney D (2010) Fossil spiders. Biol Rev 85:171–206

Selden PA, Shih C-K, Ren D (2011) A golden orb-weaver spider (Araneae: Nephilidae: *Nephila*) from the Middle Jurassic of China. Biol Lett 7:775–778

Suter RB, Stratton GE (2013) Predation by spitting spiders—elaborate venom gland, intricate delivery system. In: Nentwig W (ed) Spider ecophysiology. Springer, Heidelberg (this volume)

Tillinghast EK, Chase SF (1985) Phosphorylase activity and glycogen utilization in the aggregate duct of orb weaving spiders. Comp Biochem Physiol 81B:1053–1055

Tillinghast EK, Christenson T (1984) Observations on the chemical composition of the web of *Nephila clavipes* (Araneae, Araneidae). J Arachnol 12:69–74

Tillinghast EK, Townley MA (1994) Silk glands of araneid spiders: selected morphological and physiological aspects. Am Chem Soc Symp Ser 544:29–44

Tillinghast EK, Townley MA, Wight TN, Uhlenbruck G, Janssen E (1993) The adhesive glycoprotein of the orb web of *Argiope aurantia* (Araneae, Araneidae). Mater Res Soc Symp Proc 292:9–23

Townley MA, Bernstein DT, Gallagher KS, Tillinghast EK (1991) Comparative study of orb web hygroscopicity and adhesive spiral composition in three araneid spiders. J Exp Zool 259: 154–165

Townley MA, Tillinghast EK, Neefus CD (2006) Changes in composition of spider orb web sticky droplets with starvation and web removal, and synthesis of sticky droplet compounds. J Exp Biol 209:1463–1486

Townley MA, Pu Q, Zercher CK, Neefus CD, Tillinghast EK (2012) Small organic solutes in sticky droplets from orb webs of the spider *Zygiella atrica* (Araneae; Araneidae): β-Alaninamide is a novel and abundant component. Chem Biodivers 9:2159–2174

Uhl G (2013) Spider olfaction: attracting, detecting, luring and avoiding. In: Nentwig W (ed) Spider ecophysiology. Springer, Heidelberg (this volume)

Vetter RS (1980) Defensive behavior of the black widow spider *Latrodectus hesperus* (Araneae: Theridiidae). Behav Ecol Sociobiol 7:187–193

Vogel S (2003) Comparative biomechanics: life's physical world. Princeton University Press, Princeton

Vollrath F (1999) Biology of spider silk. Int J Biol Macromol 24:81–88

Vollrath F, Edmonds DT (1989) Modulation of the mechanical properties of spider silk by coating with water. Nature 340:305–307

Vollrath F, Tillinghast EK (1991) Glycoprotein glue beneath a spider web's aqueous coat. Naturwissenschaften 78:557–559

Vollrath F, Fairbrother WJ, Williams RJP, Tillinghast EK, Bernstein DT, Gallagher KS, Townley MA (1990) Compounds in the droplets of the orb spider's viscid spiral. Nature 345:526–528

Volsi ECFR, Mendes MA, Marques MR, dos Santos LD, Santos KS, de Souza BM, Babieri EF, Palma MS (2006) Multiple bradykinin-related peptides from the capture web of the spider *Nephila clavipes* (Araneae, Tetragnathidae). Peptides 27:690–697

Chapter 22
Cribellar Thread

Brent D. Opell

22.1 Introduction

Cribellar threads, sometimes called hackled band threads, are spun by members of 180 spider genera in 23 families (Coddington and Levi 1991; Platnick 2011). To the unaided eye, newly spun cribellar threads appear hazy and often faintly blue because they are formed of thousands of fine fibrils making them translucent and causing them to present a "soft" outline (Fig. 22.1a). These were the first highly evolved sticky prey capture threads spun by spiders, their appearance corresponding to the Triassic origin of the suborder Opisthothelae, infraorder Araneomorphae (Platnick and Gertsch 1976; Seldon et al. 1999), which contains 95% of all living spider species (Platnick 2011). This clade's basal family Hypochilidae spins cribellar threads and also shares primitive characters, such as paraxial chelicerae and two pairs of book lungs, with members of its sister infraorder, the Mygalomorphae, suggesting that the ability to spin cribellar threads was a key innovation, which contributed to araneomorph diversity. While some mygalomorphs spin webs that capture flying insects, the inclusion of cribellar threads in araneomorph aerial webs better equipped them to retain insects, giving a spider more time to subdue its prey and perhaps the ability to capture larger prey relative to its own size.

Although most araneomorph species have subsequently lost the ability to spin cribellar threads, cribellate spiders have a broad geographic distribution and are found through the araneomorph phylogenetic tree (Griswold et al. 1999, 2005). This chapter explains how cribellar threads are spun, how they adhere, and how the texture of insect surfaces affects thread adhesion. It also examines the relationship between spinning anatomy and cribellar thread stickiness and reviews how these features coevolved with web architecture.

B.D. Opell (✉)
Department of Biological Sciences, Virginia Tech, Blacksburg, VA 24061, USA
e-mail: bopell@vt.edu

W. Nentwig (ed.), *Spider Ecophysiology*,
DOI 10.1007/978-3-642-33989-9_22,

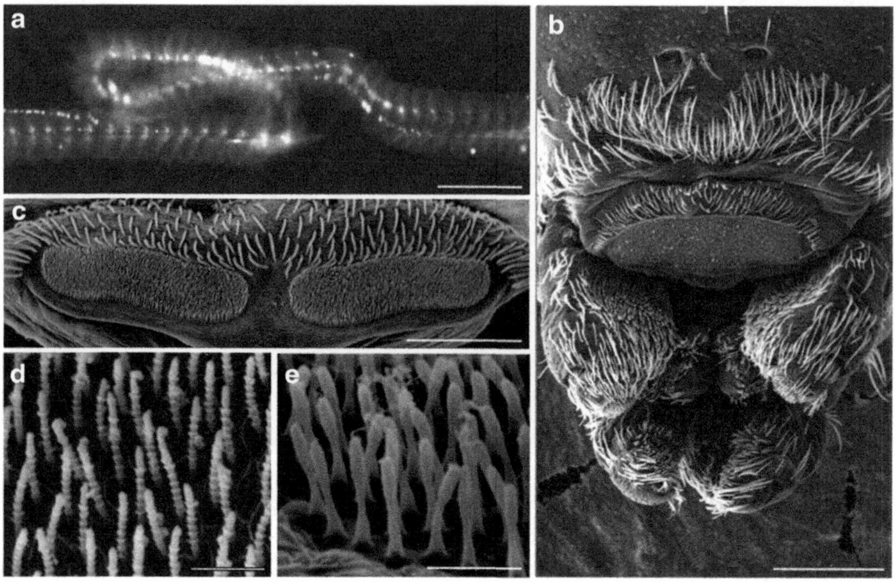

Fig. 22.1 (**a**) Cribellar thread of *Miagrammopes animotus* (Uloboridae; scale = 200 μm; Opell 1990). (**b**) Spinning anatomy of *Zosis geniculata* (Uloboridae), showing the cribellum situated anterior to the spinnerets (scale = 400 μm; Opell 1993). (**c**) Divided cribellum of *Badumna longinqua* (Desidae; scale = 200 μm; Opell 1999). (**d**) Strobilate cribellar spigots of *Waitkera waitakerensis* (Uloboridae; scale = 5 μm; Opell 1994c). (**e**) Claviform cribellar spigots of *Kukulcania hibernalis* (Filistatidae; scale = 5 μm). (**a**) Light microscopy, (**b**)–(**e**) SEM

22.2 Cribellar Thread Production

22.2.1 Cribellum and Cribellar Fibrils

The outer surface of a cribellar thread that confers stickiness is formed of thin, dry, protein fibrils, which are spun from spigots on a ventral opisthosomal spinning plate termed the cribellum, located anterior to the spinnerets (Fig. 22.1b; Peters 1984, 1992, 1995). The cribellum evolved from anterior median spinnerets, which are found today only in members of the primitive suborder Mesothelae. Although originating from a paired structure, the cribellum's plesiomorphic condition is a single median plate (Griswold et al. 1999, 2005). In some groups, the cribellum is medially divided (Fig. 22.1c) and in a few formed of four plates. Cribellar glands with short ducts are situated anterior and dorsal to the cribellum, each gland opening through a pore at the tip of one of a cribellum's hundreds to thousands of spigots (Kovoor and Peters 1988). Most spigots are strobilate, segmented and taper toward their tips and arise from cuticular sockets (Fig. 22.1d), although those found in members of the family Filistatidae are claviform or club shaped, with blunt tips and lacking sockets and segmentation (Fig. 22.1e; Griswold et al. 2005). All spigots are thought to simultaneously contribute their fibrils to the forming cribellar thread, making spigot number approximately equal to the number of fibrils included in a

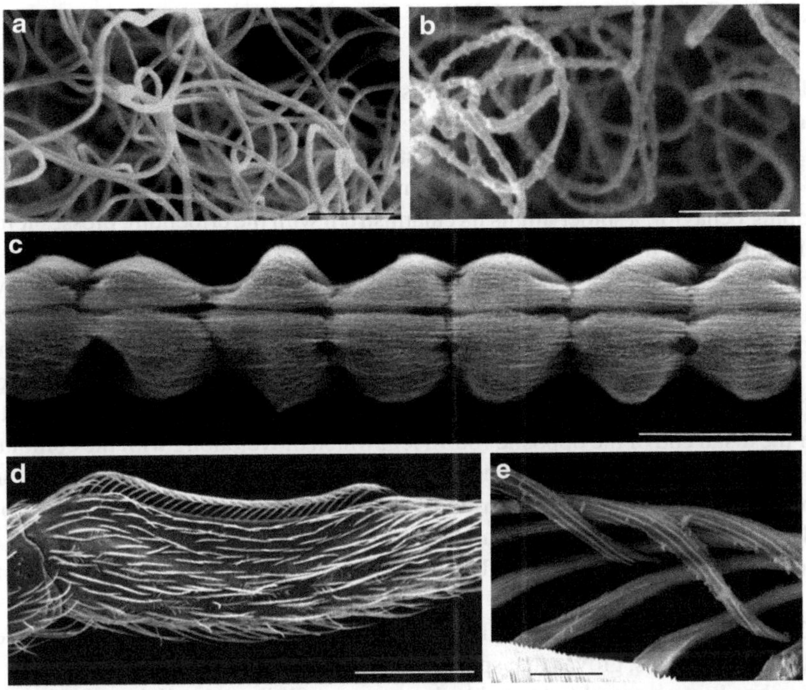

Fig. 22.2 (**a**) Cylindrical cribellar fibrils of *Hypochilus pococki* (Hypochilidae; scale = 300 nm). (**b**) Noded cribellar fibrils of *Hyptiotes cavatus* (Uloboridae; scale = 300 nm). (**c**) Cribellar thread of *Hyptiotes cavatus* (scale = 100 μm; Opell 1994b). (**d**) Calamistrum of *Miagrammopes animotus* (Uloboridae; scale = 200 μm). (**e**) Calamistrum setae of *Tangaroa beattyi* (Uloboridae; scale = 10 μm). All SEM

cribellar thread. Penultimate males posses a functional cribellum and associated combing organ, but these features and the ability to spin cribellar threads are always lost when males mature (Griswold et al. 2005).

The cribellar fibrils of the 12 hypochilid species are cylindrical with a diameter of about 25 nm (Fig. 22.2a), whereas fibrils of all other cribellate species, except members of the family Filistatidae, have regularly spaced, swollen nodes that have a diameter of about 35 nm (Fig. 22.2b; Peters 1986; Eberhard and Pereira 1993; Hawthorn and Opell 2003). Filistatid fibrils are flat and without nodes, presumably the result of being drawn from spigots with slit-like rather than circular openings (Eberhard and Pereira 1993). Cribellar fibrils remain separated and assume a looped and coiled configuration in completed threads (Fig. 22.2c), possibly held apart by static electric charges. The nodes of adjacent fibrils may interact when force is applied to a cribellar thread that has adhered to a surface, offering resistance to fibrils being pulled from the cribellar thread (Köhler and Vollrath 1995; Blackledge and Hayashi 2006).

22.2.2 Calamistrum and Fibril Combing

Cribellar fibrils are drawn from the spigots by the combing action of the calamistrum, a setal comb found on the penultimate segment of each of the spider's fourth legs (Fig. 22.2d). Most species have a calamistrum formed of a single row of setae (Peters 1984), although a double row is present in some (Griswold et al. 2005). During rapid combing behavior, a spider aligns its calamistrum with the cribellum, presses it lightly against the tips of the spigots, and sweeps the fourth leg posteriorly, drawing out a sheet of fibrils. Throughout development calamistrum length is linked to cribellum width and not the length of the fourth leg segment that bears the calamistrum (Opell 2001). Depending on the family of spiders, the tip of the fourth leg whose calamistrum is in use is supported in one of two ways: (1) It rests near the distal end of the third leg, which remains immobile as the combing leg rocks on the supporting leg, or (2) it rests near the distal end of the opposite fourth leg, which moves along with the combing leg (Eberhard 1988). Deinopidae species and orb-weaving uloborids begin combing a cribellar strand with the calamistrum of the leg nearest the outside of the web and, midway between spinning a strand that will span two radii, switch to the other calamistrum. Other cribellate spiders also alternate the use of their calamistrum, but not before finishing and attaching a cribellar strand (Eberhard 1988).

Calamistrum setae are socketed and movable. In some groups their shafts have fine toothlike extensions, groves that spiral to a narrowing tip (Fig. 22.2e), or take the form of tiny brushes, presumably to enable them to snag fibrils and pull them from cribellar spigots (Peters 1984; Griswold et al. 2005; Foelix 2011). In Uloboridae, the complicated tips of adjacent setae may temporarily lock together when in use to produce a semirigid comb (Peters 1984). Angular changes in the spider's fourth leg as it completes a combing stoke may allow the cribellar fibrils to disengage from calamistrum setae, releasing the sheet of cribellar fibrils, which will be formed around larger and stronger fibers spun from spigots on the spinnerets to form a completed cribellar thread.

22.2.3 Cribellar Thread Support Lines

The lines that support the mat of cribellar fibers differ greatly among taxa (Eberhard and Pereira 1993). Simpler arrangements, like those found in threads produced by members of the Uloboridae, have two components: (1) a pair of large, straight pseudoflagelliform or axial fibers, spun from spigots on the posterior median spinnerets (these are thought to be homologous with the flagelliform fibers of the viscous capture threads of araneoids, though the two kinds of fibers have different mechanical properties (Blackledge et al. 2009)) and (2) a coiled network of about 54 very thin paracribellar fibrils, which are spun from spigots just anterior to the cribellum and appear to form a superstructure around the axial fibers. In other spiders, such as members of the family Filistatidae, there are more support lines,

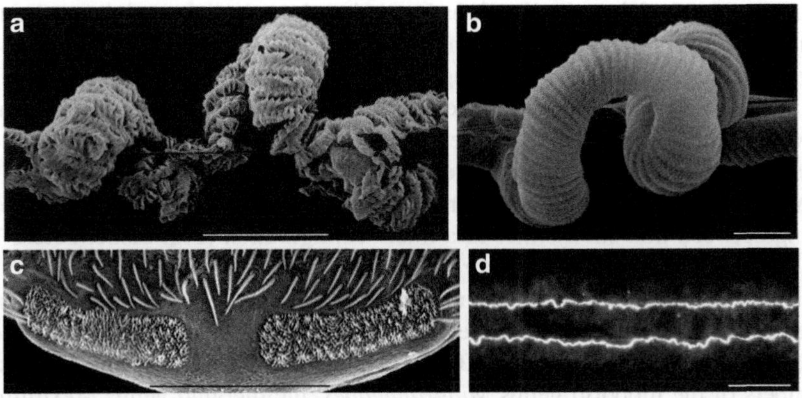

Fig. 22.3 (**a**) Cribellar thread of *Kukulcania hibernalis* (Filistatidae; scale = 800 μm; Opell 2002). (**b**) Cribellar thread of *Mexitilia trivittata* (Dictynidae; scale = 200 μm). (**c**) Cribellum of *Neolana pallida* (Neolanidae; scale = 100 μm). (**d**) Cribellar thread of *N. pallida* (scale = 40 μm). (**a**)–(**c**) SEM, (**d**) Light microscopy

which include a pair of large helical fibers, a pair of thick and curled reverse wrap fibers, and a pair of midsized and crinkled axial fibers.

The combing action of the fourth legs was originally assumed to be responsible for combining cribellar fibrils and support threads. However, among his many contributions to cribellar thread spinning, Peters (1984) hypothesized that rhythmic adductions of the posterior lateral spinnerets press the sheet of cribellar fibers around support fibers to form a cribellar thread whose width is slightly less than half that of cribellum's width. Evidence for this comes from the observation that only the posterior lateral spinnerets move laterally during cribellar thread spinning and from the regular, puffed configuration of many cribellar threads (Figs. 22.1a and 22.2c). Angular changes in the fourth combing leg may twist or reshape the cribellar mat before it is combined or as it is combined with support threads, making the process more complicated than a simple clamping action.

22.2.4 Diversity in Cribellar Thread Spinning

Median division of the cribellum occurred multiple times (Griswold et al. 1999, 2005) and even some genera, such as the Dictynidae genus *Mallos*, have species with both forms of the cribellum (Bond and Opell 1997). A divided cribellum may produce a wider array of cribellar fibrils relative to cribellum spigot number, resulting in a stickier cribellar thread. The divided cribellum of the filistatid genus *Kukulcania* produces a bipartite cribellar thread (Fig. 22.3a), whose two halves do not adhere tightly and can be artificially separated into two intact cribellar strands (Opell 2002). Most species in this family deposit intact cribellar threads on an elevated foundation line as they return to their retreat (Eberhard 1987). However, after spinning a short span of cribellar thread, species of the genera *Misionella*, *Pikelinia*, and *Pritha* contact the thread with both fourth tarsi, spread their legs, and touch each half of the cribellar thread strand to the substrate at right angles to the foundation line (Lopardo and Ramirez 2007). Cribellar fibril production stops before the thread is pulled by the legs, but the spider continues to produce support fibers, which are then attached to the foundation line a short distance further along, at which point a new cribellar thread strand is begun. This behavior produces a dense mesh of radial foundation lines with transverse cribellar threads.

Most of the cribellar threads produced by orb-weaving uloborid species are strung between non-sticky lines in a linear configuration (Fig. 22.2c; Peters 1984, 1986). In some regions of these orb webs and more extensively in modified webs constructed by other uloborid genera (*Polenecia* and *Hyptiotes*), as well as in many other cribellate spiders, linear cribellar threads are deposited along non-sticky web lines (Lubin 1986; Peters 1995; Griswold et al. 2005). Other spiders, such as members of the uloborid genus *Miagrammopes*, the dictynid genus *Mexitilia*, and filistatid species comb out cribellar thread faster than they move and then fold this thread before depositing it in a looped fashion along a foundation line (Figs. 22.1a and 22.3a, b; Opell 1990, 1999, 2002). This configuration is stickier per unit length than a linear thread and presents a wider capture area (Opell 2002). It may also bring cribellar fibrils into contact with a larger portion of an insect's cuticle and, as a loop pull free from the foundation line, help dissipate the force generated by a struggling prey before the adhesion is overcome.

22.3 Cribellar Thread Adhesion

22.3.1 Forces of Thread Adhesion

Depending on the texture of the surface that it contacts, the ambient humidity, and the type of fibrils that it contains, cribellar threads can implement three adhesive mechanisms: (1) mechanical interlock, when cribellar fibrils catch on the setae and surface irregularities of insects, much like the soft component of a Velcro® fastener catches on the stiff component; (2) van der Waals forces, when the molecules of fibrils interact with those of a surface; and (3) capillary force, when water molecules cohere to both fibrils and a surface, holding them together. Cribellar thread stickiness has been measured by pressing a 2 mm wide contact plate against a cribellar thread suspended between two supports and measuring the force required to separate the two surfaces. Thread adhesion is usually expressed as μN of force per mm length of cribellar thread contact (Table 22.1). All cribellar threads can implement mechanical interlock if the surface they contact has irregularities (see Sect. 22.3.4). At 1–2% relative humidity, cribellar threads with noded fibrils (Fig. 22.2b) generated van der Waals forces of adhesion when they contact smooth acetate surfaces. At 46% relative humidity, they doubled their stickiness by generating capillary forces but showed only an additional 20% increase in stickiness at 99% relative humidity (Hawthorn and Opell 2003). Lacking noded fibrils, threads of *Hypochilus* species (Fig. 22.2a) generated only van der Waals forces of adhesion, registering the same stickiness at 1% and 99% relative humidity. Thus,

Table 22.1 Examples of cribellum size and cribellar thread stickiness of adult female spiders as measured with a fine sandpaper contact plate (from Opell 1999)

Family	Species	Web form	Mass mg	Cribellum spigots	Stickiness μN/mm	Stickiness per mg
Filistatidae	*Kukulcania hibernalis*	Radiating lines	335	7,767	22.7	0.07
Neolanidae	*Neolana pallida*	Sheet	52	1,835	5.5	0.11
Desidae	*Matachia livor*	Radiating network	31	6,740	22.8	0.74
	Badumna insignis	Funnel	296	10,878	20.4	0.07
	Badumna longinqua	Funnel	209	9,552	13.5	0.06
Uloboridae	*Waitkera waitakerensis*	Orb	9	3,894	15.5	1.72
	Siratoba referena	Orb	4	1,800	11.7	2.93
	Uloborus glomosus	Orb	9	4,717	15.5	1.72
	Octonoba sinensis	Orb	13	4,465	17.1	1.32
	Philoponella arizonica	Orb	14	5,110	15.0	1.07
	Hyptiotes cavatus	Triangle	8	7,276	26.2	3.28
	Hyptiotes gertschi	Triangle	10	7,724	29.8	2.98
	Miagrammopes animotus	Simple	5	8,990	31.5	6.30
	Miagrammopes species	Simple	5	7,254	24.4	4.88

fibril nodes appear to be hydrophilic sites, which attract atmospheric moisture, or to provide sites where moisture collects from a surface and generates capillary forces. The impact of humidity on cribellar thread stickiness has not been examined at relative humidities between 2% and 46%. Therefore, the lower limit of capillary force adhesion has not been established.

There is no evidence that static electric charges contribute to thread stickiness. At 50% relative humidity, threads adhered to surfaces that had the same smooth texture, but different dielectric properties with the same force (Opell 1995a).

22.3.2 Role of Support Fibers

A cribellar thread's internal supporting fibers are probably crucial components of the thread's integrated adhesive system. These fibers sum and transfer the adhesive forces generated by cribellar fibrils. When cribellar threads are laid on foundation lines, the mechanical properties of the foundation lines also appear to be included in the dynamics of this system. This is suggested by the observation that the axial fibers of the supported, linear cribellar threads of *Miagrammopes* are thinner than the axial fibers in suspended, self-supporting cribellar threads from similar-sized orb-weaving uloborid species (Opell 1994a).

The low extensibility of axial fibers within cribellar threads from uloborid orb webs (Blackledge and Hayashi 2006) limits their ability to sum adhesion from all regions of a thread that has adhered to a surface. Consequently, most expressed adhesion appears to be generated by narrow bands at the outer edges of thread contact, with inner thread regions contributing little adhesion (Opell and Schwend 2008). The width of these bands has been estimated to be less than 1 μm (Hawthorn and Opell 2003). In contrast, more extensible axial fibers of viscous threads of araneoid orb weavers sum the adhesion over greater lengths of capture thread (Opell et al. 2008; Opell and Hendricks 2009). As the kinds of support fibrils and the cribellar fibril configuration differ greatly among cribellar threads (Eberhard and Pereira 1993), other cribellar threads may more effectively sum adhesion.

22.3.3 Determinants of Thread Stickiness

Cribellar thread stickiness is directly related to the number of thread fibrils that contact a surface and, therefore, to the number of spigots on the cribellum (Table 22.1; Opell 1989, 1994b,c, 1995b, 1999; Opell and Schwend 2008). However, the manner in which fibrils and support lines are combined may also result in minor differences in stickiness. Threads spun by members of the uloborid genus *Miagrammopes* achieve a slightly greater stickiness per cribellar spigot number than do other uloborid species, apparently because of differences in the density of their cribellar spinning spigots and calamistrum setae and the configuration of their cribellar thread puffs (Opell 1995b).

Thus, following the evolution of noded fibrils, the cost of achieving cribellar thread stickiness appears to have been the sum of the material cost of cribellar fibrils incorporated in a thread and the behavioral cost of spinning the capture thread (Lubin 1986), which may be related to the number of fibrils that must be pulled from the calamistrum. Looping lengths of linear cribellar threads before they are attached to a foundation line (Figs. 22.1a and 22.3a, b) further increased the stickiness of threads but also increased the cost per unit length of thread (Opell 1990).

The size of a spider does not appear to limit cribellum size or constrain thread stickiness (Opell 1999). With a mass of only 5 mg, *Miagrammopes animotus* has a cribellum that bears 8,990 spigots, 16% more than that of 335 mg female *Kukulcania hibernalis* and only a few hundred less than 209 mg *Badumna longinqua* (Desidae) (Table 22.1).

When protected, cribellar threads retain their stickiness for long periods (Eberhard 1980; Opell 1993). However, under natural conditions, stickiness begins to degrade as threads accumulate dust and pollen and are damaged by rain and contact with prey.

22.3.4 The Texture of the Surface to Which a Thread Adheres

In addition to humidity, the texture of the surface that a cribellar thread contacts determines the mechanism or mechanisms responsible for adhesion. Interspecific comparisons have measured cribellar thread stickiness in the range of 50% relative humidity with artificial surfaces of uniform texture. These include smooth acetate surfaces, which probably maximize capillary forces of adhesion or fine sandpaper, which may be held by a combination of mechanical interlock and capillary adhesion (Table 22.1). At 56–57% relative humidity, uloborid cribellar threads held beetle elytra (probably principally by capillary forces) more securely than flesh fly nota, whose stout setal covering was probably held by mechanical interlock (Opell 1994d). Bug hemelytra and flesh fly wings, surfaces with shorter setae, were held less securely than either of these surfaces. In addition to favoring mechanical interlock, setae probably also limit the area of cribellar fibril contact with the underlying cuticle, thereby reducing the adhesion generated, as documented for viscous threads (Opell and Schwend 2007). Moth wings were held the least securely of all because their scales detached and remained in the cribellar threads, as they did in viscous capture threads (Eisner et al. 1964). Thus, this predator avoidance strategy may have evolved long before viscous capture threads appeared.

Despite cribellar thread's remarkable adhesive versatility, it does not stick to a spider's calamistrum, posterior spinnerets, or first leg tips that tap the inner capture spiral loop during spiral construction. Moreover, in the process of subduing prey, cribellate spiders routinely come into contact with their own cribellar threads and either do not stick or stick only weakly to these. The reason for this is unknown.

22.4 Cribellar Thread Stickiness and Web Architecture

22.4.1 Web Forms and Costs

Cribellar threads are found in many web types, including sheet, funnel, space filling, and orb webs (Opell 1999; Griswold et al. 2005). Most cribellate spiders repair their webs and add fresh cribellar threads, but only members of the orb-weaving family Uloboridae are known to take down and replace their webs daily, ingesting silk from the old web. It is presumed that they recycle some of the material in a new web, as araneoid orb weavers do (Townley and Tillinghast 1988), but this has not been documented. Although cribellar thread enhances prey capture, its stickiness may be constrained by the cost of production, the length of thread placed in a web, the time over which this thread functions, and the potential to recycle the silk invested in it.

22.4.2 Evolution of Stickier Cribellar Thread

Cribellate orb webs constructed by members of the family Uloboridae might be viewed as reduced webs, because their cribellar thread is restricted to the capture spiral rather than being placed more densely throughout the web. The lengths of cribellar threads in webs constructed by members of other spider families have not been measured, so direct comparisons are not possible at present. However, it appears that, combined with the potential to recycle web components, the parsimonious thread used by uloborids may have reduced the total cost of cribellar thread in an orb web, permitting these spiders to spin stickier cribellar thread than non-orb species belonging to other families (Table 22.1; Opell 1999). When measured with fine sandpaper contact plates that registered the same stickiness as flesh fly wings on cribellar thread (Opell 1994b), the cribellar threads of orb-weaving uloborids were, on average, 8.4 times stickier per mg spider mass than threads produced by members of non-orb-weaving families (Table 22.1; Opell 1999). Among the non-orb weavers, *Matachia livor*, whose web superficially resembles an orb (Opell 1999), had the greatest stickiness per mg of spider mass.

Members of the uloborid sister genera *Hyptiotes* and *Miagrammopes* construct reduced webs, which they monitor more actively and manipulate during prey capture (Opell 1994b). *Hyptiotes* species build triangle webs comprised of four diverging radii between which cribellar threads are suspended. The simple webs of *Miagrammopes* species form an irregular and non-stereotypic branching network of draglines, on which cribellar threads are laid, often in a looped fashion (Fig. 22.1a). Threads spun by *Hyptiotes* were 1.8 times stickier per mg spider mass than those spun by orb-weaving uloborids, while threads spun by *Miagrammopes* were 1.8 times stickier per mg spider mass than those spun by *Hyptiotes* (Opell 1999).

The three uloborid web architectures described above look very different and contain different lengths of cribellar treads. However, in each case the web's total silk volume, its total cribellar fibril volume, its total stickiness, and its total capture area were directly related to spider mass (Opell 1996). Thus, across a wide range of web forms, spiders that produce shorter lengths of cribellar thread and have the potential to recycle this material appear to have evolved stickier cribellar threads. However, many aspects of this system, such as the behavioral cost of producing and operating these webs, still require investigation.

22.4.3 Reduced Reliance on Cribellar Thread

The cost of using cribellar thread may explain its frequent evolutionary loss. Although there are some families and even some genera that have both cribellate and ecribellate members, this loss has not been carefully examined. The funnel web-weaving *Tengella radiata* from Costa Rica may exhibit characteristics of a species on its way to cribellar thread loss (Barrantes and Madrigal-Brenes 2008). Spiderlings of this species initially spin simple sheet webs, which, as they develop, gradually take the form of the adult funnel web. However, although third instar spiderlings possess a cribellum and calamistrum, it is not until they enter the seventh instar (one molt away from maturity for males and two for females) that they start adding cribellar threads to their webs. Newly constructed adult female webs often lack cribellar threads, which are added over time. Thus, in this species cribellar thread production appears to be useful, but not required for prey capture.

One can imagine how changes in factors such as web architecture, prey capture behavior, or insect abundance might lead to loss of cribellar thread use in this and other species.

Another apparent example of reduced reliance on cribellar thread adhesion is the New Zealand species *Neolana pallida* (family Neolanidae), which spins a large sheet web of closely spaced cribellar threads and rests on a silk platform constructed above this sheet (Opell 1999). Unless the web has been damaged and recently repaired, fresh cribellar threads appear to be found mostly on the perimeter of the capture sheet, threads in the remainder of the sheet being contaminated by dust. Members of this species have a small cribellum formed of two widely spaced plates that bear widely spaced spigots (Fig. 22.3c). The straight axial fibers and coiled support fibers of their cribellar threads are covered by only a sparse array of cribellar fibrils and generate very low adhesion (Fig. 22.3d, Table 22.1).

22.5 Conclusions

Cribellar prey capture thread relies on thousands of thin, dry protein fibrils to generate mechanical interlock, van der Waals, and capillary forces of adhesion. Relative humidity and the texture of the surface contacted determine the degree to which these forces contribute to thread adhesion and the stickiness a thread registers. Cribellar fibrils are drawn from spigots on a cribellum, a ventral opisthosomal spinning plate situated anterior to a spider's spinnerets, by the calamistrum setal comb on the spider's fourth legs. These fibrils are then formed around larger supporting fibrils to produce a composite cribellar thread, which often has a regular, puffed configuration. Cribellar thread production is plesiomorphic for the large infraorder Araneomorphae and still widely distributed among this clade, where it is used in many web forms, including the most primitive orb webs. The stickiness of cribellar thread is directly related to the number of fibrils invested in a thread. Consequently, the number of thread fibrils and the length of cribellar thread in a web, along with a spider's ability to recycle these proteins, appear to determine the cost of using cribellar thread and to constrain the evolution of thread stickiness. This may explain why most araneomorphs no longer spin cribellar thread, abandoning it either in favor of other methods of prey capture or, in the case of modern orb-weaving spiders, replacing it with viscous capture thread.

Acknowledgements I thank William Eberhard for new insights into cribellar thread and helpful comments on this chapter.

References

Barrantes G, Madrigal-Brenes R (2008) Ontogenetic changes in web architecture and growth rate of *Tengella radiata* (Araneae, Tengellidae). J Arachnol 36:545–551

Blackledge TA, Hayashi CY (2006) Unraveling the mechanical properties of composite silk threads spun by cribellate orbweaving spiders. J Exp Biol 209:3131–3140

Blackledge TA, Scharff N, Coddington J, Szüts T, Wenzel JW, Hayashi CY, Agnarsson I (2009) Spider web evolution and diversification in the molecular era. Proc Natl Acad Sci USA 106: 5229–5234

Bond JE, Opell BD (1997) The functional significance of medially divided cribella in the spider genus Mallos (Araneae, Dictynidae). Bull Br Arachnol Soc 10:239–241

Coddington JA, Levi HW (1991) Systematics and evolution of spiders (Araneae). Annu Rev Ecol Syst 22:565–592

Eberhard WG (1980) Persistent stickiness of cribellar silk. J Arachnol 8:283

Eberhard WG (1987) Construction behavior of non-orb weaving cribellate spiders and the evolutionary origin of orb-webs. Bull Br Arachnol Soc 7:175–178

Eberhard WG (1988) Combing and sticky silk attachment behaviour by cribellate spiders and its taxonomic implications. Bull Br Arachnol Soc 7:247–251

Eberhard WG, Pereira F (1993) Ultrastructure of cribellate silk of nine species in eight families and possible taxonomic implications. (Araneae: Amaurobiidae, Deinopidae, Desidae, Dictynidae, Filistatidae, Hypochilidae, Stiphidiidae, Tengellidae). J Arachnol 21:161–174

Eisner T, Alsop R, Ettershank G (1964) Adhesivness of spider silk. Science 146:1058–1061

Foelix RF (2011) Biology of spiders, 3rd edn. Oxford University Press, New York

Griswold CE, Coddington JA, Platnick NI, Forster RR (1999) Towards a phylogeny of entelegyne spiders (Araneae, Entelegynae). J Arachnol 27:53–63

Griswold CE, Ramírez MJ, Coddington JA, Platnick NI (2005) Atlas of phylogenetic data for entelegyne spiders (Araneae: Araneomorphae: Entelegynae) with comments on their phylogeny. Proc Calif Acad Sci 56(Supp II):1–324

Hawthorn A, Opell BD (2003) van der Waals and hygroscopic forces of adhesion generated by spider capture threads. J Exp Biol 206:3905–3911

Köhler T, Vollrath F (1995) Thread biomechanics in the two orb-weaving spiders *Araneus diadematus* (Araneae, Araneidae) and *Uloborus walckenaerius* (Araneae, Uloboridae). J Exp Zool 271:1–17

Kovoor J, Peters HM (1988) The spinning apparatus of *Polenecia producta* (Araneae, Uloboridae): structure and histochemistry. Zoomorphology 108:47–59

Lopardo L, Ramirez MJ (2007) The combing of cribellar silk by the prithine *Misionella mendensis*, with notes on other filistatid spiders (Araneae: Filistatidae). Am Mus Novit 3563:1–14

Lubin YD (1986) Web building and prey capture in the Uloboridae. In: Shear WA (ed) Webs, behavior, and evolution. Stanford University Press, Stanford, pp 132–171

Opell BD (1989) Functional associations between the cribellum spinning plate and prey capture threads of *Miagrammopes animotus* (Araneida, Uloboridae). Zoomorphology 108:263–267

Opell BD (1990) Material investment and prey capture potential of reduced spider webs. Behav Ecol Sociobiol 26:375–381

Opell BD (1993) What forces are responsible for the stickiness of spider cribellar threads? J Exp Zool 265:469–476

Opell BD (1994a) Factors affecting the diameters of axial fibers in cribellar threads of the spider family Uloboridae. J Arachnol 22:12–18

Opell BD (1994b) Increased stickiness of prey capture threads accompanying web reduction in the spider family Uloboridae. Funct Ecol 8:85–90

Opell BD (1994c) Factors governing the stickiness of cribellar prey capture threads in the spider family Uloboridae. J Morphol 221:111–119

Opell BD (1994d) The ability of spider cribellar prey capture thread to hold insects with different surface features. Funct Ecol 8:145–150

Opell BD (1995a) Do static electric forces contribute to the stickiness of a spider's cribellar prey capture threads? J Exp Zool 273:186–189

Opell BD (1995b) Ontogenetic changes in cribellum spigot number and cribellar prey capture thread stickiness in the spider family Uloboridae. J Morphol 224:47–56

Opell BD (1996) Functional similarities of spider webs with diverse architectures. Am Nat 148: 630–648

Opell BD (1999) Changes in spinning anatomy and thread stickiness associated with the origin of orb-weaving spiders. Biol J Linn Soc 68:593–612

Opell BD (2001) Cribellum and calamistrum ontogeny in the spider family Uloboridae: linking functionally related but separate silk spinning features. J Arachnol 29:220–226

Opell BD (2002) How spider anatomy and thread configuration shape the stickiness of cribellar prey capture thread. J Arachnol 30:10–19

Opell BD, Hendricks ML (2009) The adhesive delivery system of viscous prey capture threads spin by orb-weaving spiders. J Exp Biol 212:3026–3034

Opell BD, Schwend HS (2007) The effect of insect surface features on the adhesion of viscous capture threads spun by orb-weaving spiders. J Exp Biol 210:2352–2360

Opell BD, Schwend HS (2008) Adhesive efficiency of spider prey capture threads. Zoology 112: 16–26

Opell BD, Markley BJ, Hannum CD, Hendricks ML (2008) The contribution of axial fiber extensibility to the adhesion of viscous capture threads spun by orb-weaving spiders. J Exp Biol 211:2243–2251

Peters HM (1984) The spinning apparatus of Uloboridae in relation to the structure and construction of capture threads (Arachnida, Araneida). Zoomorphology 104:96–104

Peters HM (1986) Fine structure and function of capture threads. In: Nentwig W (ed) Ecophysiology of spiders. Springer, New York

Peters HM (1992) On the spinning apparatus and structure of the capture threads of *Deinopis subrufus* (Araneae, Deinopidae). Zoomorphology 112:27–37

Peters HM (1995) *Polenecia producta* and its web: structure and evolution (Araneae, Uloboridae). Zoomorphology 115:1–9

Platnick NI (2011) The world spider catalog, version 12.0. American Museum of Natural History, online at http://research.amnh.org/iz/spiders/catalog. doi:10.5531/db.iz.0001

Platnick NI, Gertsch W (1976) The suborders of spiders: a cladistic analysis (Arachnida, Araneae). Am Mus Novit 2607:1–15

Seldon PA, Anderson JM, Anderson HM, Fraser NC (1999) Fossil araneomorph spiders from the Triassic of South Africa and Virginia. J Arachnol 27:401–414

Townley MA, Tillinghast EK (1988) Orb web recycling in *Araneus cavaticus* (Araneae, Araneidae) with an emphasis on the adhesive spiral component, GABamide. J Arachnol 16: 303–319

Part VII
Colouration

Since ever, we are fascinated by coloured arthropods because colour makes them beauties. Spiders suffered a bit from this because their overall appearance is usually described as yellowish–brownish to black. We know meanwhile that this is completely untrue: spiders have structural and chemical mechanisms for colours, arising from hairs and scales or the cuticle. Spiders apply their colouration for species recognition and not only to impress females but also to camouflage. The perception of spiders' colouration is wavelength dependent and colours may rapidly appear or disappear, thus a beautiful spider can change its appearance within a moment and different observers see a spider against its background differently. It has also to be emphasized that spiders are the only arthropods able to perform such a rapid colour change.

Chapter 23
Insect View of Orb Spider Body Colorations

I-Min Tso

23.1 Introduction

The conspicuous color patterns on body surface have recently been considered to be important in predator–prey visual interactions. Orb-weaving spiders of the families Araneidae, Nephilidae, and Tetragnathidae hunt during the day, and many of them exhibit conspicuous color patterns. To confirm whether the brightly colored markings of these diurnal orb-weaving spiders exhibit any function, it is crucial to know how these markings are viewed by organisms interacting with these spiders.

Traditionally, numerous species of brightly colored orb-weaving spiders have been regarded as diurnal predators. However, few researchers have actually followed the spiders for 24 h to investigate their temporal activity patterns. At least for some brightly colored orb-weaving spiders, their prey catching may not be limited to daytime. Some brightly colored spiders also hunt during the night and obtain a certain quantity of prey during nocturnal hunting. Therefore, does their body coloration play any role during nighttime? Recent studies showed that certain nocturnal insects have good color visions (Warrant 1999; Land and Osorio 2003). Is it possible that the color of orb web spiders play a role in nocturnal spider–insect visual interactions? An important step towards answering this question is to realize how the brightly colored spiders are viewed by insects under the nocturnal condition.

I.-M. Tso (✉)
Department of Life Science, Tunghai University, Taichung, Taiwan
e-mail: spider@thu.edu.tw

W. Nentwig (ed.), *Spider Ecophysiology*,
DOI 10.1007/978-3-642-33989-9_23, © Springer-Verlag Berlin Heidelberg 2013

23.2 Spider Coloration in the Diurnal Context

Various diurnal orb-weaving spiders exhibit brightly colored markings on their body surface, and the functions of these colorations had been debated. One group of researchers regarded the bright color patterns of these diurnal orb-weaving spiders as functioning to increase foraging success by providing attractive visual signals to prey (Craig and Ebert 1994; Hauber 2002). The camouflaging hypothesis, on the other hand, regards the bright coloration of orb-weaving spiders as functioning to conceal the spiders against the vegetation background (Zschokke 2002). This hypothesis proposes that because the reflectance spectra of the spiders' body surface are similar to those of the background vegetation, the spiders are not easily perceived by insects. To determine whether the brightly colored markings of these diurnal orb-weaving spiders function to camouflage the spiders or to visually attract prey, it is crucial to know how the colors of various body parts of these spiders are viewed by organisms interacting with these spiders.

23.2.1 *Quantifying How Colors Are Viewed by Diurnal Insects*

Most insects have a pair of compound eyes. Their visual systems are different from those of vertebrates and they can perceive light signals of much shorter wavelength (Chittka 1997). For example, the honeybee (*Apis mellifera*) has ultraviolet-, blue-, and green-sensitive photoreceptors (Chittka 1997). Insects view colors by detecting the contrasts between objects and their environments and all kinds of color receptors and signals are involved (Chittka and Menzel 1992; Vorobyev and Brandt 1997; Briscoe and Chittka 2001). Therefore, all types of receptor signals should be considered when exploring the visual interactions between predators and prey. Compared with the traditional UV approach, i.e., measuring UV reflectance of the object then infer nature of interactions, the color contrast approach is more realistic in exploring the visual interactions between spiders and insects. To quantify how spider colorations are viewed by a hymenopteran insect such as honeybee, we have to obtain the following information: spectral sensitivity of photoreceptors in the eyes of insects, reflectance spectra of spider colors, and illumination function of the habitat in which the interactions take place.

The spectral sensitivity functions of standard photoreceptors for Hymenoptera were popularly used to determine photoreceptor excitations for each measured spectra. The relative amount of light absorbed by each photoreceptor type can be expressed as

$$P = R\int_{300}^{700} Is(\lambda)S(\lambda)D(\lambda)\mathrm{d}\lambda$$

where $Is(\lambda)$ is the spectral reflectance function of the spider colorations, $S(\lambda)$ is the spectral sensitivity function of the receptor in question, and $D(\lambda)$ is the illuminating daylight spectrum (Chittka 1996). The sensitivity factor R in equation (1) can be determined by the equation

$$R = 1/\int_{300}^{700} I_{\mathrm{B}}(\lambda)S(\lambda)D(\lambda)\mathrm{d}\lambda$$

The receptors were assumed to be adapted to a background reflection function and $I_{\mathrm{B}}(\lambda)$ is the spectral reflection function of the background to which the receptors are adapted (Chittka and Menzel 1992). With this model, it is assumed that the photoreceptors display half their maximal response when stimulated by the light reflected from the adaptation background (Naka and Rushton 1966).

The quantum catch in the photoreceptors P is the input to the photoreceptors, not input to the insect brain. On a neural level, the brain performs "calculations" with graded potentials generated by receptor cells. These signals are not linearly related to the logarithm of the quantum flux that forms the input to the receptor. When the maximum excitation E_{max} of the photoreceptors is set to one, the nonlinear phototransduction process is well described by

$$E = \frac{P}{(P+1)}$$

where P is the stimulus strength, in units such that for $P = 1$, $E = 0.5$. Bees are reported to adopt achromatic vision by using green receptor signal when searching for objects far ahead (i.e., with a subtending area between 5° and 15°) and adopt chromatic vision by using green, blue, and UV receptor signals when viewing objects from a relatively short distance (i.e., with a subtending area greater than 15°) (Giurfa et al. 1997; Spaethe et al. 2001). The three excitation values in the bee's UV, blue, and green receptors can be depicted in a three-dimensional receptor excitation space or in the color hexagon (Chittka 1996). With the three receptor excitation values plotted at angles of 120°, the x and y coordinates in the color plane are given by

$$x = \sin 60°(E_{\mathrm{G}} - E_{\mathrm{UV}})$$

$$y = E_{\mathrm{B}} - 0.5(E_{\mathrm{UV}} + E_{\mathrm{G}})$$

where E_{u}, E_{B}, and E_{G} are the inputs from the three photoreceptors. Euclidean distances ΔSt between stimuli are calculated as

$$\Delta St = \sqrt{(\Delta x)^2 + (\Delta y)^2}$$

Fig. 23.1 Dorsal and ventral view of brightly colored orb web spiders from three families and color contrasts of various body parts when viewed by hymenopteran insects against vegetation background. (**a**, **b**) *Nephila pilipes* (Nephilidae). (**c**, **d**) *Argiope aemula* (Araneidae). (**e**, **f**) *Leucauge magnifica* (Tetragnathidae). *Dashed line* represents color contrast discrimination threshold of 0.05 determined for hymenopteran insects

The Euclidean distance (ΔSt) is the color contrast in the color space of organisms under consideration. Under the chromatic condition when the color contrast value is greater than the discrimination threshold of 0.05, bees can discriminate the color signals of the object from those of the background (Théry and Casas 2002).

23.2.2 Spider Coloration from Eyes of Diurnal Insects

The color contrast approach has been extensively used to quantify how the colors of various body parts of a wide array of orb web spiders are viewed by hymenopteran insects. For example, the prosoma, two yellow bands on opisthosoma, and spots on legs and ventrum of *Nephila pilipes* (Nephilidae) (Fig. 23.1a) all exhibited color contrasts significantly higher than the discrimination threshold for color contrast detection estimated for honeybees (Fig. 23.1b) (Tso et al. 2004). The dark parts of the body and spinneret had low color contrasts that were not significantly different from the discrimination threshold when viewed by chromatic vision (Fig. 23.1b). But they had color contrasts significantly higher than the threshold when only green receptor signal was used. Empirical evidence for the co-occurrence of low and high color contrast body coloration was also reported from *Argiope aemula* (Araneidae) (Fig. 23.1c, d) (Cheng and Tso 2007) and *Leucauge magnifica* (Tetragnathidae) (Fig. 23.1e, f) (Tso et al. 2006). Many diurnal orb-weaving spiders exhibit dramatic patterns of coloration with both high-contrast bright and low-contrast dark markings on their bodies. When bees view these spiders by chromatic vision, the brightly colored markings are visible but the dark parts become indistinguishable from the vegetation background. Those high-contrast bright and low-contrast dark markings may be arranged in such a way to alter the contour of the spider and generate a signal unlike that of a predator. Besides, results from both empirical and applied studies demonstrate that a yellow cue plays an important role in the signal communication between pollinator insects and flowers. Coincidently, many taxa of orb spiders exhibit various forms of yellow markings on their bodies (Yaginuma 1986). The effectiveness of yellow signals in luring pollinator prey might be one major reason for the convergent possession of such a trait in spiders of divergent phylogenetic relationships.

23.2.3 Function of Conspicuous Spider Coloration at Day

The co-occurrence of low and high color contrast body parts in these orb-weaving spiders may be an adaptive morphological trait. Break in contour due to low-contrast body colorations, plus the resource-mimicking color signals of high-contrast body color, may make it difficult for insects to associate these spiders with predation risk. Many insects such as honeybees have innate preference for symmetric and disruptive patterns (Rodriguez and Gumbert 2004). The arrangement of body color patches and their differential visual distinctiveness to insects may make these spiders resemble innate preference of pollinator insects such as symmetric and disruptive patterns.

Evidence from several studies has shown that altering the color signals of orb-weaving spiders reduced their insect catching rate (Craig and Ebert 1994; Hauber 2002), and therefore, this seems to provide direct support for the prey attraction hypothesis. However, such results could also be interpreted as being congruent with the camouflaging hypothesis because the alteration of body coloration in the treatment might have destroyed the camouflaging pattern, thus rendering the spider more visible against the background and therefore lowering the insect catching rate.

Tso et al. (2006) showed that compared with orbs without the orchid spider *L. magnifica*, those with spiders intercepted almost twice as many insects. Such a result is not congruent with the camouflaging hypothesis, which predicts a similar prey interception rate between orbs with and without spiders. The attractiveness of conspicuous orb web spider's body coloration seems to be achieved by the properties of the color signal, rather than the visibility of the spider. Tso et al. (2006) purposely used a paint exhibiting a reflectance spectrum different from that of the spiders. The color contrasts of such paint viewed against the vegetation background were significantly higher than the discrimination threshold, indicating that the paint used could be readily seen by the insects. However, given such high visibility, those painted spiders still intercepted and consumed far fewer insects than the control group (Tso et al. 2006). Such results indicate that the chromatic properties of orb-weaving spiders' body coloration are quite critical to their insect interception. The properties of their color signal have been fine-tuned by selection to achieve the best attractiveness to their prey. Once such property was altered, even though the changed coloration was still quite visible, they were no longer attractive to insects.

23.2.4 Factors Shaping Spider Coloration Pattern: Diurnal Context

The presence of low-contrast coloration to break the contour of the body and high-contrast coloration to attract prey seems to be a product of various counteracting selection pressures involved in spider–insect visual interactions. Empirical evidence from behavioral and neuroethological studies confirms that visual signal design of brightly colored orb web spiders reflects a compromise between efficaciously increasing foraging benefit and strategically reducing predation cost. In a field experiment, Fan et al. (2009) used a dummy approach to assess how visual luring signals of giant wood spiders *N. pilipes* traded off opposing pressures of feeding and surviving. They created dummies made of cardboard to test how changing size of conspicuous signal affected attractiveness to prey and predators. Fan et al. (2009) found that a spiders' color signal alone was sufficient to lure prey in the natural condition. Moreover, uniformly yellow-colored dummies attracted significantly more prey than those dummies which mimicked the color pattern of *N. pilipes*. However, yellow-colored dummies also attracted far more hymenopteran predators. Therefore, the current body coloration pattern of *N. pilipes* seems to exhibit the benefit of attracting prey as well as the cost of increased predation pressure. Many orb spiders from a diverse array of families all exhibit similar body coloration patterns. The contrasting conspicuous and inconspicuous parts may generate a contour-breaking disruptive coloration pattern (Cuthill et al. 2005) to reduce detection by predators (Zschokke 2002; Hoese et al. 2006) and lower the cost of this morphology-associated foraging trait. Such commonly seen coloration pattern might reflect orb spiders' adaptation in response to high predation pressure from predators such as wasps. (Blackledge et al. 2003).

23.3 Spider Coloration in the Nocturnal Context

While discussing the color signaling between animals, most focus is placed on the diurnal system. The first reason for such bias is that the daylight intensity is higher than that of nighttime. Secondly, the dim light environment in the nighttime has very high noise-to-signal ratio (Warrant 2004). Therefore, it is easier to identify and analyze various forms of color signaling in the diurnal condition. Many spiders only hunt during nighttime and their colorations are usually dark, gray, or brown reducing the spiders' visibility during daytime (Oxford and Gillespie 1998). However, the ventrum of various species of *Neoscona* and *Araneus* exhibits two to four bright spots. During nighttime the spiders will sit in the center of the orb web and expose the bright ventrum spots. Why does a sit-and-wait predator which is only active during the night have such conspicuous body coloration? Is it possible that the conspicuous ventrum spots of nocturnal orb spiders also serve as visual lures to attract visually orientated prey under the dim light condition, as does the conspicuous body coloration of those diurnal orb-weaving spiders? Besides, some researchers conducted a round-the-clock survey and they found that in addition to diurnal hunting, certain conspicuous orb web spiders such as *Nephila* and *Leucauge* also actively hunted for prey during the night (Chuang et al. 2007; Tso et al. 2007). Since the body coloration of these conspicuous orb web spiders are effective visual lure during daytime, will these visual signals play a role in spider–insect visual interactions under the dim light condition? In this section I will first describe the neuroethological methods used to quantify how colorations of spiders are viewed by nocturnal moth. Then results of empirical studies manipulating color signal of spiders in the nocturnal condition will be reviewed. The factors shaping the design of spider visual lure in the nocturnal condition will also be discussed.

23.3.1 Quantifying How Colors Are Viewed by Nocturnal Insects

Many nocturnal insects have specialized eyes which enable them to discriminate color stimuli (Kelber et al. 2002) and to detect food resource at night (Raguso and Willis 2005). The superposition compound eyes of numerous nocturnal insects combine the light signal received by hundreds of ommatidia. The signal intensity can be greatly magnified thereby solving the problem of low light intensity in dim light environments (Kelber et al. 2003). In addition, the rhabdoms of superposition eyes are longer than those of apposition eyes and thus can help to improve the noise-to-signal ratio (Kelber and Roth 2006). The visual sensitivity of nocturnal insects is furthered structurally by wide pupil aperture and physiologically by spatial/temporal summation of visual channel neural outputs (Warrant 1999). Nocturnal insects such as moths generally exhibit UV, blue and green photoreceptors, and similar types of receptors are also found in numerous diurnal insects (Briscoe and Chittka 2001). However, while moths can distinguish color signals in the night, diurnal insects are color-blind under dim light conditions (Kelber et al. 2003). Therefore, we cannot use the diurnal neuroethological models commonly used in numerous studies to quantify how the coloration of an organism is viewed under dim light conditions. A solution was provided by Johnsen et al. (2006). These researchers developed a neuroethological model based upon the visual system of a hawkmoth to calculate nocturnal color contrasts. In this model, information such as reflectance spectra of spiders, illumination function of the environment, and parameters regarding the visual system of the hawk moth are needed to calculate nocturnal color contrast. In the neuroethological model developed by Johnsen et al. (2006), reflection is expressed as the percentage ($\%R_\lambda$) relative to the reflection from a standard reference:

$$\%R_\lambda = \frac{S_\lambda - D_\lambda}{R_\lambda - D_\lambda} \times 100\%$$

where S is the sample intensity at wavelength λ, D is the dark intensity at wavelength λ, and R is the reference intensity at wavelength λ.

The following equation can be used to calculate the quantum catches of one ommatidium of moth (Warrant and Nilsson 1998):

$$N = 1.13(\pi/4)n\Delta P^2 D^2 \Delta t \int_{350}^{700} \kappa\tau(1 - e^{-kRi(\lambda)l})L(\lambda)\mathrm{d}\lambda$$

where n is the effective facets in the superposition, ΔP is the photoreceptor acceptance angle, D is the diameter of a facet lens, Δt is the integration time of a photoreceptor, κ is the quantum efficiency of transduction, τ is the fractional transmission of the eye media, k is the absorption coefficient of the rhabdom, l is the rhabdom length doubled by tapetal reflection, R_i (λ) are the absorbance spectra of each photoreceptor, and $L(\lambda)$ is the color signal of the object, which is the multiplication of reflectance spectra of objects and that of nocturnal light environment. The difference of an object of interest and the background, the achromatic contrast, can be estimated by

$$C = \frac{N_X - N_{\text{green}}}{N_X + N_{\text{green}}}$$

where N_x is quantum catches of object and N_{green} is quantum catches of green vegetation background (Johnsen et al. 2006).

To calculate the nocturnal chromatic contrasts, first the quantum catch values (N) of UV (uv), blue (b), and green (g) photoreceptors are each estimated to generate N_{uv}, N_{b}, and N_{g}. Then q_{uv}, q_{b}, and q_{g}, the relative quantum catches of each type of photoreceptor, can be calculated by

$$q_{\text{uv}} = \frac{N_{\text{uv}}}{N_{\text{uv}} + N_{\text{b}} + N_{\text{g}}}$$

$$q_{\text{b}} = \frac{N_{\text{b}}}{N_{\text{uv}} + N_{\text{b}} + N_{\text{g}}}$$

$$q_{\text{g}} = \frac{N_{\text{g}}}{N_{\text{uv}} + N_{\text{b}} + N_{\text{g}}}$$

Then values of each stimulus were used to calculate relative distances in the color triangle by

$$X_1 = \frac{1}{\sqrt{2}}(q_g - q_b)$$

$$X_2 = \frac{\sqrt{2}}{\sqrt{3}}\left(q_{uv} - \frac{q_g + q_b}{2}\right)$$

X_1 and X_2 are the distances on the x-axis and y-axis, which represent the relative intensity of three types of photoreceptors in the 2D color space. The distance of two color stimuli in the color space is the nocturnal chromatic color contrast (Johnsen et al. 2006). So far, the theoretical discrimination threshold value for the nocturnal chromatic neuroethological model is still not available. In some studies, researchers compared the nocturnal achromatic and chromatic contrast values of various body parts of spiders and those averaged from petals of flowers by ANOVA tests and LSD mean comparisons (Chuang et al. 2008). Results of these tests can help determine whether the ventrum spots are more conspicuous than other body parts and whether they are similar to those of the resources the nocturnal moths were searching for.

23.3.2 Spider Coloration from Eyes of Nocturnal Insects

Members of the nocturnal orb spider genera *Neoscona* and *Araneus* hunt actively during the night, but during daytime they hide in vegetation or under loose bark, fallen timber, or rocks. The coloration of spiders' dark gray or brown opisthosomal dorsum matches well with the background they perch during daytime (Fig. 23.2a). The coloration of *Neoscona* spider's dorsum and legs seems to effectively conceal the spiders because from the eyes of hymenopteran insects, these spiders' major predators during daytime, the color contrasts are smaller than the discrimination threshold (Fig. 23.2b) (Chuang et al. 2008). To determine whether ventrum spots mimicked the color signals of the resources of spiders' prey, Chuang et al. (2008) measured the reflectance spectra of flower petals of three species of herb plants frequently visited by moths and found that the nocturnal color contrasts of spots and flowers were similar. Therefore, nocturnal orb spiders seem to exploit the color vision of nocturnal insects by exhibiting visual lures resembling the color signal of flowers open at night.

In the previous section it is known that when the brightly colored orb-weaving spiders are viewed by diurnal hymenopteran, the dark or green parts of a spider's body are indistinguishable from the vegetation background but the conspicuous parts are highly visible to insects. This phenomenon also occurs when these spiders are viewed by nocturnal insects under dim light conditions. Through nocturnal

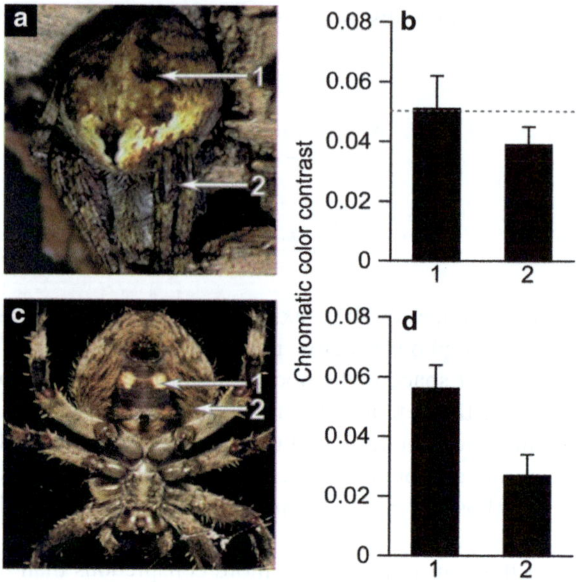

Fig. 23.2 (**a**) Dorsal view of the strictly nocturnal orb web spider *Neoscona punctigera* (Araneidae). (**b**) Diurnal color contrasts of its dorsum when viewed against bark by hymenopteran insects. *Dashed line* represents discrimination threshold of 0.05 determined for hymenopteran insects. (**c**) Ventral view of *Neoscona punctigera* with the brightly colored spots exposed. (**d**) Nocturnal color contrasts of bright and dark parts of ventrum when viewed by lepidopteran insects

achromatic vision of moth, when conspicuously colored orb web spiders such as *N. pilipes* are viewed in the dim light environment, the conspicuous body parts are more visible than the dark parts (Chuang et al. 2007). The nocturnal achromatic as well as chromatic contrasts of conspicuous yellow body parts of *N. pilipes* when viewed against vegetation background by lepidopteran insects were significantly higher than those of the black body parts. The combination of high- and low-contrast body colorations might make the appearance of spiders unlike that of a predator but rather like some form of resource. Many pollinator insects have an innate preference for symmetric and disruptive patterns (Rodriguez and Gumbert 2004). Moreover, floral guides, stingless bee nest entrances, and insectivorous pitchers all exhibit a similar dark center, radiating stripes, and peripheral dots (Biesmeijer et al. 2005). The arrangement of body color patches of *N. pilipes* and their differential visual distinctiveness to insects may resemble the global visual attributes of a pollinator's resource.

23.3.3 Function of Conspicuous Spider Coloration at Night

Chuang et al. (2008) evaluated when spiders sit on webs with conspicuous ventrum spots fully exposed if they would serve as deceptive color signals to lure visually orientated nocturnal prey. In the field, webs with *Neoscona punctigera* intercepted significantly more insects than those without. When the color signal of ventrum spots was altered by paint, webs' prey interception rates decreased significantly. Blamires et al. (2012) monitored, using infrared video cameras, the prey attraction rates of *N. punctigera* dummies to determine whether the coloration of the ventral spots of these nocturnal spiders acts as a prey lure. Blamires et al. (2012) placed various form of dummies on vacant spider webs in the field at night. The results of field experiments showed that the coloration of the spots functions to lure prey. The standard dummies, mimicking adult female *N. punctigera* in coloration, size, and shape, attracted more prey than the dummies with gray spots, the entirely black dummies, and webs without spiders. These results demonstrate that even in the nocturnal condition certain terrestrial organisms exhibit visual lures to attract prey. Unexpectedly, however, the entirely yellow dummies attracted fewer prey than the standard dummies. The ecological significance of such finding will be discussed in Sect. 23.3.4.

On the other hand, while the orchid spider *L. magnifica* hunts during the day and is generally regarded as a diurnal predator, it also hunts actively during the night. Tso et al. (2007) demonstrated that the bright body coloration of *L. magnifica* serves as a visual lure to nocturnal insects. When the color signals of the ventral yellow stripes of *L. magnifica* were altered by paint, the moth interception rate was reduced significantly. Results of a field survey showed that the number of actively hunting *L. magnifica* during the night was two to three times higher than that during the day. A comparison of diurnal and nocturnal hunting performance of *L. magnifica* recorded by video cameras revealed that they obtained higher prey intake from nocturnal hunting, and nocturnal hunting seems to be their primary source of prey intake (Tso et al. 2007). Similar results were reported from the brightly colored *N. pilipes* (Chuang et al. 2007). The prey interception rate during diurnal hunting was significantly lower than that in nocturnal hunting. In the diurnal hunting lepidopteran insects comprised less than 10% of prey consumed. However, between 33% and 60% of nocturnal prey were lepidopterans. The average size of the intercepted nocturnal prey was significantly larger than that of the diurnal prey. Results of these studies indicate that large lepidopteran insects seem to be the major target of certain colorful orb web spiders during nocturnal hunting.

23.3.4 Factors Shaping Spider Coloration Pattern: Nocturnal Context

Many nocturnal insects have superposition eyes and these eyes are extremely sensitive to very dim light signals. Such extraordinary sensitivity is achieved structurally by wide pupil aperture and physiologically by spatial/temporal summation of visual channel neural outputs (Warrant 1999). The sensitivity of superposition eyes is so high (theoretically estimated to be 10,000 times that of locust, Warrant 1999) that they can even perceive far away point light source (Warrant 2006). Although the opisthosomal spots of *N. punctigera* are small, the color signal reflected from them should be easily perceived by nocturnal insects with superposition eyes. On the other hand, the size of ventrum spots might be constrained by strong predation pressure. During daytime, when the major predators of orb spiders such as parasitoid insects (Blackledge et al. 2003) are most abundant, *Neoscona* or *Araneus* spiders usually perch on bark with their bright ventrum spots well concealed. If the size of ventral spots is too large, they will be difficult to conceal, and this will consequently increase the conspicuousness of spiders to predators.

In the study conducted by Blamires et al. (2012), comparisons between the yellow-spotted dummies and the entirely yellow dummies and the extremely low number of observed predation events suggest that, unlike diurnal spiders, the ventral spots of nocturnal spiders such as *Neoscona* species do not represent a compromise between prey attraction and predator avoidance. Perhaps the spots are exploiting a key, as yet unidentified, chromatic or achromatic visual cue used by foraging or navigating insects. The symmetry of the spots may be implicit in the luring of insects as they may resemble pattern symmetries that insects use to identify flower parts (Dafni and Kevan 1996). The paired spots may resemble nectar guides used by insects to locate food. Alternatively, they may be the most energy-efficient way, given the available resources, for nocturnal spiders to invest in prey-enticing coloration (Kelsh et al. 2009) or are an evolutionary or ontogenetic remnant that is costly to discard and happens to lure prey (Kemp et al. 2005). Experiments should be performed to test the significance of spot size, shape, and position in the functioning of such nocturnal visual lure.

23.4 Conclusions

Color vision requires the integration of information from all the primary receptors. Many studies have overemphasized the UV component of the spider and background light signals to infer the nature of insect–spider visual interactions. Insects see by detecting the contrasts between objects and their environments and all kinds of color receptors and signals are involved (Chittka and Menzel 1992; Vorobyev and Brandt 1997; Briscoe and Chittka 2001). Therefore, all types of receptor signals should be considered when exploring the visual interactions between predators and prey. On the other hand, the results of recent studies examining spider–insect interactions on a 24-h basis indicate that while studying animal communications, we should have a comprehensive view of the timing as well as the visual systems of all organisms involved in the interaction. If the color signaling of one organism functions in a range of light conditions but research is conducted only in a subset of them, the conclusions subsequently made might be biased. To date, almost all empirical studies on foraging behaviors of so-called "diurnal" web spiders only investigated the diurnal hunting. In the past, in the terrestrial ecosystem, color signals were generally considered to be used only by diurnal organisms. However, more and more studies have demonstrated that various nocturnal organisms utilize color signals to locate food resources and mates (Kelber and Roth 2006). In the case of *L. magnifica* and *N. pilipes*, their conspicuous body coloration actually attracts much more nocturnal than diurnal prey. Since orb-weaving spiders are sit-and-wait predators, they are under strong selection pressure to evolve ways to make prey orient toward them. Therefore, the selection pressures of effectively exploiting the color vision of large nocturnal prey might be one of the major forces driving the evolution of orb spider's body coloration. I suggest that the selection pressure of effectively exploiting the color vision of insects in the nocturnal condition could be one major force driving the evolution of spider coloration.

References

Biesmeijer JC, Giurfa M, Koedam D, Potts SG, Joel DM, Dafni A (2005) Convergent evolution: floral guides, stingless bee nest entrances, and insectivorous pitchers. Naturwissenschaften 92:444–450

Blackledge TA, Coddington JA, Gillespie RG (2003) Are three-dimensional spider webs defensive adaptations? Ecol Lett 6:13–18

Blamires SJ, Lai CH, Cheng RC, Liao CP, Shen PS, Tso IM (2012) Body spot coloration of a nocturnal sit-and-wait predator visually lures prey. Behav Ecol 23:69–74

Briscoe AD, Chittka L (2001) The evolution of color vision in insects. Annu Rev Entomol 46: 471–510

Cheng RC, Tso IM (2007) Signaling by decorating webs: luring prey or deterring predators? Behav Ecol 18:1085–1091

Chittka L (1996) Optimal sets of color receptors and opponent processes for coding of natural objects insect vision. J Theor Biol 181:179–196

Chittka L (1997) Bee color vision is optimal for coding flower colors, but flower colors are not optimal for being coded – why? Isr J Plant Sci 45:115–127

Chittka L, Menzel R (1992) The evolutionary adaptation of flower colors and the insect pollinators' color vision. J Comp Physiol A 171:171–181

Chuang CY, Yang EC, Tso IM (2007) Diurnal and nocturnal prey luring of a colorful predator. J Exp Biol 210:3830–3837

Chuang CY, Yang EC, Tso IM (2008) Deceptive color signaling in the night: a nocturnal predator attracts prey with visual lures. Behav Ecol 19:237–244

Craig CL, Ebert K (1994) Color and pattern in predator–prey interactions: the bright body colors and patterns of a tropical orb-spinning spider attract flower-seeking prey. Funct Ecol 8: 616–620

Cuthill IC, Stevens M, Sheppard J, Maddocks T, Parraga CA, Troscianko TS (2005) Disruptive coloration and background pattern matching. Nature 434:72–74

Dafni A, Kevan PG (1996) Floral symmetry and nectar guides: ontogenetic constraints from floral development, color pattern rules and functional significance. Bot J Linn Soc 120:371–377

Fan CM, Yang EC, Tso IM (2009) Hunting efficiency and predation risk shapes the color-associated foraging traits of a predator. Behav Ecol 20:808–816

Giurfa M, Vorobyev M, Brandt R, Posner B, Menzel R (1997) Discrimination of colored stimuli by honeybees: alternative use of achromatic and chromatic signals. J Comp Physiol A 180:235–243

Hauber ME (2002) Conspicuous coloration attracts prey to a stationary predator. Ecol Entomol 27:686–691

Hoese FJ, Law EAJ, Rao D, Herberstein ME (2006) Distinctive yellow bands on sit-and-wait predator: prey attractant or camouflage? Behaviour 143:763–781

Johnsen S, Kelber A, Warrant E, Sweeney AM, Widder EA, Lee RL, Javier HA (2006) Crepuscular and nocturnal illumination and its effects on color perception by the nocturnal hawkmoth *Deilephila elpenor*. J Exp Biol 209:789–800
Kelber A, Roth LSV (2006) Nocturnal color vision – not as rare as we might think. J Exp Biol 209: 781–788
Kelber A, Balkenius A, Warrant EJ (2002) Scotopic color vision in nocturnal hawkmoths. Nature 419:922
Kelber A, Balkenius A, Warrant EJ (2003) Color vision in diurnal and nocturnal hawkmoths. Integr Comp Biol 43:571–579
Kelsh RN, Harris ML, Colanesi S, Erickson CA (2009) Stripes and belly-spots – a review of pigment cell morphogenesis in vertebrates. Semin Cell Dev Biol 20:90–104
Kemp DJ, Rutowski RL, Mendoza M (2005) Color pattern evolution in butterflies: a phylogenetic analysis of structural ultraviolet and melanic markings in North American sulphurs. Evol Ecol Res 7:133–141
Land MF, Osorio DC (2003) Color vision: coloring the dark. Curr Biol 13:83–85
Naka KI, Rushton WAH (1966) S-potentials from color units in the retina of fish (Cyprinidae). J Physiol (Lond) 185:536–555
Oxford GS, Gillespie RG (1998) Evolution and ecology of spider coloration. Annu Rev Entomol 43:619–643
Raguso RA, Willis MA (2005) Synergy between visual and olfactory cues in nectar feeding by wild hawkmoths, *Manduca sexta*. Anim Behav 69:407–418
Rodriguez I, Gumbert A (2004) Symmetry is in the eye of the 'beeholder': innate preference for bilateral symmetry in flower-naive bumblebees. Naturwissenschaften 91:374–377
Spaethe J, Tautz J, Chittka L (2001) Visual constraints in foraging bumblebees: flower size and color affects search time and flight behavior. Proc Natl Acad Sci USA 98:3898–3903
Théry M, Casas J (2002) Predator and prey views of spider camouflage. Nature 415:133
Tso IM, Lin CW, Yang EC (2004) Colorful orb-weaving spiders through a bee's eyes. J Exp Biol 207:2631–2637
Tso IM, Liao CP, Huang JP, Yang EC (2006) Function of being colorful in web spiders: attracting prey or camouflaging oneself? Behav Ecol 17:606–613
Tso IM, Huang JP, Liao CP (2007) Nocturnal hunting of a brightly colored sit-and-wait predator. Anim Behav 74:787–793
Vorobyev M, Brandt R (1997) How do insect pollinators discriminate colors. Isr J Plant Sci 45: 103–113
Warrant EJ (1999) Seeing better at night: life style, eye design and the optimum strategy of spatial and temporal summation. Vision Res 39:1611–1630
Warrant EJ (2004) Vision in the dimmest habitats on earth. J Comp Physiol A 190:765–789
Warrant EJ (2006) Invertebrate vision in dim light. In: Warrant E, Nilsson DE (eds) Invertebrate vision. Cambridge University Press, Cambridge
Warrant EJ, Nilsson DE (1998) Absorption of white light in photoreceptors. Vision Res 38:195–207
Yaginuma T (1986) Spiders of Japan in color. Hoikusha Publishing Company, Osaka
Zschokke S (2002) Ultraviolet reflectance of spiders and their webs. J Arachnol 30:246–254

Chapter 24
Structural Colors in Spiders

Rainer F. Foelix, Bruno Erb, and David E. Hill

24.1 Introduction

> The finely color'd Feathers of some Birds, and particularly those of Peacock Tails, do, in the very same part of the feather, appear of several Colours . . . after the very same manner than thin Plates were found to do . . . and therefore their Colours arise from the thinness of the transparent parts of the Feathers. . . . and to the same purpose it is, that the Webs of some Spiders, by being spun very fine, have appear'd color'd . . . and by varying the Position of the Eye, do vary their Colours.
> Isaac Newton (1730)

It is quite remarkable that almost 300 years ago Isaac Newton already knew that certain colors are not due to pigments but are caused by the microstructure of an object. The "thin plates" he was referring to were later found to be an essential part for light interference, which is mostly responsible for such "structural colors." Newton's correct assumption is even more astonishing, if we consider that those "thin plates" are in the nanometer range (or the wavelengths of visible light) and therefore could not be resolved with a light microscope.

Interference of light often occurs in stacks of thin plates, mostly in alternating materials with different optical densities (refractive index). At the border of each layer, a light ray is transmitted but also reflected; due to their different path lengths, a phase shift occurs (varies as a function of wavelength), and the reflected light will appear of a different color than the incoming light (Land 1972; Kinoshita et al. 2008).

Structural colors may also be produced if a surface is scored by fine parallel ridges (and grooves) which act as diffraction gratings. Incident light is separated

R.F. Foelix (✉) • B. Erb
Neue Kantonsschule Aarau, Biology Department, Aarau, Switzerland
e-mail: r.foelix@gmx.ch

D.E. Hill
213 Wild Horse Creek Drive, Simpsonville, SC 29680, USA

W. Nentwig (ed.), *Spider Ecophysiology*,
DOI 10.1007/978-3-642-33989-9_24, © Springer-Verlag Berlin Heidelberg 2013

Fig. 24.1 Colorful theraphosids. (**a**) A deep-blue Cyriopagopus resting on a tree trunk (Photo: Rick West). (**b**) A juvenile *Ephebopus cyanognathus* with the typical blue chelicerae, a green-golden opisthosoma, and yellow fringes on the distal femora (Photo: Bastian Rast)

into its component wavelengths and reflected into different directions (Parker and Martini 2006). This effect is commonly seen on the surface of some beetle wings (Hinton 1968) and also on compact disks.

The most conspicuous structural (or physical) colors are known as *iridescence*, which means the colors change, depending on the angle of the incident light—or the angle an observer is viewing an iridescent object (Doucet and Meadows 2009; Meadows et al. 2009). Iridescent colors are caused on layered or crystalline nanostructures by coherent scattering (Prum 2006). Well-known biological examples of iridescence are certain feathers of birds (e.g., in peacocks and hummingbirds), certain insect wings (e.g., in butterflies and beetles), sea shells (mother of pearl), and even plant petals (Glover and Whitney 2010). Inanimate objects such as minerals (e.g., opal and labradorite), thin oil films on water, or soap bubbles may also display distinct interference colors (Kinoshita 2008; Vukusic 2004).

Among animals, structural colors have been extensively studied in birds (e.g., Durrer 1977; Prum 2006) and in insects (e.g., Ghiradella 1998, 2010; Prum et al. 2006), but spiders have received relatively little attention (Foelix et al. 2009; Ingram et al. 2009, 2011; Parker and Hegedus 2003). It is true that most spiders are rather drab and inconspicuously colored, yet there are notable exceptions. For instance, many jumping spiders (Salticidae) are quite colorful and even the nocturnal theraphosid may have brightly colored species (Fig. 24.1). This chapter will focus on structural colors found in these two families. Structural colors in these spiders reside mostly in special hairs and scales but may occur also in the solid body cuticle, for instance, of the chelicerae and the lenses of the eyes (Homann 1985).

24.2 Structural Colors in Hairs of Theraphosid Spiders

Even most arachnologists are not aware that there are quite a few brightly colored theraphosids; yet among pet spider keepers, these are highly prized, for instance, *Avicularia versicolor*, *Cyriopagopus* spp., *Ephebopus cyanognathus*, *Euathlus* spp.,

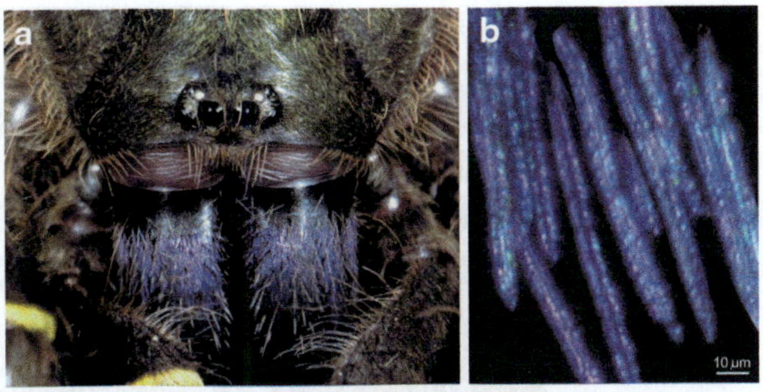

Fig. 24.2 Blue hairs on the chelicerae of *E. cyanognathus*. (**a**) The blue color resides entirely in the hairs, not in the cuticle. (**b**) Epi-illumination with a weak LED light source yields deep-blue and purple color hues (from Foelix et al. 2009)

Haplopelma lividum, *Heterothele gabonensis*, *Holothele* spp., *Iridopelma* spp., *Pamphobeteus* spp., *Phormictopus* spp., *Poecilotheria metallica*, and *Xenesthis immanis*, to name just a few. Since most of them are iridescent, it is clear that the coloration is due (at least in part) to structural colors. But where exactly are the structural colors of those theraphosids localized, and how are they produced? In the few species we have examined so far, the bright colors were always restricted to specialized hairs and were not found in the body cuticle. The two species which we studied closely were *E. cyanognathus* and *P. metallica*.

24.2.1 Ephebopus cyanognathus

This spider from Brazil, Guyana, and Suriname is known under the vernacular name "blue fang tarantula" because of its metallic blue chelicerae (Figs. 24.1b and 24.2). A closer inspection under a dissecting microscope reveals that the blue color comes from deep-blue hairs covering the cheliceral basal segment (Fig. 24.2). Under epi-illumination the color may change from a deep violet blue to lighter blue or even turquoise, depending on the direction of the incident light. Under transmitted light these hairs appear completely transparent and colorless. They measure 100–200 µm in length and only 5–6 µm in diameter. The hair shaft bears several rows of fine extensions near the base, which fuse into solid laterally projecting fins along the distal hair shaft. Cross sections of the hair shaft appear star-shaped; only the central core consists of solid cuticle, whereas the peripheral part is layered (Fig. 24.3). The solid core probably absorbs non-reflected light and may function like the dark backing of a mirror. The peripheral lamellation is barely visible with the scanning electron microscope (Fig. 24.3a) but is clearly seen with the transmission electron microscope: about six cuticular lamellae (50–70 nm) alternate with narrow air spaces (60–120 nm) (Fig. 24.3 b, c). Due to the different refractive indices of cuticle (1.58) and air (1), incoming light rays will be split into refracted and reflected rays at each border line of cuticle and air. Depending on the different distances, these light waves have traveled into this air/cuticle stack, the reflected rays will show phase shifts, and the resulting interference shows up in different colors (Fig. 24.4a). Many thin layers (lamellae) cause "constructive

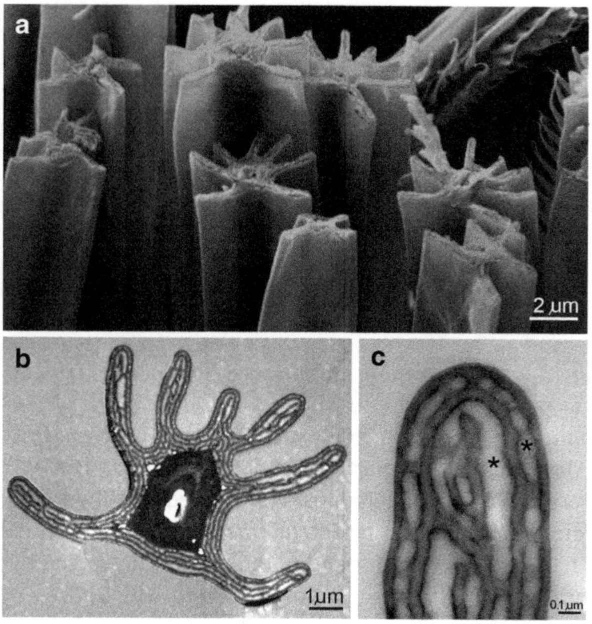

Fig. 24.3 Fine structure of the blue cheliceral hairs in *E. cyanognathus* (Theraphosidae). (**a**) Hair shafts, clipped with a razor blade, exhibit a typical cogwheel shape in cross section. The lamellar structure of the hair wall is barely visible in this scanning electron micrograph. (**b**) Cross section of a blue hair as seen in a thin section in the transmission electron microscope. Only the center of the hair shaft consists of solid cuticle, whereas the lateral fins are distinctly layered. (**c**) High magnification of a lateral fin showing several cuticular lamellae of 50–70-nm thickness, alternating with equally narrow air spaces (*asterisk*) (from Foelix et al. 2009)

interference," that is, the intensity of the interference colors is enhanced. About ten alternating layers yield an almost complete reflection of a given wavelength (Land 1972); in the case of the blue hairs in *Ephebopus,* we found 5–6 alternating layers (Fig. 24.3b, c). The color of the reflected light is mostly a deep blue, with shades of violet and green. This is confirmed, when the theoretical reflectance spectra are calculated, based on measurements of the optical thickness and the geometrical dimensions of the lamellae (Huxley 1968; Fig. 24.4b). In addition to a strong UV reflectance, there is a distinct peak in the blue and violet (maximum at 430 nm) and a smaller one in the green (maximum at 540 nm) in the calculated spectrum.

24.2.2 *Poecilotheria metallica*

Another deeply blue-colored theraphosid is *P. metallica* from India, the "Gooty Sapphire Ornamental Tree Spider," or simply "Gooty Sapphire," which is among the most treasured spiders among pet spider keepers. All legs, palps, and chelicerae are covered by deep-blue hairs, and the proximal tibiae show a patch of bright

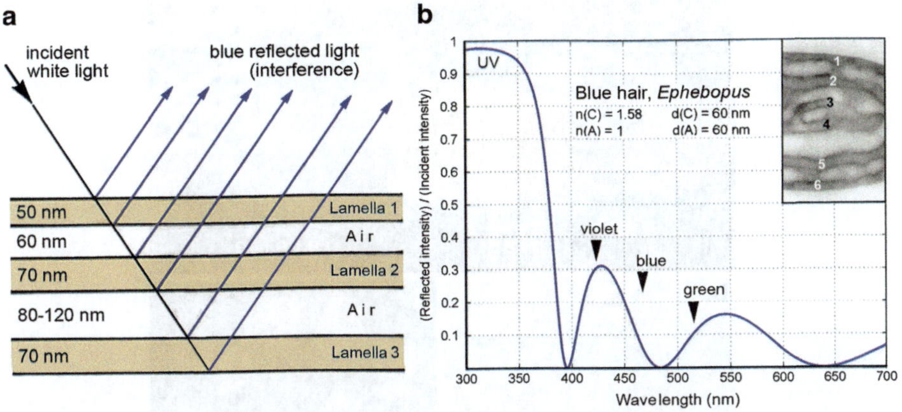

Fig. 24.4 Blue iridescent colors in theraphosid hairs. (**a**) Diagram showing the reflection of incoming *white light* at the border lines of cuticular lamellae and air spaces. Measurements were taken from cheliceral hairs in *E. cyanognathus* (Fig.24.3c). (**b**) Theoretical spectral reflectance of six cuticular lamellae of high refractive index n(C) = 1.58 separated by five gaps of low refractive index n(A) = 1, based on measurements taken from the *inset* picture. Note that the largest peaks lie in the UV, *violet* and *blue part* of the spectrum, and a smaller one in the *green* (courtesy of Rolf Thieleczek)

yellow hairs (Fig. 24.5a). Whereas the blue hairs are clearly iridescent, the yellow hairs hardly change their hue when the angle of the incident light varies.

24.2.2.1 Blue Hairs

We selected the proximal tibia, where blue and yellow hairs are juxtaposed (Fig. 24.5a). Unexpectedly, cross sections of blue and yellow hairs often appear very similar in the light microscope (Fig. 24.5b). This is because the blue hairs have the same spiny extensions along the basal hair shaft as the yellow hairs do. The more distal part of the shaft in blue hairs lacks these spines and has only smooth longitudinal ridges (Fig. 24.6a). Cross sections show about nine such ridges (Figs. 24.5c and 24.6c) which are made up of a cuticular meshwork. High magnification reveals a solid cuticle in the core of the hair shaft and a distinct lamellation in the periphery (Fig. 24.7a). This lamellation of 6–7 stacked cuticular layers and intervening air spaces is well developed in the longitudinal ribs but changes into an irregular meshwork toward the center of the hair shaft. Interestingly, the same irregular meshwork is present in the spiny extensions of the basal hair shaft. This is most likely the reason why these hairs appear blue over the entire length of the hair shaft. Calculations of the spectral reflectance give essentially the same results as shown for the blue hairs in *Ephebopus* in Fig. 24.4b; the peak for violet blue has the maximum at 445 nm, and the smaller green peak at 540 nm (13 layers of cuticle/air, 65-nm thickness).

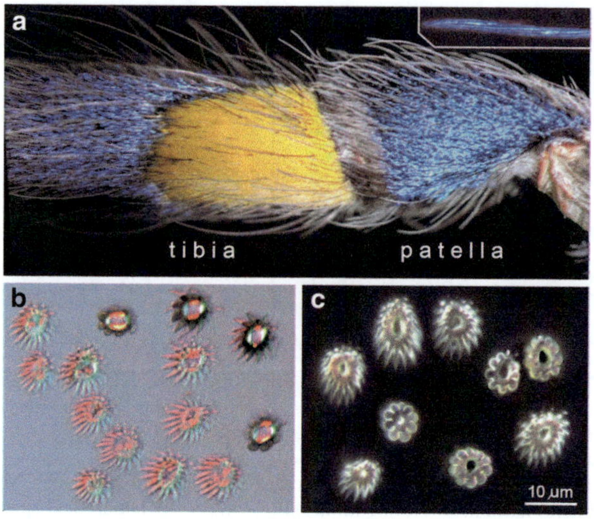

Fig. 24.5 Blue and yellow hairs of the theraphosid *P. metallica*. (**a**) Note the difference of the blue color hues on tibia and patella, depending on the angle of the reflected light. *Inset*: higher magnification of a blue hair showing longitudinal ridges along the hair shaft. (**b**) Cross sections of ten spiny (yellow) hairs, which appear pinkish under polarized light; four darker (blue) hairs are seen in the upper right. (**c**) Cross sections of nine blue hairs under dark-field illumination. The hair shafts bear spines toward the hair base, but are smooth with longitudinal ridges in the distal part

24.2.2.2 Yellow Hairs

The yellow hairs occur on all the legs (more pronounced on legs 1 and 2) but are restricted to lateral patches on the proximal tibia (Fig. 24.5a). Each hair is brushlike, with 10–14 rows of cuticular extensions along the entire hair shaft (Fig. 24.6b). Cross sections of the hair shaft show only an irregular cuticular meshwork (Fig. 24.6d; 24.7b) but no lamellation as in the blue hairs. This meshwork is probably responsible for the yellow interference color. Unfortunately, no precise measurements could be obtained from this irregular meshwork. Nevertheless, using an approximation of 140 nm for the lamellar thickness, the calculated reflectance curve showed a peak in the yellow (maximum at 560 nm) and also a strong reflection in the UV. No pigment is present in these hairs, because sections appear colorless under transmitted white light. This is surprising, since a yellow coloration is generally based on ommochrome pigments (Holl 1987).

24.3 Structural Colors in Scales of Jumping Spiders (Salticidae)

The most colorful spiders are certainly found among the jumping spiders. Their structural colors arise mostly from modified hairs (scales) but also from solid cuticle (see Sect. 23.4). Scales are defined as flattened hairs (setae) that have a pedicel bent

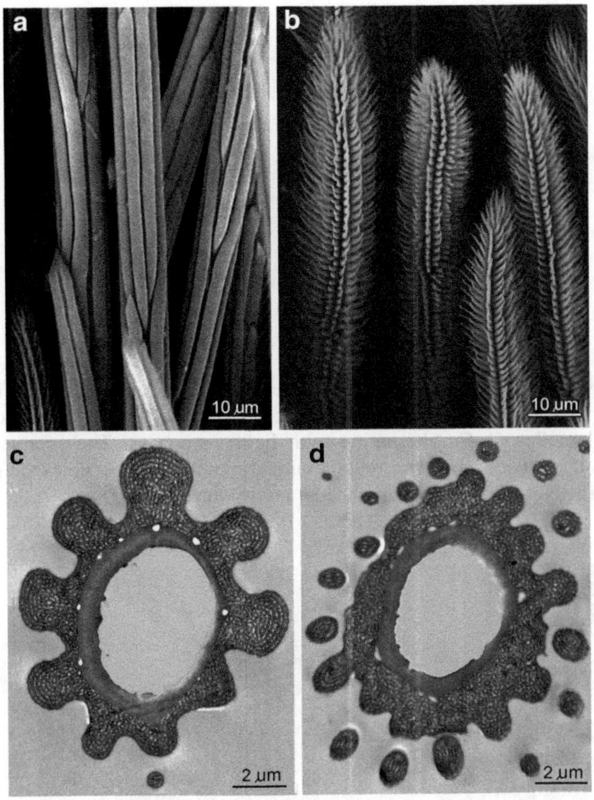

Fig. 24.6 Fine structure of blue and yellow tibial hairs in *P. metallica* (Theraphosidae). (**a**) The blue hairs have longitudinal ridges along the distal hair shaft. (**b**) The yellow hairs bear many fine extensions along the entire hair shaft. (**c**) Cross-section of a blue hair (TEM) showing a hollow center, surrounded by a thin cuticular tube and nine distinct ridges that consist of a highly lamellated cuticle. (**d**) Cross-section of a yellow hair showing a solid cuticle ring in the center surrounded by a loose cuticular meshwork that is also present in the peripheral extensions

near the socket, so that the scale lies close and parallel to the surface cuticle (Hill 1979; Townsend and Felgenhauer 1998). Although the size and shape of scales vary a great deal among the 5,000 species of salticids, they may be roughly grouped into three types: lanceolate, spatulate, and lamelliform (Roth 1993). Most studies on salticid scales have covered morphological, functional, or phylogenetic aspects, but only a few dealt with the structural colors of these scales (Land et al. 2007; Taylor and McGraw 2007). We shall focus here on two examples, the golden scales of *Habronattus hallani* and the blue scales of *Maratus splendens* and *Maratus volans*.

24.3.1 *Habronattus hallani*

The genus *Habronattus* contains about 100 different species, the males of which are very colorful. We picked *H. hallani* (from Arizona, USA) which is renowned for

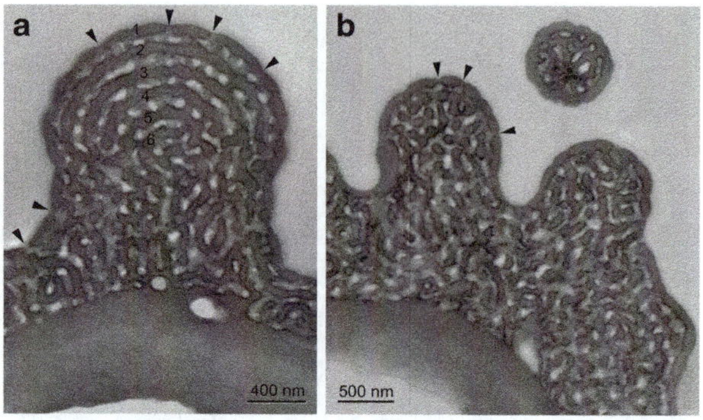

Fig. 24.7 Cross sections of a blue hair (a) and a yellow hair (b) in *P. metallica* (Theraphosidae) at high magnification. (**a**) Only the innermost part of the hair shaft is solid; the outer cuticle consists of six regular lamellae (1–6), alternating with light air spaces. They connect through pores (arrowheads) to the outside. (**b**) An irregular cuticular meshwork occupies the peripheral part of yellow hair shafts. Note the pores (arrowheads) on the surface

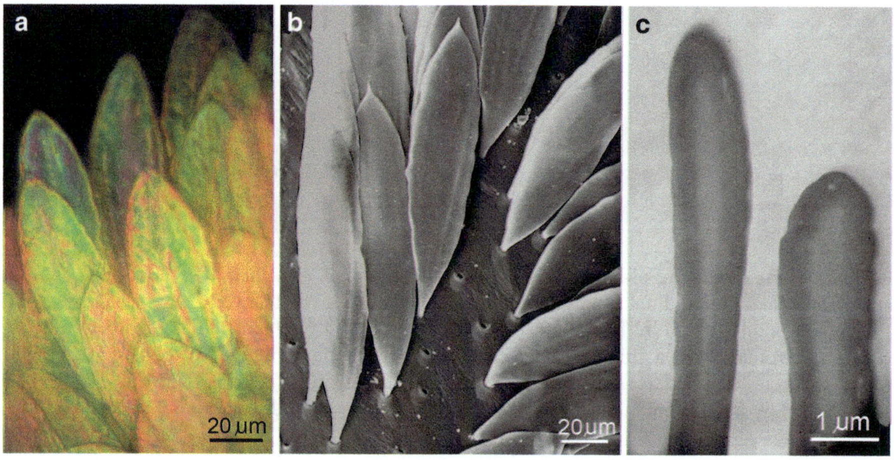

Fig. 24.8 Golden iridescent scales in the salticid *Habronattus hallani*. (**a**) The femora are densely covered by overlapping, golden scales (Photo: Lisa Taylor). (**b**) These scales appear very smooth, with hardly any surface structure. (**c**) This is confirmed in thin sections; the wall thickness of a scale is only 0.2–0.3 µm

having iridescent scales on the front legs and also on the prosoma and clypeus. The femora and patellae are covered by overlapping lanceolate scales that have a distinct golden shine (Fig. 24.8a). The surface of these scales is very smooth, even at high magnification under the scanning electron microscope (Fig. 24.8b). Each scale measures only about 1 µm in thickness, the actual wall being around 300 nm thick, which leaves a narrow air space in the center (Fig. 24.8c). A calculation of the reflectance spectrum shows a peak in the green (at 545 nm) and another one in the red

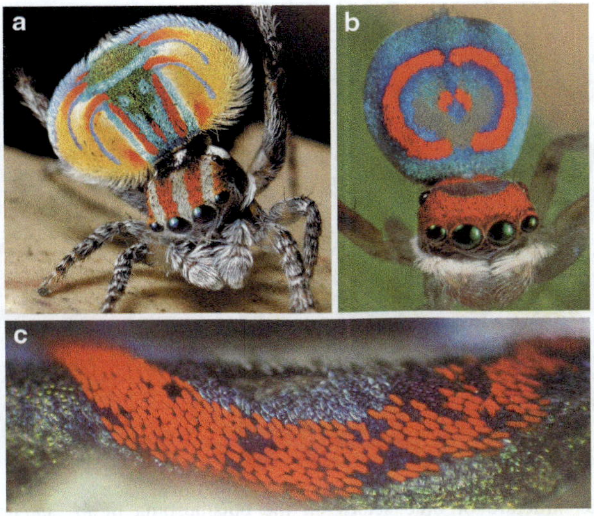

Fig. 24.9 The most colorful jumping spider is the Australian salticid genus *Maratus*. (**a**) A courting male of *M. volans*, displaying its iridescent opisthosomal fan (Photo: Jürgen Otto). (**b**) A courting male of *M. splendens* demonstrating his blue and red opisthosomal flap; the *blue scales* may change to green, depending on the angle of illumination. The red hairs do not vary in color (Photo: Jürgen Otto). (**c**) Detail from the lateral opisthosoma of *M. splendens* showing a band of red brush hairs, flanked on both sides by blue or green scales

(at 700 nm), which together may account for the observed golden hue. Overall, the structure of these golden scales is surprisingly simple and resembles that of UV-reflecting scales of the jumping spider *Cosmophasis* (Land et al. 2007).

24.3.2 Maratus splendens and M. volans

Some of the most colorful jumping spiders belong to the genus *Maratus* from Australia. The vernacular name of these tiny salticids is "peacock spiders," because the males elevate and extend an opisthosomal flap to display their striking colors (Fig. 24.9a). From our preliminary observations, it seems that blue and green are due to a variety of iridescent scales, whereas red (and perhaps yellow) is due to pigmented, brushlike hairs (Fig. 24.9b). The blue scales—which may also appear green depending on the angle of illumination (Fig. 24.9c)—are leaf-shaped, about 40 μm long and 15 μm wide (Fig. 24.10a). At high magnification the scale surface reveals a regular pattern of fine cuticular ridges (Fig. 24.10b) that probably act as a diffraction grating (Parker and Hegedus 2003). Sections of these thin scales (1 μm) show a double-layered cuticle of different densities below the surface ridges (Fig. 24.10c). Another set of fine ribs lines the inner side of the scale wall bordering the narrow, air-filled central lumen. These blue scales are structurally much more

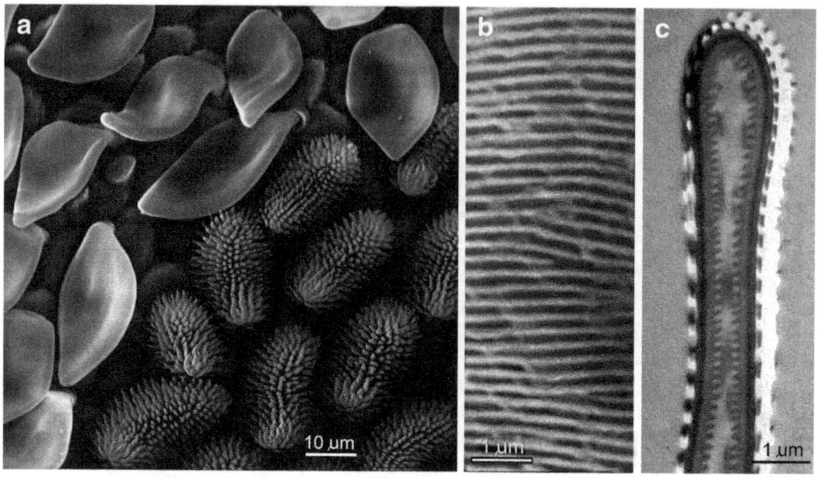

Fig. 24.10 Fine structure of blue scales on the opisthosoma of *M. splendens* (Salticidae). (**a**) Leaflike (blue) scales are seen on the upper left and (red) hairs on the lower right. (**b**) High magnification of a blue scale showing fine ridges on the surface. (**c**) Thin sections reveal a narrow scale wall with surface ridges and a cuticular meshwork inside (*M. volans*)

complex than the golden scales in *H. hallani*. We surmise that the iridescent blue color results from a combination of surface diffraction and interference effects within the layered scale wall.

24.4 Iridescence in the Integument

Many male spiders, again mostly among salticids, are known for having shiny, colorful chelicerae (Fig. 24.11). These interference colors are caused by a multilayered body cuticle rather than by hairs or scales (Ingram et al. 2009, 2011). Among jumping spiders the genus *Phidippus* is probably best known for its green-to-blue (and also red to violet) chelicerae; for most *Phidippus* species, even juveniles and females exhibit colorful chelicerae, sometimes in different colors than in the males (Foelix et al. 2010). Ingram et al. (2009, 2011) assumed that the cheliceral cuticle lamellae contained intermittent air spaces that would provide a multi-reflector system. We did not see evidence for any air spaces and believe that the variation of refractive index between adjacent cuticular layers would be sufficient to cause interference.

Many close-up pictures of jumping spiders show the large eyes (front row) in either a green or red color hue (Fig. 24.9b). This color effect is due to interference in the multilayered lens cuticle, as was nicely demonstrated by Homann (1985). When eye preparations were soaked in lactic acid, the cuticle layers would slightly swell,

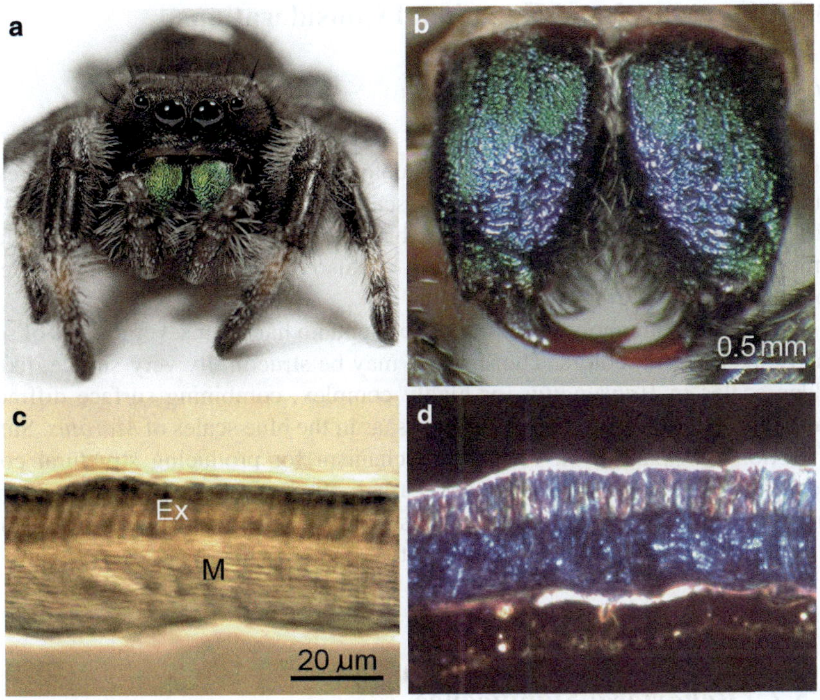

Fig. 24.11 Colorful chelicerae in jumping spiders of the genus *Phidippus*. (**a**) A female *Phidippus regius* with dark-green chelicerae (Photo: Bastian Rast). (**b**) Even juveniles exhibit brightly colored chelicerae (exuvium). Note the change of color from green to blue to purple. (**c**) Semi-thin section of the cheliceral cuticle in *Phidippus audax*, unstained. Note the brown exocuticle (Ex) and the lighter, multilayered mesocuticle (M). (**d**) Same cuticle as in (**c**) but stained with methylene blue, dark-field illumination. Note the different optical properties of exo- and endocuticle

and a color shift from green to red could be induced. Similarly, when cuticular layers are compressed, a color change towards shorter wavelengths will result (Steinbrecht 1985). The observation that iridescence in a live spider (or in an exuvium) may change over time, or even disappear, can be explained by a slight change in the dimensions of the cuticle layers (or air spaces, if present).

Finally, another type of structural coloration should be mentioned, namely, the white or silvery reflection seen in several spider families (e.g., Araneidae, Tetragnathidae, Theridiidae). This is caused by guanine deposits in the peripheral cells of the midgut (guanocytes). If the stored crystals are small and cuboid, a matt white will result, but if the guanine forms very thin platelets (0.1 µm), a shiny silver reflection is produced (Oxford 1998). It seems that the matt guanine represents the ancestral type, whereas the silver form is phylogenetically derived (Oxford and Gillespie 1998) (see also Wunderlin and Kropf 2013).

24.5 Structural and Functional Considerations

24.5.1 Structural Variation

Although we have looked at only a few colorful theraphosids, it is our impression that they have only special hairs—but no scales—that cause structural colors. These hairs may vary morphologically, but the basic design of having a lamellar structure in the nanometer range seems consistent. We also did not find any body cuticle that would exhibit structural colors.

In contrast, in araneomorph spiders we found no hairs but several types of scales that yield structural colors. Those scales may be structurally very simple, like the golden scales in *Habronattus*, or highly complex, combining surface diffraction gratings with internal multilayer reflectors, as in the blue scales of *Maratus*. Surface diffraction is apparently a very old mechanism for producing structural colors, already known from the Cambrian (Parker 1998). There may be many more variations on structural colors in spiders—only a few species have been studied so far and the picture is far from complete.

24.5.2 Behavioral Significance

Until now we have not yet posed the decisive question: what is the biological significance of structural colors in spiders? In the case of diurnal spiders with good color vision (as in salticids), the obvious answer is that these colors ought to play a role in communication, for example, during male–male contests or courtship (Fig. 24.9a; Hill and Otto 2011). There is indeed good evidence that body coloration is important during the courtship display of some spiders, regardless whether the colors are due to pigments or to interference (Lim et al. 2007; Taylor and McGraw 2007). Both immature and adult *Phidippus* flash their iridescent chelicerae with rapid movements of their pedipalps, while facing either sighted prey or a human observer (Fig. 24.12). During male–male contests or courtship, these chelicerae are displayed, but not flashed. The limited directional impact of this intense flickering color may insure that the signal is sent only to the intended recipient and not to undetected predators nearby.

It is much more difficult to interpret structural colors in the nocturnal theraphosids. Since their communication does not rely on visual signals, it is hard to say what kind of advantage any coloration might afford. Under twilight conditions it is conceivable that they may have an aposematic function, for instance, when *Poecilotheria* assumes its typical defense posture, raising its front legs and exposing the blue and yellow undersides. However, it may also be that bright colors in theraphosids are of no particular advantage (e.g., neutral)—perhaps comparable to colorful organisms living in the abyss of the oceans.

Fig. 24.12 Three sequential images of a subadult female *Phidippus pulcherrimus* (Salticidae), flashing its iridescent blue-green chelicerae at the photographer at a rate of 2–4/s. This palp flicking also occurs when stalking certain kinds of prey

24.6 Conclusions

Structural colors are described and analyzed in theraphosid and salticid spiders. Some theraphosids are brightly blue: this is caused by special hairs with a lamellated wall that causes an interference of the incoming light. The incident white light is then reflected as a deep blue. In some cases, the hairs are bright yellow: there the hair wall exhibits a fine cuticular meshwork of slightly different dimensions than in the blue hairs but also results in interference, and the reflected light appears yellow. In salticids, iridescent colors are produced by flattened hairs (scales). Golden scales have a rather simple structure of two thin cuticular layers on the outside and a narrow air space in between. Blue iridescent scales are more complex, with multilayered scale walls and fine ridges on the surface that act as a diffraction grating. The body cuticle (e.g., chelicerae and eye lenses) may also be brightly colored due to light interference on many thin cuticular layers. The biological significance of structural colors in spiders is well understood in the diurnal salticids where optical signals are exchanged in courtship, usually in bright daylight. In contrast, theraphosids are mostly active at night and visual communication hardly plays any role. Their coloration may be of advantage at dawn during their defensive behavior, when the brightly colored blue and yellow legs are raised toward an aggressor.

Acknowledgements We are grateful to several colleagues who have helped us with this study: Bastian Rast and Benno Wullschleger, for supplying us with theraphosid material, and Judith Kastenmeier, Jürgen Otto, and Lisa Taylor for sending us salticids. Rolf Thieleczek kindly calculated the reflectance spectra of iridescent spider hairs and scales for us; Jerome Rovner and Benno Wullschleger critically read our manuscript. The Neue Kantonsschule Aarau generously let us use their electron microscope facilities. Digital photographs with the transmission electron microscope were kindly taken by Karin Boucke and Sherry Vinsant.

References

Doucet SM, Meadows MG (2009) Iridescence: a functional perspective. J R Soc Interface 6: 115–132
Durrer H (1977) Schillerfarben der Vogelfeder als Evolutionsproblem. Denkschr Schweiz Naturforsch Ges 91:1–127
Foelix R, Erb B, Wullschleger B (2009) Worauf beruht die Blaufärbung gewisser Vogelspinnenarten? Arachne 14(3):4–12
Foelix R, Erb B, Kastenmeier J (2010) Morphologische Besonderheiten der Springspinne *Phidippus regius* C. L. Koch, 1846. Arachne 15(6):4–11
Ghiradella H (1998) Hairs, bristles and scales. In: Locke M (ed) Microscopic anatomy of invertebrates. Insecta, vol 11A. Wiley, New York, pp 257–287
Ghiradella H (2010) Insect cuticular surface modifications: scales and other structural formations. Adv Insect Physiol 38:135–180
Glover BJ, Whitney HM (2010) Structural colour and iridescence in plants: the poorly studied relations of pigment colour. Ann Bot 105:505–511
Hill DE (1979) The scales of salticid spiders. Zool J Linn Soc 65:193–218
Hill DE, Otto JC (2011) Visual display of male *Maratus pavonis* (Dunn 1947) and *Maratus splendens* (Rainbow 1896) (Araneae: Salticidae: Euophryinae). Peckhamia 89.1:1–41
Hinton HE (1968) An electron microscope study of the diffraction gratings of some carabid beetles. Insect Physiol 15:959–962
Holl A (1987) Coloration and chromes. In: Nentwig W (ed) Ecophysiology of spiders. Springer, Berlin
Homann H (1985) Interferenzfarben an Spinnenaugen. Biologie in unserer Zeit 15:90–91
Huxley AF (1968) A theoretical treatment of the reflexion of light by multilayer structures. J Exp Biol 48:227–245
Ingram AL, Ball AD, Parker AR, Deparis O, Boulenguez J, Berthier S (2009) Characterization of the green iridescence on the chelicerae of the tube web spider, *Segestria florentina* (Rossi 1790) (Araneae, Segestriidae). J Arachnol 37:68–71
Ingram AL, Deparis O, Boulenguez J, Kennaway G, Berthier S, Parker AR (2011) Structural origin of the green iridescence on the chelicerae of the red-backed jumping spider, *Phidippus johnsoni* (Salticidae: Araneae). Arthropod Struct Dev 40:21–25, Epub 2010 Oct 29
Kinoshita S (2008) Structural colors in the realm of nature. World Scientific Publishing, London
Kinoshita S, Yoshioka S, Miyazaki J (2008) Physics of structural colors. Rep Prog Phys 71: 076401. doi:10.1088/0034-4885/71/7/076401
Land MF (1972) The physics and biology of animal reflectors. Prog Biophys Mol Biol 24:75–106
Land MF, Horwood J, Lim ML, Li D (2007) Optics of the ultraviolet reflecting scales of a jumping spider. Proc R Soc London B 74:1583–1589
Lim MLM, Land MF, Li D (2007) Sex-specific UV and fluorescence function as courtship signals in jumping spiders. Science 315:481
Meadows MG, Butler MW, Morehouse NI, Taylor LA, Toomey MB, McGraw KJ, Rutkowski RL (2009) Iridescence: views from many angles. J R Soc Interface 6:107–113
Newton I (1730) Opticks: or, a treatise of the reflections, refractions, inflections and colours of light. Book II, 4th edn. W. Innys, London, pp 226/227
Oxford GS (1998) Guanine as a colorant in spiders: development, genetics, phylogenetics and Ecology. Proc 17th Eur Coll Arachnol Edinburgh 1997:121–131
Oxford GS, Gillespie RG (1998) Evolution and ecology of spider coloration. Annu Rev Entomol 43:619–643
Parker AR (1998) Colour in Burgess Shale animals and the effect of light on evolution in the Cambrian. Proc R Soc B 265:967–972
Parker AR, Hegedus Z (2003) Diffractive optics in spiders. J Opt A Pure Appl Opt 5:111–116
Parker AR, Martini N (2006) Structural colour in animals – simple to complex optics. Opt Laser Technol 38:135–180

Prum RO (2006) The anatomy, physics and evolution of avian structural color. In: Hill GE, McGraw KJ (eds) Bird coloration, vol. 1: mechanisms and measurements. Harvard University Press, Cambridge, MA, pp 295–353

Prum RO, Quinn T, Jones RH (2006) Anatomically diverse butterfly scales all produce structural colors by coherent scattering. J Exp Biol 209:748–765

Roth VD (1993) Spider genera of North America with keys to families and genera, and a guide to literature. Am Arachnol Soc, Gainesville, Florida

Steinbrecht A (1985) Fine structure and development of the silver and golden cuticle in butterfly pupae. Tissue Cell 17:745–762

Taylor LA, McGraw KJ (2007) Animal coloration: sexy spider scales. Curr Biol 17:592–593

Townsend VR, Felgenhauer BE (1998) Cuticular scales of spiders. Invertebr Biol 117:318–330

Vukusic P (2004) Natural photonics. Phys World 17:35–39

Wunderlin J, Kropf C (2013) Rapid colour change in spiders. In: Nentwig W (ed) Spider ecophysiology. Springer, Heidelberg (this volume)

Chapter 25
UV and Camouflage in Crab Spiders (Thomisidae)

Marie E. Herberstein and Felipe M. Gawryszewski

25.1 Colour and Colour Vision Models

The perceived colour of an object is the result of difference in photoreceptor responses to the light that reaches the animal's eyes and further computation in the brain (Endler 1990; Kelber et al. 2003). The number and type of photoreceptors vary between and sometimes within animal species (Osorio and Vorobyev 2008). Humans, for instance, have three photoreceptors, with sensitivity peaks in the regions we see as blue, green and red (Osorio and Vorobyev 2008). Therefore, human vision is outside the ultraviolet (UV) range of the light spectrum. The human blue photoreceptor actually reaches the UV, but the crystalline lens in human eyes filters wavelengths smaller than 400 nm. Most other animal taxa, however, do have a photoreceptor with a sensitivity peak in the UV region (Osorio and Vorobyev 2008). Bees and most insects, for instance, have three photoreceptors, one in the UV range, other in the blue and a third in the green region of the light spectrum (Briscoe and Chittka 2001).

Because of the differences in visual systems, organisms may perceive the same colour patch differently. What seems conspicuous to one organism could seem highly cryptic to another. Therefore, it is extremely useful to estimate how different animals perceive an object (Endler 1990). In response to that, several visual models that try to calculate how colour patches are perceived by animals have been developed (Endler 1990; Backhaus 1991; Chittka 1992; Vorobyev and Osorio 1998; Endler and Mielke 2005). Despite being extremely useful, all of them are only estimations because many aspects of how light is processed by these visual systems are simply unknown. Generally speaking, these models take into consideration the photoreceptor sensitivity curves, the irradiance of the environment (e.g. sunlight at midday), the transmittance of the medium (e.g. air, water), the reflectance of the environmental

M.E. Herberstein (✉) • F.M. Gawryszewski
Department of Biological Sciences, Macquarie University, North Ryde 2109, NSW, Australia
e-mail: marie.herberstein@mq.edu.au

W. Nentwig (ed.), *Spider Ecophysiology*,
DOI 10.1007/978-3-642-33989-9_25, © Springer-Verlag Berlin Heidelberg 2013

background and the reflectance of the colour patch (Endler 1990; Chittka 1992; Endler and Mielke 2005). Using this information, models may, for instance, calculate the degree of colour contrast between an organism colour and its background. The bee colour hexagon (Chittka 1992) is the most common model used in the crab spider literature. Nonetheless, other models could also be used, and it may actually be valuable to use more than one model.

These models have in common that all photoreceptors present in the animal's eyes are equally important in generating the perceived colour. For bees, for instance, there is no evidence that the UV photoreceptor is more important than the blue or green photoreceptors (Kevan et al. 2001). Considering a crab spider sitting on a white flower, the degree of conspicuousness—colour contrast—from a bee's perspective mainly depends on variation in UV. Nevertheless, if both spider and flower cause the same UV and blue photoreceptor excitation levels but vary in green photoreceptor excitation, the spider would also be perceived as a different colour from the flower. Therefore, while in this chapter we focus on the UV reflectance of crab spiders, we also consider other regions of the spectrum as contributing to the final spider colour.

25.2 The Biology of Crab Spiders

Crab spiders are members of the Thomisidae family, which currently consists of 177 genera and 2,152 species (Platnick 2011). The medium-sized spiders are relatively easily recognised by their elongated first two pairs of legs, which they extend laterally. Their tendency to walk sideways gives them the appearance of a crab and hence the name (Foelix 2011). A recent phylogeny has excluded the Philodromidae but included the Aphantochilidae as a thomisid subfamily (Benjamin et al. 2008); however, taxonomy and systematics of crab spiders is a work in progress with many species yet to be identified.

Crab spiders are sit-and-wait predators that do not employ webs for prey capture. Silk is utilised for securing prey after capture, construction of cocoons, dispersal and as draglines (Foelix 2011). Crab spiders wait motionless on a substrate for prey to arrive, which they then slowly approach, quickly grab with their front legs and envenomate with their chelicerae (Morse 2007). Crab spiders can be found on various substrates including bark, leaves and flowers. The foraging behaviour of flower-dwelling crab spiders has been studied intensively, and in some species, such as *Misumena vatia*, we have an excellent account of their foraging ecology (Morse 2007). Crab spiders attack any prey within their vicinity and on flowers that often means pollinating insects such as honeybees and bumblebees, or syrphid flies (Morse 1981, 2010).

Pollinators do respond to the predatory risk imposed by crab spiders. For example, honeybees are less likely to land on a flower where a bee has been recently killed or attacked (Dukas 2001). Furthermore, honeybees apparently communicate this information to the hive via a modified waggle dance (Abbott and Dukas 2009). Other pollinators, such as bumblebees, are also less likely to visit patches where crab spiders are present (Dukas and Morse 2003).

Fig. 25.1 *Top left*, *Zygometis lactea* awaits its prey (photo by Mayra Amboni). *Top right*, *Sidymella longipes*. *Bottom*, yellow and white morphs of *Thomisus spectabilis* (all Thomisidae)

Flower-dwelling thomisids are well recognised for their colourful appearance and their ability to reversibly change colour over several days (Gabritschevsky 1927; Oxford and Gillespie 1998; Schmalhofer 2000; Fig. 25.1). The best-studied models of colour change are *M. vatia* and *Thomisus onustus*. Colour change is frequently interpreted in terms of background matching and crypsis: by matching their body colour to the flower colour (Théry 2007), spiders are less likely to be detected by predators or prey (Chittka 2001; Théry and Casas 2002, 2009). However, crypsis in these thomisids may be far from perfect (Defrize et al. 2010) and pollinators avoid flowers harbouring crab spiders irrespective of the degree of colour matching (Brechbühl et al. 2010). Crab spider colour, colour change and prey response is clearly a complex phenomenon that we discuss in greater detail in the following sections.

25.3 Distribution of UV Reflectance Among Crab Spiders

The availability and affordability of spectrophotometers that measure light in the UV (below 400 nm) has led to the discovery that UV reflection is present in many animals, including birds (Mullen and Pohland 2008), fish (Losey et al. 1999), lizards (Whiting et al. 2006), insects (Wilts et al. 2011) and spiders (Lim et al. 2007).

Fig. 25.2 *Thomisus spectabilis* (Thomisidae) on a white daisy, (**a**) under light visible to humans and (**b**) the same animal photographed using only UV light (below 400 nm). Figure from Heiling and Herberstein (2004); reprinted with kind permission of Highwire Press

UV reflection in crab spiders has only recently become a focus of research, with the discovery of an Australian crab spider (*Thomisus spectabilis*) that reflects substantial amount of UV light (as well as blue and green light) (Heiling et al. 2003). Reflecting UV light per se is nothing extraordinary, but this spider perches on large open flowers such as daisies, which do not reflect UV light (Heiling et al. 2005b; Fig. 25.2). As a consequence, the colour contrast created between the spider and the flower is substantial. The colour contrast is large enough for a honeybee (and probably any insect with a UV receptor) to detect the crab spider on the flower, and behavioural experiments have confirmed that several different species of pollinators can detect the spider on the flower (Heiling and Herberstein 2004; Llandres et al. 2011).

This discovery has two broad implications. First, the existence of strongly contrasting crab spiders challenges the generalisation that crab spider colour is selected for maximal crypsis, reducing the contrast between the spider and flower. Second, because flowers generally reflect little UV light (but for UV markers) (Kevan et al. 1996), a UV-reflecting crab spider is likely to be contrasting against most flowers. However, is UV reflection a common trait amongst crab spiders, or an isolated curiosity?

The first mention of UV reflectivity in crab spiders in the literature comes from Sato (1987). While this account has been cited in a major review of spider colour (Oxford and Gillespie 1998), no further cases of UV reflection have been reported until almost 10 years ago in the Australian *T. spectabilis*. At about the same time, a colour analysis of *T. onustus* confirmed the lack of UV reflection in this European crab spider (Théry and Casas 2002). A recent comparative study sampled several Australian and European crab spiders to investigate if UV reflection is a phenomenon exclusive to Australian species (Herberstein et al. 2009). Of the four additional Australian species [*Diaea evanida, Runcinia* sp., *Sidymella* sp. and *Diaea lactea* (now transferred to *Zygometis*)], all reflected light below 400 nm. The three European species (in addition to *T. onustus*: *Synema globosum, M. vatia* and *Xysticus* sp.) however showed no evidence of UV reflection even though over

100 individuals in total were measured (Herberstein et al. 2009). Based on these rather preliminary samples, it seems that the tendency to reflect UV light is predominantly found in Australian species but not in European species. This raises several exciting evolutionary questions, such as whether UV reflection is ancestral or derived, whether it has evolved once or several times independently and whether it is unique to the Australian crab spider fauna.

It is likely that UV reflection is not unique to the Australian fauna but is broadly distributed amongst crab spiders: at least one Malaysian crab spider species (Gawryszewski 2011) revealed considerable UV reflection. Furthermore, a recent study of an Indian *Thomisus* sp. also documented that this species reflects UV light thereby creating a colour contrast against its flower background that is detectable by hymenopteran prey (Bhaskara et al. 2009).

While there is increasing evidence that UV reflection is not unique to Australian crab spiders, there may be a geographic signature in the distribution of this trait. Depending on the function and mechanism of UV reflectance, there may be selection on this trait in certain regions due to local environmental or ecological conditions. To fully resolve this question, we require detailed morphological and colourimetric studies of a wider selection of species and regions.

25.4 Mechanisms of Colour Production and Colour Change

The proximate cause of animal colouration has been traditionally divided into pigmentary and structural colours. Pigmentary colours are being produced by molecules that absorb part of the electromagnetic spectrum and structural colours produced by structures that interfere with electromagnetic spectrum at the wavelength scale (Kinoshita 2008; see also Foelix et al. 2013). However, this division is often misleading because the colour we observe is frequently the result of a combination of both pigmentary and structural components (Grether et al. 2004). Furthermore, not all components that contribute to the final colour can be easily categorised into pigmentary or structural.

For the crab spider opisthosoma, three components interact to form the perceived spider colour: cuticle, hypodermis and guanocytes (Fig. 25.3). The cuticle absorbs part of the light spectrum, especially wavelengths below 350 nm (Gawryszewski 2011). Therefore, the cuticle, although often looks transparent to our eyes, functions as a pigment. It absorbs part of the UV light, which affects the final crab spider colouration.

Below the cuticle lies a single layer of tightly connected cells, the hypodermis (Fig. 25.3). The hypodermis has granules and crystals that, similarly to the cuticle, absorb the light spectrum (Insausti and Casas 2008, 2009; Riou and Christidès 2010; Gawryszewski 2011). The granules contain ommochromes and its precursors, such as kynurenine and 3-OH-kynurenine (Seligy 1972; Riou and Christidès 2010). The crystals found in the hypodermis are of unknown nature but are likely to be biochemically related to ommochromes (Gawryszewski 2011).

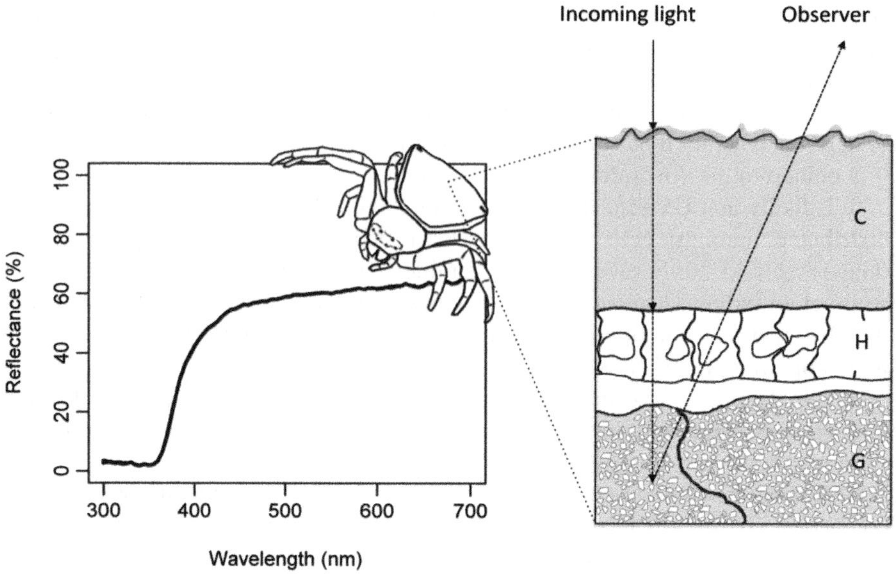

Fig. 25.3 On the *left*: a typical spectrum of a UV-reflective *Thomisus spectabilis* (Thomisidae). On the *right*: a cross section of the spider's opisthosoma depicting the three components contributing to the observed crab spider colouration: cuticle (C), hypodermis (H) and guanocytes (G). Incoming light is absorbed by the cuticle and by pigments or crystals present in the hypodermis. The light that reaches the guanocytes is then reflected back to the observer. UV-reflective individuals have very few pigments and crystals in the hypodermis, making it transparent to the UV light that is reflected by guanine crystals

The colour change in crab spiders is the consequence of changes in the components within these granules and crystals (Insausti and Casas 2008, 2009; Gawryszewski 2011). Yellow spiders, for instance, have a high concentration of 3-OH-kynurenine, whereas white, non-UV-reflective, spiders have a high concentration of kynurenine (Riou and Christidès 2010).

The innermost layer is composed of guanocytes (Fig. 25.3). Guanocytes are storage cells that contain guanine crystals. The guanine crystals found in crab spiders are irregular and randomly arranged (Oxford 1998, Gawryszewski 2011). To the human observer, they have a matt white appearance. This happens because they are highly reflective and have a flat reflectance spectrum from 400 to 700 nm. Nonetheless, they are also highly UV reflective from 330 to 400 nm. From 300 to 330 nm, guanine crystals absorb almost 100 % of the light. UV and non-UV crab spider individuals analysed so far posses the same UV-reflective guanine crystals (Gawryszewski 2011).

The difference between UV- and non-UV-reflective individuals is found in the hypodermis. UV-reflective individuals have a hypodermis without or with very few granules and crystals. This makes the hypodermis transparent to the UV light that passed through the cuticle. The UV light therefore reaches the guanocytes and is reflected back by the guanine crystals (Gawryszewski 2011). Non-UV-reflective individuals (e.g. white non-UV and yellow phenotypes) have pigments or crystals in the hypodermis that filter the UV light. In this case no UV light reaches the guanine crystals to be reflected back. Thus, in order to become UV reflective, a non-UV-reflective individual needs simply to catabolise or remove pigments and crystals from the hypodermis (Gawryszewski 2011).

25.5 The Function of UV Reflectance in Crab Spiders

When examining colours, visual functions are often proposed without considering possible nonvisual alternatives. The research on colour in crab spiders has been extremely fruitful and exciting over the last few decades, but the function of colour is mostly considered in terms of whether or not it can be detected by prey or predators. Similarly, the function of UV reflectance has been tested exclusively in the context of visual signals to either prey or predators. Few alternative (but perhaps not mutually exclusive) functional hypotheses have been proposed, the most prominent of which suggests that the colour-creating mechanisms (pigments) may protect underlying cells from photo-damage (Théry and Casas 2009).

The effect of a UV-reflective spider body against a mostly non-UV-reflecting background is a colour contrast that should be detectable by a receiver (a predator or a prey) that is sensitive to UV light, such as most insects and birds. If the spider is then visible, prey or predator may react in one of three possibilities: avoid the spider, be attracted to the spider or be indifferent to the spider. These behavioural predictions were tested comprehensively in Australian UV-reflective crab spiders and more recently in an Indian species.

When offered two flowers, one with a UV-reflective spiders and one without, European honeybees (*Apis mellifera*) preferentially landed on the flower containing the spider, both in *T. spectabilis* (Heiling et al. 2003; Llandres and Rodriguez-Girones 2011) and *D. evanida* (Herberstein et al. 2009). When the amount of UV light reflected was artificially eliminated in *T. spectabilis* by using sunscreen, European honeybees avoided flowers with spiders (Heiling et al. 2005a). European honeybees have only recently (last 200 years) been introduced into Australia and have thus not had an opportunity to coevolve with these crab spiders. Similar experiments on Australian native bees (*Trigona carbonaria* and *Austroplebeia australis*) showed that native bees are still attracted to UV-reflective crab spiders but only approach the flowers. They are less likely to land on flowers occupied by crab spiders, suggesting that they have evolved a mechanism to recognise and avoid the spiders (Heiling and Herberstein 2004; Llandres et al. 2011). Similar experiments on an Indian thomisid (*Thomisus* sp.) showed that *Apis cerana* approached spiders highly contrasting in the UV more frequently than less contrasting spiders (Bhaskara et al. 2009).

While the accumulating evidence that UV reflection is attractive to pollinators is mounting, we still have very little idea of how it affects predator behaviour. What we can assume is that if spiders create stronger contrast against a background, they will also be under stronger predation pressure from predators that can detect that contrast. To date, no experiments that test such an idea have been published, but this is clearly an important next step in the investigation of UV reflection in crab spiders.

The structures responsible for UV reflection (guanine crystals) may have evolved primarily to prevent photo-damage to the underlying cells: harmful UV light is reflected before it can reach vital cells. The surrounding pigments, ommochromes, are known to absorb UV light and function as protection in insect eyes (Théry and Casas 2009), but perhaps they are more costly to produce than guanine crystals that are a by-product of digestion. Thus, especially in geographic regions where the incident of UV irradiation is high, these protective mechanisms are selected for. Nonetheless, the spider cuticle is not completely transparent to UV light and therefore already offers the first barrier against photo-damage (Gawryszewski 2011). The effect on pollinator behaviour may subsequently be co-incidental, but together both effects (photo-protection and increased prey capture) could maintain this trait in these geographic regions. Clearly, photo-protection as a function is very possible but still requires more work to confirm this idea.

25.6 Evolution of UV Reflectance

There are two main questions regarding the evolution of UV reflectance among crab spiders: (1) when and how many times has UV reflectance evolved and (2) is UV reflectance an adaptation of flower-dwelling species to capture pollinating insects? In order to answer the first question we have to consider the mechanism resulting in UV-reflective crab spiders. As explained above, the proximate cause of UV reflectance suggests that its evolution does not require the appearance of any new structure but only the removal of UV-filtering pigments or crystals in the hypodermis (Gawryszewski 2011). Interestingly, this implies that the production of UV reflectance is less costly than the lack of UV reflectance. Despite that, at least for Australian crab spiders ($n = 28$ taxa), the presence of UV reflectance (measured as 10 % or more maximum reflectance between 300 and 400 nm) was less common both within and between species (9 out of 28 taxa) than other regions of the light spectrum (blue 400–500 nm, 19/28 taxa; green 500–600 nm, 22/28 taxa; and red 600–700 nm, 25/28 taxa) (Gawryszewski 2011). A possible explanation why UV reflectance is relatively rare is that the substrata where spiders perch do not reflect UV light but the other regions of visible light. In this scenario, UV-reflective spiders would suffer the cost of greater conspicuousness to predators that can detect UV light. For flower-dwelling species, this could be the case, because flowers rarely reflect UV light (Chittka et al. 1994; Dyer 1996; Kevan et al. 1996). However, this may not be necessarily the case of other components of the environment.

A phylogenetic analysis of the same crab spiders suggested that UV reflectance evolved at least five times within this family. This analysis also found that the most likely ancestral state for most crab spider clades is the absence of UV reflectance.

In contrast, reflectance of other regions of the light spectrum was the most likely ancestral state for most crab spider clades (Gawryszewski 2011). However, it is often difficult to be sure that a species does or does not have the ability to reflect UV light because of limited sample sizes. Therefore, an increase of sample size for some of these species could change some of the results presented here. In addition, the data on the evolution of UV reflectance are almost exclusively based on Australian crab spiders and our conclusions could shift once spiders from other regions are included.

Is UV reflectance in flower-dwelling spiders used to attract pollinators? If this is the case, we would expect UV reflectance to be prevalent in flower-dwelling spiders but less so in species that perch on leaves or bark. An initial analysis of Australian species suggests that UV reflectance is more frequently found in flower-dwelling than non-flower-dwelling species (Gawryszewski 2011). Other regions of the light spectrum (e.g. blue and green), however, were more or less evenly distributed between flower- and non-flower-dwelling spiders species. Furthermore, the evolution of UV reflectance seems to be correlated with the evolution of the flower-dwelling habit (Gawryszewski 2011). Therefore, it is likely that UV reflectance in flower-dwelling species is an adaptation for capturing pollinating insects. Nevertheless, the ability to reflect UV light was also present in some non-flower-dwelling species where it may serve other function, such as improving crypsis against a UV-bright background. For example, some individuals of the bark-dwelling species *Stephanopis* cf. *scabra* reflect UV (12 % maximum reflectance from 300 to 400 nm), which is likely to improve colour matching against UV-reflective barks (Gawryszewski 2011).

25.7 Conclusions

Our understanding of the function, prevalence and evolution of UV reflection in crab spiders is only in its infancy with many exciting questions still unanswered. It is nevertheless already clear that crab spider colour is unexpectedly complex and dynamic. Invoking crypsis and camouflage for crab spiders untested is likely to underestimate the true complexity of this fascinating phenomenon.

References

Abbott KR, Dukas R (2009) Honeybees consider flower danger in their waggle dance. Anim Behav 78:633–635

Backhaus W (1991) Color opponent coding in the visual system of the honeybee. Vision Res 31: 1381–1397

Benjamin SP, Dimitrov D, Gillespie RG, Hormiga G (2008) Family ties: molecular phylogeny of crab spiders (Araneae: Thomisidae). Cladistics 24:708–722

Bhaskara RM, Brijesh CM, Ahmed S, Borges RM (2009) Perception of ultraviolet light by crab spiders and its role in selection of hunting sites. J Comp Physiol A 195:409–417

Brechbühl R, Casas J, Bacher S (2010) Ineffective crypsis in a crab spider: a prey community perspective. Proc R Soc B Biol Sci 277:739–746

Briscoe AD, Chittka L (2001) The evolution of color vision in insects. Annu Rev Entomol 46: 471–510

Chittka L (1992) The colour hexagon: a chromaticity diagram based on photoreceptor excitations as a generalized representation of colour opponency. J Comp Physiol A 170:533–543

Chittka L (2001) Camouflage of predatory crab spiders on flowers and the colour perception of bees (Aranida: Thomisidae/Hymenoptera: Apidae). Entomol Gen 25:181–187

Chittka L, Shmida A, Troje N, Menzel R (1994) Ultraviolet as a component of flower reflections, and the colour perception of Hymenoptera. Vision Res 34:1489–1508

Defrize J, Thery M, Casas J (2010) Background colour matching by a crab spider in the field: a community sensory ecology perspective. J Exp Biol 213:1425–1435

Defrize J, Lazzari CR, Warrant EJ, Casas J (2011) Spectral sensitivity of a colour changing spider. J Insect Physiol 57:508–513

Dukas R (2001) Effects of perceived danger on flower choice by bees. Ecol Lett 4:327–333

Dukas R, Morse DH (2003) Crab spiders affect flower visitation by bees. Oikos 101:157–163

Dyer AG (1996) Reflection of near-ultraviolet radiation from flowers of Australian native plants. Aust J Bot 44:473–488

Endler J (1990) On the measurement and classification of colour in studies of animal colour patterns. Biol J Linn Soc 41:315–352

Endler J, Mielke P (2005) Comparing entire colour patterns as birds see them. Biol J Linn Soc 86: 405–431

Foelix RF (2011) Biology of spiders. Oxford University Press, Oxford

Foelix RF, Erb B, Hill DE (2013) Structural colors in spiders. In: Nentwig W (ed) Spider ecophysiology. Springer, Heidelberg (this volume)

Gabritschevsky E (1927) Experiments on color changes and regeneration in the crab spider, *Misumena vatia*. J Exp Zool 42:117–267

Gawryszewski FM (2011) Colouration in crab spiders (Thomisidae): mechanism, function and evolution. Ph.D. Thesis, Macquarie University, Australia

Grether G, Kolluru G, Nersissian K (2004) Individual colour patches as multicomponent signals. Biol Rev 79:583–610

Heiling AM, Herberstein ME (2004) Predator–prey coevolution: Australian native bees avoid their spider predators. Proc Biol Sci 271:S196–S198

Heiling AM, Herberstein ME, Chittka L (2003) Crab-spiders manipulate flower signals. Nature 421:334

Heiling AM, Cheng K, Chittka L, Goeth A, Herberstein ME (2005a) The role of UV in crab spider signals: effects on perception by prey and predators. J Exp Biol 208:3925–3931

Heiling AM, Chittka L, Cheng K, Herberstein ME (2005b) Colouration in crab spiders: substrate choice and prey attraction. J Exp Biol 208:1785–1792

Herberstein ME, Heiling AM, Cheng K (2009) Evidence for UV-based sensory exploitation in Australian but not European crab spiders. Evol Ecol 23:621–634

Insausti TC, Casas J (2008) The functional morphology of color changing in a spider: development of ommochrome pigment granules. J Exp Biol 211:780–789

Insausti TC, Casas J (2009) Turnover of pigment granules: cyclic catabolism and anabolism of ommochromes within epidermal cells. Tissue Cell 41:421–429

Kelber A, Vorobyev M, Osorio D (2003) Animal colour vision—behavioural tests and physiological concepts. Biol Rev 78:81–118

Kevan P, Giurfa M, Chittka L (1996) Why are there so many and so few white flowers? Trends Plant Sci 1:252

Kevan P, Chittka L, Dyer AG (2001) Limits to the salience of ultraviolet: lessons from colour vision in bees and birds. J Exp Biol 204:2571–2580

Kinoshita S (2008) Structural colors in the realm of nature. World Scientific, Singapore
Lim MLM, Land MF, Li D (2007) Sex-specific UV and fluorescence signals in jumping spiders. Science 315:481
Llandres AL, Rodriguez-Girones MA (2011) Spider movement, UV reflectance and size, but not spider crypsis, affect the response of honeybees to Australian crab spiders. PLoS One 6:e17136
Llandres AL, Gawryszewski FM, Heiling AM, Herberstein ME (2011) The effect of colour variation in predators on the behaviour of pollinators: Australian crab spiders and native bees. Ecol Entomol 36:72–81
Losey GS, Cronin TW, Goldsmith TH, Hyde D, Marshall NJ, McFarland WN (1999) The UV visual world of fishes: a review. J Fish Biol 54:921–943
Morse DH (1981) Prey capture by the crab spider *Misumena vatia* (Clerck) (Thomisidae) on three common native flowers. Am Midl Nat 105:358–367
Morse DH (2007) Predator upon a flower. Harvard University Press, Cambridge
Morse DH (2010) Effect of prey size on growth of newly emerged crab spiderlings *Misumena vatia*. J Arachnol 38:309–312
Mullen P, Pohland G (2008) Studies on UV reflection in feathers of some 1000 bird species: are UV peaks in feathers correlated with violet-sensitive and ultraviolet sensitive cones? Ibis 150:59–68
Osorio D, Vorobyev M (2008) A review of the evolution of animal colour vision and visual communication signals. Vision Res 48:2042–2051
Oxford G (1998) Guanine as a colorant in spiders: development, genetics, phylogenetics and ecology. In: Proceedings of the 17th European Colloquium of Arachnology, Edinburgh, 1997
Oxford GS, Gillespie RG (1998) Evolution and ecology of spider coloration. Annu Rev Entomol 43:619–643
Platnick NI (2011) The world spider catalog, version 12.0. American Museum of Natural History. Online at http://research.amnh.org/iz/spiders/catalog. doi:10.5531/db.iz.0001
Riou M, Christidès JP (2010) Cryptic color change in a crab spider (*Misumena vatia*): identification and quantification of precursors and ommochrome pigments by HPLC. J Chem Ecol 36: 412–423
Sato Y (1987) A spider *Thomisus labefactus*, changing colour by reflected UV rays. Insectarium 24:118–226
Schmalhofer VR (2000) Diet-induced and morphological color changes in juvenile crab spiders (Araneae, Thomisidae). J Arachnol 28:56–60
Seligy V (1972) Ommochrome pigments of spiders. Comp Biochem Physiol A 42:699–709
Théry M (2007) Colours of background reflected light and of the prey's eye affect adaptive coloration in female crab spiders. Anim Behav 73:797–804
Théry M, Casas J (2002) Predator and prey views of spider camouflage. Nature 415:133
Théry M, Casas J (2009) The multiple disguises of spiders: web colour and decorations, body colour and movement. Philos Trans R Soc B 364:471–480
Vorobyev M, Osorio D (1998) Receptor noise as a determinant of colour thresholds. Proc R Soc B 265:351–358
Whiting MJ, Stuart-Fox DM, O'Connor D, Firth D, Bennet NC, Blomberg SP (2006) Ultraviolet signals ultra-aggression in a lizard. Anim Behav 72:353–363
Wilts BD, Pirih P, Stavenga DG (2011) Spectral reflectance properties of iridescent pierid butterfly wings. J Comp Physiol A 197:693–702

Chapter 26
Rapid Colour Change in Spiders

Judith Wunderlin and Christian Kropf

26.1 Introduction

Among the various ways of camouflage in the animal kingdom, rapid colour change belongs to the most impressive ones. In contrast to slow morphological colour change, the so-called physiological colour change occurs relatively fast (Holl 1987; Stuart-Fox and Moussalli 2009). It may take some hours but can also be finished within parts of a second.

Physiological colour change is generally enabled by movements of pigment granules within cells or by modifications of the morphology of the pigment-containing cells (Holl 1987). When pigment granules move within chromatophores, they either disperse or concentrate (Stuart-Fox and Moussalli 2009). Relatively well-known examples of animals being capable of physiological colour change are chameleons and cephalopods (Stevens and Merilaita 2009). Among the circa 1.1 million described (and presumably up to 3.7 million really existing arthropod species; Hamilton et al. 2010), there are only few examples of physiological colour change. Some crustaceans (Auerswald et al. 2008) and insects (Key and Day 1954; Hinton and Jarman 1972) are able to change their colour rather slowly. Rapid colour change (i.e. a physiological colour change within parts of a second) within arthropods is only known from spiders (Araneae).

These spiders show mostly a white or at least light opisthosomal pattern, made up by specialized midgut cells beneath the hypodermis, the so-called guanocytes (Millot 1926; Seitz 1972). These cells store the excretory product guanine that appears white and acts as a colourant (Oxford 1998). In all reported cases of rapid

J. Wunderlin • C. Kropf (✉)
Institute of Ecology and Evolution, University of Bern, Bern, Switzerland

Natural History Museum, Bern, Switzerland
e-mail: judith.wunderlin@gmail.com; christian.kropf@iee.unibe.ch

W. Nentwig (ed.), *Spider Ecophysiology*,
DOI 10.1007/978-3-642-33989-9_26,

colour change in spiders, the white guanine markings on the opisthosomal surface diminish in size or disappear more or less completely. As a consequence, the general colouration of the spider's opisthosoma changes to the darker or greyish brown of the digestive mass. Recovery of the original colour pattern requires a few or several minutes (e.g. Blanke 1975; Holl 1987). Rapid colour change in spiders is always related to disturbances, especially when the spider has to drop from the web (Oxford and Gillespie 1998).

Up to now, it is not clear how this impressive phenomenon works. In a study on the araneid *Cyrtophora cicatrosa*, the importance of guanocytes for rapid colour change of spiders was recognized (Blanke 1975). This author proposed that movement of guanine within the guanocytes rather than the contraction of the cells is responsible for the colour change. On the other hand, Edmunds and Edmunds (1986) assumed that there should be contractile elements in the guanocytes in *Argiope flavipalpis*, also an araneid. However, their efforts to find such elements failed. In this context, Oxford and Gillespie (1998) concluded that guanine contraction or retraction seems to be the basis of all physiological colour change in spiders.

Bristowe (1941) was the first who documented the rapid colour change of the linyphiid spider *Floronia bucculenta* that is native to temperate regions of Europe and Russia. An ongoing study on this species gives new insights into the underlying mechanism of rapid colour change.

26.2 Spiders Capable of Rapid Colour Change

Up to now 21 spider species (most of them web-building) from five different families are known to be able to change their colour rapidly. They mainly belong to the Tetragnathidae (ten species) and Araneidae (eight species).

26.2.1 Tetragnathidae

Three species of the genus *Leucauge* are able to change colour rapidly. The name refers to the Greek *leukos* meaning "white" and points to the white guanine pattern on the opisthosoma of *Leucauge* species. Uyemura (1957) observed large flecks on the opisthosoma of *Leucauge subgemmea* becoming smaller and more separated when the spider is picked or shaken strongly. In the genus *Tetragnatha*, many Hawaiian species perform rapid colour change when dropping from the web (R. Gillespie, personal communication), so the species list given in Table 26.1 may have to be extended in the future.

Table 26.1 List of spider species with rapid colour change (adapted from Oxford and Gillespie 1998, distribution mainly from Platnick 2012)

Species	Family	Distribution	References
Leucauge blanda	Tetragnathidae	Russia, China, Korea, Taiwan, Japan	Feng (1990)
Leucauge celebesiana	Tetragnathidae	India to China, Laos, Japan, Sulawesi, New Guinea	Yaginuma (1986), Feng (1990); both sub *Leucauge magnifica*
Leucauge subgemmea	Tetragnathidae	Russia, China, Korea, Japan	Uyemura (1957)
Tetragnatha eurychasma	Tetragnathidae	Hawaii	Gillespie (personal communication)
Tetragnatha filiciphilia	Tetragnathidae	Hawaii	Gillespie (personal communication)
Tetragnatha paludicola	Tetragnathidae	Hawaii	Gillespie (personal communication)
Tetragnatha sp. A, B, C	Tetragnathidae	Hawaii	Oxford and Gillespie (1998)
Tylorida striata	Tetragnathidae	China to Australia	Feng (1990)
Argiope flavipalpis	Araneidae	Africa, Yemen	Edmunds and Edmunds (1986)
Argiope reinwardti	Araneidae	Malaysia to New Guinea	Bristowe (1976)
Argiope sp.	Araneidae	Malaysia or Sumatra	Bristowe (1976)
Argiope sp.	Araneidae	W. Africa	Bell (1893)
Cyrtophora cicatrosa	Araneidae	Pakistan to Northern Territory	Blanke (1975)
Gasteracantha "fornicata"	Araneidae	Sumatra	Bristowe (1976)
Gea heptagon	Araneidae	USA to Argentina, South Pacific Islands, Australia	Sabath (1969)
Phonognatha graeffei	Araneidae	Australia	Roberts (1936, sub *Araneus wagneri*)
Chrysso scintillans	Theridiidae	Myanmar, China, Korea, Japan, Philippines	Uyemura (1957, sub *Argyria venusta*)
Floronia bucculenta	Linyphiidae	Europe, Russia	Bristowe (1941, 1958)
Philodromus spinitarsis	Philodromidae	Russia, China, Korea, Japan	Ikeda (1989)

Gasteracantha "fornicata" is probably misidentified as this species should occur only in Queensland (Platnick 2012).

26.2.2 *Araneidae*

In the genus *Argiope*, two identified (*A. flavipalpis*, *A. reinwardti*) and two unidentified species are able to change colour rapidly. Edmunds and Edmunds (1986) assumed contractile elements to be responsible for the contraction of guanocytes in *A. flavipalpis*. They observed the darkening of the opisthosoma in the moment when the spider dropped from the web. The spiders remain dark during thanatosis but start to become bright soon when climbing back into the web. Recovery of the original white pattern needs considerably more time than darkening.

One of the first observed spider species with rapidly changing colour was *Phonognatha graeffei* where the creamy mottled pattern disappears when it falls out of the web (Roberts 1936). *Gasteracantha "fornicata"* (probably misidentified, see, Table 26.1) changes its colour from red to black; this should be caused by a contraction of pigment-containing cells (Bristowe 1976). *Gea heptagon* is patterned with bright dots and immediately turns to brown when dropping (Sabath 1969); recovery of the original white markings takes several minutes. In the same way, *Cyrtophora cicatrosa* changes its colour (Blanke 1975). Blanke found new white guanine dots after the recolouring, not observed previously to colour change, and therefore, he proposed a guanine relocation to be the underlying mechanism of the colour change.

26.2.3 Other Families

The theridiid spider *Chrysso scintillans* changes the abdominal pattern from a continuous golden yellow colouration to starlike flecks when disturbed (Uyemura 1957). The opisthosomal pattern of the philodromid *Philodromus spinitarsis* is less conspicuous; the colour changes partially from pale brown to almost black when the spider drops (Ikeda 1989).

26.3 Guanocytes

The midgut fills a great part of the spider's opisthosoma with its folded diverticula. Midgut tissue basically consists of four cell types: secretory cells, resorption cells, basal cells, and guanocytes (Foelix 2011). Guanocytes are specialized midgut cells whose main function is to store the purine excretory product guanine (Millot 1926). They support the tissues responsible for excretion as they absorb purine-containing metabolic products (Seitz 1972, 1987).

Guanine is stored in anhydrous crystalline form and therefore made nonhazardous. Most guanine is present just after the reproductive phase. Subsequently, guanine starts to be discarded and the so-called guanine-storing minimum is reached at the end of the life cycle (Seitz 1972).

Guanocytes in spiders may cause a matt white or a silvery colouration. In the first case, the cells are packed with prismatic guanine crystals. In the second case, plates of doublet crystals are formed within guanocytes that are stacked with cytoplasm between the plates. The two crystals of each plate are cemented by layers of amorphous guanine. Crystal morphology and stack dimensions determine the light-reflective properties of a guanine stack (Levy-Lior et al. 2010).

Already Millot (1926) discerned the function of guanine as a colourant. However, Seitz (1972) disputed this in the case of *Araneus diadematus*. Nowadays, the importance of guanine as a biochrome in spider colouration is generally acknowledged (Holl 1987; Oxford 1998). The white or silvery colour shines through the thin and transparent opisthosomal cuticle. The amounts of white, for example, range from only few stripe-like markings (e.g. in *Linyphia* species) to a totally white opisthosoma as found in the thomisid *Misumena vatia*. In many spider species, conspicuous colours like red are brightened by the white guanine, whereas in dark spider species, guanine plays no role for the colouration of the integument (Holl 1987).

Guanine storage obviously is one precondition for developing colour change in spiders (Oxford 1998). Almost all colour-changing spider species are representatives of the families with many guanine-storing species: Araneidae, Tetragnathidae, Theridiidae, and Philodromidae. However, maintaining stored guanine is costly (Oxford 1998), and this might be a reason why so few species developed rapid colour change.

26.4 Colour Change in *Floronia bucculenta* (Linyphiidae)

26.4.1 Life Observations

The colour pattern of *F. bucculenta* consists of many white dots on a brownish background. The white dots are relatively diverse in their shape; only few are truly circular. Dots close to the dorso-median line are bigger than lateral ones. During colour change, the dots seemingly become very small or disappear completely (Fig. 26.1). *F. bucculenta* changes its colour from white to a dark brown the moment it drops from the web. The change can also be induced when the spider is shaken in a tube. During colour change, the white dots are almost instantaneously pulled inwards. Often, the spider will exhibit thanatosis after dropping. This presumably makes it even more difficult for predators to find it.

The darkening of the opisthosoma takes place no matter if the spider lands on a dark or a light surface. In the lab, the spiders changed their colour to dark although they landed on a white ground and were even better visible then. Therefore, the spider probably is not able to control the reaction. Under natural conditions, the spider lands on the dark forest floor, where it is well camouflaged.

Whereas the disappearance of the guanine dots happens instantaneously, the dots reappear in approximately one to two minutes. This looks as if the white dots were slowly coming out of the deep. Similar observations were made in other colour-changing spiders (see Sect. 26.2).

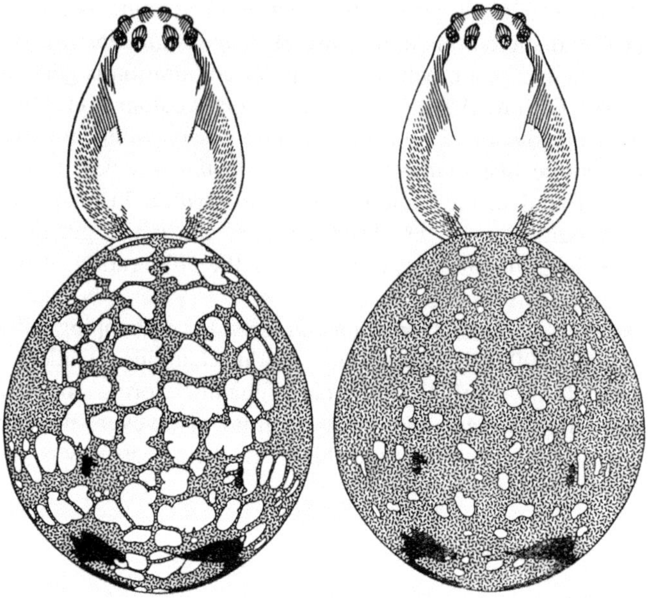

Fig. 26.1 Two states of colouration in *Floronia bucculenta* (taken from Bristowe 1958). *Left*: normal colouration with a lot of extended white dots. *Right*: dark state just after colour change

26.4.2 Guanocytes

The white dots on the surface of the opisthosoma of *Floronia bucculenta* have different forms. Therefore, the shape of guanine accumulations varies also in histological sections. Generally, they are about 50 μm in diameter and up to 100 μm in length (Fig. 26.2). Towards the outside, the white accumulations are normally wider than at the proximal end. One guanine accumulation that is visible from the outside as one white point consists in fact of several elongate guanocytes with a diameter of 5–10 μm and a length of 50–100 μm.

The shape of the guanocytes and consequently the form of the whole guanine accumulations of *F. bucculenta* are different as compared to other species. Generally, guanocytes in other spiders are about 30–50 μm long and show an elongated cubic form which varies from species to species (Millot 1926). For example, *Araneus diadematus* has relatively broad clubbed guanocytes that are about 60 μm long (Seitz 1972). In *A. diadematus,* there are much larger parts made out of normal resorption cells fitted in between the proximal ends of the guanocytes than it is the case in *F. bucculenta. Misumena vatia* shows guanocytes with a cubic form (Insausti and Casas 2008). The possible function of the elongated shape of the guanocytes in *F. bucculenta* probably has to do with their ability to retract (see below) but needs to be explored in more detail.

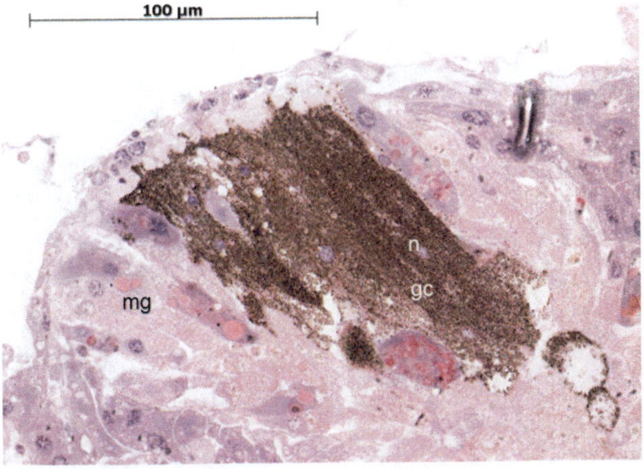

Fig. 26.2 Longitudinal section of a midgut diverticulum of *Floronia bucculenta* with the longish guanocytes. Staining: hematoxylin–eosin. *gc* guanocytes, *mg* midgut, *n* nucleus

In *F. bucculenta*, no tight connection between guanocytes and hypodermis can be found, which seems to be a precondition for a rapid colour change (see below). However, according to Seitz (1972), the guanocytes in *A. diadematus* (that is not able to change its colour) are tightly interconnected to the hypodermis by protuberances and invaginations of cell membranes. These connections may be important for the stability of the colour pattern as the possible signal function should be independent of the feeding state of the spider.

26.4.3 Muscles

Fine striated muscles lie above the guanine-containing parts of the midgut diverticula. Looking at a midgut diverticulum from above, the muscle fibres are arranged like a dense grid (Fig. 26.3a). Each muscle strand is about 2–3 µm wide. The muscles are mainly present at the apical parts of the diverticula. Further inside, their number decreases rapidly (Fig. 26.3b). Here, the muscles lie mainly at the side of the guanine accumulations. At the proximal ends of the guanocytes and of the whole guanine accumulation, no muscles are visible any more. The position of the muscle grid within the surrounding tissues is illustrated in a model drawing (Fig. 26.4). A contraction of the muscle grid in *Floronia bucculenta* will likely lead to the observable colour change during which the white guanocytes retreat within the spider's opisthosoma. However, the exact mechanism is still part of ongoing studies.

The guanine-storing areas of another linyphiid spider, *Linyphia triangularis*, were investigated. This species shows white guanine markings on its opisthosoma too; however, it is not able to change its colour pattern. *L. triangularis* has broader and shorter guanocytes than *F. bucculenta*, and no muscles connected to the midgut cells or the guanocytes are visible.

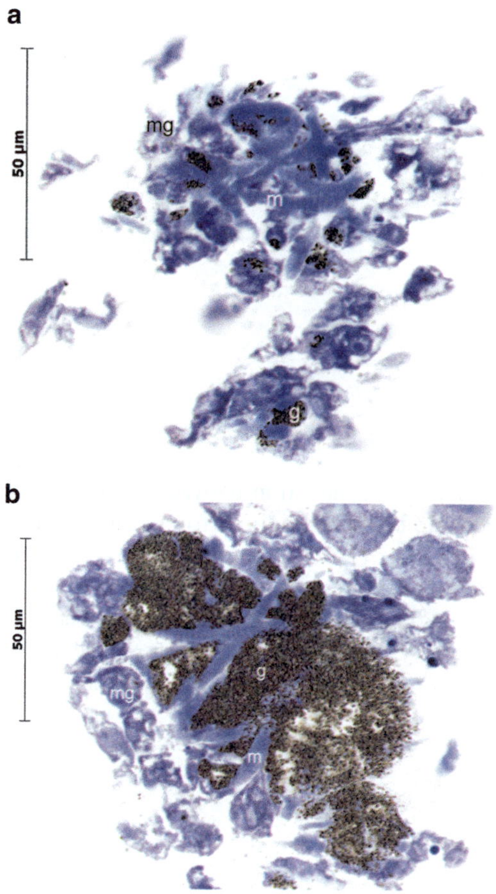

Fig. 26.3 Cross section of the outer part of a midgut diverticulum of *Floronia bucculenta* with guanine and striated muscles. Staining: toluidine blue. (**a**) Section right under the hypodermis, note the dense muscle grid and little guanine. (**b**) Section from further inside, less muscles and more guanine are visible. *g* guanine, *m* muscle, *mg* midgut tissue

At first glance, the described muscle grid may be confused with the abdominal sac (Whitehead and Rempel 1959), especially in longitudinal sections of midgut diverticula. However, the abdominal sac consists of smooth muscles, whereas the newly found muscle grid is clearly striated.

Physiological colour change was proposed to be a process by either migration of chromatic inclusions within cells or alteration of the chromatocyte morphology (Blanke 1975; Holl 1987). The findings in *F. bucculenta* are compatible with the second mechanism, a change of form and position of guanocytes. Such a mechanism was also proposed by Edmunds and Edmunds (1986) and by Oxford and Gillespie (1998). Our study is the first to present morphological support for this idea.

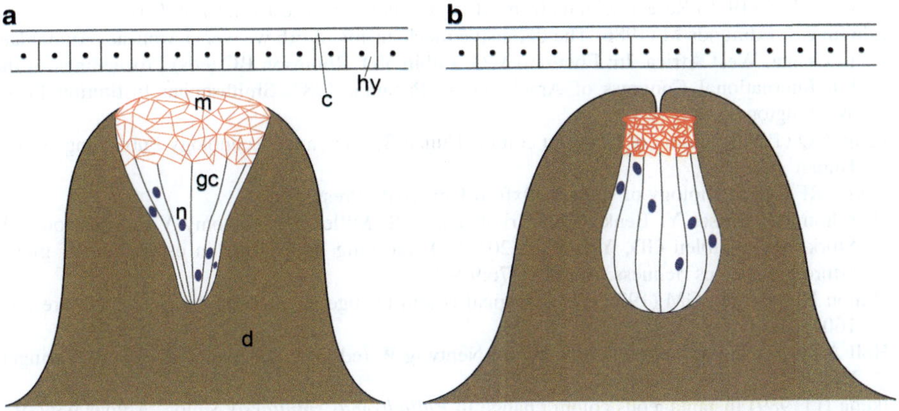

Fig. 26.4 Model of a midgut diverticulum in *Floronia bucculenta*. The white guanine is visible through the hypodermis and cuticle. Fine muscles (in red) cover the guanocytes like a grid. (**a**) Muscle grid relaxed, guanocytes beneath hypodermis; (**b**) muscle grid contracted, guanocytes retreated. *c* cuticle, *d* midgut diverticulum (brown), *gc* guanocytes, *hy* hypodermis, *m* muscles, *n* nucleus

26.5 Conclusions

Rapid physiological colour change is a rare phenomenon in animals. Among arthropods, it is known only from 21 spider species. Guanine markings are always involved when colour change can be observed. However, no studies dealing with the mechanism of the colour change have existed until now. In *Floronia bucculenta,* a set of fine striated muscles was discovered that is associated with the guanocytes. Contraction of these muscles probably leads to colour change by a retreat of the white guanocytes within parts of a second. No similar muscles are present in a related species that is not able to change its colour. The exact mechanism has still to be explored.

Acknowledgements Thanks go to R. Gillespie (Berkeley, USA) for giving valuable unpublished information and to H. Ono (Tokyo, Japan) for translating a Japanese text.

References

Auerswald L, Freier F, Lopata A, Meyer B (2008) Physiological and morphological colour change in Antarctic krill, *Euphausia superba*: a field study in the Lazarev Sea. J Exp Biol 211: 3850–3858

Bell HHJ (1893) Notes on a spider. Nature 47:557–558

Blanke R (1975) Die Bedeutung der Guanozyten für den physiologischen Farbwechsel bei *Cyrtophora cicatrosa* (Arachnida: Araneidae). Ent Germ 2:1–6

Bristowe WS (1941) The comity of spiders II. Ray Society, London

Bristowe WS (1958) The world of spiders. Collins, London

Bristowe WS (1976) Rare arachnids from Malaysia and Sumatra. J Zool 178:7–14
Edmunds J, Edmunds M (1986) The defensive mechanisms of orb weavers (Araneae: Araneidae) in Ghana, West Africa. In: Eberhard WG, Lubin YD, Robinson BC (eds) Proceedings of the 9th International Congress of Arachnology, Panama, 1983. Smithsonian Institution Press, Washington
Feng Z-Q (1990) Spiders of China in colour. Hunan Science and Technology Publishing House, Hunan
Foelix RF (2011) Biology of spiders. Oxford University Press, Oxford
Hamilton AJ, Basset Y, Benke KK, Grimbacher PS, Miller SE, Novotny V, Samuelson GA, Stork NE, Weiblen GD, Yen JDL (2010) Quantifying uncertainty in estimation of global arthropod species richness. Am Nat 176:90–95
Hinton HE, Jarman GM (1972) Physiological colour change in the Hercules beetle. Nature 238: 160–161
Holl A (1987) Coloration and chromes. In: Nentwig W (ed) Ecophysiology of spiders. Springer, Berlin
Ikeda H (1989) Instantaneous colour change in *Philodromus spinitarsis* Simon. Atypus 93:7–9
Insausti TC, Casas J (2008) The functional morphology of color changing in a spider: development of ommochrome pigment granules. J Exp Biol 211:780–789
Key KHL, Day MF (1954) The physiological mechanism of colour change in the grasshopper *Kosciuscola tristis* (Sjöst) (Orthoptera, Acrididae). Aust J Zool 2:340–363
Levy-Lior A, Shimoni E, Schwartz O, Gavish-Regev E, Oron D, Oxford G, Weiner S, Addadi L (2010) Guanine-based biogenic photonic-crystal arrays in fish and spiders. Adv Funct Mater 20:320–329
Millot J (1926) Contributions à l'histophysiologie des Aranèides. Bull Biol France Belg Suppl 7:1–238
Oxford GS (1998) Guanine as a colorant in spiders: development, genetics, phylogenetics, and ecology. In: Selden PA (ed) Proceedings of the 17th European Colloqium of Arachnology, Edinburgh, 1997. British Arachnological Society, Burnham Beeches, Bucks
Oxford GS, Gillespie RG (1998) Evolution and ecology of spider coloration. Annu Rev Entomol 43:619–643
Platnick NI (2012) The world spider catalog, version 13.0. American Museum of Natural History. Online at http://research.amnh.org/iz/spiders/catalog. doi:10.5531/db.iz.0001
Roberts NL (1936) Colour change in the leaf-curling spider (*Araneus wagneri*). Proc R Zool Soc NSW 1936:28–29
Sabath LE (1969) Color change and life history observations of the spider *Gea heptagon* (Araneae: Araneidae). Psyche 76:367–374
Seitz KA (1972) Elektronenmikroskopische Untersuchungen an den Guanin-Speicherzellen von *Araneus diadematus* Clerck (Araneae, Araneidae). Z Morph Tiere 72:245–262
Seitz KA (1987) Excretory organs. In: Nentwig W (ed) Ecophysiology of spiders. Springer, Berlin
Stevens M, Merilaita S (2009) Animal camouflage: current issues and new perspectives. Phil Trans R Soc B 364:423–427
Stuart-Fox D, Moussalli A (2009) Camouflage, communication and thermoregulation: lessons from colour changing organisms. Phil Trans R Soc B 364:463–470
Uyemura T (1957) Colour change of two species of Japanese spiders. Acta Arachnol 15:1–10
Whitehead WF, Rempel JG (1959) A study of the musculature of the black widow spider, *Latrodectus mactans* (Fabr). Can J Zool 37:831–870
Yaginuma T (1986) Spiders of Japan in color—new edition. Hoikusha, Osaka

Part VIII
Nutrition

For centuries, spiders were regarded as exclusively predacious, feeding on a wealth of arthropods, which guarantee their high-quality nutrition. Today we know that many prey items are just low-quality food and a well-balanced uptake of nutrients is of upmost importance. This modern view adds malnutrition and starvation as important threats to the survival of spiders. Occasionally, spiders also take up plant-derived products, mainly nectar and pollen, perhaps not only for energetic reasons. The cultivation of spiders, especially mass-rearing, still suffers from the need to supply living food, but there are important approaches to design an artificial diet for spiders.

Chapter 27
Nutritional Aspects of Spider Feeding

Søren Toft

27.1 Introduction

The prey of spiders provides energy, nutrients and water needed for maintenance, growth and reproduction, but the acquisition of these nutriments also implies costs for capture, digestion and metabolism as well as for handling of indigestible fractions and toxins in the prey. Until recently there was consensus that a diet of animal tissues is nutrient rich and well-balanced and that predators consume food of high nutritional quality (Slansky and Scriber 1985). Food stress in these animals was equated with low prey availability and starvation, and optimal foraging models explained predator behaviour in terms of optimizing energy intake. Studies over the last two decades have effectively rejected these views for generalist predators of several invertebrate and vertebrate groups, including spiders. Hunger is not the only way in which the nutritional state of spiders may be stressful. Whether prey is scarce or rich, the available prey species may have an imbalanced nutrient composition compared with the spiders' requirements, and the available prey may contain toxins and deterrents in amounts that make them unsuitable as staple food. All these conditions have consequences not only for the single individual but also for the part of the biological community that it interacts with. For long, the pioneering paper of Greenstone (1979) stood alone to claim a role for nutrients in predator ecology. This review summarizes recent advances in our understanding of interactions between spiders and their prey from a nutritional viewpoint. An extensive bibliography of the topic can be found in Wilder (2011).

S. Toft (✉)
Department of Bioscience, Aarhus University, Aarhus, Denmark
e-mail: soeren.toft@biology.au.dk

W. Nentwig (ed.), *Spider Ecophysiology*,
DOI 10.1007/978-3-642-33989-9_27, © Springer-Verlag Berlin Heidelberg 2013

27.2 Nutritional Quality of Prey

The food quality of a prey can be defined as its potential contribution to the fitness of the spider. It is a central concept because fitness maximization is assumed to govern the evolution of prey choice and handling as well as food utilization. Prey quality is usually determined in performance experiments that compare fitness parameters when the spider is fed diets including the test prey compared with diets that differ only in lacking this prey. In single-prey experiments starved spiders form the control group, and a group fed a high-quality standard "comparison prey" (usually fruit flies, house flies, crickets or springtails, depending on the size of the spider) is included to provide a scale on which the new prey can be compared with previously tested prey. Results of single-prey and mixed-prey experiments may supplement each other; e.g. aphids are consistently low-quality or toxic food when presented in monotypic diets but may enhance fitness as part of mixed diets (Toft 2005).

Simultaneous tests of multiple prey have revealed a continuum of prey qualities, ordinally classified as "high-quality", "intermediate-quality", "low-quality", "poor-quality" and "toxic" prey (Toft and Wise 1999; Toft 2009). High-quality prey allows completion of the full life cycle with low mortality on a monotypic diet; intermediate- and low-quality prey allows substantial or some growth, respectively, but development stops before maturity. Poor quality is useless to the spiders whose survival and growth are not statistically distinguishable from that of starved spiders. Toxic prey results in no growth and faster death than in starved spiders. Some prey groups (e.g. aphids) are consistently of very low food quality, while others (e.g. Collembola, Diptera) may cover the whole spectrum of food qualities depending on the species.

Though knowledge of the specific toxins and nutrient imbalances of prey will ultimately be important for understanding the details of spider–prey interactions, food quality cannot be determined by chemical analysis but only by performance experiments. The reasons for this are that even a perfect composition of nutrients can be useless if the prey also contains toxins or deterrents, the same nutrient composition may be good for one species and not for another, and the optimal nutrient composition may change within a species according to season and life-cycle phase. Likewise, toxicity is no chemical trait but a characteristic of consumer physiology; the same prey may be toxic to spiders and high-quality prey for other predators. For example, the collembolan *Folsomia candida* is toxic to spiders but high-quality prey for generalist predatory soil mites.

Though most studies have concerned typical euryphagous wolf spiders, only a minority of prey species included in the tests were of high food quality. This suggests that most prey in nature may be of low food quality to spiders. The notion of the "generalist predator" does not imply that all potential prey are valuable and substitutable but only that prey with these qualities come from diverse taxa (classes, orders); this contrasts with specialized stenophagous spiders whose focal prey are usually from a single taxon.

27.2.1 Role of Nutrient Deficiencies and Toxins for Prey Quality

Spider food is live individuals of other animal species which makes detailed manipulation of food composition difficult. Tests of prey quality and the factors influencing this have used two approaches: comparison of different prey species or comparison of nutritionally manipulated groups of the same non-toxic prey species. The nutritional manipulations have been done by enriching the growth media of the prey, either as a broadscale (multi-nutrient) enrichment or as enrichment with a single type of nutrient (e.g. protein, lipid, vitamins). Mayntz and Toft (2001) confirmed that nutritional enrichment of the prey via its growth medium was due to a modified composition of the prey biomass and not to the prey's gut contents.

How do toxins and nutrient deficiencies contribute to create the continuum of prey qualities? Experiments with nutrient-deficient prey show that these are initially eaten in about the same amounts as high-quality prey; the spiders at first grow normally, and negative effects may take weeks to develop. Thus, nutrient deficiency may explain the "intermediate" but not the more inferior categories. Toxins and deterrents prevent feeding immediately or very quickly through induced aversions (Toft 1997); they are associated with low consumption and high death rates and thus may account for the low food quality categories ("low-quality" to "toxic" prey). Weak toxins and nutrient deficiency cannot be directly separated by performance experiments, though, unless it is possible to test the same prey with and without the toxins. The masking of a prey's nutritional value by a toxin is repeated in mixed diets that include a toxic prey, preventing utilization of the diet's high-quality prey. It is unknown to what extent spiders can avoid toxic prey in the field.

27.2.2 Effects of Broadscale Nutrient Enrichment

Broadscale nutrient additions are appropriate for analyzing the effects of overall nutritional condition on behaviour or performance if the specific mechanisms of these effects are of minor importance. In most cases the enrichment consisted of adding dog food to the growth medium of the prey. Such manipulations have been applied successfully to fruit flies, house flies, crickets and collembolans. Experiments using broadscale nutritional enrichments have demonstrated several nutrition-dependent phenomena. Enriched prey may lead to increased predation rate (Bressendorff and Toft 2011); enhanced survival, growth and development (Mayntz and Toft 2001); better utilization of prey biomass (Mayntz and Toft 2006); increased courtship activity and mating success of males (Lomborg and Toft 2009); enhanced egg maturation (Wilder and Rypstra 2008); and higher propensity for cannibalism (including sexual cannibalism) (Mayntz and Toft 2006; Wilder and Rypstra 2008). All these improvements appeared though prey was unlimited also in the groups fed deficient prey, i.e. they are results of enhanced nutrient balance and not because more energy was available.

27.2.3 Effects of Specific Nutrient Enrichments

Manipulation of single nutrients or nutrient groups provides information on the mechanistic basis of deficiency effects. It should be kept in mind that since the manipulations are done on the prey media, they do not directly control the nutrient composition of the prey; nutrients added to the prey's media may be biochemically transformed in the prey. It is known, however, that the levels of some fatty acids (Pollierer et al. 2010) are increased in consumers at least two trophic levels above the enrichment. Focus has so far been mostly on the macronutrients protein and lipid and their effect on feeding behaviour and life histories.

Addition of multiple amino acids, the single amino acid methionine or mixed fatty acids + cholesterol to the basic *Drosophila* medium either enhanced survival, increased growth or both of wolf spiders (*Pardosa amentata*) fed the resulting flies compared with spiders fed flies raised on the standard medium. In contrast, addition of multiple vitamins had no effect on any parameter (Mayntz and Toft 2001). In the orb weaver, *Argiope keyserlingi*, protein addition increased growth and web decorations (Blamires et al. 2009). Testing flies with a range of lipid to protein (L:P) ratios Jensen et al. (2011b) found a positive correlation between protein content and rates of development and growth. Salomon et al. (2008, 2011) found that lipid—but not protein-enriched prey—promoted development towards adulthood in a field population of *Stegodyphus lineatus* but that protein enrichment via matriphagy enhanced juvenile survival. We still have no coherent picture of how spiders' nutrient demands change over the life cycle and the seasons.

27.3 Nutrient Balancing in Spiders

The negative effects of nutrient deficiency on behaviour, physiology and overall performance provide selection pressure for the evolution of nutrient self-selection. A nutritional theory of optimal foraging has been formulated as the so-called geometric framework (Simpson et al. 2004). It posits that animals seek to obtain a certain amount of food of a specific nutrient composition (Fig. 27.1a). This "intake target" can be reached by eating a single prey of optimal nutrient composition or by mixing prey of imbalanced but complementary composition. The central assumption of the theory is that the self-selected intake target reflects the nutrient composition that results in maximal fitness. Available prey may not allow the spider to reach the intake target, however. Therefore, other regulatory mechanisms must be involved in attaining the optimal body composition, some behavioural (acting mostly prior to or during prey ingestion) and others physiological (acting mostly post-ingestively). The ability of spiders to actively balance their intake of macronutrients was first demonstrated by Mayntz et al. (2005). Two balancing mechanisms were revealed. Groups of wolf spiders (*Pardosa prativaga*) were fed imbalanced prey to both sides of the assumed protein–lipid target, and the amounts

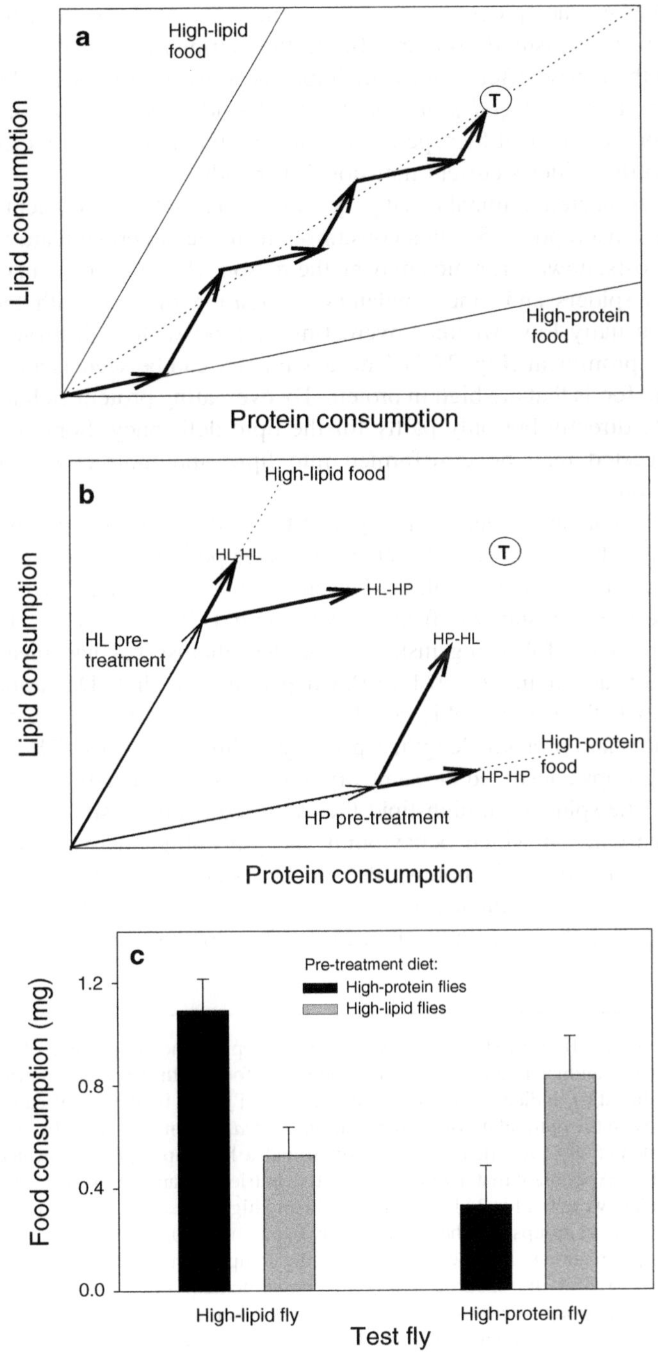

Fig. 27.1 (**a**) In the geometric framework, food intake is displayed in an n-dimensional coordinate system with each relevant nutrient forming an axis. Each prey type can be represented by a *straight line*

consumed when the spiders had access to either the same or the complementary prey type were measured. Spiders offered the complementary fly type consumed more food than those offered the same imbalanced fly as previously (Fig. 27.1b, c). Similarly pretreated *Stegodyphus lineatus* was found to extract more of the deficient nutrient from subsequent flies, perhaps indicating that release of digestive enzymes depends on the spider's current nutritional demands.

Spiders subjected to imbalanced prey may compensate for deficient nutrients by increasing consumption. Surplus consumption of the superabundant nutrient may also have costs, however, which may be the reason why this mechanism is not fully utilized. In spiders and other predators, overeating of food with too high-lipid content is usually low, whereas overeating of food with high protein content is much more prominent (Fig. 27.2). This asymmetry can be seen as an adaptation of predators to foods that are high in protein. By overeating protein-rich foods, spiders compensate directly but only partly for the lipid deficiency. Some of the surplus protein ingested may be transformed into lipid and indirectly provide further compensation.

These mechanisms cannot fully regulate the spider's nutrient composition when confronted with an unbalanced diet for a longer period. Concomitant changes in body composition also take place. Jensen et al. (2011b) fed *Pardosa prativaga* through the second instar with fruit flies varying in L:P ratio between 0.10 and 0.89. At the start of the following instar, the spiders themselves had mean L:P ratios ranging from approximately 0.1 to 0.6 depending on diet. Deviations from the optimal prey L:P ratio (probably near 0.25) had costs in terms of reduced growth in lean dry mass and prosoma length, especially on high-lipid food. This was the case though all groups consumed the same dry mass of food and had the same total mass gain. Thus, the spiders fed high-lipid flies suffered from obesity effects.

The timescale on which behavioural and physiological responses to nutrient imbalances take place is noteworthy. In the studies of Mayntz et al. (2005) (Fig. 27.1b, c), pretreatment times were only 24 h. Jensen et al. (2011b) found feeding responses by the fifth fly (Fig. 27.2). These behavioural responses appeared

---

Fig. 27.1 (continued) from the origin whose slope depicts the proportion of nutrients. It is assumed that the animal requires a certain amount of food with a specific nutrient ratio (here lipid to protein ratio), indicated by the "intake target" (T). The intake target can be reached by eating a prey with the optimal nutrient composition (*dotted line* through T) or by combining meals of two complementary prey (here a high-protein and a high-lipid prey; indicated by *arrows*). (**b**) Design of an experiment that demonstrated active nutrient balancing (Mayntz et al. 2005). Two groups of spiders were fed for 24 h with high-lipid and high-protein fruit flies, respectively. Each of these pretreatment groups was then divided in two; one subgroup from each was fed the same fly type as in the pretreatment, and the other was fed the complementary fly type. Nutrient balancing would be indicated if the subgroups offered complimentary prey consumed more than the corresponding subgroups that were again offered the same prey. (**c**) Results demonstrating balancing of lipid and protein intake by the wolf spider *Pardosa prativaga* as predicted in (**b**). Spiders consumed more of lipid-rich flies if they were pretreated with protein-rich flies, and they consumed more of protein-rich flies if pretreated with lipid-rich flies (from Mayntz et al. 2005)

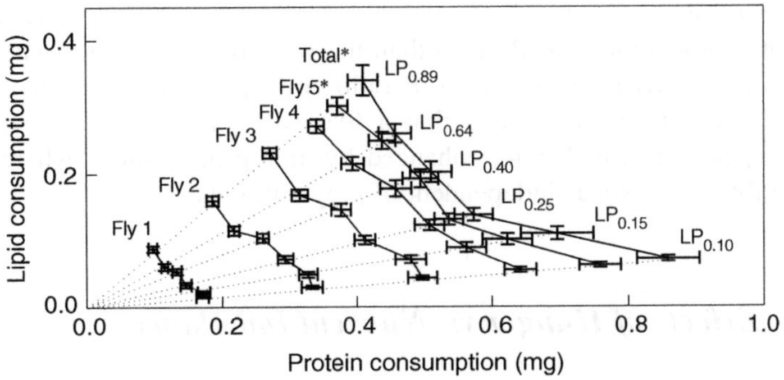

Fig. 27.2 Results of an experiment in which six groups of *Pardosa prativaga* were fed fruit flies with different lipid to protein (L:P) ratios throughout the second instar. L:P ratios are indicated at the end of each fly line. The *points* show cumulative protein and lipid intake from the first through the fifth fly and the total amount eaten during the instar. Notice that data for consumption of fly 1–4 each lie on a *straight line* with a slope of -1 (indicating consumption independent of nutrient composition) but that data for fly 5 and total consumption deviates significantly from this (indicated by *asterisk*) and that the deviation is due to excessive consumption of the two most protein-rich diets (from Jensen et al. 2011b)

much earlier than the 5 weeks it took to obtain significant detrimental growth effects (Mayntz and Toft 2001). This indicates that behavioural and physiological regulation is a continuously ongoing process, an integrated part of day-to-day foraging activities rather than a response to an already threatening situation. The adaptive value of this may be that the costs of regulatory mechanisms are likely to be exaggerated in highly imbalanced spiders which are clearly weakened by their condition. For example, long-time imbalanced wolf spiders show reduced rates of predation and cannibalism, thus failing to make use of behavioural options that might have been used to redress their nutritional imbalance (Mayntz and Toft 2006; Wilder and Rypstra 2008).

27.3.1 Nutrient Balancing and Life Cycles

Spiders' life cycles may be regulated in different ways. Some species grow faster than others; some show highly synchronous development and fixed cycles, while others have great flexibility in life-cycle duration, and the population may consist of individuals from several instars. Such differences are likely to influence a species' nutritional demands. Juvenile growth implies high requirements for protein; a high growth rate may therefore be associated with a larger risk of experiencing protein limitation than a low growth rate. Likewise, a fixed life-cycle forces the animals to complete development even under suboptimal nutritional conditions, i.e. they are more at the mercy of the prey available than spiders with a flexible life cycle. Jensen et al. (2011a) compared two species of *Pardosa* wolf spiders and confirmed that the species with slower growth and a flexible life cycle (*P. prativaga*) was better able to regulate its body content (L:P ratio) than the fast-growing species with fixed life cycle (*P. amentata*) in which the L:P ratio corresponded closely to that of the consumed prey. They also found that a fixed life cycle was associated with a nutrient-dependent growth rate, enhanced by high-protein diets, whereas growth in the flexible species was independent of prey nutrients.

27.3.2 Effects of Hunger vs. Nutrient Imbalance

Starvation and nutrient deficiency are two stressful conditions, and spiders may suffer from both. Experiments that varied them in a factorial design have revealed aspects of behaviour, web structure and physiology that are affected by one or the other, or both. In the araneid *Zygiella x-notata,* orb-web size increased under food limitation, while low prey quality reduced the number of radii (Mayntz et al. 2009). Life history parameters and feeding may also be affected; some parameters were found to change in the same direction (e.g. increased development time, reduced adult mass); some were affected by one but not the other (juvenile survival, egg sac production reduced by nutrient deficiency; cannibalism propensity increased by starvation); and some were affected in opposite directions (food consumption and number of instars in males increased by starvation and reduced by nutrient deficiency) (Mayntz et al. 2003; Mayntz and Toft 2006). Imbalanced spiders are also less tolerant to toxic prey (aphids), while starvation has no effect on this. Surprisingly, in all these studies there was no evidence of significant interactions between the two factors. Thus, starvation and nutrient deficiency effects are independent.

A body condition index (BCI, residuals from a regression of body mass on a measure of body size; Jakob et al. 1996) has been devised to characterize the feeding condition of animals, in particular their lipid stores. Wilder and Rypstra (2008) and Lomborg and Toft (2009) found that BCI in wolf spiders was unaffected by nutrient imbalance which modifies the spiders' body composition without reducing their mass. The latter authors concluded that BCI is suitable only for expressing prey limitation, but not nutrient deficiencies. A functional definition of condition is the ability to mobilize resources in appropriate situations. In this sense, condition is positively correlated with BCI when the diet is balanced, but negatively when the diet is imbalanced and the animal obese. Unfortunately, we have no simple way to express whether an individual from nature suffers from nutrient imbalance.

27.4 Spider Nutrition and Ecological Webs

Nutritional demands have implications for several aspects of spider community ecology, in particular their role in food webs and in the natural control of prey populations. Given the food quality continuum and the suggestion that most potential prey species are in the lower end of the food quality scale, the notion that alternative prey species can substitute each other in the diet of generalist predators according to their abundance may be of limited validity. Instead it has been argued that the prey responsible for growth and reproduction is much less diverse than indicated by prey lists and perhaps restricted to a few high-quality prey species (Toft and Wise 1999). These are not necessarily of high abundance, and spiders may starve in spite of apparent prey abundance, just like herbivores may starve in a green world. This may be true even for "generalist predators" with highly diverse diet lists. Spiders may develop a specific aversion against an abundant low-quality prey (Toft 1997), but these prey will not be deleted completely from the menu because aversions last for only a short time. Thus, spiders maintain a highly diverse prey list even if the prey species are not nutritionally substitutable and even if most of the prey types contribute little to the spiders' fitness.

Predators, including spiders, have higher nitrogen content than their herbivorous prey. From stoichiometric theory, it has therefore been argued that predators in nature should be limited by nitrogen and that the frequent occurrence of cannibalism and intra-guild predation among predators has evolved in response to a high nitrogen demand. There is no direct test of this hypothesis. Available evidence generally confirms enhanced performance of spiders given high-protein prey. However, geometric framework theory predicts a fitness peak at an intermediate L:P ratio of the food (the intake target, Fig. 27.1a), implying a reduced performance when the protein content of the food is very high. This has been confirmed for a carabid beetle (Jensen et al. 2012) and must be expected to hold for spiders too. Compensatory overingestion of protein-rich flies (Fig. 27.2) indicates that it will. Reduced nutrient extraction from males (high-protein prey) during sexual cannibalism compared with normal (high-lipid) cricket prey (Wilder and Rypstra 2009) also indicates that spiders are as likely to be limited by a low as by a high L:P ratio.

The nutrient composition of the prey affects spider performance through its effect on the spider's own nutrient balance. One aspect of modified performance is reduced ability to consume prey unbalanced in the same direction as the spider, especially if the imbalance is to the lipid side (Fig. 27.2). A consequence of this may be that the spiders' functional response is reduced if the nutrient composition of the prey is far from the spider's intake target. Using a wolf spider–fruit fly system, Bressendorff and Toft (2011) found a normal type 2 functional response when the prey was of near-optimal nutrient composition, but this turned into the dome-shaped response (i.e. with reduced killing rate at high prey densities) if the prey was imbalanced. Interestingly, when tested in spring, the dome-shaped response appeared when the flies had a high L:P ratio, while in late autumn it was induced by a low ratio. This reflected the high-lipid demand of spiders prior to hibernation and higher protein demand in the reproductive period. The ability to reduce the population of a prey species is lowered if the prey is nutritionally imbalanced compared to the spider's intake target, which possibly is the case with most non-predatory prey. Notice that most laboratory-reared prey is deficient if reared on their optimal media.

To what extent can information on food quality from the laboratory be used to predict the most important food web connections of a predator? We would expect that, all else being equal, a high-quality prey is more actively hunted than a low-quality prey and that spiders would have a stronger impact on the high-quality prey. One piece of evidence is available to confirm this. Wise (2004) manipulated populations of *Schizocosa* and found one group of Collembola that responded positively to spider removal, i.e. Tomoceridae. These collembolans were found to be nutritionally complete prey for *Schizocosa* by Toft and Wise (1999). In contrast, predators are expected to be poor controllers of prey species with effective chemical defences because of a low killing rate. Several studies have indicated, however, that generalist predators including spiders are sometimes effective against aphids (Symondson et al. 2002). Thus, substantial reduction of a prey population may be possible even if the prey is of low food quality.

27.5 Conclusion

Contrary to previous beliefs, potential prey species of spiders are not all of high food quality but form a continuum from complete to useless and even detrimental food. Possibly, the majority of prey species are of low nutritional quality even to generalist predators. Low food quality is primarily due to defensive toxins in the prey, whereas nutrient deficiency leads only to suboptimal food value. Whether a prey is deficient or not depends on the current demands of the spider. Together, two deficient but complementary prey may form a complete diet, but a toxic prey in a mixed diet may render even high-quality prey useless. Dependence on nutrient-deficient prey reduces all aspects of spider performance like prey capture, growth, development, reproductive success, survival and cannibalism. Generally, prey with high-protein content promotes performance, but lipids may also be limiting. As stress factors, hunger and nutrient imbalances have very different effects on individual performance. Hunger increases individual activity and aggression against prey and thus enhances the possibilities for recovery, whereas nutrient imbalance weakens the animal and reduces its possibilities for relief. Spiders can balance their intake of the macronutrients protein and lipid in an optimal ratio that may depend on season and phase of the life cycle. Balancing may be accomplished by differential feeding on prey items with complementary composition or by selective nutrient extraction. The nutrient composition of prey influences spiders' role as biocontrol agents because predation efficiency is reduced for prey with a nutrient composition that differs from their demand.

References

Blamires SJ, Hochuli DF, Thompson MB (2009) Prey protein influences growth and decoration building in the orb web spider *Argiope keyserlingi*. Ecol Entomol 34:545–550

Bressendorff BB, Toft S (2011) Dome-shaped functional response induced by nutrient imbalance of the prey. Biol Lett 7:517–520

Greenstone MH (1979) Spider feeding behaviour optimizes dietary essential amino acid composition. Nature 282:501–503

Jakob EM, Marshall SD, Uetz GW (1996) Estimating fitness: a comparison of body condition indices. Oikos 77:61–67

Jensen K, Mayntz D, Toft S, Raubenheimer D, Simpson SJ (2011a) Prey nutrient composition has different effects on *Pardosa* wolf spiders with dissimilar life histories. Oecologia 165:577–583

Jensen K, Mayntz D, Toft S, Raubenheimer D, Simpson SJ (2011b) Nutrient regulation in a predator, the wolf spider *Pardosa prativaga*. Anim Behav 81:993–999

Jensen K, Mayntz D, Toft S, Clissold FJ, Hunt J, Raubenheimer D, Simpson SJ (2012) Optimal foraging for specific nutrients in predatory beetles. Proc R Soc B 279:2212–2218. doi:10.1098/rspb.2011.2410

Lomborg JP, Toft S (2009) Nutritional enrichment increases courtship intensity and improves mating success in male spiders. Behav Ecol 20:700–708

Mayntz D, Raubenheimer D, Salomon M, Toft S, Simpson SJ (2005) Nutrient-specific foraging in invertebrate predators. Science 307:111–113

Mayntz D, Toft S (2001) Nutrient composition of the prey's diet affects growth and survivorship of a generalist predator. Oecologia 127:207–213

Mayntz D, Toft S (2006) Nutritional value of cannibalism and the role of starvation and nutrient imbalance for cannibalistic tendencies in a generalist predator. J Anim Ecol 75:288–297

Mayntz D, Toft S, Vollrath F (2003) Effects of prey quality and availability on the life history of a trap-building predator. Oikos 101:631–638

Mayntz D, Toft S, Vollrath F (2009) Nutrient balance affects foraging behaviour of a trap-building predator. Biol Lett 5:735–738

Pollierer MM, Scheu S, Haubert D (2010) Taking it to the next level: trophic transfer of marker fatty acids from basal resource to predators. Soil Biol Biochem 42:919–925

Salomon M, Mayntz D, Lubin Y (2008) Colony nutrition skews reproduction in a social spider. Behav Ecol 19:605–611

Salomon M, Mayntz D, Toft S, Lubin Y (2011) Maternal nutrition affects offspring performance via maternal care in a subsocial spider. Behav Ecol Sociobiol 65:1191–1202

Simpson SJ, Sibly RM, Lee KP, Behmer ST, Raubenheimer D (2004) Optimal foraging when regulating intake of multiple nutrients. Anim Behav 68:1299–1311

Slansky F, Scriber JM (1985) Food consumption and utilization. In: Kerkut GA, Gilbert LI (eds) Comprehensive insect physiology, biochemistry and pharmacology, vol 4. Pergamon, Oxford

Symondson WOC, Sunderland KD, Greenstone MH (2002) Can generalist predators be effective biocontrol agents? Annu Rev Entomol 47:561–594

Toft S (1997) Acquired food aversion of a wolf spider to three cereal aphids: intra- and interspecific effects. Entomophaga 42:63–69

Toft S (2005) The quality of aphids as food for generalist predators: implications for natural control of aphids. Eur J Entomol 102:371–383

Toft S (2009) Effects of prey quality on the outcome of dietary mixing in a generalist predator, the wolf spider *Pardosa prativaga* (L. Koch). Contrib Nat Hist 12:1387–1401

Toft S, Wise DH (1999) Growth, development, and survival of a generalist predator fed single- and mixed-species diets of different quality. Oecologia 119:191–197

Wilder SM (2011) Spider nutrition: an integrative perspective. Adv Insect Physiol 40:87–136

Wilder SM, Rypstra AL (2008) Diet quality affects mating behaviour and egg production in a wolf spider. Anim Behav 76:439–445

Wilder SM, Rypstra AL (2009) Males make poor meals: a comparison of nutrient extraction during sexual cannibalism and predation. Oecologia 162:617–625
Wise DH (2004) Wandering spiders limit densities of a major microbi-detritivore in the forest-floor food web. Pedobiologia 48:181–188

Chapter 28
Herbivory in Spiders

Dirk Sanders

28.1 Introduction

Though spiders are commonly regarded as "generalist predators" (Nentwig 1986; Wise 1993), there is a long reaching evidence that spiders also directly feed on resources provided by plants, such as nectar (Vogelei and Greissl 1989) and pollen (Smith and Mommsen 1984). For example, several species of wandering spiders take up nectar as a supplement to animal prey (Jackson et al. 2001), and orb-web spiders ingest regularly pollen when recycling their webs (Smith and Mommsen 1984). Meehan et al. (2009) even discovered an extreme case of herbivory in a Neotropical jumping spider that feeds primarily and deliberately on the tips of an ant-acacia leaflet. Most cases of herbivory in spiders, however, seem rather occasional but have the potential to play a role in spiders. The arising question is whether herbivory only occurs in certain life stages and in a very few species or is a more general phenomenon which has been largely ignored. If this is the case, many spiders need to be seen as omnivores rather than carnivores. To shed some light on this issue, I look at the existing literature and sketch some expectations about the actual frequency of this kind of resource use. Is there evidence for herbivory in spiders being associated with age and hunting strategy? How can we improve our understanding about actual resource use of spiders in the field including herbivory?

28.2 Pollen Feeding

Smith and Mommsen (1984) found that regularly dusting birch pollen in orb webs of *Araneus diadematus* (Araneidae) doubled the life expectancy of spiderlings and altered their web-spinning behaviour, so that they spun more frequently than did

D. Sanders (✉)
Institute of Ecology and Evolution, University of Bern, Bern, Switzerland
e-mail: dirk.sanders@iee.unibe.ch

W. Nentwig (ed.), *Spider Ecophysiology*,
DOI 10.1007/978-3-642-33989-9_28, © Springer-Verlag Berlin Heidelberg 2013

Fig. 28.1 Pollen in the orb web of *Aculepeira ceropegia* (Araneidae). When recycling the web, the spider will also feed on the pollen

fasting control spiders. This suggests that orb-web spiders as they regularly recycle their webs can use pollen as an additional food resource (Fig. 28.1) and even adapt their behaviour in response to pollen availability. This may be important for the spiders especially in times when insect prey is scarce and aerial plankton, such as pollen and fungus spores, is abundant; pollen in combination with fungal spores is found throughout the vegetation period in the web of *A. diadematus* (del Fiol et al. 2007). The horizontal webs of sheet-web spiders (Linyphiidae) also have the potential to intercept pollen grains, and in laboratory feeding trials, 80 % and 90 % of individuals of the two linyphiid spider species *Frontinella communis* and *Tennesseellum formicum,* respectively, fed on pollen grains dusted on their webs (Peterson et al. 2010). Spiderlings of *Thomisus onustus* (Thomisidae) were starved or fed on pollen, and similar to the study by Smith and Mommsen (1984), the access to pollen as resource doubled their life expectance (Table 28.1, Vogelei and Greissl 1989). In contrast to web-building spiders where pollen feeding can be seen as more or less accidentally, wandering spiders have to search actively for pollen making this an intentional act of herbivory.

These examples highlight that pollen feeding as a supplementary resource use may be an important component of the feeding biology of many spiders in general, with evidence from members of two major families of web builders, Araneidae and Linyphiidae, and for some wandering spiders. There is, however, also evidence for some spiders to ignore pollen in their web (Carrel et al. 2000), and nothing is known about the frequency of this interaction for spiders in general. Anyway, for all spiders that recycle their webs, it is likely that they use pollen to some degree. Pollen feeding might also be a pathway exposing spiders to *Bt*-crop pollen (for details, see Meissle 2013). The importance of fungivory is still unclear as fungal spores may have a deleterious effect on spiders. Smith and Mommsen (1984) reported such deleterious effects of fungal spores on spiderlings in their experiment, though they tested only one kind of spore (*Cladosporium herbarum*). But fungi

Table 28.1 Survival time of *Thomisus onustus* (Thomisidae) spiderlings in a feeding experiment with a starvation control group (no food), two treatments with pollen feeding, provision of sucrose solution to simulate nectar, and insect prey (Vogelei and Greissl 1989)

		Spiderling survival time (in days)			
			Quartiles		
Feeding treatment	*n*	Mean	75 %	50 % (median)	25 %
Starvation control group	12	21.4	20	21	22
Pollen of *Erigeron annuus*	12	34.9	30	35	43
Pollen of *Bellis perennis*	15	43.9	36	44	53
Sucrose solution (30 %)	13	129.4	92	128	156
Drosophila melanogaster	12	No deaths occurred during the experiment			

have different organic chemical composition and content of toxic metabolites; consequently, many species can be nutritionally useful and some can be deleterious (del Fiol et al. 2007; see also Evans 2013).

28.3 Nectar Feeding

There is more literature on nectar use by spiders than on pollen feeding. By being an exceptionally rich source of sugar, and often containing significant quantities of amino acids and other nutrients, nectar may be an especially rewarding addition to the diet of predatory arthropods (Jackson et al. 2001). Along with studies on thomisids (Beck and Connor 1992; Pollard et al. 1995), salticids (Jackson et al. 2001), and several wandering spiders (Taylor and Foster 1996), the study by Chen et al. (2010) suggests that nectar feeding is indeed a widespread strategy in spiders. Chen et al. (2010) tested for the presence of plant sugar (fructose) in wide range of spider species from different families using cold-anthrone tests. The cold anthrone-sulphuric acid method can be used to determine the concentration of fructose in spiders to judge whether they were feeding on nectar (Taylor and Pfannenstiel 2008). A test for fructose can ensure that the sugar is from plants, distinguishing it from glucose, which also occurs in the blood and lymphatic elements of spiders. In the study of Chen et al. (2010), 19 % of all spiders (745 individuals) tested positive for fructose including the families Oxyopidae, Thomisidae, Pisauridae, Salticidae, Lycosidae, Tetragnathidae, Araneidae, Nephilidae, and Agelenidae. In contrast, there was no evidence for nectar feeding in Linyphiidae, Clubionidae, and Theridiidae. For the species *Ebrechtella tricuspidata* (Thomisidae), 216 individuals were included into the study allowing to test for differences based on sex and age. Significantly more females were tested positive than males (87.5 % versus 42.9 %), and immatures tested positive at a lower rate than adults (26.5 % and 66.7 %, respectively). In contrast, nectar feeding was frequent in male crab spiders of the species *Misumenoides formosipes* and increased their longevity, while females of this species rarely take in nectar (Pollard et al. 1995). Therefore, these data suggest that nectar feeding might be more important for adult spiders, but this has to be taken with some precautions, as this conclusion is based on single species, and there is strong evidence for juvenile spiders as well.

Field and laboratory observations confirmed nectar feeding for each of the 90 jumping spider species (Salticidae) studied by Jackson et al. (2001). This suggests that the use of nectar as a resource is a widespread, if not routine, feeding supplement at least for the early instars of salticids. As nectar is taken in as a liquid, it might seem relevant to ask whether the salticids are truly feeding, instead of simply drinking, from flowers. However, in spiders, drinking and feeding are overlapping processes anyway (Jackson et al. 2001). In the same study on salticids, choice tests demonstrated that each species spent longer time taking up sucrose solutions than distilled water.

Especially for active hunters, such as the salticids, nectar can be an important source of energy, which is probably less needed for sit and wait predators. There is strong evidence that nectar feeding increases activity and survival of wandering spiders (Taylor and Foster 1996; Taylor and Pfannenstiel 2008, 2009; Taylor and Bradley 2009) and that they use olfactory cues to locate nectar sources (Patt and Pfannenstiel 2008). For example, Patt and Pfannenstiel (2008) revealed that the spider *Hibana futilis* (Anyphaenidae) can recognize and remember particular chemical stimuli associated with nectar. Following ingestion of minute amounts of sugar, these spiders exhibited counterturning and other local searching behaviours that increased their chances of finding more nectar. In another study ingestion of honey solution significantly increased the survival and shortened the development time of *Ebrechtella tricuspidata*. Female spiders that fed on honey solution had a shorter pre-oviposition period and laid more eggs than those given only water. This study suggests that nectar could be a high-quality supplementary food to maintain normal growth and metabolism in spiderlings and adult female spiders (Wu et al. 2011).

Access to nectar might enhance longevity because of the amino acids, lipids, vitamins, and minerals normally found in nectar in addition to sugars. Vogelei and Greissl (1989) and Taylor and Foster (1996) both showed that spiderlings given access to a simulated nectar source survived longer than spiders given access to water alone. Wu et al. (2011) were interested whether crab spiders of the species *Ebrechtella tricuspidata* show an active preference for honey solution of a certain concentration. The results showed that the number of females feeding on the 10 %, 20 %, and 30 % honey solution was significantly higher than those feeding on water alone, but there was no difference in the number of spiders feeding on the three different concentrations.

28.4 Exploitation of an Ant-Plant Mutualism

The species *Bagheera kiplingi*, a Neotropical jumping spider (Salticidae), exploits a well-studied ant-plant mutualism (Meehan et al. 2009). With evidence from behavioural field observations and stable isotope analyses, the authors show that the main diet of this host-specific spider comprises specialized leaf tips (Beltian food bodies) from *Vachellia* spp. ant-acacias, structures traded for protection in the plant's coevolved mutualism with *Pseudomyrmex* spp. ants that inhabit its hollow thorns. The spiders not only use the plant resources but also feed on ant larvae. *Bagheera kiplingi* is able to circumvent the well-known defences of the acacia's *Pseudomyrmex* ant inhabitants, which keep the plant free of most herbivores and encroaching vegetation. These spiders occur almost exclusively on ant-occupied acacias, where they breed year-round and generally build their nests at the distal tips of older leaves that have low rates of ant patrol (Meehan et al. 2009).

28.5 Spider-Plant Mutualism

The provision of nectar as a reward for predators that keep away herbivores is a common mutualistic interaction between plants and ants. Ruhren and Handel (1999) examined a similar nectar-mediated mutualistic interaction between a plant and spider, namely, between *Chamaecrista nictitans* (Caesalpiniaceae) and jumping spiders in New Jersey, USA. These jumping spiders (*Eris* sp. and *Metaphidippus* sp.) on *C. nictitans* collect nectar in addition to feeding on herbivores, ants, bees, and other spiders. When given a choice between plants with or without active extrafloral nectaries, they prefer plants with nectar. They are indeed involved in the same kind of mutualism shown for ants and plants because *C. nictitans* plants with resident jumping spiders set significantly more seeds than plants with no spiders, indicating a beneficial effect of predator presence on plant fitness. This example suggests that if nectar use is more frequent in spiders, plants that allow access to nectar may indeed have higher density of wandering spiders which can increase trophic control of herbivores, a likely mechanism increasing plant fitness. As ants are often engaged in ant-aphid mutualism which is certainly not beneficial for the plant, attracting spiders instead seems a good idea.

28.6 Conclusion

There is a growing body of evidence for a wide range of spider families that they use nectar as supplementary resource, while pollen feeding in web-building spiders has been studied in the laboratory including a few species only. In one study the simulated nectar intake by feeding with sucrose solution increased life span much longer than pollen feeding implying that a supplementary sugar diet supplies energy and indeed increases survival in times when prey is scarce (Table 28.1). To judge the overall importance of pollen and nectar resources for spiders, however, we need data on the proportion incorporated into the body in relation to prey catches as they are rarely the only source. A promising method is the analysis of stable isotopes to understand long-term resource use in spiders, as it can reveal the proportions of different food sources in the diet of spiders, that is pollen and insect prey. Considering that many spiders use nectar provided by plants, we can assume that there is a greater potential for mutualistic interaction between plants and spiders, as currently known in the literature.

Acknowledgements I wish to thank Wolfgang Nentwig, Eva Knop, and Sebastian Schuch for the valuable discussion on herbivory in spiders.

References

Beck MW, Connor EF (1992) Factors affecting the reproductive success of the crab spider *Misumenoides formosipes*: the covariance between juvenile and adult traits. Oecologia 92:287–295
Carrel JE, Burgess HK, Shoemaker DM (2000) A test of pollen feeding by a linyphiid spider. J Arachnol 28:243–244
Chen X, Chen Y, Wu L, Peng Y, Chen J, Liu F (2010) A survey of nectar feeding by spiders in three different habitats. Bull Insect 63:203–208
del Fiol F, Solveig T, Riccardo G (2007) Fungal spores and pollen as potential nutritional additives for the cross spider *Araneus diadematus* Clerck (Araneae, Araneidae). Bol Micológ 22:47–50
Evans HC (2013) Fungal pathogens of spiders. In: Nentwig W (ed) Spider ecophysiology. Springer, Heidelberg (this volume)
Jackson RR, Pollard SD, Nelson XJ, Edwards GB, Barrion AT (2001) Jumping spiders (Araneae: Salticidae) that feed on nectar. J Zool 255:25–29
Meehan CJ, Olson EJ, Reudink MW, Kyser TK, Curry RL (2009) Herbivory in a spider through exploitation of an ant-plant mutualism. Curr Biol 19:R892–R893
Meissle M (2013) Side effects of *Bacillus thuringiensis* toxins on spiders. In: Nentwig W (ed) Spider ecophysiology. Springer, Heidelberg (this volume)
Nentwig W (1986) Non-webbuilding spiders: prey specialists or generalists? Oecologia 69:571–576
Patt JM, Pfannenstiel RS (2008) Odor-based recognition of nectar in cursorial spiders. Entomol Exp Appl 127:64–71
Peterson JA, Romero SA, Harwood JD (2010) Pollen interception by linyphiid spiders in a corn agroecosystem: implications for dietary diversification and risk-assessment. Arthropod Plant Interact 4:207–217
Pollard SD, Beck MW, Dodson GN (1995) Why do male crab spiders drink nectar? Anim Behav 49:1443–1448
Ruhren S, Handel SN (1999) Jumping spiders (Salticidae) enhance the seed production of a plant with extrafloral nectaries. Oecologia 119:227–230
Smith RB, Mommsen TP (1984) Pollen feeding in an orb-weaving spider. Science 226:1330–1332
Taylor RM, Bradley RA (2009) Plant nectar increases survival, molting, and foraging in two foliage wandering spiders. J Arachnol 37:232–237
Taylor RM, Foster WA (1996) Spider nectarivory. Am Entomol 42:82–86
Taylor RM, Pfannenstiel RS (2008) Nectar feeding by wandering spiders on cotton plants. Environ Entomol 37:996–1002
Taylor RM, Pfannenstiel RS (2009) How dietary plant nectar affects the survival, growth, and fecundity of a cursorial spider *Cheiracanthium inclusum* (Araneae: Miturgidae). Environ Entomol 38:1379–1386
Vogelei A, Greissl R (1989) Survival strategies of the crab spider *Thomisus onustus* Walckenaer 1806 (Chelicerata, Arachnida, Thomisidae). Oecologia 80:513–515

Wise D (1993) Spiders in ecological webs. Cambridge University Press, Cambridge
Wu L, Yun Y, Li J, Chen J, Zhang H, Peng Y (2011) Preference for feeding on honey solution and its effect on survival, development, and fecundity of *Ebrechtella tricuspidata*. Entomol Exp Appl 140:52–58

Chapter 29
Artificial Diets for Spiders

Divina M. Amalin

29.1 Introduction

Spiders are one of the ubiquitous groups of predaceous organisms in the animal kingdom. They are exceptionally well adapted to survive in nature. They are widely distributed and occupying all terrestrial habitats. Surveys conducted by several researchers have shown that spiders are a large and frequent part of the generalist predatory arthropod fauna (Kagan 1943; Whitcomb et al. 1963a, b; Aguilar 1968; Leigh and Hunter 1969; Howell and Pienkowski 1971; Wheeler 1973; Yeargan and Dondale 1974; Dondale et al. 1979; Dean et al. 1982; Mansour et al. 1982; Culin and Yeargan 1983; Barrion and Litsinger 1994; Plagens 1985; Amalin and Barrion 1992; Amalin et al. 2009). They have an important influence in the dynamics of the arthropod pests' population in natural and agricultural landscapes. For instance, in the natural landscape, Van Hook (1971) found that the wolf spider, *Lycosa* sp. (Lycosidae), consumed 21 % of the total arthropods in a grassland ecosystem. An experiment on a forest floor system by Clarke and Grant (1968) indicates that spiders managed the populations of Collembola and other insects. Amalin et al. (2009) reported the diversity of spiders and low insect pest populations in the agricultural buffer zone of the Everglades National Park in Florida. Spiders' potential as biological control of agricultural pests has been recognized to help solve problems associated with the use of chemical pesticides. Several studies on the spider fauna in agricultural landscapes have been documented. Spiders in agricultural fields limit habitation of insect pests. They are considered "farmer's best friend" particularly by the small-scale farmers because they consume large numbers of insect pests as prey.

The use of spiders as biological control agents depends mainly on the conservation and increases in numbers through the use of agricultural practices such as

D.M. Amalin (✉)
Department of Science and Technology, Bicutan, Taguig City, Philippines
e-mail: dmamalin@gmail.com

W. Nentwig (ed.), *Spider Ecophysiology*,
DOI 10.1007/978-3-642-33989-9_29, © Springer-Verlag Berlin Heidelberg 2013

selective spraying (Marc and Canard 1997; Wisniewska and Prokopy 1993; Amalin et al. 2001a) and habitat manipulation rather than mass rearing and periodic release. A very popular example of spider augmentation in the field was in China, where farmers build straw or bamboo shelters for spiders and then move these shelters to whichever paddies are experiencing pest outbreaks. This method of spider augmentation led to a 60 % reduction in pesticide use (Riechert and Bishop 1990; Marc et al. 1999). In Japan, spider populations are maintained and enhanced by the release of *Drosophila* fruit flies into fields when pest insects are not abundant (Marc et al. 1999). Augmentation by releasing laboratory-reared spiders has not been done due to difficulty of mass rearing spiders. The need to rear different insect prey species makes it especially difficult to culture spiders in the laboratory. Formulation of artificial diets would greatly facilitate laboratory rearing of spiders. Few attempts to rear spiders on artificial diets had been done (Peck and Whitcomb 1968; Amalin et al. 1999, 2001b; Smith 2008). This book chapter reviews some of the attempts to rear spiders using artificial diets and discuss important factors that will contribute to advancing in mass rearing of spiders using artificial diet.

29.2 Development of Artificial Diet for Spiders

The global campaign for sustainable agriculture, to sustain the economic viability of farm operations and enhance the quality of life for farmers and society as a whole, shifted the emphasis in pest control utilizing biological control agents, such as predatory arthropods (insect, mites, and spiders), which call for the development of more efficient and economical methods of mass production by artificial diets. Currently, several species of insects are mass produced using artificial diets. About 50 species of beneficial insects have been successfully mass produced and utilized in the field for biological control through augmentation. Advances in rearing beneficial insects in countries like China, Holland, the United Kingdom, the USA, and the former Soviet Union are well documented (Singh 1982). However, laboratory rearing of spiders using artificial diet is still a big challenge. Spider rearing is primarily by confining a single spider in a container provided with water and live insects as food. It appears that most spiders must feed on a variety of insect prey species to obtain the optimum nutrition for survival and reproduction (Greenstone 1979; Uetz et al. 1992). The need to rear different insect prey species makes it especially difficult to culture spiders in the laboratory. Formulation of artificial diets would greatly facilitate laboratory rearing of spiders.

Rearing of spiders on artificial diets has been reported with some degree of success. Peck and Whitcomb (1968) fed *Cheiracanthium inclusum* (Clubionidae) and *Gladicosa gulosa* (Lycosidae) on artificial diet consisting of homogenized insects. Smith (2008) reared araneid spiders in the genus *Poltys* on a nonliving food mix based on houseflies, pollen, egg, and moths. Amalin et al. (1999, 2001b) reared three species of hunting spiders, *C. inclusum* (Clubionidae), *Trachelas volutus* (Corinnidae), and *Hibana velox* (Anyphaenidae), on three different food mixes consisting of milk, egg yolk, and soybean, as follows: (1) milk + egg yolk, (2) soybean liquid, and (3) combination diet (milk + egg yolk + soybean). Peck and Whitcomb (1968) and Smith (2008) food mixes are quite laborious since additional rearing activity is required to have continuous source of prey insects. The artificial diets formulated by Amalin et al. (2001b) are easier since they were made of food ingredients readily accessible in the grocery. Formulation of the artificial diets by Amalin et al. (1999, 2001b) was inspired from the finding that some species of wandering spiders are facultative nectar feeders (Taylor and Foster 1996).

The three diets were used to investigate the survival and development of three species of hunting spiders reared under laboratory conditions. Among the three diets, the combination diet was found superior based on survival and development of the three spider species. The combination diet provided more complete nutritional or dietary requirements (Table 29.1) for the three spider species. The effectiveness of the combination diet was confirmed when compared with natural diet. Females of these three spider species that matured using the combination diet and were fertilized in captivity produced 1–3 egg masses. Oviposition took place 2–7 days after mating, *Cheiracanthium inclusum* had an average of 57 eggs per egg mass, whereas *Hibana velox* and *Trachelas volutus* had an average of 110 and 56 eggs per egg mass, respectively. Thus far, there is no documented data on the reproduction in field populations of these three hunting spiders to compare data, but these complete development data showed that the combination diet formulated by Amalin et al. (2001b) revealed that an artificial diet is adequate for these spiders under laboratory conditions. Further attempts towards mass rearing of these spiders using artificial diets should be pursued. Clearly the proportions of the various ingredients in combination diet must be evaluated to optimize spider survival and reproduction.

29.3 Factors to Be Considered in Formulating Artificial Diet for Spider Rearing

Development of artificial diets for arthropod species is not a young scientific endeavor. Several artificial diets have been developed particularly for insects both phytophagous and entomophagous. However, the development of an artificial diet for spiders is still on a juvenile stage. Similar to entomophagous insects, the development of artificial diets for spiders should address some important factors such as (1) attractiveness of the diet and behavioral cues from the prey to initiate attack and feeding; (2) essential nutrients, feeding on a variety of insect prey species for optimum nutritional requirement (see also Toft 2013); (3) easy food mix, formulated artificial diets with readily accessible ingredients; (4) rearing setup, feeding arena and confinement; and (5) economics of production, affordability of the mass-reared individuals.

Table 29.1 Nutritional composition of the different diets based on the manufacturers' nutritional analyses (mg per 100 ml) (according to Amalin et al. 2001b)

Nutrient composition	Milk + egg yolk	Soybean liquid	Combination (milk + egg yolk + soybean)
Total fat	3,040	1,300	4,340
Saturated fat	1,300	0	1,300
Cholesterol	97.83	0	97.83
Sodium	82.61	39.13	121.74
Total carbohydrates	6,090	10,870	17,040
Sugars	5,220	6,520	12,990
Protein	6,040	2,610	8,650
Potassium	0	126.09	126.09
Vitamin A	348 IU	0	348 IU
Thiamin (B1)	0	0.05	0.05
Riboflavin (B2)	0	0.03	0.03
Niacin (B3)	0	0.52	0.52
Pantothenic acid (B5)	0	0.35	0.35
Pyridoxine hydrochloride (B6)	0	0.05	0.05
Folate (B9)	0	0.02	0.02
Vitamin C	0.52	0	0.52
Vitamin D	43.48 IU	0	43.48 IU
Biotin (vitamin H)	0	0.3	3
Calcium	140	30	170
Iron	0.31	0.31	0.62
Phosphorus	110	40	150
Magnesium	0	17.4	17.4
Zinc	0	0.26	0.26

29.3.1 *Attractiveness of the Diet*

The key in formulating artificial diet is the acceptance or recognition of the diet by the spiders, that is, getting the diet to emulate the actual texture of natural prey. Even diet containing the correct nutritional requirements could remain unacceptable because of the absence of the proper chemical attractants or stimulants to incite the satisfactory feeding responses (House 1961). Advancement in searching for artificial diets that will satisfy the voracious appetite of beneficial insects showed that some store-bought ingredients can trigger feeding by predatory arthropods, like cooked egg and meat paste from ground beef and beef liver (Hardin and Garcia 2001; Senft 2007). Cooked hens' eggs in particular provide the mimicry of the actual texture of the natural prey. Amalin et al. (1999, 2001b) included egg yolk but in a raw form, which probably was the reason why the three hunting spiders were able to locate and feed on the artificial diet. The raw egg yolk was mixed with the other ingredients, cow's milk and soybean milk, to form a liquid diet. Green food color was added to the diet to serve as an indicator if the diet has been consumed since the color of the spiders' opisthosoma changes depending on the food they consumed. The spiders held singly in a glass vial were offered with the liquid diet soaked in a cotton swab. The spiders perched on the stick as it fed on the diet and the opisthosoma color changes to green, which clearly indicated that the diet contains the chemical cues or olfactory cues needed by the hunting spiders to fully accept the diet. Hostettler and Nentwig (2006) found that the wandering spider, *Cupiennius salei*, uses olfactory cues to detect and choose their prey species. Similar phenomenon probably applies to the three hunting spiders used by Amalin et al. (2001b) raised on artificial liquid diet.

29.3.2 Essential Nutrients

Spiders are known to feed on a variety of insect prey species to obtain the optimum nutrient or chemical requirements. The general nutritional requirements for arthropods include carbohydrates, protein, vitamins, cholesterol, and minerals. Carbohydrates are known to be the major energy source important for survival and longevity of any arthropod species. They are usually added to artificial diets in the form of glucose, sucrose, starch, and dextrose. Protein is for growth and reproduction. The common protein supplements are casein, sodium caseinate, and casein hydrolysate given alone or supplemented with amino acid mixes depending on the suitability to a certain species. Vitamins are associated to female fecundity and its eggs' viability as well as its growth and development. Different strains of yeast which vary in vitamin contents were used in different artificial diets. Cholesterol is a common sterol and a precursor of ecdysone, the molting hormone, essential for growth and development. Egg yolk is an excellent source of cholesterol. Minerals in artificial diets are reported essential for growth, longevity, and female fecundity. Salts such as potassium hydrogen phosphate and magnesium sulfate are a good source of minerals. The artificial diet formulated by Amalin et al. (2001b) contains all of the abovementioned nutritional requirements (Table 29.1). Although it gave a satisfactory result in rearing three species of hunting spiders, the nutrient quantities and proportions should be evaluated for a more stable rearing up to several generations.

29.3.3 Easy Food Mix

Common artificial diets for spiders consisted of food mixes with ground insects, which command additional activity in rearing the prey insects. It would be more convenient to have an artificial diet consisting of food ingredients easily accessible from the supermarket.

29.3.4 Rearing Setup

One of the big challenges in mass rearing of spiders is the design of the feeding arena, how to confine them without cannibalizing. Rearing of spiders is usually in single confinement to avoid cannibalism. The primary drawback is the labor-intensive care regime. Rearing container size should be adequate to avoid crowding and allow food search and mating. Construction material should be nontoxic, ventilated, and secured to avoid escape.

29.3.5 Economics of Production

Production of spiders for biological control must be cost-effective to be an integral component of the pest management program in agricultural fields. For instance, the new diet formulated by the US Agricultural Research Service scientist Allen Cohen (Hardin and Garcia 2001; Senft 2007), for predatory insects with pure store-bought ingredients, costs about $50/kg of diet to produce 60,000 big-eyed bug adults or up to 20,000 lacewing adults that will yield about six million eggs during their life span compared to the older diet formulation which cost $600/kg of diet to produce similar number of individuals.

29.4 Conclusion

Spiders are important biological control agents regulating insect populations. They cluster in sites where many prey species are maintained at a high density (Schmitt 1987) and insects represent over 99 % of their diet. For this reason, spiders are a favorable biological control agent in the agricultural ecosystem. Despite the known importance of spiders as biological control agents, its augmentation in the agricultural fields is limited to conservation and increase in numbers through the use of agricultural practices such as selective spraying rather than on mass rearing and release. Their widespread use has been hampered by inadequate methods to rear them on artificial diets or by the costs of rearing prey for them to feed on.

Improvement of the earlier attempts to mass rear spiders on artificial diets (Peck and Whitcomb 1968; Amalin et al. 2001b; Smith 2008) may enable their use in agriculture for augmentation of field populations. Results of Amalin et al. (2001b) showed that an artificial diet is adequate for the survival and development of three species of hunting spiders under laboratory conditions. Attempts towards mass rearing of these spiders using artificial diets should be further pursued. Likewise, successful mass rearing using artificial diets of other biological control agents (parasitoids and predators) can be tried on and modified for spiders. Modification should be based on the attractiveness of spiders on the diet to initiate attack and feeding, the right proportion of the ingredients to satisfy the nutrient requirements, the appropriate rearing arena which can accommodate mass production, and the cost-effective rearing system to understand the spiders' feeding biology and nutritional chemistry. Advancements in mass rearing spiders on artificial diets may enable their use in agriculture for augmentation of field populations.

References

Aguilar F (1968) Araneida frecuentes en el algodonero. Rev Peru Entomol 11:92–95
Amalin DM, Barrion AA (1992) Spiders of potato in the Philippines. Philipp Agric 73:179–184
Amalin DM, Reiskind J, Mcsorley R, Peña J (1999) Survival of the hunting spider, *Hibana velox* (Araneae, Anyphaenidae) raised on different artificial diets. J Arachnol 27:692–696
Amalin DM, Peña JE, Mcsorley R, Browning HW, Crane JH (2001a) Comparison of different sampling methods and effect of pesticide application on spider populations in lime orchards in South Florida. Environ Entomol 30:1021–1027
Amalin DM, Pena JE, Reiskind J, Mcsorley RM (2001b) Comparison of the survival of the three species of sac spiders on natural and artificial diets. J Arachnol 29:253–262
Amalin DM, Pena JE, Duncan R, Leavenwood J, Koptur S (2009) Effects of pesticides on the arthropod community in the agricultural areas near the Everglades National Park. Proc Fla State Hort Soc 122:429–437
Barrion AT, Litsinger JA (1994) Taxonomy of rice insect pests and their arthropod parasites and predators. In: Heinrichs SA (ed) Biology and management of rice insects. Wiley Eastern Limited, London
Clarke RD, Grant PR (1968) An experimental study of the role of spiders as predators in forest litter community. Ecology 49:1152–1154
Culin JD, Yeargan KV (1983) Comparative study of spider communities in alfalfa and soybean ecosystems: foliage dwelling spiders. Ann Entomol Soc Am 76:825–831
Dean DA, Sterling WL, Horner NV (1982) Spiders in eastern Texas cotton fields. J Arachnol 13: 111–120
Dondale CD, Parent B, Pitre D (1979) A 6-year study of spiders (Araneae) in a Quebec apple orchard. Can Entomol 111:377–380
Greenstone MH (1979) Spider feeding behavior optimizes dietary essential amino acid composition. Nature 282:501–503
Hardin B, Garcia J (2001) High performance insect diets. Agr Res (2):16–17
Hostettler S, Nentwig W (2006) Olfactory information saves venom during prey-capture of the hunting spider Cupiennius salei (Araneae: Ctenidae). Funct Ecol 20:369–375
House HL (1961) Insect nutrition. Annu Rev Entomol 6:13–26
Howell JO, Pienkowski RL (1971) Spider populations in alfalfa, with notes on spider prey and effect of harvest. J Econ Entomol 64:163–168
Kagan M (1943) The Araneida found on cotton in central Texas. Ann Entomol Soc Am 36: 257–258
Leigh TF, Hunter RE (1969) Predaceous spiders in California cotton. Calif Agric 23:4–5
Mansour F, Ross JW, Edwards GB, Whitcomb WH, Richman DB (1982) Spiders of Florida citrus groves. Fla Entomol 65:514–552
Marc P, Canard A (1997) Maintaining spider biodiversity in agroecosystems as a tool in pest control. Agric Ecosyst Environ 62:229–235
Marc P, Canard A, Ysnel F (1999) Spiders (Araneae) useful for pest limitation and bioindication. Agric Ecosyst Environ 74:229–273
Peck WB, Whitcomb WH (1968) Feeding spiders on artificial diet. Entomol News 79:233–236

Plagens MJ (1985) The corn field spider community: composition, structure, development, and function. PhD dissertation. University of Florida, Gainesville, 206 pp
Riechert SE, Bishop L (1990) Prey control by an assemblage of generalist predators in a garden test system. Ecology 71:726–736
Schmitt RJ (1987) Indirect interactions between prey: apparent competition, predator aggregation and habitat segregation. Ecology 68:1887–1897
Senft D (2007) Mass reared insect gets fast food. Available at http://www.ars.usda.gov/is/ar/archive/jun97/bugdiets0697.htm
Singh P (1982) The rearing of beneficial insects. N Z Entomol 7:306–310
Smith HM (2008) Some notes on rearing Poltys (Araneae: Araneidae) in captivity. J Arachnol 36: 207–209
Taylor RM, Foster WA (1996) Spider nectarivory. Am Entomol 42:82–86
Toft S (2013) Nutritional aspects of spider feeding. In: Nentwig W (ed) Spider ecophysiology. Springer, Heidelberg (this volume)
Uetz GW, Bischoff J, Rover J (1992) Survivorship of wolf spiders (Lycosidae) reared on different diets. J Arachnol 20:207–221
Van Hook RI (1971) Energy and nutrient dynamics of spider and orthopteran populations in grassland ecosystems. Ecol Monogr 41:1–26
Wheeler AG (1973) Studies on the arthropod fauna of alfalfa. V. Spiders (Araneida). Can Entomol 105:425–432
Whitcomb WH, Exline H, Hite M (1963a) Comparison of spider populations of ground stratum in Arkansas pasture and adjacent cultivated field. Proc Ark Acad Sci 17:6
Whitcomb WH, Exline H, Hite M (1963b) Spiders of the Arkansas cotton field. Ann Entomol Soc Am 56:653–660
Wisniewska J, Prokopy R (1993) Spiders that feed on leafhoppers and leafminer. Fruit Notes 58:1–2
Yeargan KV, Dondale CD (1974) The spider fauna of alfalfa fields in northern California. Ann Entomol Soc Am 67:681–682

Part IX
Ecotoxicology

With their high species diversity and abundance in virtually all terrestrial habitats, spiders are a perfect group for purposes of biomonitoring, for example in an ecotoxicological context. How do spiders suffer from heavy metal or synthetic pesticide contamination of their environment? Can they be used as indicator species? Since insecticides derived from the neem tree are less toxic to spiders than synthetic pesticides, wouldn't this give a good argument against the latter? Since *Bacillus thuringiensis* toxins from genetically engineered crop do not harm spiders, couldn't this be seen as a further good argument in favour of such crops?

Chapter 30
Effects of Heavy Metal Contamination

Paweł Migula, Grażyna Wilczek, and Agnieszka Babczyńska

30.1 Introduction

Heavy metals are naturally occurring constituents of each environment. Most of them are essential to spiders (such as Ca, Fe, Mg, Mn, Zn or Cu), some only in trace levels (Co, Mo, Ni, V) and just a few are recognised as biologically nonessential xenobiotics (As, Cd, Hg, Pb or Sr). Xenobiotic metals are always harmful, while negative effects of essential metals are dose dependent. Spiders, as carnivores, have been recognised as potent biological indicators of heavy metal contamination (Maelfait 1996). Due to a specific form of feeding, even biogenic metals may reach their body at too high concentrations. If the regulatory mechanisms supporting adequate internal metal levels and their proper cellular distribution are insufficient, spiders may suffer from various health alterations easily seen at the cellular level (Wilczek 2005; Wilczek et al. 2008). At the organism level, negative effects appear later as metabolic disorders, inhibition of growth or changes in life history traits (Babczyńska et al. 2011a; Chen et al. 2011; Eraly et al. 2011). At the population level, heavy metals act negatively on population dynamics, density, survival or reproduction (Hendrickx et al. 2003a; Chen et al. 2011; Eraly et al. 2011). Heavy metals may affect species richness and the guild structure (Ferreira 2010).

In nature, spiders may use two main mechanisms to cope with excessive heavy metal burdens: (1) tolerance through phenotypic, morphological, biochemical or behavioural plasticity or (2) resistance. If adaptations are heritable, adapted organisms may have lower fitness than non-adapted ones (Morgan et al. 2007). Under metal pollution conditions only those spider populations are able to survive which developed various tolerance mechanisms to stressors. Genetic adaptations of spiders to heavy metals were not proven yet. Increased tolerance could be reached through physiological adaptations involving trade-off mechanisms leading to

P. Migula (✉) • G. Wilczek • A. Babczyńska
University of Silesia, Katowice, Poland
e-mail: pawel.migula@us.edu.pl

W. Nentwig (ed.), *Spider Ecophysiology*,
DOI 10.1007/978-3-642-33989-9_30, © Springer-Verlag Berlin Heidelberg 2013

different life history traits (Wilczek et al. 2004; Eraly et al. 2010). In this chapter we review the recent knowledge concerning the effects of heavy metals on spiders at various levels of biological organisation. Factors deciding on their distribution in spiders and the potential of spiders to cope with excess of metals will be discussed.

30.2 Factors Determining the Uptake of Heavy Metals

Spiders take up heavy metals mostly from their food which is restricted mainly to the soft tissues of arthropods. Prey types and numbers hunted by spiders at a given location depend on the environmental, climatic and prey population dynamics at a given site. Heavy metal levels might change between years and sites, hunting behaviour and activity (Wilczek and Babczyńska 2000; Ramirez et al. 2011). A high metal body burden of spiders is rather a result of the specificity of metal excretion and storage than the quality of consumed prey (Hopkin 1989; Wilczek and Babczyńska 2000).

Differences between species in accumulation of heavy metals result from physiological and ecological variability between species. In a comparative study on *Pirata piraticus* (Lycosidae) and *Clubiona phragmitis* (Clubionidae), Tojal et al. (2002) showed that both species differ in foraging behaviour and prey composition. The first species feeds on the soil-dwelling insects, while the second forages on flying insects that have lower concentrations of metals.

Studies at the species level have shown differences in the accumulation of heavy metals between juveniles and adults, due to differences in their physiology or temporal differences in their exposure to metals by ingestion (Maelfait 1996; Hendrickx et al. 2003a). In nature, body burdens of metals are often higher in males, irrespective of species (Table 30.1). They have a shorter lifespan and invest less energy to cope with toxic substances. The longer-living females are characterised by a slow rejection of metals, and they should invest more energy into an efficient regulation of metal levels, being responsible for egg production and nursing of the spiderlings (Wilczek et al. 2004). Some laboratory feeding experiments on spiders sampled from polluted areas, however, showed that body concentrations were gender independent, as in the cases of *Pardosa lugubris* (Lycosidae) or *Agelena labyrinthica* (Agelenidae) (Table 30.2).

At specific areas, such as intertidal sites, metal bioavailability for spiders depends on the cation exchange capacity of the sediments. *P. piraticus* often occurs in freshwater and brackish water banks of the rivers and feeds on organisms which also live in close contact with the intertidal sediments. Du Laing et al. (2002) showed that contents of Cd, Cu and Zn in their body were higher than in the sediment in sites situated closer to the river mouth and were characterised by a higher salinity. Chloride content in sediments correlated well with the Cd, Cu and Zn contents in the spiders' body.

Table 30.1 Maximum body burdens of four heavy metals in unpolluted and polluted sites in selected species of spiders

	Unpolluted sites (μg/g body weight)				Polluted sites (percentage of unpolluted site)				
Species	Cd	Pb	Zn	Cu	Cd	Pb	Zn	Cu	Reference
Araneus diadematus	8.2	10	613	62	256	290	309	152	1, 3, 5
Linyphia triangularis	7.1	5.2	694	43	296	519	157	137	1, 3, 5
Agelena labyrinthica midgut glands	31.5	16.8	824	36	240	89	285	183	1
Agelena labyrinthica females	17	2.7	817	129	16	163	73	100	2
Agelena labyrinthica males	2.2	0.4	749	92	2,273	1,600	106	168	2
Metellina segmentata	5.4	7.1	462	41.7	407	239	204	216	3, 5
Pardosa palustris	6	5	512	54	367	240	186	576	3
Pardosa sp.	0.4	0.3	128	14.8	325	200	154	88	4
Pirata piraticus	10.9	30	444	157	41	90	89	64	6

References refer to (1) Babczyńska et al. (2011b), (2) Wilczek et al. (2003), (3) Wilczek and Migula (1996), (4) Van Straalen et al. (2001), (5) Wilczek et al. (1997), (6) Tojal et al. (2002).

Table 30.2 Cd concentrations (mean ± standard deviation) measured in experimentally exposed and control offspring originating from different populations

		Cd concentration (μg/g)	
Species	Material	Not Cd exposed	Cd exposed
Pardosa saltans (1)	Reference site	0.39 ± 0.37	61.1 ± 13.3
	Polluted site	0.45 ± 0.53	57.7 ± 14.5
Pardosa lugubris (2)	Reference site		
	Females	5.9 ± 2.2	141.0 ± 25.0
	Males	6.7 ± 2.6	116.0 ± 58.0
Agelena labyrinthica (3)	Reference site		
	Juveniles	8.2 ± 1.6	339.0 ± 56.3
	Females	1.8 ± 1.7	44.2 ± 29.3
	Males	2.8 ± 0.5	176.6 ± 67.8
	Polluted site		
	Juveniles	28.0 ± 4,3	228.2 ± 25.7
	Females	21.9 ± 10.4	276.7 ± 46.9
	Males	26.5 ± 6.8	278.5 ± 28.7

Data sources: (1) Eraly et al. (2010), (2) Babczyńska and Migula (2002), (3) Babczyńska et al. (2011a), source 1 refers to wet weight; sources 2–3 to dry weight.

30.3 Bioconcentration and Bioaccumulation of Heavy Metals

Metals reach the body through the alimentary tract mostly as sucked liquid digest or mashed pulp of the captured prey (Hopkin 1989). Contamination of spider webs with heavy metals may be an additional source of metal toxicity for orb-web species that regularly consume old webs (Hose et al. 2002).

Ground-dwelling wolf spiders better reflect the pollution level of their biotope than web-building spiders (Peterson et al. 2003; Wilczek et al. 2005; Eraly et al. 2010). Internal Pb, Cd and Cu concentrations increased with the soil concentration for most spiders (as well as for most other taxonomic groups) in the order Pb > Cd > Cu. Body concentrations of Zn were quite constant over a large range of soil concentrations (Heikens et al. 2001), but Jung and Lee (2012) showed different results for *Pardosa astrigera*. The metal content in the soil was in the order of Zn > Pb > Cu > Cd, while in spiders, the order of Zn > Cu > Cd > Pb was identified. The bioconcentration factors (BCF: metal body content/metal soil content) were 28.1, 0.14, 2.7 and 6.3 for Cd, Pb, Zn and Cu, respectively. Based on the ratio of assimilated to eliminated amounts of metals, spiders in general are regarded as macro-concentrators of heavy metals, but they can better regulate biogenic metals. High accumulation of metals (Ni, Pb, Zn) characterised spiders from moderately contaminated lowland floodplain along the river Rhine (Schipper et al. 2008). Their body burdens reached up to ten times higher levels than coleopterans.

In areas such as Ni-rich serpentine sites, spiders demonstrated some Ni regulatory possibilities. Concentration of Ni in spiders exceeded 100 μg/g dry weight, and their insect prey also contained Ni in even much higher concentrations than in spiders (Peterson et al. 2003; Boyd 2009). The bioaccumulation factor (metal body burdens of spiders/metal in prey) for two spiders (the thomisid *Misumena vatia* and the pholcid *Pholcus phalangioides*) fed with Ni-rich mirid bugs (*Melanotrichus boydi*) was low and reached the value of 0.54 and 0.60, respectively (Boyd 2009). In this case, Ni was not magnified along the food chain. The wolf spider *P. lugubris* demonstrated also some Cd regulatory possibilities. Spiderlings fed with Cd-contaminated flies until maturation accumulated less Cd than their prey (Table 30.2).

Bioaccumulation rates are metal dependent. For example, *P. astrigera* exposed for 8 weeks reached saturation levels of Pb and Hg within 2 weeks (ca 3–4 μg/spider), while the saturation level for Cd (10 μg/spider) was only reached after 6 weeks (Jung et al. 2005). The excretion rate of Cd is very slow as it was shown experimentally for *Pirata piraticus* fed with Cd-contaminated flies (Hendrickx et al. 2003b).

Although *Agelena labyrinthica* is a web-building species, its ecological niche is similar to wandering spiders, and it accumulates higher amounts of metals than other web-building species (Wilczek and Babczyńska 2000; Wilczek et al. 2005). The bioaccumulation factor of Cd depends on the developmental stage, gender and preadaptation to heavy metals under natural conditions (Table 30.2).

30.4 Internal Metal Distribution and Loss of Heavy Metals

In six wandering and web-building spider species levels of Cd, Pb, Cu and Zn in the midgut glands were higher than in their gonads (Wilczek and Babczyńska 2000). The spiders protect their reproductive functions and thus continue to occur in heavy metal-contaminated sites. Metals are deposed mainly in the digestive cells of the midgut glands in metal-rich mineral granules, which were already described by

Hopkin (1989). Three types (A, B and C) are localised inside and the forth (D) outside the cells. Type A contains Ca, Zn and Pb; type B granules include Cu, Hg, Cd, Zn, Pb or Fe; type C contains Fe; and type D comprises calcium carbonate granules. The structures grow with spider age and with the duration of exposure. Metals from these mineral granules can be transported into the gut lumen and eventually are lost with faeces after every meal. Elimination of Cd with the exuviae is low. In exuviae of Cd-intoxicated *A. labyrinthica,* Cd concentration constituted from 0.2 % in juveniles to 10 % in males of the opisthosomal concentration. Cu losses with the exuviae of this spider are low in males (3 %) but high (36 %) in females (Babczyńska et al. 2011a).

30.5 Subcellular Effects of Heavy Metals

30.5.1 Metallothioneins

As most animals, also spiders are able to synthesise metallothioneins which are small (3–14 kDa) cysteine-rich metal-binding proteins. The level of metallothioneins in web-building spiders reflects the soil pollution with heavy metals. Babczyńska et al. (2011a) showed that in *Araneus diadematus* (Araneidae) from polluted sites, metallothionein synthesis was found to be induced by metals unlike in spiders of this species from the reference site. The response of other examined species [*Linyphia triangularis* (Linyphiidae) and *Agelena labyrinthica*] was less pronounced. The level of metallothioneins is not only species specific but also metal dependent. High metallothionein concentrations follow high concentrations of Cd and Pb, but body concentrations of essential Cu and Zn and metallothionein levels were not correlated. This suggests that metallothioneins play a regulatory role in relation to biogenic metals if their concentrations exceed the tolerable levels. In case of Cd and Pb, the role of metallothioneins shifts rather towards defence than to regulation and storage.

A significant increase of metallothioneins was also indicated in feeding experiments on *A. labyrinthica* fed Cd- or Cu-contaminated fruit flies (Babczyńska et al. 2011b). The protective role of the proteins demonstrated itself by a relative increase of metallothionein-positive cells in the midgut glands and increased total concentration of metallothioneins in this organ. In a majority of cases, the variability of this parameter reflects the degree of pollution in natural habitats of spiders, their developmental stage and gender.

The level of metallothioneins does not increase immediately. Eraly et al. (2010), studying the response of *Pardosa saltans* intoxicated with Cd under laboratory conditions, demonstrated that the synthesis of the proteins is energetically costly and starts only when metals in the spiders' body exceed a certain level. Thus in spiders, metallothioneins as protective molecules in metal intoxication might play rather a complementary role to the other efficient mechanism of metal neutralisation, such as binding of metals and their storage as granular concretions in cells of the midgut glands.

30.5.2 Defensive Mechanisms Against Heavy Metals

Metals affect the cellular integrity and functioning of the organism and therefore evoke various defensive mechanisms that repair damages and restore homeostasis. Metals, if not bound to any ligand or sequestered in granules, often attack mitochondria causing disturbances in the respiratory chain and uncoupling of oxidative phosphorylation, thus reducing energy available for defensive processes. Analysis of the cellular effects caused by Cd and Cu indicated a series of alterations in mitochondria of the midgut gland cells of *Agelena labyrinthica* (Babczyńska et al. 2011b). Mitochondria may play a strategic role in cell death initiation, and therefore, a decrease of mitochondrial transmembrane potential in midgut gland cells indicates early stages of apoptosis (Wilczek 2005).

One of the most important negative effects of heavy metals in the organism is to deplete the ATP concentration. The ADP/ATP ratio used as a measure of the energetic status of cells indirectly informs whether the cells proliferate (<0.1), overcome apoptosis (0.1–1) or suffer from necrosis (>1), and this defines the level of ATP depletion. Experiments with *A. labyrinthica*, dietarily exposed to an excess of Cd or Cu, indicated that their energy status was not disturbed. While ATP remained at similar levels in control and intoxicated spiders, the ADP/ATP ratio was two to six times higher for metal-exposed spiders and was age and gender dependent. ATP in females was kept at high levels, and the increased ADP/ATP ratio reflected an increased ADP pool, not a depletion of ATP, suggesting enhanced anabolic processes (Babczyńska et al. 2011b). Such an energy status of females was probably due to activated anaerobic pathways. An enhanced activity of enzymes involved in anaerobic reactions was already documented earlier for females of *A. diadematus*, *Metellina segmentata* (Tetragnathidae), *Linyphia triangularis* and *Pardosa palustris* sampled at heavily polluted sites (Marczyk et al. 1993). The adenylate energy charge index (= ATP + 0.5ADP/ATP + ADP + AMP) used in these spiders to quantify the metabolic energy held in the adenylate pool was kept at a physiologically optimal level (i.e. above 0.75), what can result from the enhancement of anaerobic pathways of energy production (Wilczek 1996).

Heavy metals are responsible for an enhanced production of reactive oxygen species and the inhibition of antioxidant activities. Intracellular accumulation of reactive oxygen species can damage biological macromolecules. The degree of the damage depends on the efficiency of the antioxidative defence which includes nonenzymatic scavengers, such as glutathione or heat-shock proteins, and major antioxidative enzymes: superoxide dismutase, catalase or glutathione peroxidases which are selene dependent or selene independent. Enzymatic antioxidative reactions in spiders living in metal-polluted habitats are species specific. In males of *P. lugubris,* collected from sites along a heavy metal pollution gradient, positive correlations between the activity of glutathione S-transferase and body burdens of Pb and Zn and between both peroxidase types and Cu were found. These spiders also maintained high glutathione concentration. Web-building *A. labyrinthica* collected from the same habitats had lower levels of these parameters (Wilczek et al. 2004).

Similar types of enzymatic relations were identified between catalase and superoxide dismutase and heavy metals in the body of *P. palustris* from metal-contaminated areas. Another examined species, *L. triangularis,* also kept high activities of these enzymes, although the metal level in their body was low (Wilczek and Migula 1996).

The activities of both types of glutathione peroxidases were also higher in *Xerolycosa nemoralis* (Lycosidae) than in *L. triangularis* and *A. labyrinthica* when isolated midgut glands were analysed. Heat-shock proteins (Hsp70) play an important cytoprotective role. In the above-mentioned species from heavily polluted areas, the percentage of Hsp70 positive cells of midgut glands was species, gender and site dependent. They were the highest in *X. nemoralis* and the lowest in *L. triangularis* (Wilczek et al. 2008).

Interspecies differences in detoxifying responses of spiders from heavily polluted habitats were identified studying activity levels of enzymes involved in the neutralisation of xenobiotics during the first and second phase of detoxification. Wolf spiders (*P. palustris*, *P. lugubris*) maintain high hydrolytic processes catalysed by carboxylesterases and the conjugation with glutathione led by glutathione S-transferases. The intensity of these reactions in web-building spiders (*Metellina segmentata, L. triangularis, A. diadematus, A. labyrinthica*) was lower than in wolf spiders (Wilczek and Migula 1996; Wilczek et al. 2003, 2004).

Responses of spiders living in habitats chronically polluted by heavy metals to chemical stress are often gender dependent. Maintaining a high glutathione level is an important element in the defence process of males against high metal loads and their pro-oxidative activity. The antioxidative reactions in females are species specific and depend mainly on high glutathione peroxidase and catalase activities, although the levels of these enzymes are species specific (Wilczek et al. 2008). It cannot be excluded that under an increased impact of pollutants, female spiders are able to utilise various detoxifying mechanisms and strategies, even if energetic costs are high, to protect their genetic material. The main direction in the use of the described parameters by different spider species is schematically presented in Fig. 30.1.

30.6 Effects of Heavy Metals on Spider Populations and Communities

30.6.1 Heavy Metals and Spider Populations

Often a cascade of ecological effects can be seen under heavy metal pollution. In the case of spiders, this could be measured as reduced secondary productivity due to restrictions in the energy budget caused by such factors as reduced prey availability or increased predation risk leading to reduced fitness. This, in turn, affects both reproductive output and survivorship.

The relationship between density and richness of ground-dwelling spider species along a heavy metal pollution gradient was documented by Koponen and Koneva

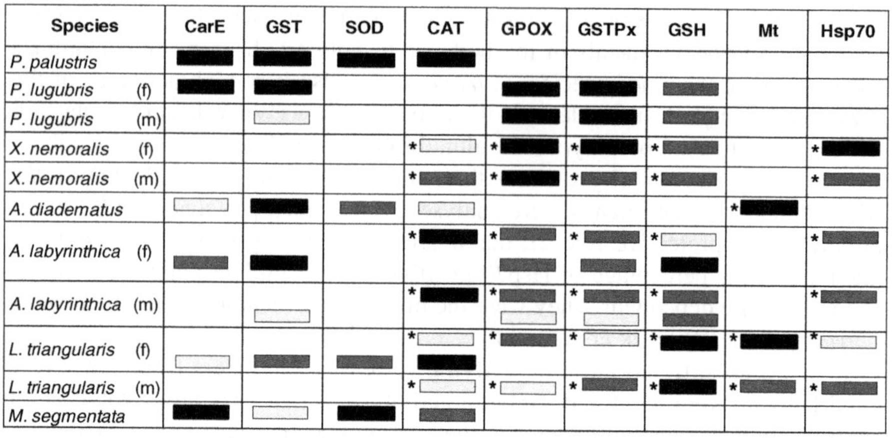

Fig. 30.1 Relative levels of *CarE* carboxylesterase, *GST* glutathione S-transferase, *SOD* superoxide dismutase, *CAT* catalase, *GPOX* and *GSTPx* glutathione peroxidases, *GSH* glutathione contents, *Mt* metallothioneins and *Hsp70* heat-shock proteins in the body of actively hunting and web-building spiders from heavily polluted environments. *Black shaded* high, *grey shaded* medium, *white shaded* low, *asterisk* measured in midgut glands, *f* females, *m* males

(2005) for the areas around the Severonickel smelter (Kola Peninsula). In the nearest belt, 2.5 km from the smelter, the spider density was low (6 individuals/m^2) with a diversity of only two species. Also at a distance of 5 km, only 3 individuals/m^2 and up to 10 km (heavily polluted and afforested zone) 8 individuals/m^2 were collected. At a distance of 30 km, the density reached nearly 60 individuals/m^2. In another Ni-contaminated area the density of spiders representing 18 species was 20 individuals/m^2, while in only slightly polluted sites it was six times higher and represented by 58 species. Linyphiid spiders were most abundant in slightly polluted sites (64 % of all species and 60 % of the individuals). At heavily polluted sites their contribution was reduced to 23 % and 38 %, respectively, but they still were the most abundant family among identified ten families of spiders (Koponen 2011).

Changes in growth, reproduction and survival are sensitive factors, indicating sublethal effects caused by the metal pollution at population level. The overall effects on reproductive traits depend on trade-off mechanisms of energy investments between inactivation or elimination of metals and reproduction. Pollutants are limiting the availability of energy which could be allocated for reproduction. In such conditions, less but larger offspring is favoured. Such a strategy is used by the wolf spider *Pirata piraticus*. Hendrickx et al. (2003a) studied populations from four polluted sites located on the banks of the river Schelde and one unpolluted inland reference site. Reproductive output and fecundity of spiders from populations with high metal loads was strongly reduced as they allocated energy to detoxification processes. The size of the females was positively correlated with metal levels and negatively with fecundity and reproductive output, but they produced larger eggs.

If the reproductive output is neglected, such a strategy indicates adaptive trade-off mechanisms between egg size and number of produced eggs.

Such a strategy, as described above, is not common in spiders. Effects of heavy metals on both growth and reproduction could be negative and are species dependent. For example, in laboratory conditions, the total number of eggs laid by females of Pb and Zn-stressed *Pardosa astigera* was sharply reduced (Chen et al. 2011). In another wolf spider (*Pardosa saltans*) experimentally exposed to Cd (100 μg/g wet body weight) also resulted in prolonged development, reduced growth and reproduction (Eraly et al. 2010). The authors concluded that the main defence mechanism against heavy metals is the up-regulated expression of metallothionein-like proteins (MTLP). Further field studies on this species indicated that metals affected life history and physiological traits in four out of six *P. saltans* populations (Eraly et al. 2011). Cd concentrations correlated negatively with the adult size and positively with the egg mass. Reproduction and the reproductive output (fecundity, egg mass, cocoon weight) were the lowest or even postponed in population with the highest Cd and Zn body burdens. At the most polluted site the highest number of females carrying cocoons appeared later in the season than in other sites. These changes were not related with an increase of MTLP, and the authors stated that under field conditions, physiological defence plays a lower role than indirect and synergistic effects of heavy metal pollution.

Fluctuating asymmetry (random differences in the development of both sides of a bilateral symmetrical character) is often used as an indicator of environmental and/or genetic stress. In *P. piraticus* Maelfait and Hendrickx (1998) recorded a significant positive correlation between the distance between two dorsal spines on tibia IV and internal concentrations of Cd. Further studies on the same species showed that in populations adapted to metal stress fluctuating asymmetry was negatively correlated with clutch mass but positively with the egg size. Selection obviously eliminated suffering spiders and favours females producing larger eggs. Thus, diminished clutch masses are compensated by a production of larger eggs (Hendrickx et al. 2003c). A stronger relationship between fluctuating asymmetry and fitness demonstrated possible selection against developmentally unstable individuals with the evolution of adaptive responses to environmental stress and the resources devoted to reproduction.

30.6.2 Heavy Metals and Spider Communities

Species composition in areas around the contamination sources is directly and indirectly influenced not only by heavy metals but also by many other factors, such as alterations of vegetation structure, properties of soils and their acidification or deacidification by liming (Maelfait 1996; Ormerod and Rundle 1998).

The composition of spider species may differ distinctly between highly polluted and unpolluted sites. Read et al. (1998) analysed the relations between spider species and a suite of environmental variables in woodlands. They showed that the heavy metal pollution near the main pollution source affected the complete invertebrate community, causing a lack of lycosid spiders in highly polluted woods. This vacant niche was filled by large agelenid species which seem to be more adaptable to such pollution levels.

More difficult is to demonstrate differences in spider communities between moderately polluted and unpolluted sites. Jung et al. (2008), using a Shannon diversity index, showed that a moderate contamination with heavy metals did not affect communities of ground-dwelling spiders, although the diversity index was higher in the unpolluted site, and species diversity was negatively correlated with Pb levels in soil. Using discrimination techniques, the authors stated that among the dominant spider families, only the composition and structure of Linyphiidae separated unpolluted and moderately polluted sites. Moreover, some lycosid species (*Pardosa laura*, *P. astrigera*) were highly abundant at moderately polluted sites and could even be used as indicators of metal bioaccumulation.

The use of guilds (groups of organisms that explore the same resource, e.g. generalist feeders of arthropods that differ in hunting behaviour) is one of the functional ways of community structure analysis. Guilds may reflect functional diversity and alterations of the integrity of biological communities. Ferreira (2010) assessed the impact of Cu and Zn in surface soils on the abundance and community structure of ground-dwelling spider communities in an area of 2 km around a copper mine. The abundance of spiders increased with increased distance from the contamination source, but Cu in the body was not correlated with Cu in soils. The author stated that the abundance of the ground hunter guild responded better to soil contamination, while specialist web-building spiders responded better to atmospheric deposition.

30.7 Conclusions

The results presented above indicate that under strong environmental pressure as a consequence of high concentrations of heavy metals in soils and prey, spiders are able to protect themselves and their genetic material using various, efficient strategies. The strategies differ between species. Hunting spiders in general intensify the activity of detoxifying enzymes, including antioxidative defence, whereas web-building species with their moderate levels of protective mechanisms seem better adapted to the unfavourable conditions in postindustrial or ore-bearing areas.

References

Babczynska A, Migula P (2002) Cadmium—fenitrothion interaction in spider *Pardosa lugubris* and the fruit fly *Drosophila melanogaster*. Bull Environ Contam Toxicol 69:589–592

Babczyńska A, Wilczek G, Szulińska E, Franiel I (2011a) Quantitative immunodetection of metallothioneins in relation to metals concentration in spiders from variously polluted areas. Ecotoxicol Environ Saf 74:1498–1503

Babczyńska A, Wilczek G, Wilczek P, Szulińska E, Witas I (2011b) Metallothioneins and energy budget indices in cadmium and copper exposed spiders *Agelena labyrinthica* in relation to their developmental stage, gender and origin. Comp Biochem Physiol 154C:161–171

Boyd RS (2009) High-nickel insects and nickel hyperaccumulator plants: a review. Insect Sci 16:19–31

Chen XQ, Zhang ZT, Liu R, Zhang XL, Chen J, Peng Y (2011) Effects of the metals lead and zinc on the growth, development, and reproduction of *Pardosa astrigera* (Araneae: Lycosidae). Bull Environ Contam Toxicol 86:203–207

Du Laing G, Bogaert N, Tack FMG, Verloo MG, Hendrickx F (2002) Heavy metal contents (Cd, Cu, Zn) in spiders (*Pirata piraticus*) living in intertidal sediments of the river Scheldt estuary (Belgium) as affected by substrate characteristics. Sci Total Environ 289:71–81

Eraly D, Hendrickx F, Bervoets L, Lens L (2010) Experimental exposure to cadmium affects metallothionein-like protein levels but not survival and growth in wolf spiders from polluted and reference populations. Environ Pollut 158:2124–2131

Eraly D, Hendrickx F, Backeljau T, Bervoets L, Lens L (2011) Direct and indirect effects of metal stress on physiology and life history variation in field populations of a lycosid spider. Ecotoxicol Environ Saf 74:1489–1497

Ferreira JG (2010) Impact of mine pollution on the abundance and community structure of ground-dwelling spiders (Araneae): potential use as bioindicators. MSc Thesis, University of Lisboa, Lisboa

Heikens A, Peijnenburg WJG, Hendriks AJ (2001) Bioaccumulation of heavy metals in terrestrial invertebrates. Environ Pollut 113:385–393

Hendrickx F, Maelfait JP, Speelmans M, Van Straalen NM (2003a) Adaptive reproductive variation along a pollution gradient in a wolf spider. Oecologia 134:189–194

Hendrickx F, Maelfait JP, Langenbick F (2003b) Absence of cadmium excretion and high assimilation result in cadmium biomagnification in a wolf spider. Ecotoxicol Environ Saf 55:287–292

Hendrickx F, Maelfait JP, Lens L (2003c) Relationship between fluctuating asymmetry and fitness within and between stressed and unstressed populations of the wolf spider *Pirata piraticus*. J Evol Biol 16:1270–1279

Hopkin SP (1989) Ecophysiology of metals in terrestrial invertebrates. Elsevier, London

Hose GC, James JM, Graz MR (2002) Spider webs as pollution indicators. Environ Pollut 120:725–733

Jung CS, Lee SB, Jung MP, Lee JH, Lee SW, Lee SH (2005) Accumulated heavy metal content in wolf spider, *Pardosa astrigera* (Araneae: Lycosidae), as a bioindicator of exposure. J Asia Pac Entomol 8:185–192

Jung MP, Kim ST, Kim S, Lee JH (2008) Species diversity and community structure of ground-dwelling spiders in unpolluted and moderately heavy metal-polluted habitats. Water Air Soil Pollut 195:15–22

Jung MP, Lee JH (2012) Bioaccumulation of heavy metals in the wolf spider, *Pardosa astrigera* L. Koch (Araneae: Lycosidae). Environ Monit Assess 184:1773–1779

Koponen S (2011) Ground-living spiders (Araneae) at polluted sites in the Subarctic. Arachnol Mitt 40:80–84

Koponen S, Koneva G (2005) Spiders along a pollution gradient (Araneae). Acta Zool Bulg Suppl 1:131–136

Maelfait JP (1996) Soil spiders and bioindication. In: Van Straalen NM, Krivolutsky DA (eds) Bioindicator systems for soil pollution. Kluwer, Dordrecht

Maelfait JP, Hendrickx F (1998) Spiders as bio-indicators of anthropogenic stress in natural and semi-natural habitats in Flanders (Belgium): some recent developments. In: Selden A (ed) Proceedings of the 17th European Colloquium of Arachnology, Edinburgh

Marczyk G, Migula P, Trzcionka E (1993) Physiological responses of spiders to environmental pollution in the Silesian Region (Southern Poland). Sci Total Environ 2:1315–1321

Morgan A, Kille P, Sturzenbaum S (2007) Microevolution and ecotoxicology of metals in invertebrates. Environ Sci Technol 41:1085–1096
Ormerod SJ, Rundle SD (1998) Effects of experimental acidification and liming on terrestrial invertebrates: implications for calcium availability to vertebrates. Environ Pollut 103:183–191
Peterson LR, Trivett V, Baker AJM, Aguiar C, Pollard AJ (2003) Spread of metals through an invertebrate food chain as influenced by a plant that hyperaccumulates nickel. Chemoecology 13:103–108
Ramirez MG, McCallum J, Landry JM, Vallin VA, Fukui SA, Gerqus HE, Torres JD, Sy CL (2011) Relationships between physiological characteristics and trace metal body burdens of banded garden spiders *Argiope trifasciata* (Araneae, Araneidae). Ecotoxicol Environ Saf 74:1081–108
Read H, Martin MH, Rayner JMV (1998) Invertebrates in woodlands polluted by heavy metals—an evaluation using canonical correspondence analysis. Water Air Soil Pollut 106:17–42
Schipper AM, Wijnhoven S, Leuven RS, Ragas A, Hendrickx AJ (2008) Spatial distribution and internal metal concentrations of terrestrial arthropods in a moderately contaminated lowland floodplain along the Rhine River. Environ Pollut 151:17–26
Tojal C, Hendrickx F, Tack FMG, Maelfait JP, Bogaert N, Willems K, Vernaillen P, Mertens J, Verloo MG (2002) Heavy metal concentrations in the spiders *Pirata piraticus* (Clerck, 1757) and *Clubiona phragmitis* (C.L. Koch, 1843) along the Scheldt Estuary (Belgium). Sci World J 2:978–982
Van Straalen NM, Butovsky RO, Pokarzhevskii AD, Zaitsev AS, Verhoef CS (2001) Metal concentrations in soil and in invertebrates in the vicinity of a metallurgical factory near Tula (Russia). Pedobiologia 45:451–466
Wilczek G (1996) Adenine nucleotides and adenylate energy charge as measures of physiological condition of spiders exposed to environmental stress. Stud Soc Sci Torun 4:137–142
Wilczek G (2005) Apoptosis and biochemical biomarkers of stress in spiders from industrially polluted areas exposed to high temperature and dimethoate. Comp Biochem Physiol 141C:194–206
Wilczek G, Babczyńska A (2000) Heavy metals in the gonads and hepatopancreas of spiders (Araneae) from variously polluted areas. Ekologia (Bratislava) 19:283–292
Wilczek G, Migula P (1996) Metal body burdens and detoxifying enzymes in spiders from industrially polluted areas. Fresenius J Anal Chem 354:643–647
Wilczek G, Majkus Z, Migula P, Bednarska K, Świerczek E (1997) Heavy metals and detoxifying enzymes in spiders from coal and metallurgic dumps near Ostrava (Czech Republic). In: Żabka M (ed) Proceedings of the 16th European Colloquium of Arachnology, Siedlce
Wilczek G, Babczyńska A, Migula P, Wencelis B (2003) Activity of esterases as biomarkers of metal exposure in spiders from the metal pollution gradient. Pol J Environ Stud 12:765–771
Wilczek G, Babczyńska A, Augustyniak M, Migula P (2004) Relations between metals (Zn, Pb, Cd and Cu) and glutathione-dependent detoxifying enzymes in spiders from a heavy metal pollution gradient. Environ Pollut 132:453–461
Wilczek G, Babczyńska A, Majkus Z (2005) Body burdens of metals in spiders from the Lidice coal dump near Ostrava (Czech Republic). Biologia (Bratisl) 60:599–605
Wilczek G, Babczyńska A, Wilczek P, Doleżych B, Migula P, Młyńska H (2008) Cellular stress reactions in female and male spiders from areas variously polluted with heavy metals. Ecotoxicol Environ Saf 70:127–137

Chapter 31
Side Effect of Synthetic Pesticides on Spiders

Stano Pekár

31.1 Introduction

Spiders are natural enemies occurring in many agroecosystems. Indeed, they are often the most abundant and diversified natural enemies, which contribute to the reduction of several pests. However, certain pest management practices, such as the application of pesticides, can disrupt their role in pest control. There are a number of different synthetic insecticide and acaricide classes with varying effects on the nontarget arthropods. Most synthetic insecticides such as organophosphates, cyclodienes, pyrethroids, carbamates, and organochlorines are neurotoxic and have a highly negative effect. A few are insect growth regulators, antifeedants, or microbial pesticides, which are usually less detrimental, at least in terms of acute toxicity. Unexpectedly, even herbicides and fungicides or their additives can have serious detrimental effects.

Research on the ecotoxicology of spiders has received rather limited attention in comparison with other natural enemies, for example, parasitoids (Theiling and Croft 1988). Early papers investigated the side effects of pesticides in the field by focusing on the abundance of the spider population and species richness of the community. Later toxicological studies mainly evaluated the direct toxicity, specifically mortality at different pesticide concentrations and/or doses, with a main emphasis on the concentration/dose recommended for pest control. Formulations that caused low mortality were considered harmless and recommended for use. However, such acute toxicity tests did not consider side effects on other life-history traits, such as foraging, defence, mating, or migration, which may severely impair spider pest control abilities. Recent research, therefore, focuses more on the mechanisms behind the intoxication of surviving individuals and, in particular, on so-called sublethal effects.

S. Pekár (✉)
Department of Botany and Zoology, Masaryk University, Brno, Czech Republic
e-mail: pekar@sci.muni.cz

W. Nentwig (ed.), *Spider Ecophysiology*,
DOI 10.1007/978-3-642-33989-9_31,

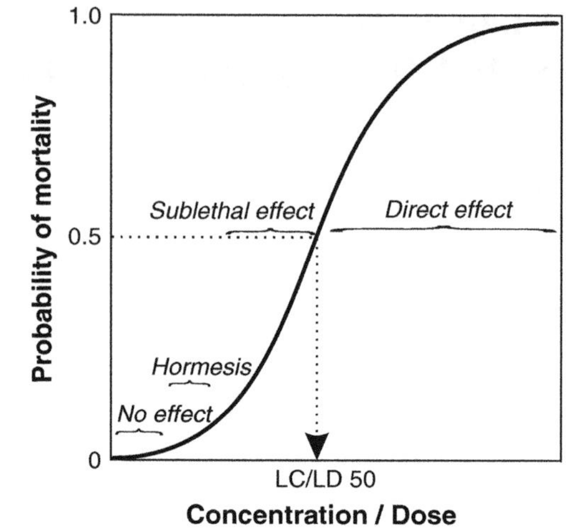

Fig. 31.1 Logit response curve of mortality showing the variety of effects in relation to the concentration or dose (LC or LD_{50})

The type of effect is a function of the concentration and/or dose (Fig. 31.1). At relatively high concentrations/doses, contact with a formulation causes high mortality. This is known as a direct effect. At concentrations/doses lower than LC_{50} (i.e., lethal concentration) or LD_{50} (i.e., lethal dose)—that is, concentration/dose at which more than 50 % of individuals survive—sublethal effects are observed. These are evaluated as behavioural or physiological changes in an individual that survived exposure to a pesticide. Yet even lower concentrations/doses may have a stimulating effect on behavioural and physiological functions. Such effects are called hormesis. Very low concentrations or doses have negligible or no effect. For example, contact with herbicide did not affect the walking activity of two lycosid species, *Pardosa milvina* and *Hogna helluo* (Evans et al. 2010), or courtship, mating, prey capture, or escape from predators in another lycosid *Pardosa palustris* (Michalková and Pekár 2009). The application of carbamate did not alter web structure in an araneid *Larinioides sclopetarius* (Lengwiler and Benz 1994). Oviposition and fecundity did not decrease in the lycosid *Pirata piratoides* after insect growth regulator application (Deng et al. 2008). Rate of development was not extended in the lycosid *Pardosa amentata* after contact with pyrethroid and organophosphate (Toft and Jensen 1998). Diacylhydrazine did not affect the functional response in the philodromid *Philodromus cespitum* (Řezáč et al. 2010).

31.2 Direct Effect

Mortality resulting from contact with a pesticide, administered topically, orally, or via residues, differs among pesticides, formulations, spider species, developmental stage, and sex and is also influenced by current abiotic and biotic conditions. In general, highest mortality was caused by standard doses of insecticides and acaricides and lowest mortality by fungicides and herbicides (Theiling and Croft 1988). The causes of mortality differ among pesticide classes. Little is known of their effect in spiders. Neurotoxic formulations, such as type I pyrethroids, have a quick knockdown effect. The symptoms of type II pyrethroids include ataxia, convulsions, contractions, and paralysis. Mortality is caused secondarily through disturbing the water balance. Besides the effect on neurotransmitters, organophosphates tend to accumulate in cell membranes and modify their permeability. It has been shown in the sparassid *Polybetes pythagoricus* that contact with organophosphate altered the lipid dynamics and caused diminished capacity for oxygen binding (Cunningham et al. 2002). Pyrethroids can further cause dysfunction in myocardial cells and alter heart rate (Desneux et al. 2007).

31.3 Sublethal Effects

Sublethal effects have been studied in 21 species of spiders using 26 formulations (Table 31.1). The frequency of encountering sublethal concentrations/doses for spiders in the field is assumed to be higher than that of lethal doses. Due to the enhanced stability of synthetic pyrethroids in the environment, spiders may contact sublethal doses for several days after spraying (Baatrup and Bayley 1993). Sublethal doses also occur when the pesticide is diluted in water (such as dew) or because of the filter effect of plants or spray drift, which give rise to surfaces with sparse contamination. Such situations are frequent in many types of crops. A further situation arises because of webs. Three-dimensional webs of theridiid and dictynid spiders can reduce contact of the resident spider with pesticide droplets (Pekár 1999).

Different classes of pesticides can cause different sublethal effects. These effects of intoxication interact with numerous life-history traits and can be measured by investigating changes in physiology and behaviour. Behavioural responses often reflect changes at the physiological level. Activities such as movement, prey capture, reproduction, development, and defence are highly sophisticated, governed by complex neural interactions, and particularly disrupted by neurotoxic formulations. However, in contrast to direct effects, recovery from sublethal effects is possible over several days. For example, *P. amentata* recovered in 3–4 days (Baatrup and Bayley 1993).

31.3.1 Enzymatic Activity

Pesticides can have a dramatic effect on many physiological processes. Organophosphorous and carbamate compounds inhibit cholinesterases, enzymes in the central nervous system, as has been shown for the lycosid *Anoteropsis hilaris* following organophosphates application (Van Erp et al. 2002) and the linyphiid *Hylyphantes graminicola* following both organophosphate and pyrethroid applications (Peng et al. 2010). In the latter species, reduced activity of cholinesterases pertained even to offspring.

Table 31.1 List of spider species and synthetic formulations that have been used to evaluate sublethal effects

Family/species	Insecticide (acaricide)	Fungicide	Herbicide	Reference
Araneidae	–	–	–	–
Alpaida veniliae	–	–	Glyphosate	Benamú et al. (2010)
Araneus diadematus	Cypermethrin, paraffin oil	Prochloraz, triadimenol	–	Samu and Vollrath (1992)
Larinioides sclopetarius	Deltamethrin, diazinon, dicofol, pirimicarb	–	–	Lengwiler and Benz (1994)
Neoscona pratensis	Endosulfan, spinosad	–	–	Benamú et al. (2007)
Linyphiidae	–	–	–	–
Erigone atra	Fenvalerate	–	–	Dinter et al. (1998)
Frontinella communis	Malathion	–	–	Tietjen and Cady (2007)
Hylyphantes graminicola	Dimethoate, fenvalerate, methamidophos	–	–	Deng et al. (2006, 2007) and Peng et al. (2010)
Oedothorax apicatus	Deltamethrin, fenvalerate	–	–	Everts et al. (1991), Jagers op Akkerhuis et al. (1995; 1997) and Dinter et al. (1998)
Tenuiphantes tenuis	Cypermethrin	–	–	Shaw et al. (2005)
Lycosidae	–	–	–	–
Anoteropsis hilaris	Chlorpyrifos, diazinon	–	–	Van Erp et al. (2002)
Pardosa amentata	Cypermethrin, dimethoate, λ-cyhalothrin	–	–	Baatrup and Bayley (1993), Toft and Jensen (1998), Nielsen et al. (1999) and Shaw et al. (2006)
Pardosa milvina	–	–	Glyphosate	Evans et al. (2010) and Wrinn et al. (2012)
Pardosa palustris	Chlorpyrifos + cypermethrin, deltamethrin	–	Clomazone, glyphosate	Pekár and Beneš (2008) and Michalková and Pekár (2009)
Pardosa pseudoannulata	Imidacloprid, methamidophos	–	–	Widiarta et al. (2001) and Wang et al. (2006a, b)
Pardosa prativaga	Dimethoate	–	–	Pedersen et al. (2002)

Pirata piratoides	Buprofezin	–	–	Deng et al. (2008)
Rabidosa rabida	Malathion	–	–	Tietjen (2006) and Tietjen and Cady (2007)
Schizocosa ocreata	Malathion	–	–	Tietjen and Cady (2007)
Philodromidae	–	–	–	–
Philodromus cespitum	Acetamiprid, azadirachtin, chlorpyrifos + cypermethrin, deltamethrin, diflubenzuron, methoxyfenozide, spinosad	–	Clomazone	Pekár and Beneš (2008) and Řezáč et al. (2010)
Salticidae	–	–	–	–
Salticus scenicus	Malathion	–		Tietjen and Cady (2007)
Sparassidae	–	–	–	–
Polybetes pythagoricus	Fenitrothion	–	–	Cunningham et al. (2002)

Acetylcholinesterase is an enzyme decomposing the neurotransmitter acetylcholine. Inhibition of acetylcholinesterase would lead to hyperactivity and general perturbation in all systems. This can result in death. Spiders can resist neurotoxic effects by producing detoxification enzymes, such as glutathione S-transferase and glutathione peroxidase. The absence or low activities of detoxification enzymes increase their susceptibility. In *Pardosa amentata* Nielsen et al. (1999) found that detoxification enzymes are present in the bodies of spiders throughout the year. The activity of glutathione S-transferase was affected only slightly by pyrethroid application, while the activity of glutathione peroxidase was strongly induced. Similarly, in another study, glutathione S-transferase was not induced by an organophosphorous formulation applied on *Anoteropsis hilaris* (Van Erp et al. 2002). Glutathione peroxidase, unlike glutathione S-transferase, is thus considered an important system used to combat unpredictable exposure to toxins. Alternatively, resistance to intoxication can be achieved by the overproduction of acetylcholinesterase, which was found to be affected by the nutritional state of spiders. Pedersen et al. (2002) found a strong synergistic effect of nutrition on the regenerative ability of acetylcholinesterase after organophosphate application in the lycosid *Pardosa prativaga*.

Not only enzymes in the nervous system were found to be affected by pesticides. Wang et al. (2006a) reported the inhibition of protease activity in the gut of spiders following application of a high dose of an organophosphate on the lycosid *Pardosa pseudoannulata*.

31.3.2 Water Loss

It is known that contact with neurotoxic formulations (organophosphorous, carbamate, cyclodiene, and pyrethroid) causes accelerated water loss (Everts et al. 1991), which can lead to mortality, as has been observed in the linyphiid *Oedothorax apicatus* (Jagers op Akkerhuis et al. 1997). Using this species as a model, the following scenario has been predicted. Contact with pyrethroid will cause abnormal signalling intensity of the humidity in the cuticle. Such false signalling will cause the spiders to stop searching for a more favourable humid environment, which is also achieved via locomotion disruption. After penetration of pyrethroid into the haemolymph, diuretic hormone is produced, which will cause active water excretion (Jagers op Akkerhuis et al. 1997). Indeed, pyrethroids have been shown to increase diuresis by anal excretion (Everts et al. 1991). If passive water loss and water excretion increases with time, it can lead to mortality, particularly in a dry environment (Jagers op Akkerhius et al. 1995). In *O. apicatus*, high susceptibility was found at high temperatures and low humidity. High humidity, on the other hand, reduced intoxication.

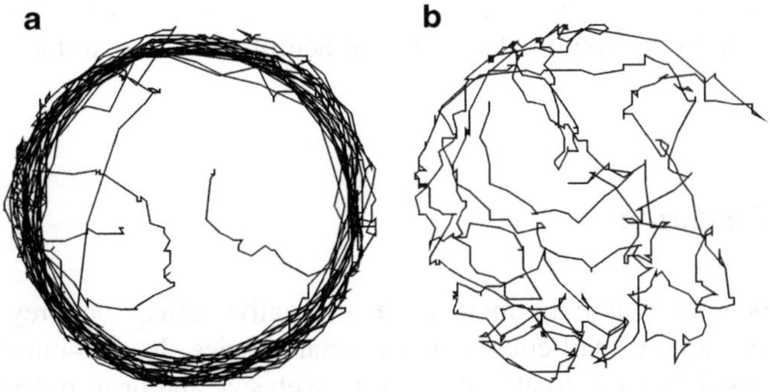

Fig. 31.2 Path tracks by *Pardosa* spiders when exposed to control surface (**a**) and pyrethroid residues (**b**) for a period of 2 h

31.3.3 Movement

Locomotion is a fundamental behavioural property of spiders. It reflects interaction with the environment; therefore, walking activity is the easiest behaviour in which to detect sublethal effects, particularly in cursorial species. At relatively higher doses or concentrations, movement is usually reduced, while at relatively lower doses or concentrations, it is induced. The application of pyrethroids on *P. amentata* caused ataxia and paralysis of the fourth legs for 3 days, probably due to relaxation of the flexor muscles (Shaw et al. 2006). It increased quiescence for 12 h, and recovery was observed after 24 h (Baatrup and Bayley 1993). In two linyphiids, *Tenuiphantes tenuis* and *Oedothorax apicatus*, pyrethroids reduced levels of movement, which lasted for several days (Shaw et al. 2005; Everts et al. 1991). This is because high velocities require greater neural control and energy, which the spider is likely deprived of. Contact with surfaces treated with organophosphate shifted the circadian rhythm in the lycosid *Rabidosa rabida* and the salticid *Salticus scenicus* (Tietjen and Cady 2007). The spiders were active earlier than normal. The authors suggest that organophosphate affected the interaction between the light receptors and the circadian clock. The lycosid *Pardosa palustris* not only moved less on organophosphate- and pyrethroid-treated surfaces than on control (Fig. 31.2) but exhibited an uncoordinated walking pattern (Pekár and Beneš 2008).

If movement is induced, it is probably an indication of avoidance and subsequent dispersal. Avoidance is induced by repulsion. In pyrethroids, the active ingredient is often highly repellent, whereas in other pesticides repulsion is caused by the additives (Desneux et al. 2007). Whether pesticides induce dispersal has not yet been studied. However, results from two studies suggest that it might. The lycosid *Pardosa milvina* increased its speed of movement on a herbicide-treated surface and thus minimised exposure and risk (Evans et al. 2010). Contact with a surface treated with an organophosphate elevated activity in the lycosids *Schizocosa ocreata* and *R. rabida*, in the linyphiid *Frontinella communis*, and in the *S. scenicus* presumably because the spiders tried to avoid it (Tietjen and Cady 2007). Alternatively, in lycosid and salticid species, increased activity could be caused by impaired visual processing due to the abnormal function of neurotransmitters in the protocerebral ganglion.

31.3.4 Predation

Reduced locomotor activity must have a negative effect on prey search and frequency of capture, particularly in cursorial species. In web-building species, such inhibition leads also to alteration of the web size and/or web design. Furthermore, some pesticides may reduce olfactory capacity and disrupt the detection of kairomones from prey and, if a pesticide has an antifeedant property, even decrease consumption (Desneux et al. 2007).

Reduced prey capture frequency lasting for several days was reported for several species: *Pardosa pseudoannulata* after neonicotinoid application (Widiarta et al. 2001); the araneid *Neoscona pratensis* following spinosyn application (Benamú et al. 2007); and *Pardosa amentata* and two linyphiids, *Erigone atra* and *Oedothorax apicatus*, following the application of pyrethroids (Dinter et al. 1998; Shaw et al. 2006). The araneid *Alpaida veniliae* rejected prey intoxicated with herbicide for 4 days probably because of aversion (Benamú et al. 2010). Surprisingly, even insect growth regulators caused reduced prey consumption in *P. piratoides*, but the mechanism is not known (Deng et al. 2008).

A more sophisticated approach to investigating sublethal effects on predation is to study the functional response, that is, the relationship between prey capture and prey density, as it allows two components of predation to be estimated: searching efficiency and handling time. Only in a few studies, the functional response has been investigated. Typically, spiders show type 2 response. In *Philodromus cespitum* exposure to benzoylurea, spinosyn, and neonicotinoid reduced type 2 functional response (Fig. 31.3) due to an increase in handling time (Řezáč et al. 2010). In females of *Hylyphantes graminicola*, prey capture was reduced for 24 h following topical application of an organophosphate, after which it changed from type 2 to type 1 response (Deng et al. 2007).

The effect on web design has been studied mainly in orb-weaving species probably because changes to the web architecture are easier to detect and quantify than in species constructing three-dimensional webs. Topical application of pyrethroids dramatically reduced web frequency and size in the araneid *Araneus diadematus* (Samu and Vollrath 1992). In another araneid, *Larinioides sclopetarius*, application of organophosphate and organochlorine only slightly altered web architecture, but the application of pyrethroid delayed web construction and reduced its size by more than 70 % (Lengwiler and Benz 1994). Oral application of herbicide altered web production in terms of number of radii and spiral threads (Fig. 31.4) in *Alpaida veniliae* (Benamú et al. 2010). Sublethal doses of spinosyn and a pyrethroid reduced the frequency of web building in *Neoscona pratensis* (Benamú et al. 2007). Reductions in the size of webs were also observed in the sheet-web-building linyphiid *Tenuiphantes tenuis* following pyrethroid application (Shaw et al. 2005).

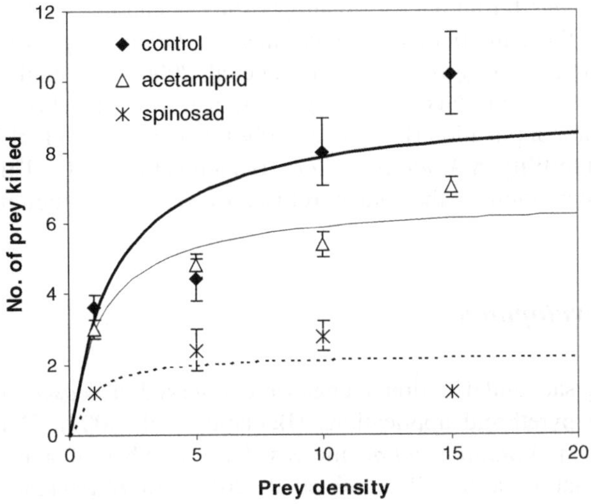

Fig. 31.3 Functional response curves of type 2 in *Philodromus* spiders and *Drosophila* flies offered as prey following application of two insecticides (neonicotinoid and spinosyn) and the control. Points are means; whiskers are standard errors of the mean

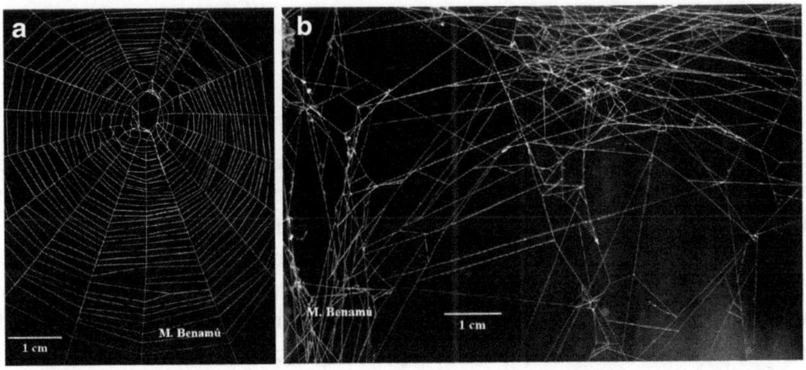

Fig. 31.4 Comparison of webs produced by *Alpaida veniliae* in the laboratory (**a**) constructed under control conditions and (**b**) constructed under glyphosate treatment applied orally via prey (photos: Benamú)

31.3.5 Reproduction

Reproduction involves a series of processes coordinated by the nervous and hormonal systems, namely, mate finding, chemical or sound communication, courtship, mating, egg sac production, spermatogenesis/oogenesis, and brood care. All these processes are prone to being affected particularly by neurotoxic substances, yet evidence is scarce. Contact with surfaces treated with an organophosphate lowered the rate of mating in *Rabidocosa rabida* (Tietjen 2006). In a few other cases the inhibitory effect seems to result from reduced prey consumption prior to oviposition. A pyrethroid and/or organophosphates caused a reduction in egg sac production in three linyphiids, *Erigone atra*, *Oedothorax apicatus* (Dinter et al. 1998), and *Hylyphantes graminicola* (Deng et al. 2006; Peng et al. 2010). Females of *Pirata piratoides* produced fewer eggs and had lower fertility following insect growth regulator application (Deng et al. 2008). Application of herbicide decreased fecundity and fertility in *Alpaida veniliae* (Benamú et al. 2010). This was probably the result of starvation, as the spiders refused to consume contaminated prey.

31.3.6 Development

Abnormal egg sacs and dehydrated eggs were observed in *Neoscona pratensis* after spinosyn and pyrethroid applications (Benamú et al. 2007). Herbicide applied through prey in *Alpaida veniliae* increased the number of abnormal eggs and prolonged instar duration (Benamú et al. 2010). In *Hylyphantes graminicola*, offspring from mothers exposed to pyrethroid took longer to develop and had lower mass (Peng et al. 2010). Organophosphate application resulted in smaller body size in *H. graminicola* (Deng et al. 2006). Whether this was a direct effect of the formulation or an indirect effect of reduced feeding remains to be investigated.

31.3.7 Defence

Secondary defence strategies, such as escape, are based on movement abilities. In cursorial species, reduced speed of movement would increase susceptibility to predation. In web-building species, for example, linyphiids, spraying causes them to leave their webs. As a result, they become more exposed to predators. Everts et al. (1991) proved that reduced speed of movement in *Oedothorax apicatus* females following contact with pyrethroid caused higher predation by carabids.

Pesticides can also interrupt signalling from kairomones produced by predators. Spiders thus would not be able to avoid areas signalling the presence of predators. Wrinn et al. (2012) tested whether *Pardosa milvina* can recognise cues from its intraguild predators when the surface is treated with glyphosate and found that the response to kairomones from one predator was not dramatically altered by the herbicide but elevated to cues from another predator.

31.4 Hormesis

As outlined above, low doses or concentrations of pesticides can cause improved performance. This could help spiders to suppress pests if the magnitude of improvement was marked. Evidence for hormesis in spiders is still scarce and insufficient chiefly because most toxicological studies have centred on higher concentrations/

doses. A low dose of organophosphates stimulated predation in *Hylyphantes graminicola* (Deng et al. 2007), which killed more prey (but did not consume them), and in two lycosids, *Pardosa pseudoannulata* and *Pardosa amentata* (Toft and Jensen 1998; Wang et al. 2006b). Detailed analysis of the functional response revealed that it was due to an improvement in the searching efficiency of the predator. Physiological processes are stimulated by low doses too. Organophosphates induced, though only slightly, the hatching rate in *H. graminicola* (Deng et al. 2006). When *P. pseudoannulata* was treated with a low dose of an organophosphate, the protease activity increased (Wang et al. 2006a). Strangely enough, growth rate was higher and body size was larger in *Pirata piratoides* after application of an insect growth regulator at LD50 than after application at LD10 (Deng et al. 2008). The occurrence and magnitude of hormesis in spiders thus requires additional investigation.

31.5 Conclusions

Sublethal concentrations/doses affect a variety of traits. Although evidence of sublethal effects has been gathered for several spider species and formulations, it is still insufficient with respect to all possible effects, species, formulations, and concentrations/doses. Many effects—for example, on dispersal, defence, or fecundity—are only presumed and remain to be investigated.

Due to the current development of plant protection measures towards the use of selective chemical substances with limited direct effect on natural enemies, the toxicology of spiders should be centred on sublethal effects. New protocols and guidelines for risk assessment are needed to achieve fast and reliable quantification of the effects. This should be paralleled by the detailed study of a selected agrobiont species and estimation of the relative importance of particular life-history traits prone to sublethal doses. For example, predation and reproduction represent two different life-history components. Predation is a daily activity, while reproduction usually takes place only once in a spider's life. Thus, failure to capture prey has an immediate effect on biological control, whereas failure to reproduce would affect the control potential of the next generation.

What is the influence of sublethal effects at the population level? It is difficult to estimate because the consequences of effects at physiological levels are often unknown (Desneux et al. 2007). However, it is expected that even minor disruptions caused by sublethal doses can render spiders ineffective for biological control and, on a longer-term basis, can have an impact on the intrinsic rate of population increase.

References

Baatrup E, Bayley M (1993) Effects of the pyrethroid insecticide cypermethrin on the locomotor activity of the wolf spider *Pardosa amentata*: quantitative analysis employing computer-automated video tracking. Ecotoxicol Environ Saf 26:138–152

Benamú MA, Schneider MI, Pineda S, Sanchez NE, Gonzalez A (2007) Sublethal effects of two neurotoxican insecticides on *Araneus pratensis* (Araneae: Araneidae). Commun Agric Appl Biol Sci 72:557–559

Benamú MA, Schneider MI, Sánchez NE (2010) Effects of the herbicide glyphosate on biological attributes of *Alpaida veniliae* (Araneae, Araneidae), in laboratory. Chemosphere 78:871–876

Cunningham M, Garcia CF, González-Baró MR, Garda H, Pollero R (2002) Organophosphorous insecticide fenitrothion alters the lipid dynamics in the spider *Polybetes pythagoricus* high density lipoproteins. Pestic Biochem Physiol 73:37–47

Deng L, Dai J, Cao H, Xu M (2006) Effects of an organophosphorous insecticide on survival, fecundity, and development of *Hylyphantes graminicola* (Sundevall) (Araneae: Linyphiidae). Environ Toxicol Chem 25:3073–3077

Deng L, Dai J, Cao H, Xu M (2007) Effects of methamidophos on the predating behavior of *Hylyphantes graminicola* (Sundevall) (Araneae: Linyphiidae). Environ Toxicol Chem 26:478–482

Deng L, Xu M, Cao H, Jiayin D (2008) Ecotoxicological effects of buprofezin on fecundity, growth, development, and predation of the wolf spider *Pirata piratoides* (Schenkel). Arch Environ Contam Toxicol 55:652–658

Desneux N, Decourtye A, Delpuech J-M (2007) The sublethal effects of pesticides on beneficial arthropods. Annu Rev Entomol 525:81–106

Dinter A, Mausch H, Brauckhoff U (1998) Risk assessment for the linyphiid spiders *Erigone atra* and *Oedothorax apicatus* and the pyrethroid Sumicide 10 resulting from laboratory dose–response studies. IOBC Bull 21:77–87

Evans SC, Shaw EM, Rypstra AL (2010) Exposure to glyphosate-based herbicide affects agrobiont predatory arthropod behaviour and long-term survival. Ecotoxicology 19:1249–1257

Everts JW, Willemsen I, Stulp M, Simons L, Aukema B, Kammenga J (1991) The toxic effect of deltamethrin on linyphiid and erigonid spiders in connection with ambient temperature, humidity, and predation. Arch Environ Contam Toxicol 20:20–24

Jagers op Akkerhuis GAJM, Rossing WAH, Piet GJ, Everts JW (1997) Water depletion, an important cause of mortality in females of the spider *Oedothorax apicatus* after treatment with deltamethrin: a simulation study. Pestic Biochem Physiol 58:63–76

Jagers op Akkerhius GAJM, Westerhof R, van Straalen NM, Koeman JH (1995) Water balance, respiration and immobilisation in relation to deltamethrin poisoning and physical conditions in the epigeal spider *Oedothorax apicatus*. Pestic Sci 44:123–130

Lengwiler U, Benz G (1994) Effects of selected pesticides on web building behaviour of *Larinioides sclopetarius* (Clerck) (Araneae, Araneidae). J Appl Entomol 117:99–108

Michalková V, Pekár S (2009) How glyphosate altered the behaviour of agrobiont spiders (Araneae: Lycosidae) and beetles (Coleoptera: Carabidae). Biol Control 51:444–449

Nielsen SA, Toft S, Clausen J (1999) Cypermethrin effects on detoxication enzymes in active and hibernating wolf spider (*Pardosa amentata*). Ecol Appl 9:463–468

Pedersen L-F, Dall LG, Soerensen BC, Mayntz D, Toft S (2002) Effects of hunger level and nutrient balance on survival and acetylcholinesterase activity of dimethoate exposed wolf spiders. Entomol Exp Appl 103:197–204

Pekár S (1999) Foraging mode: a factor affecting the susceptibility of spiders (Araneae) to insecticide applications. Pestic Sci 55:1077–1082

Pekár S, Beneš J (2008) Aged pesticide residues are detrimental to agrobiont spiders (Araneae). J Appl Entomol 132:614–622

Peng Y, Shao X-L, Hose GC, Chen J (2010) Dimethoate, fenvalerate and their mixture affects *Hylyphantes graminicola* (Araneae: Linyphiidae) adults and their unexposed offspring. Agric For Entomol 12:343–351

Řezáč M, Pekár S, Stará J (2010) The negative effect of some selective insecticides on the functional response of a potential biological control agent, the spider *Philodromus cespitum*. BioControl 55:503–510

Samu F, Vollrath F (1992) Spider orb web as bioassay for pesticide side effects. Entomol Exp Appl 62:117–124

Shaw EM, Wheater CP, Langan AM (2005) The effects of cypermethrin on *Tenuiphantes tenuis* (Blackwall, 1852): development of a technique for assessing the impact of pesticides on web building in spiders (Araneae: Linyphiidae). Acta Zool Bulg suppl 1:173–179

Shaw EM, Waddicor M, Langan AM (2006) Impact of cypermethrin on feeding behaviour and mortality of the spider *Pardosa amentata* in arenas with artificial 'vegetation'. Pest Manag Sci 62:64–68

Theiling KM, Croft BA (1988) Pesticide side-effect on arthropod natural enemies: a database summary. Agr Ecosyst Environ 21:191–218

Tietjen WJ (2006) Pesticides affect the mating behavior of *Rabidosa rabida* (Araneae, Lycosidae). J Arachnol 34:285–288

Tietjen WJ, Cady AB (2007) Sublethal exposure to a neurotoxic pesticide affects activity rhythms and patterns of four spider species. J Arachnol 35:396–406

Toft S, Jensen AP (1998) No negative sublethal effects of two insecticides on prey capture and development of a spider. Pestic Sci 52:223–228

Van Erp S, Booth L, Gooneratne R, O'Halloran K (2002) Sublethal responses of wolf spiders (Lycosidae) to organophosphorous insecticides. Environ Toxicol 17:449–456

Wang Z, Song D-X, Fu X-Q, Li Y-L (2006a) Effect of methamidophos on midgut proteinase activity in the wolf spider, *Pardosa pseudoannulata* (Araneae: Lycosidae) assayed with piezoelectric bulk acoustic wave impedance analysis method. Acta Entomol Sin 49:700–704

Wang Z, Song D-X, Zhu M-S (2006b) Functional response and searching behavior to the brown planthopper, *Nilaparvata lugens* by the wolf spider, *Pardosa pseudoannulata* under low dose chemical pesticides. Acta Entomol Sin 49:295–301

Widiarta IN, Matsumura M, Suzuki Y, Nakasuji F (2001) Effects of sublethal doses of imidacloprid on the fecundity of green leafhoppers, *Nephotettix* spp. (Hemiptera: Cicadellidae) and their natural enemies. Appl Entomol Zool 36:501–507

Wrinn KM, Evans SC, Rypstra AL (2012) Predator cues and an herbicide affect activity and emigration in an agrobiont wolf spider. Chemosphere 87(4):390–396. doi:10.1016/j.chemosphere.2011.12.030

Chapter 32
Side Effects of *Bacillus thuringiensis* Toxins on Spiders

Michael Meissle

32.1 Introduction

In the first half of the twentieth century, researchers discovered that intracellular crystalline protein inclusions (parasporal crystals) produced by the bacterium *Bacillus thuringiensis* (*Bt*) during sporulation show insecticidal properties. At the same time, first insecticidal spray products based on sporulating bacterial suspensions of *Bt* were commercialized in France. Since then hundreds of different subspecies, producing different combinations of insecticidal proteins, have been discovered. Biopesticides based on *Bt* often contain viable bacterial spores, parasporal crystals that consist of crystal (Cry) and cytotoxic proteins, and formulation components that are added to improve product characteristics. Cry proteins, however, are considered the main active ingredients of *Bt* sprays. The most widely used subspecies in commercial insecticides has been *Bt kurstaki*, which kills a wide range of lepidopteran pest species. Its products have been mainly used in forestry and in organic farming to protect vegetables, field crops, tree fruits, berries and ornamentals. Also, stored commodities can be protected from pest infestation. Chrysomelid beetles, such as the Colorado potato beetle *Leptinotarsa decemlineata*, can be controlled with products based on *Bt tenebrionis*. *Bt israelensis* shows high activity against dipterans and has been used for the control of mosquito and blackfly larvae. Today, *Bt* sprays are the most widely used biopesticides in the world and comprise 1–2 % of the global insecticide market (Glare and O'Callaghan 2000).

In 1996, genetically engineered (GE) plants transformed with genes from *Bt* became commercially available. Those GE maize and cotton varieties have been protected from the most serious lepidopteran pests by expression of specific Cry proteins, Cry1Ab or Cry1Ac, respectively. Since then, *Bt* crops have been

M. Meissle (✉)
Agroscope ART, Zurich, Switzerland
e-mail: michael.meissle@art.admin.ch

W. Nentwig (ed.), *Spider Ecophysiology*,
DOI 10.1007/978-3-642-33989-9_32, © Springer-Verlag Berlin Heidelberg 2013

Table 32.1 Overview of plants genetically engineered with *Bacillus thuringiensis* (*Bt*) genes, their target pests and the insecticidal proteins that they express

Plant	Target pests	*Bt* protein
Maize	Corn borers (Lepidoptera)	Cry1Ab
		Cry1Ab + Vip3A
	Corn rootworms (Col.: Chrysomelidae)	Cry3Bb1
Cotton	Several Lepidoptera species	Cry1Ac
		Cry1Ab
		Cry1Ac + Cry2Ab
		Cry1Ac + Cry1F
		Cry1Ac + Cry1Ab
		Vip3a
Potato	Colorado potato beetle (Col.: Chrysomelidae)	Cry3Aa
Rice	Several Lepidoptera species	Cry1Ab
		Cry1Ab/Cry1Ac fusion protein
Eggplant	Colorado potato beetle (Col.: Chrysomelidae)	Cry3Bb

Only plants for which data on spiders are available are listed (Peterson et al. 2011).

cultivated on ever increasing areas. The most important crop transformed with *Bt* genes is maize, with a global area of 43 million hectares, followed by cotton, with 23 million hectares in 2011 (James 2011). This represents one quarter of the global maize production area and two thirds of the cotton area, which illustrates the importance of *Bt*-transgenic varieties for modern agriculture on a global scale. Today's *Bt* plants can express a variety of *cry* genes that target lepidopteran or coleopteran pests (Table 32.1). In addition, genes encoding for vegetative insecticidal *Bt* proteins (*vip*) have been engineered into crops. For better protection of a plant against one pest complex (e.g., different lepidopteran species) and for protection against several different pests (e.g., lepidopteran and coleopteran species), *Bt* proteins are often combined (stacked) in one plant (James 2011).

Insecticidal Cry proteins have a relatively narrow spectrum of activity against certain groups of herbivores, while other herbivores remain unaffected. Therefore, it is important to maintain a healthy natural enemy complex which helps to keep populations of non-target herbivores below economic injury thresholds. In an integrated pest management context, *Bt* products including GE crops represent highly specific tools, which efficiently solve the main pest problems and allow combination with other measures, including biological control, to help control other, minor pests. In the risk assessment process of GE plants and pesticides, which precedes registration and commercial use, studies on beneficial non-target species, such as pollinators, decomposers and natural enemies, play an important role. Spiders are amongst the most numerous arthropod predators in agricultural crops, and it is widely acknowledged that they contribute to the biological control of pests (Nyffeler and Sunderland 2003; Peterson et al. 2011). This chapter provides an overview of the knowledge on the routes and consequences of exposure of spiders to *Bt* Cry proteins.

32.2 Mode of Action of Cry Proteins

To be effective against target pests, *Bt* Cry proteins need to be ingested. In the case of *Bt* sprays, this happens during feeding on (sprayed) plant surfaces. In the insect gut, the parasporal crystals are solubilised and processed to active toxins by proteases. In *Bt* crops, the plant tissue produces specific Cry proteins in a soluble and often truncated or modified form, which shortcuts the solubilisation and processing steps in the insect gut after ingestion. After traversing the peritrophic membrane, the Cry proteins bind to specific receptors on the brush border of the columnar cells lining the midgut. Subsequently, they insert irreversibly into the plasma membrane of the gut cells and form oligomers. This results in the formation of pores that function as ion channels, leading to a loss of transmembrane potential, cell lysis and ultimately death of the insect. Susceptibility of an arthropod to a certain Cry protein is only given when the midgut environment is suitable and when appropriate receptors are present in the membrane of the midgut cells. In general, lepidopterans and dipterans have an alkaline gut environment. While lepidopterans show susceptibility to Cry1A and Cry2A class toxins, dipterans are susceptible to Cry2A, Cry4 and Cry11B class toxins. Chrysomelid beetles have a neutral to acidic midgut pH and are susceptible to Cry3 class toxins. The complex mode of action of insecticidal Cry proteins and the narrow spectrum of activity were key to the attractiveness of *Bt* for use as biopesticide and for genetic engineering (Knowles 1994; Glare and O'Callaghan 2000).

32.3 Exposure of Spiders to *Bt* Proteins

Bt sprays are applied once or several times per growing season, depending on pest pressure. The crystals and spores are deposited on the plant surface. They are highly sensitive to degradation by UV light and are easily washed off by rain. In contrast, GE crops produce *Bt* proteins in high concentrations over the whole growing season. The insecticidal proteins, which are produced in most tissues, are contained within the plant. Those differences indicate that exposure of arthropods to *Bt* toxins differs between fields sprayed with *Bt* formulations and fields planted to *Bt* crops. In sprayed fields, spiders can be exposed when hit directly by the spray, when feeding on prey that is covered with *Bt* suspension, or when recycling their own web that intercepted *Bt* suspension after spraying (Fig. 32.1). In sprayed fields and in *Bt* crops, spiders can be exposed when feeding on prey which has consumed *Bt* proteins, e.g., herbivores or other predators. In the case of *Bt* crops, spiders may also feed directly on GE plant material, such as pollen (Fig. 32.1). While no data are available for exposure of spiders to *Bt* spray formulations, exposure of spiders to Cry proteins in *Bt* crops via prey and via pollen has been studied in detail. Those pathways are discussed below.

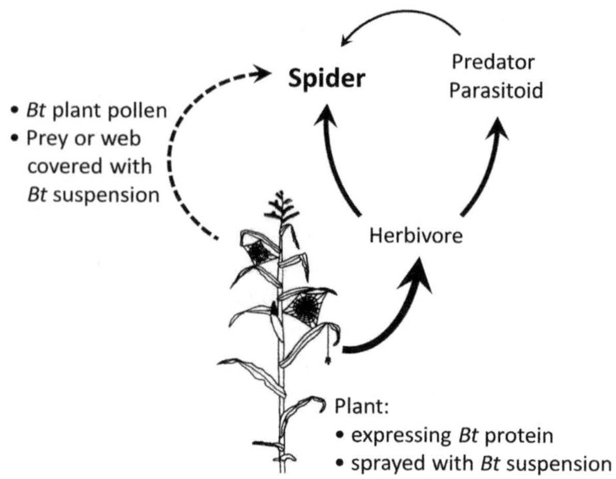

Fig. 32.1 Major routes of exposure of spiders to *Bacillus thuringiensis* (*Bt*) proteins in crops sprayed with *Bt* formulations or genetically engineered crops expressing *Bt* proteins. *Solid arrows* show the transfer of *Bt* protein from plants to spiders via prey. The fact that *Bt* proteins get diluted when transferred to higher levels is symbolized by the thickness of the *arrows*

32.3.1 Exposure via Prey

Most spiders have a rather broad prey spectrum, which is influenced by the crop and its dominant arthropod fauna; the landscape context of the field, i.e., proximity to other habitats such as open water bodies, forests, or grasslands; and the ecology and biology of the spider species, e.g., body size, hunting and web-building behaviour, or habitat. One example of a European web-building spider that occurs frequently in maize fields is *Phylloneta impressa* (Theridiidae). Mainly herbivores, such as aphids, thrips, mirid bugs, leafhoppers and chrysomelid beetles have been recollected from the webs of this species (Meissle and Romeis 2009). Those insect groups have also been reported to be major prey organisms for various spider species in different crops worldwide, together with flies and midges, parasitoids and bees, other spiders, butterflies and moths, as well as collembolans (Nyffeler and Sunderland 2003; Li et al. 2011). Field studies revealed that the prey spectra of spiders in *Bt* maize and *Bt* rice were not different to those in non-*Bt* crops (Árpás et al. 2005; Li et al. 2011; Tian et al. 2012).

Exposure of spiders to *Bt* proteins via prey depends on the actual amount of the insecticidal protein contained in the prey species. Herbivores feeding on *Bt* plant tissue, such as thrips, mirid bugs and chrysomelid beetles, contained relatively high concentrations of Cry protein, almost reaching the level in the leaves (Harwood et al. 2005; Meissle and Romeis 2009) (Fig. 32.2). In contrast, phloem-feeding species, such as aphids, contained no or at best traces of Cry protein, which is not transported in the phloem (Romeis and Meissle 2011). Concentrations of Cry protein in predatory species, e.g., ladybird beetles, lacewings, predatory bugs, or hoverfly larvae, are generally lower than in the tissue-feeding herbivores because the *Bt* protein is diluted when it is transferred to higher trophic levels (Fig. 32.2). Thus, environmental exposure concentrations of spiders in a *Bt* crop may occasionally approach the expression level in leaf tissue. The average exposure, however, is considerably lower.

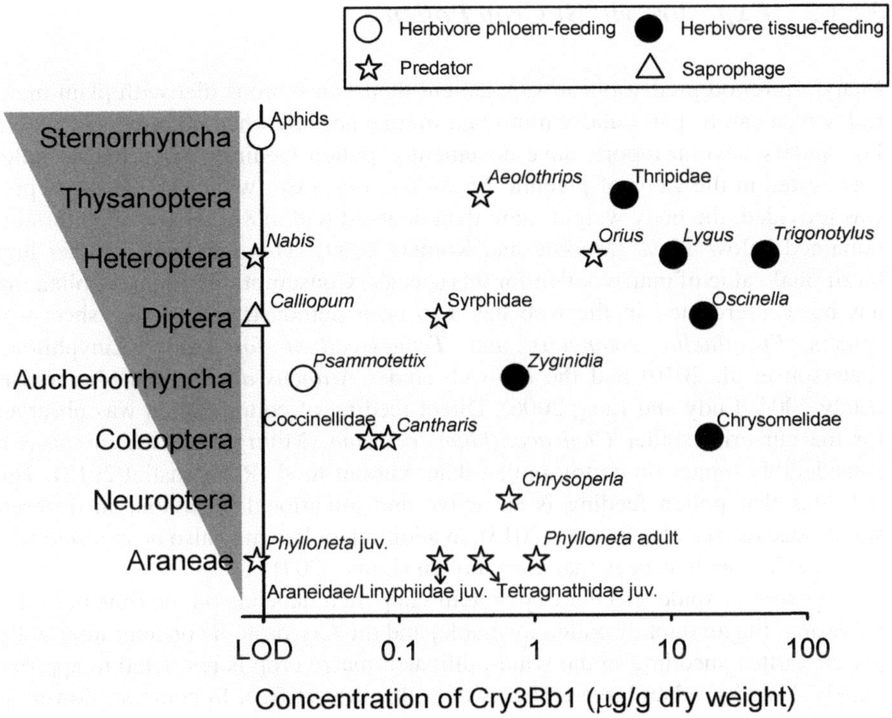

Fig. 32.2 Concentrations of the *Bacillus thuringiensis* (*Bt*) protein Cry3Bb1 in arthropod groups included in the prey spectrum of *Phylloneta impressa*, a web-building spider that is common in European maize fields. The grey triangle symbolizes the importance of the taxonomic order as prey for the spider. LOD denotes the limit of detection of the immunological assay (ELISA) used to quantify the Cry3Bb1 concentration. Data have been summarized and simplified from Meissle and Romeis (2009)

Cry protein measurements of spiders collected in *Bt* crops confirmed that they actually did contain *Bt* protein. In Cry3Bb1-expressing maize, adult *P. impressa* contained ca. 0.5 % of the concentration in the leaves (Meissle and Romeis 2009). Juveniles collected after anthesis, however, contained no measurable Cry3Bb1. In contrast, juveniles of the families Araneidae, Linyphiidae and Tetragnathidae contained Cry3Bb1, but concentrations were low (Meissle and Romeis 2009) (Fig. 32.2). Measurable concentrations of Cry1Ab were also reported from spiders (mainly Linyphiidae) collected in *Bt* maize in the USA (Harwood et al. 2005). In the laboratory, the Cry protein transfer from plants via prey to spiders and the dilution effect along the food chain have been confirmed (Chen et al. 2009; Meissle and Romeis 2009; Tian et al. 2010, 2012).

32.3.2 Exposure via Bt Crop Pollen

Many arthropod predators can supplement their carnivorous diet with plant material, which can be particularly important in time periods when no prey is available. For spiders, several reports have documented pollen feeding. When maize pollen was dusted in the webs of juvenile *Phylloneta impressa*, while no additional prey was provided, the body weight more than doubled within 8 weeks, while mortality remained below 15 % (Meissle and Romeis 2009). This indicates a rather high nutritional value of maize pollen for this species. Consumption of maize pollen that has been intercepted in the web has also been demonstrated for the sheet-web spiders *Frontinella communis* and *Tennesseellum formicum* (Linyphiidae) (Peterson et al. 2010) and the orb-web spider *Araneus diadematus* (Araneidae) (Ludy 2004; Ludy and Lang 2006). Direct feeding of cotton pollen was observed for the cursorial spider *Cheiracanthium inclusum* (Miturgidae), which survived considerably longer on cotton pollen than without food (Pfannenstiel 2012). This indicates that pollen feeding is an active and intentional behaviour in different spider species (see also Sanders 2013). In addition, pollen may also be ingested with prey species, such as bees that carry pollen (Ludy 2004).

Exposure of spiders to the Cry protein via pollen depends on the time period of flowering, the amount of pollen available, and the Cry protein concentration in the pollen. Pollen shedding in the wind-pollinated maize crop is restricted to approximately 2 weeks, when large amounts of pollen are set free. In contrast, flowers in the arthropod-pollinated cotton crop are present for a longer time (ca. 6 weeks), but each flower is available for 1 day only, relatively small amounts of pollen are produced per flower, and the pollen is not dispersed by wind. Expression of Cry protein in pollen is depending on the transformation event of the GE plant and the genetic construct used. For example, maize plants with the transformation events Event 176 (which is no longer in commercial use) and MON88017 express a relatively high concentration of Cry1Ab against corn borers and Cry3Bb1 against corn rootworms, respectively (Ludy and Lang 2006; Meissle and Romeis 2009). *Bt* maize events currently grown to control corn borers (events *Bt*11and MON810) contain only trace amounts of Cry1Ab in pollen (US EPA 2001). Similarly, *Bt* cotton varieties that are grown today express very low concentrations of Cry proteins in pollen (OGTR 2002).

When MON88017 pollen was applied to the web, *P. impressa* spiderlings contained ca. 3 % of the Cry3Bb1 concentration in pollen (Meissle and Romeis 2009). After juvenile *A. diadematus* ingested webs dusted with Event 176 pollen, Cry1Ab was detected in 65 % of the spiderlings, but concentrations in the spiders' gut were below the quantification limit (Ludy and Lang 2006). Consequently, pollen availability in current *Bt* crops is limited, Cry concentrations in pollen are often low, and Cry uptake by spiders via pollen seems to be low even in plants with relatively high expression levels. This indicates that exposure of spiders to Cry proteins via pollen is likely to occur, but less relevant than exposure via prey.

32.4 Consequences of Exposure to *Bt* Protein

Despite the fact that spiders are very common predators in crop fields, only few studies investigated potential effects of *Bt* proteins on life-table parameters of spiders in the laboratory. More abundant are data on population densities in the field.

32.4.1 Laboratory Studies

No contact activity of Novodor (*Bt tenebrionis*) was observed for six spider species from different families that were kept on treated filter paper (Pekár and Haddad 2005) and when the formulation was directly sprayed on *Phylloneta impressa* (Pekár 2002). Similarly, no contact activity was evident when *Hibana velox* (Anyphaenidae) was kept on leaf discs treated with Dipel (*Bt kurstaki*) (Amalin et al. 2000). With the known mode of action of Cry proteins that requires oral ingestion for toxicity, the absence of contact activity is not surprising. Data on potential effects of *Bt* formulations after ingestion by spiders, however, are not available.

In contrast, effects of plant-expressed Cry proteins on spiders via prey or pollen consumption have been studied. To examine the toxicity of Cry3Bb1, adult and juvenile *P. impressa* were fed with Cry protein-containing food (prey or maize pollen) for 8 weeks in the laboratory. No differences in mortality, weight development or offspring production were observed compared to spiders provided with Cry3Bb1-free food (Meissle and Romeis 2009). No effects of Event 176 maize pollen containing Cry1Ab were detected on weight increase, survival, moult frequency, reaction time, and various web variables of *Araneus diadematus* (Ludy and Lang 2006). When the ground-dwelling spider *Ummeliata insecticeps* (Linyphiidae) was fed with prey that had been reared on two *Bt* transgenic rice varieties expressing Cry1Ab/Cry1Ac and Cry1Ab, respectively, no negative effects were observed on survival and development (Tian et al. 2010). Feeding on prey that had consumed Cry1Ab-expressing rice did also not affect survival, developmental time, fecundity and predation in *Pardosa pseudoannulata* (Lycosidae) (Tian et al. 2012). Similarly, survivorship and fecundity of another lycosid, *Pirata subpiraticus* feeding on prey from Cry1Ab-expressing *Bt* rice, were not significantly different from those feeding on non-*Bt* rice prey. However, developmental time of the spider was significantly (ca. 25 %) longer in the *Bt* rice treatments (Chen et al. 2009). The prey organism used by Chen et al. (2009) was the lepidopteran target pest of Bt rice, the leaffolder *Cnaphalocrocis medinalis*. This species showed reduced feeding on *Bt* rice, demonstrating sublethal damage. Furthermore, binding studies indicated that the spiders lack receptors for Cry1Ab. Those facts suggest that indirect effects due to reduced prey quality rather than direct toxic effects led to the observed effect on developmental time (Chen et al. 2009).

Table 32.2 Studies on spray products based on *Bacillus thuringiensis* (*Bt*) that examined effects on spiders in the field

Taxon studied	*Bt* product (*Bt* subspecies)	System studied (country)	Sampling method	Reference
Pardosa milvina (Lycosidae)	Javelin WG (*Bt kurstaki*)	*Brassica oleracea* plots (USA)	Pitfalls	Muckenfuss and Shepart (1994)
Oxypes spp., *Lysosa* spp.	Cutlass, Delfin, Bactec (*Bt kurstaki*)	Cotton plots (India)	Visual counts	Patel and Vyas (2000)
12 families, 62 species combined	Dipel 4L, Thuricide 16B (*Bt kurstaki*)	Spruce-fir forests (USA)	Pitfalls	Hilburn and Jennings (1988)
12 families, 30 species combined	Dipel 2X, MVP[a] (*Bt kurstaki*)	Apple orchards (USA)	Beating	Bajwa and Aliniazee (2001)
Araneae combined	Dipel DF (*Bt kurstaki*), XenTari DF (*Bt aizawai*)	*Brassica oleracea* plots (USA)	Plant sampling	Maxwell and Fadamiro (2006)
Araneae combined	Dipel 4L (*Bt kurstaki*)	Mixed mesophytic forest (USA)	Pitfalls, litter collection	Rieske and Buss (2001)
Araneae combined	Not specified (*Bt kurstaki*)	Maize plots (India)	Visual counts	Srinivas and Panwar (2003)
Araneae combined	Not specified (*Bt galleriae*)	Rice plots (India)	Visual counts	Saikia and Parameswaran (2002)
Araneae combined	Halt (*Bt kurstaki*)	Eggplant plots (India)	Visual counts	Jyoti and Goud (2008)

None of the listed studies revealed a consistent negative effect of *Bt* products on spiders compared to untreated controls.
[a]Product based on *Bt* crystal proteins produced in genetically engineered *Pseudomonas fluorescens*.

32.4.2 Field Studies

Data on spiders from studies in arable crops, vegetable fields, forests, or orchards treated with *Bt* formulations are summarized in Table 32.2. What all studies have in common is that no consistent deleterious effects on spider populations were observed after application of different products based on different *Bt* strains.

More data on spiders have been published for GE crops expressing *Bt* proteins. The available field data have recently been reviewed and subjected to meta-analyses by Peterson et al. (2011). Untreated *Bt* crops were compared to untreated conventional crops as well as to insecticide-treated conventional crops. Altogether, 60 studies that contain spider data were identified from maize, cotton, potato, rice and eggplant. Meta-analysis is a useful tool to analyze data if a large number of studies are available, as it can reveal overall trends that are not apparent from the results of individual studies. For maize, 22 studies on plants expressing Cry1Ab,

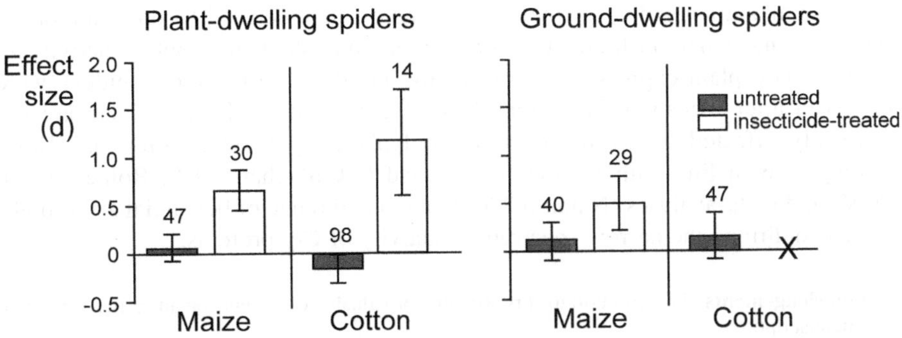

Fig. 32.3 Effects of *Bt* maize and *Bt* cotton compared to untreated and insecticide-treated conventional crops on plant-dwelling (*left panel*) and ground-dwelling (*right panel*) communities of spiders. Positive effect size indicates that spider abundance in *Bt* crops is favoured compared to the *untreated* or *insecticide-treated* conventional non-*Bt* crop. Significance is assumed if error lines, which represent 95 % confidence intervals, do not cross the null line. The number of observations is noted above each *bar*. Redrawn from Peterson et al. (2011)

Cry1Ab + Vip3A, or Cry3Bb1 were included in the meta-analysis. No effect of *Bt* maize on the abundance of foliage and ground-dwelling spiders compared to untreated conventional maize was evident (Fig. 32.3). When compared to insecticide-treated conventional maize, however, *Bt* maize harboured more spiders. For cotton, 10 studies were suitable for meta-analysis, with most data collected from Cry1Ac-expressing plants but also from plants expressing Cry1Ab, Cry1Ac + Cry2Ab, Cry1Ac + Cry1F, Cry1Ac + Cry1Ab, or Vip3A. The abundance of foliar spiders in *Bt* cotton was marginally lower relative to untreated non-*Bt* fields (Fig. 32.3). The effect size, however, was small compared to the positive effect of *Bt* cotton versus insecticide-treated conventional cotton. No difference in spider abundance was evident between *Bt* and untreated non-*Bt* cotton for the community of ground-dwelling spiders (Fig. 32.3).

Field studies on potato expressing Cry3Aa (four studies), rice expressing Cry1Ab or fusion protein of Cry1Ab and Cry1Ac (five studies), and eggplants expressing Cry3Bb1 (one study) reported no deleterious effects on spiders (Peterson et al. 2011; Tian et al. 2010).

32.5 Conclusions

Spiders living in agricultural fields that are sprayed with *Bt* formulations or cropped with *Bt* plants are likely to ingest insecticidal *Bt* proteins. The main route of exposure to Cry proteins in *Bt* crops is via prey and depending on the prey spectrum of the individual species. With the exception of phloem feeders, herbivorous species tend to contain high concentrations of Cry protein. When transferred to higher trophic levels, the Cry protein gets diluted. There is evidence that spiders ingest small amounts of Cry protein when feeding on pollen of *Bt* crops. Laboratory bioassays have demonstrated that *Bt* formulations show no contact activity to spiders. For plant-expressed Cry proteins, no direct effects on spiders were observed after ingestion. Numerous field studies confirmed that spiders are not adversely affected by *Bt* formulations and plant-expressed *Bt* proteins. Those findings are in line with reviews (Glare and O'Callaghan 2000; Romeis et al. 2006) and meta-analyses (Naranjo 2009) on a broad range of beneficial arthropods, which confirmed the narrow spectrum of activity of Cry proteins.

Acknowledgements I am grateful to Jörg Romeis for valuable comments on an earlier version of the manuscript.

References

Amalin DM, Peña JE, Yu SJ, McSorley R (2000) Selective toxicity of some pesticides to *Hibana velox* (Araneae: Anyphaenidae), a predator of citrus leafminer. Fla Entomol 83:254–262

Árpás K, Tóth F, Kiss J (2005) Foliage-dwelling arthropods in *Bt*-transgenic and isogenic maize: a comparison through spider web analysis. Acta Phytopathol Entomol Hung 40:347–353

Bajwa WI, Aliniazee MT (2001) Spider fauna in apple ecosystem of Western Oregon and its field susceptibility to chemical and microbial insecticides. J Econ Entomol 94:68–75

Chen M, Ye GY, Liu ZC, Fang Q, Hu C, Peng YF, Shelton AM (2009) Analysis of Cry1Ab toxin bioaccumulation in a food chain of *Bt* rice, an herbivore and a predator. Ecotoxicology 18:230–238

Glare TR, O'Callaghan M (2000) *Bacillus thuringiensis*: biology, ecology and safety. Wiley, Chicester

Harwood JE, Wallin WG, Obrycki JJ (2005) Uptake of *Bt* endotoxins by nontarget herbivores and higher order arthropod predators: molecular evidence from a transgenic corn agroecosystem. Mol Ecol 14:2815–2823

Hilburn DJ, Jennings DT (1988) Terricolous spiders (Araneae) of insecticide-treated spruce-fir forests in West-Central Maine. Great Lakes Entomol 21:105–114

James C (2011) Global status of commercialized biotech/GM crops: 2011. ISAAA Brief No. 43, ISAAA, Ithaca, NY, USA

Jyoti DP, Goud KB (2008) Safety of organic amendments and microbial pesticides to natural enemies in brinjal ecosystem. Annu Plant Prot Sci 10:12–16

Knowles BH (1994) Mechanism of action of *Bacillus thuringiensis* insecticidal δ-endotoxins. Adv In Insect Phys 24:275–308

Li K, Tian J, Wang Q, Chen Q, Chen M, Wang H, Zhou Y, Peng Y, Xiao J, Ye G (2011) Application of a novel method PCR-ligase detection reaction for tracking predator—prey trophic links in insect-resistant GM rice ecosystem. Ecotoxicology 20:2090–2100

Ludy C (2004) Intentional pollen feeding in the spider *Araneus diadematus* Clerck, 1757. Newslett Br Arachnol Soc 101:4–5

Ludy C, Lang A (2006) Bt maize pollen exposure and impact on the garden spider, *Araneus diadematus*. Entomol Exp Appl 118:145–156

Maxwell EM, Fadamiro HY (2006) Evaluation of several reduced-risk insecticides in combination with action threshold for managing lepidopteran pests of cole crops in Alabama. Fla Entomol 89:117–126

Meissle M, Romeis J (2009) The web-building spider *Theridion impressum* (Araneae: Theridiidae) is not adversely affected by *Bt* maize resistant to corn rootworms. Plant Biotechnol J 7:645–656

Muckenfuss AE, Shepart BM (1994) Seasonal abundance and response of diamondback moth, *Plutella xylostella* (L.) (Lepidoptera: Plutellidae), and natural enemies to esfenvalerate and *Bacillus thuringiensis* subsp. *kurstaki* Berliner in coastal South Carolina. J Agric Entomol 11:361–373

Naranjo SE (2009) Impacts of *Bt* crops on non-target invertebrates and insecticide use patterns. CAB Rev: Perspect Agricult, Vet Sci, Nutr Nat Resour 4(011):1–23

Nyffeler M, Sunderland KD (2003) Composition, abundance and pest control potential of spider communities in agroecosystems: a comparison of European and US studies. Agr Ecosyst Environ 95:579–612

OGTR (2002) Risk assessment and risk management plan—commercial release of Bollgard II cotton. Office of the Gene Technology Regulator, Australia. http://www.ogtr.gov.au/internet/ogtr/publishing.nsf/content/dir012-3/$FILE/dir012finalrarmp.pdf. Accessed 26 Apr 2012

Patel MC, Vyas RN (2000) Field bioefficacy of *Bacillus thuringiensis* var. *kurstaki* and neem based formulations against cotton bollworms. Indian J Plant Prot 28:78–83

Pekár S (2002) Susceptibility of the spider *Theridion impressum* to 17 pesticides. J Pestic Sci 75:51–55

Pekár S, Haddad CR (2005) Can agrobiont spiders (Araneae) avoid a surface with pesticide residues? Pest Manag Sci 61:1179–1185

Peterson JA, Romero SA, Harwood JD (2010) Pollen interception by linyphiid spiders in a corn agroecosystem: implications for dietary diversification and risk-assessment. Arthropod Plant Interact 4:207–217

Peterson JA, Lundgren JG, Harwood JD (2011) Interactions of transgenic *Bacillus thuringiensis* insecticidal crops with spiders (Araneae). J Arachnol 39:1–21

Pfannenstiel RS (2012) Direct consumption of cotton pollen improves survival and development of *Cheiracanthium inclusum* (Araneae: Miturgidae) spiderlings. Ann Entomol Soc Am 105:275–279

Rieske LK, Buss LJ (2001) Effects of gypsy moth suppression tactics on litter- and ground-dwelling arthropods in the central hardwood forests of the Cumberland Plateau. For Ecol Manage 149:181–195

Romeis J, Meissle M (2011) Non-target risk assessment of *Bt* crops—Cry protein uptake by aphids. J Appl Entomol 135:1–6

Romeis J, Meissle M, Bigler F (2006) Non-target risk assessment of *Bt* crops—Cry protein uptake by aphids. Nat Biotechnol 24:63–71

Saikia P, Parameswaran S (2002) Eco-friendly strategies for the management of rice leaffolder, *Cnaphalocrocis medinalis* Guenee. Annu Plant Prot Sci 10:12–16

Sanders D (2013) Herbivory in spiders. In: Nentwig W (ed) Spider ecophysiology. Springer, Heidelberg (this volume)

Srinivas PS, Panwar VPS (2003) Efficacy of neem and *Bt* formulations against *Chilo partellus* (Seinhoe) infestation in maize. Pestic Res J 15:131–133

Tian JC, Chen Y, Li ZL, Li K, Chen M, Peng YF, Hu C, Shelton AM, Ye GY (2012) Transgenic Cry1Ab rice does not impact ecological fitness and predation of a generalist spider. PLoS One 7:e35164

Tian JC, Liu ZC, Chen M, Chen Y, Chen XX, Peng YF, Hu C, Ye GY (2010) Laboratory and field assessments of prey-mediated effects of transgenic *Bt* rice on *Ummeliata insecticeps* (Araneida: Linyphiidae). Environ Entomol 39:1369–1377

US EPA (2001) Biopesticides Registration Action Document—*Bacillus thuringiensis* plant-incorporated protectants. US EPA Washington DC, USA http://www.epa.gov/opp00001/biopesticides/pips/bt_brad.htm. Accessed 26 Apr 2012

Chapter 33
Effects of Neem on Spiders

John D. Stark

33.1 Introduction

Spiders are an abundant and important group of predators that inhabit many ecosystems and play a major role in the regulation of pest species (Riechert and Lockley 1984). Because many species inhabit agricultural ecosystems or their surroundings, they have the potential to be exposed to pesticides. Spiders are in general quite susceptible to pesticides, especially synthetic insecticides (Stark et al. 1994; Pekár 2013). However, much less work has been published on the effects of pesticides on spiders compared to insects. Certain natural insecticides, including those of plant origin, appear to be less toxic to spiders compared to synthetic insecticides. Pesticides from the neem tree fall into this category.

The neem tree, *Azadirachta indica*, is native to Southern Asia and has been known for centuries to have medicinal and pest control properties. More recently, commercial pesticides have been developed from the neem tree, and today, neem pesticides are used to control various pests throughout the world. In the developed world, neem-based pesticides are often used by organic farmers. Neem pesticides are produced primarily from the seed kernels which are crushed and extracted with water or alcohol. Neem oil, which is obtained by cold pressing the seeds, also has pesticidal properties. Neem soap is another product that is used as a pesticide.

More than 70 chemicals have been identified in the neem tree. The major insecticidal component is azadirachtin (Fig. 33.1), a triterpenoid compound that exhibits insect growth regulatory properties. Azadirachtin interferes with molting by inhibiting production of the insect hormone, ecdysone. Neem products also act as antifeedants causing insects to stop feeding after ingestion. Furthermore, neem can act as an egg-laying deterrent, whereby volatile compounds from neem are repulsive to some insects, and this stops them from laying eggs on plant surfaces.

J.D. Stark (✉)
Washington State University, Puyallup, Washington, USA
e-mail: starkj@wsu.edu

W. Nentwig (ed.), *Spider Ecophysiology*,
DOI 10.1007/978-3-642-33989-9_33, © Springer-Verlag Berlin Heidelberg 2013

Fig. 33.1 The chemical structure of azadirachtin, the main active compound of the neem tree (Wikipedia Commons)

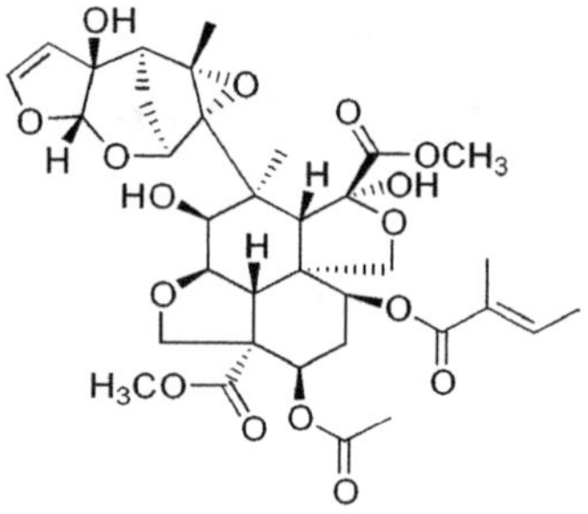

Neem pesticides are considered to be broad-spectrum insecticides because they are toxic to most major groups of insects. Furthermore, they work by contact and/or ingestion and are systemic, being taken into plants via leaves and roots. For a list of neem products and pest insect species controlled by neem, see the following web page: http://web.pppmb.cals.cornell.edu/resourceguide/mfs/08neem.php.

In this chapter, I will discuss the literature pertaining to the effects of neem pesticides on spiders. I have limited this review to include papers published only in peer-reviewed scientific journals. Papers presented at meetings, published meeting proceedings, book chapters, theses/dissertations, or reports published in newsletters are not considered. Furthermore, only studies where spiders were evaluated separately and not lumped in with predators in general are discussed. Furthermore, I have also separated out laboratory and field studies.

33.2 Laboratory Studies

A number of laboratory studies have been conducted to determine the effects of neem on spiders. One of the earliest studies to evaluate the effects of a neem-based pesticide on spiders was published by Saxena et al. (1984). In this study, topical application of 50 μg of a neem seed kernel extract had no effect on the lycosid *Pardosa pseudoannulata*. In another study, Mansour et al. (1986) evaluated the effects of a 2.5 % extract of neem seed prepared with several different solvents. They found that certain extracts were not toxic to the miturgid *Cheiracanthium mildei* but that 4 % extracts made with pentane, acetone, and ethanol resulted in 71 %, 54 %, and 33 % mortality, respectively.

Mansour and Nentwig (1988) evaluated the toxicity of 30 pesticides, including neem, to juvenile and adult web-building and hunting spiders from the tropics, Europe and the Middle East, in laboratory studies. The philodromid hunting spider, *Philodromus* sp., was not susceptible to any of the pesticides evaluated. The web-building spiders, *Argiope* sp. (Araneidae) and *Linyphia* sp. (Linyphiidae), and the hunting spider, *Cheiracanthium* sp., were moderately to highly susceptible to these pesticides. Botanical insecticides, herbicides, and fungicides were not very toxic to these spider species, while most acaricides were highly toxic.

The effects of three neem pesticides, Margosan-O, Azatin, and RD9-Repelin, were evaluated on the pest mite, *Tetranychus cinnabarinus*, the predacious mite, *Typhlodromus athiasae*, and the predatory spider, *C. mildei*, in laboratory studies (Mansour et al. 1993). None of these products were toxic to the spider species.

Two neem pesticides, Neemgard, an acaricidal and fungicidal formulation obtained from neem seed kernels, and Neemix 4.5 were tested as controls for the pest mite, *T. cinnabarinus,* the predacious mite, *Phytoseiulus persimilis,* and the predatory spider, *C. mildei* (Mansour et al. 1997). Neemgard was found to be highly toxic to *T. cinnabarinus,* but was not toxic to *C. mildei* or *P. persimilis.* Neemix 45 was repellent but caused no mortality in *T. cinnabarinus.*

Punzo (1997) studied the effects of azadirachtin on the wolf spider *Schizocosa humilis* (sub *S. episina*). In this study, azadirachtin's effects on mortality, survivorship, and immunological competency were evaluated. Adults and spiderlings were fed prey that had been injected with several concentrations of azadirachtin. Significant mortality over a 30-day period was observed when spiders ingested prey injected with 1.0 and 10.0 ppm azadirachtin. Additionally, ingestion of azadirachtin-treated prey led to significant negative effects on spiderlings body mass and the width of the prosoma as well as a significant decrease in total hemocyte counts.

Amalin et al. (2000) tested the toxicity of 14 pesticides used in the production of Tahiti lime to the anyphaenid spider, *Hibana velox*, in the laboratory. Many of the pesticides evaluated were highly toxic to this spider. However, exposure to azadirachtin resulted in the lowest impact on *H. velox* with less than 20 % mortality at the highest concentration tested.

The effects of several pesticides on the survival and web-building potential of the eresid spider, *Stegodyphus sarasinorum*, were evaluated in laboratory studies (Kumar and Yashkamal 2009). Neem pesticides were the least toxic to this spider species compared to the other insecticides tested.

The toxicity of five insecticides, including neem (NeemAzal) on the functional response of the spider, *Philodromus cespitum*, was evaluated in a laboratory study (Řezáč et al. 2010). Exposure of this spider to NeemAzal resulted in less than 10 % mortality. However, NeemAzal exposure resulted in significantly lower predation rates compared to the control.

33.3 Field Studies

In 1988 and 1989, two field experiments were conducted to study the effects of different neem products and the synthetic insecticide, monocrotophos, on rice pests and the spider, *Pardosa pseudoannulata* (Mohan et al. 1991). The neem products were effective as controls for rice pests. After application of neem, there were initial reductions in spider populations, but spiders recolonized neem-treated plots while populations remained low in monocrotophos-treated plots.

Stark (1992) found that applications of the commercial neem pesticide Margosan-O had no significant effect on spiders inhabiting turf grass. In a follow-up study, Stark and Crawford (2005) found that the synthetic insecticide chlorpyrifos significantly reduced spider species diversity, whereas Margosan-O had no significant negative effect on these spiders inhabiting turf grass.

Baitha et al. (2000) evaluated the efficacy of several insecticides on control of insect pests in rice. They found that a neem seed kernel extract was the least effective at controlling pest species but also had the lowest impact on spiders.

A neem seed extract was evaluated along with a synthetic insecticide for control of insect pests of cotton in Australia (Ma et al. 2000). The neem pesticide was moderately effective as a control for cotton pests and resulted in higher control yields than the control. Spiders and other predators were found to be unaffected by the neem extract.

Chakraborti (2001) evaluated neem pesticides along with application of the synthetic insecticide, phosphamidon, for their potential to control insect pests of mustard. Neem cake was applied to soil; a neem seed kernel extract and phosphamidon were applied to mustard foliage. Effects on aphids, diamond back moth, and several predators were recorded. This combination treatment effectively controlled the pest species but was safe for spiders, coccinellids, and syrphid flies.

Three neem pesticides—NeemAzal, Rakshak Gold, and ICIPE Neem—were evaluated against beneficial arthropods in a cotton agroecosystem (Mann and Dhaliwal 2001). Numbers of spiders were found to be higher in the NeemAzal treatment of 1 l/ha and the same as the control in a higher rate of NeemAzal (2 l/ha).

The effects of two concentrations of neem on the pest species, the alfalfa weevil *Hypera variabilis* (Coleoptera: Curculionidae) and aphids and the predators (coccinellids, anthocorids, chrysopids, and spiders) were compared to malathion in field experiments (Yardim et al. 2001). Malathion was more effective at reducing weevil numbers than neem, but the neem treatments significantly reduced weevil numbers and were more effective than malathion at controlling aphids. The number of predators was not significantly affected by the neem treatments.

The effectiveness of a sustainable management program targeting sucking pests that transmit leaf curling in chilli was conducted in West Bengal, India (Chakraborti 2004). Both neem cake and foliar applications of neem oil and azadirachtin were applied. These treatments were effective controls for the pest species and were safe to coccinellids, syrphids, and spiders compared to conventional chemical control.

The efficacy of neem oil and the synthetic insecticides, imidacloprid and carbosulfan, as a seed dress to control sucking insect pests of okra was conducted in the field (Indira Gandhi et al. 2006). Neem oil provided good control of the pest species. Additionally, numbers of coccinellids and the spiders, *Oxyopes* sp. (Oxyopidae) and *Clubiona* sp. (Clubionidae), were higher in neem-treated plots than in okra treated with synthetic pesticides.

A study was conducted to evaluate the effects of various plant extracts including neem alone and in combination with synthetic insecticides for control of bollworms in cotton in the field (Sinzogan et al. 2006). All treatments, except neem alone, significantly reduced populations of lady beetles, ants, and spiders compared to the control.

Ishtiyaq and Shaw (2008) studied the effects of several plant extracts on the host location and acceptance by the egg parasitoid, *Trichogramma japonicum*, parasitizing the yellow stem borer, *Scirpophaga incertulas,* in a Chhattisgarh agroecosystem in India. They found that applications of a neem extract (0.5 %) resulted in reductions in spider populations in the field.

Ravi et al. (2008) compared the efficacy of sequential applications of the natural pesticides, nucleopolyhedrovirus, of the cotton bollworm *Helicoverpa armigera* (*Ha*NPV @ 1.5 × 10^{12} OB/ha), *Bacillus thuringiensis* var. *kurstaki* (Delfin® 25 WG @1 kg/ha), spinosad 45 SC ((@ 75 g a.i./ha), and neem (neemazol 1.2 EC @ 1,000 ml/ha) as controls for *H. armigera* to sequential application of synthetic insecticides and untreated control on tomato F1 hybrid Ruchi. The sequential applications of the natural pesticides were as effective as the synthetic products for control of *H. armigera*. Furthermore, higher numbers of predators, including spiders, were found in the plots that received applications of the natural products.

A field study was conducted to evaluate the effectiveness of synthetic insecticides, biopesticides/bio-agents, botanicals, and their combinations, cartap hydrochloride 4 G @ 0.75 kg a.i./ha, *Beauveria bassiana* @ 2.5 kg/ha, *B. thuringiensis* @ 1.5 kg/ha, imidacloprid 17.8 SL @ 0.05 %, *T. japonicum* @ 100,000 eggs/ha, neem gold @ 5 ml/l, and neem gold @ 3 ml/l + *T. japonicum* @ 75,000 eggs/ha for control of the yellow stem borer (Singh et al. 2008). The synthetic pesticides were most effective for control of the pest species but were detrimental to spider populations. The biopesticides, including neem, conserved spider populations.

Anis Joseph et al. (2010) conducted a field study to evaluate the toxicity of several insecticides, including neem to two major tetragnathid spiders found in rice fields, *Tetragnatha mandibulata* and *T. maxillosa*. These species were highly susceptible to synthetic insecticides, but exposure to a neem seed kernel extract resulted in the least mortality.

Applications of neem oil at 1 %, 2 %, and 3 % applied at 10 days intervals were used to control tukra mealy bug *Maconellicoccus hirsutus* and leaf webber *Diaphania pulverulentalis* infestations on mulberry (Ravikumar et al. 2010). The neem treatments were effective at reducing the pest species and populations of coccinellids, spiders, and soil macrofauna in mulberry ecosystem increased.

A field study was conducted to evaluate the effectiveness of seven botanical pesticides, including neem oil, on pest and beneficial insect species inhabiting eggplant and okra in Ghana (Mochiah et al. 2011). All of the pesticides evaluated significantly reduced pest insect species in both crops compared to the control. These pesticides were relatively benign to the natural enemies, including spiders and lady beetles.

Tiwari et al. (2011) evaluated the effects of neem and the synthetic insecticides, imidacloprid and cypermethrin, on populations of the natural enemies, braconid wasps, coccinellids, and predatory spiders, in eggplants. They found that neem pesticides were less harmful to the natural enemies than the synthetic insecticides.

33.4 Conclusions

The majority of the laboratory and field studies with neem have shown that pesticides derived from this tree have little impact on spiders. Some neem products and certain spider species are sensitive to neem. However, compared to synthetic insecticides, neem pesticides are much less damaging to spiders and other beneficial insects.

References

Amalin DM, Pena JE, Yu SJ, Mcsorley R (2000) Selective toxicity of some pesticides to *Hibana velox* (Araneae: Anyphaenidae), a predator of citrus leafminer. Florida Entomol 83:254–262

Anis Joseph R, Premila KS, Nisha VG, Rajendran S, Sarika Mohan S (2010) Safety of neem products to tetragnathid spiders in rice ecosystem. J Biopestic 3:88–89

Baitha A, Hameed SF, Singh R (2000) Effectiveness and economics of various treatments and their impact on spider population in rice ecosystem. Annu Plant Prot Sci 8:13–16

Chakraborti S (2001) Neem-based management approaches for mustard insect pests. J Entomol Res 25:213–220

Chakraborti S (2004) Sustainable management of apical leaf curling in chilli. J Appl Zool Res 15:34–36

Indira Gandhi P, Gunasekaran K, Sa T (2006) Neem oil as a potential seed dresser for managing homopterous sucking pests of okra (*Abelmoschus esculentus* (L.) Moench). J Pestic Sci 79:103–111

Kumar D, Yashkamal K (2009) Effect of pesticides on the survival and web building activity of social spider *Stegodyphus sarasinorum* (Araneae: Eresidae). Indian J Entomol 71:275–278

Ishtiyaq A, Shaw SS (2008) Kairomonal effects of bio-active plant extracts on spider population and parasitization of yellow stem borer eggs by *Trichogramma japonicum*. J Biol Control 22:341–345

Ma DL, Gordh G, Zalucki MP (2000) Toxicity of biorational insecticides to *Helicoverpa* spp. (Lepidoptera: Noctuidae) and predators in cotton field. Int J Pest Manag 46:237–240

Mann GS, Dhaliwal GS (2001) Impact of neem-based insecticides on beneficial arthropods in cotton ecosystem. Annu Plant Prot Sci 9:225–229

Mansour FA, Asher KRS, Omari N (1986) Toxicity of neem, *Azadirachta indica* seed kernel extracts prepared with different solvents on the spider *Cheiracanthium mildei*. Phytoparasitica 14:73–76

Mansour FA, Nentwig W (1988) Effects of agrochemical residues on four spider taxa: laboratory methods for pesticide tests with web-building spiders. Phytoparasitica 16:317–325

Mansour FA, Ascher KRS, Omari N (1993) Effects of Margosan-o™, Azatin™ and RD9-repelin® on spiders, and on predacious and phytophagous mites. Phytoparasitica 21:205–211

Mansour FA, Ascher KRS, Abo-Moch F (1997) Effects of Neemgard on phytophagous and predacious mites and on spiders. Phytoparasitica 25:333–336

Mochiah MB, Banful B, Fening KN, Amoabeng BW, Offei Bonsu K, Ekyem S, Braimah H, Owusu-Akyaw M (2011) Botanicals for the management of insect pests in organic vegetable production. J Entomol Nematol 3:85–97

Mohan K, Gopalan M, Balasubramanian G (1991) Studies on the effects of neem products and monocrotophos against major pests of rice and their safety to natural enemies. Indian J Plant Prot 19:23–30

Pekár S (2013) Side-effect of synthetic pesticides on spiders. In: Nentwig W (ed) Spider ecophysiology. Springer, Heidelberg (this volume)

Punzo F (1997) Effects of azadirachtin on mortality, growth, and immunological function in the wolf spider, *Schizocosa episina* (Araneae: Lycosidae). Bull Environ Contam Toxicol 58: 415–421

Ravi M, Santharam G, Sathiah N (2008) Ecofriendly management of tomato fruit borer, *Helicoverpa armigera* (Hubner). J Biopestic 1:134–137

Ravikumar J, Samuthiravelu P, Qadri SMH, Hemanthkumar L, Jayaraj S (2010) Integrated Pest Management (IPM) module for Tukra mealy bug, *Maconellicoccus hirsutus* (Green) and leaf webber, *Diaphania pulverulentalis* (Hamp.) in mulberry. J Biopestic 3:354–357

Riechert SE, Lockley T (1984) Spiders as biological control agents. Annu Rev Entomol 29: 299–320

Řezáč M, Pekár S, Stará J (2010) The negative effect of some selective insecticides on the functional response of a potential biological control agent, the spider *Philodromus cespitum*. BioControl 55:503–510

Saxena RC, Justo HD, Epino PB (1984) Evaluation and utilization of neem cake against the rice brown planthopper, *Nilaparvata lugens* (Homoptera: Delphacidae). J Econ Entomol 77: 502–507

Singh S, Prasad CS, Nath L, Tiwari GN (2008) Eco-friendly management of *Scirpophaga incertulas* (Walk.) and *Cnaphalocrocis medinalis* (Guen.) in Basmati rice. Annu Plant Prot Sci 16:11–16

Sinzogan AAC, Kossou DK, Atachi P, van Huis A (2006) Participatory evaluation of synthetic and botanical pesticide mixtures for cotton bollworm control. Int J Trop Insect Sci 26:246–255

Stark JD (1992) Comparison of the impact of Margosan-O and Chlorpyrifos on non-target invertebrates inhabiting turf grass. Pestic Sci 36:293–299

Stark JD, Jepson PC, Thomas CFG (1994) Effects of pesticides on spiders from the lab to the landscape. Rev Pestic Toxicol 3:83–110

Stark JD, Crawford R (2005) Short-term effects of the neem insecticide, Margosan-O and chlorpyrifos on a spider (Araneida: Linyphiidae) community inhabiting turf grass. Biopestic Int 1:71–81

Tiwari G, Prasad CS, Nath L (2011) Effect of insecticides, bio-pesticides and botanicals on the population of natural enemies in brinjal ecosystem. Vegetos 24:40–44

Yardim EN, Ozgen I, Kulaz H (2001) Effects of neem-based and chemical insecticides on some arthropods in alfalfa. Meded Rijksuniv Gent Fak Landbouwkd Toegep Biol Wet 66:519–524

Part X
Applications

"What can this be used for?" is a typical question which biologists usually are trained not to ask. Here we give a short overview on applications increasingly arising from spiders. The hydraulic spider leg mechanism is an excellent model for gripper fingers, robot arms, legs of walking robots and even as a sonde in medical technology. Leg scopulae of spiders are great models for biomimetic transfer when sticking to smooth surfaces is requested. The applications of spider silk are numerous, ranging from medical technology, textile industry to shock absorbance, just to mention a few. Similar is the situation with venom compounds where a wide range of potential applications, including the pharmaceutical and agrochemical markets, are under investigation.

Chapter 34
Modelling and Application of the Hydraulic Spider Leg Mechanism

Lena Zentner

34.1 Introduction

Use of the knowledge about biological systems for mechanical design affords much potential in solving various interesting technical problems like for gripping, manipulation and locomotion tasks. The mechanics of the motion of the spider leg shall be considered in this context. The mobility of the limbs of the spider leg is provided by a highly developed musculature and, on the other hand, by different structure of the joints. Numerous muscles pass through the corresponding parts of the leg. These muscles can act as flexors or as extensors. The exceptions can be seen concerning for example of femur-patella joint and tibia-metatarsus joint at the theraphosid spider *Phrixotrichus roseus*; these joints have the flexors only.

The axis of rotation of the hydraulic joint goes through the peripheral point of the leg cross-section. Hence, there are no extensors in such a design (Figs. 34.1 and 34.2). In this case, the stretching occurs hydraulically. Like most of the invertebrate animals, spiders also have an open blood-vascular system. The arteries run from the heart through the whole body. Haemolymph flows from the arteries freely into the tissue and gathers in lacunae (cavities between the tissues) (see also Wirkner and Huckstorf 2013). If the muscles in the prosoma contract, then the pressure of the haemolymph in the lacunae is increased, and the legs are stretched (Foelix 2011, see also Kropf 2013).

L. Zentner (✉)
Technical University Ilmenau, Ilmenau, Germany
e-mail: lena.zentner@tu-ilmenau.de

W. Nentwig (ed.), *Spider Ecophysiology*,
DOI 10.1007/978-3-642-33989-9_34, © Springer-Verlag Berlin Heidelberg 2013

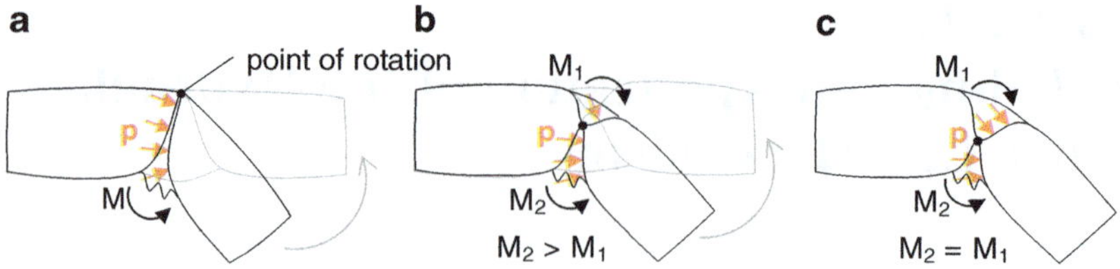

Fig. 34.1 Schematic representation of different joints of the spider leg: (**a**) pure hydraulic joint; (**b**) partial hydraulic joint; (**c**) non-hydraulic joint

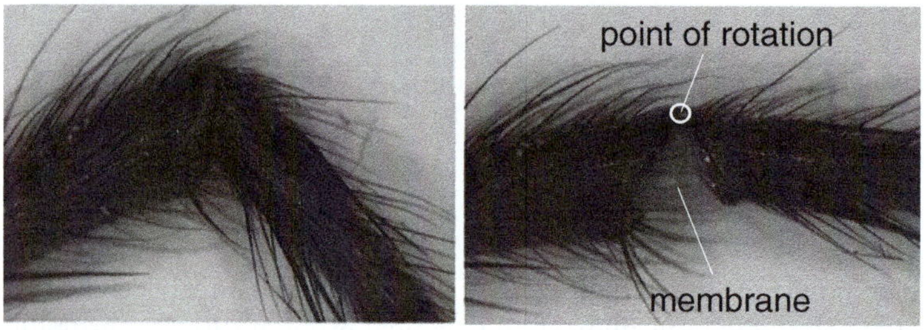

Fig. 34.2 A typical hydraulic joint of the spider leg

34.2 Parameter Identification

For the parameter identification we will consider a simple model of the movement in the femur-patella joint (Zentner 2003). The values of the characteristic parameters are determined analytically using the results of the experimental studies of the spider leg movement. For this investigation the femur part is fixed horizontally. The other parts of the leg are bonded together to perform a circular movement (Fig. 34.3a, b, left). The parameter φ is an angle between the x-axis and the moving part of the leg with the length l_1. In reality, the lacunae have a complicated shape and indefinable position at the cross-section of the leg (Fig 34.3a, right). However, the model would be simplified if it is assumed that only three lacunae with the radius a_1 are in the middle of the leg with the radius r_1, and the area of the lacuna cross-section remains constant during the movement (Fig 34.3b, right). Let the pressure in lacunae at the beginning of the femur part, which has the length l_0, be p_0 and in the femur-patella joint p_1. The pressure loss in the femur part is determined by Poiseuille's formula:

$$p_0 - p_1 = \frac{8\mu}{\pi a_0^4} l_0 \dot{V}. \tag{34.1}$$

The dynamic viscosity of the haemolymph μ is assumed to be equal to the dynamic viscosity of the water. The volume inside the membrane depends on the angle φ. Therefore, the volume flow is a function of the angle velocity and calculated as the following:

$$\dot{V} = (\pi r_1^2)\, r_1\, \dot{\varphi}. \tag{34.2}$$

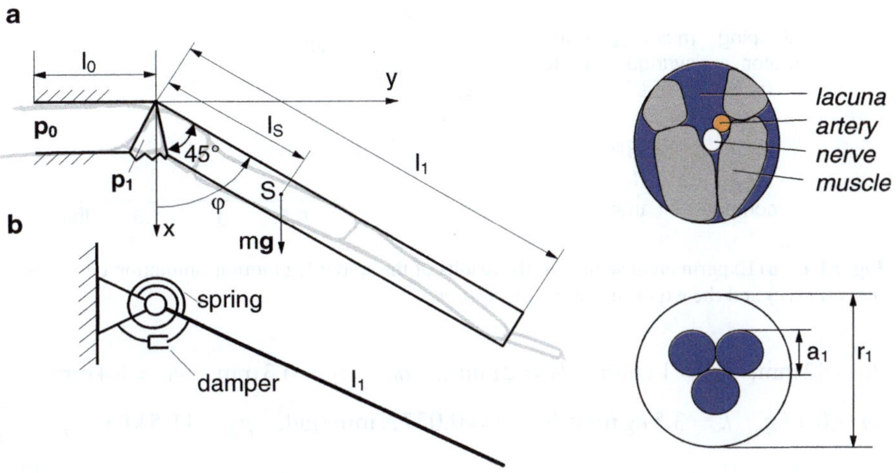

Fig. 34.3 (**a**) A schematic illustration of the fixed leg with a simplified form (*left*) and its rough illustration of the cross-section (*right*); (**b**) model of the moved leg (*left*) and simplified cross-section for the model of hydraulic spider leg mechanism (*right*)

Let us assume that the extension of the leg occurs only by increase of the pressure in the leg. The passive behaviour of the muscles-tissue-complex is modelled with the help of the system of spring and damper elements, which have the spring stiffness c and the damping coefficient k. The motion of the leg part with the mass m, the lacuna area of $A = 3\pi a_1^2$ and the mass moment of inertia J can be modelled according to Euler's law:

$$J\ddot{\varphi} = -mgl_S \sin \varphi + p_1(3\pi a_1^2)r_1\sqrt{2} - k\dot{\varphi} - M_C. \quad (34.3)$$

The distance between the mass centre and the centre of rotation is l_S, and p_1 is defined by (34.1). The moment of the spring element is

$$M_C = c(\varphi - \varphi_0). \quad (34.4)$$

The differential equation of motion (34.3) describes the dynamic behaviour of the spider leg which has only one rotational axis and moves under the inner pressure. The corresponding experiment has been carried out in the following way. A spider leg, namely the fourth leg of the theraphosid *Phrixotrichus roseus*, was fixed on the microsyringe at the femur part (Fig. 34.4a). The syringe was driven by the step drive. At the beginning of the femur part the pressure was measured. A video of the leg motion was made which allowed us to determine $\varphi(t)$ function for definite time points. The following parameters of the model were measured or calculated on the basis of this experiment:

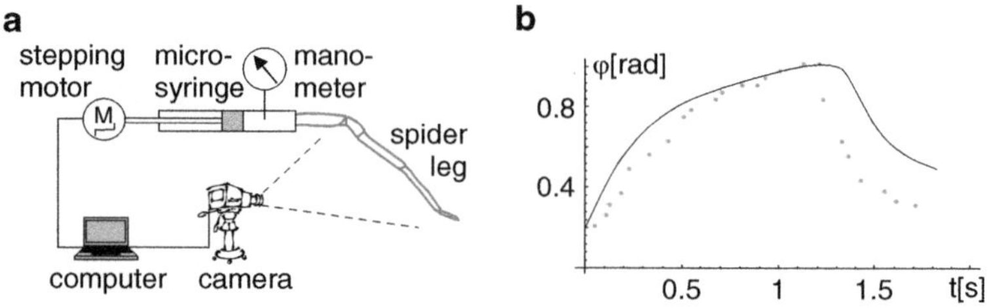

Fig. 34.4 (**a**) Experimental setup; (**b**) the results of the spider leg motion simulation on the model basis (*curve*) and the experimental results (*points*)

$$l_1 = 42\,\text{mm},\quad l_S = 14\,\text{mm},\quad l_0 = 21\,\text{mm},\quad a_0 = a_1 = 0.3\,\text{mm},\quad r_1 = 1.4\,\text{mm},$$
$$m = 0.17\,\text{g},\quad k = 3.5\,\text{kg}\,\text{mm}^2/\text{s},\quad c = 0.037\,\text{N}\,\text{mm}/\text{rad},\quad p_0 = 11.5\,\text{kPa},\quad \varphi_0 = \frac{\pi}{3}. \tag{34.5}$$

Figure 34.4b shows the results of leg motion modelling compared with the experimental results. Both results have the same qualitative dependency for $\varphi(t)$. Therefore, the found characteristic parameters were used for further modelling.

34.3 Modelling of the Spider Jump

Most spiders have the ability to jump. Jumping from the ground is carried out with the help of the third or fourth leg pair. Some spiders use both pairs simultaneously. Lifting the front legs the spider shows its jump intention. Directly before the jump, it fixes a safety thread to the ground and pulls up the jump legs symmetrically. Pressure impulse is induced in the prosoma under which the haemolymph flows in the lacunae of the legs. Then these are stretched rapidly and the jump is carried out. In order to overcome obstacles or to bridge fast short distances, the spiders often switch from short jumps to the normal run. On the contrary, theraphosids jump to overpower their prey unexpectedly.

The spider's jump can be subdivided into three phases. The first phase (pushing from the ground) begins with the start position of the spider for the jump and ends with the loss of the contact with the ground. Subsequently the second phase (the ballistic phase) begins which ends with restoring of the contact of spider legs with the ground. Finally, the third phase (landing phase) is initiated. We will consider only the first two phases. The following prerequisites should be observed for the model building of the first phase (Zentner 2003):

- Only three joints are moved (trochanter-femur joint, femur-patella joint and tibia-metatarsus joint). These are the most supple joints of the spider leg.
- The stretching of the legs occurs purely hydraulically without the effect of muscles.

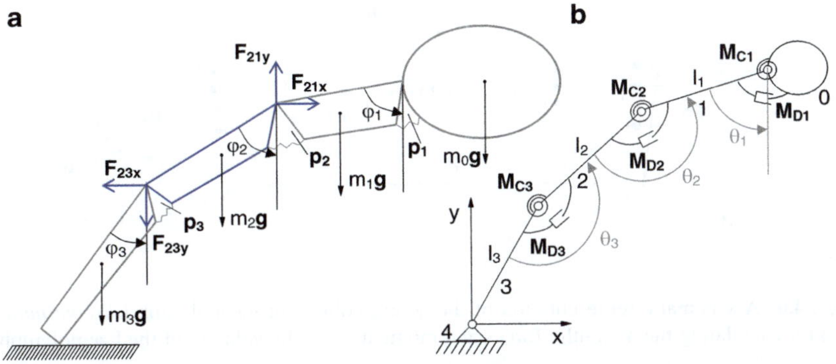

Fig. 34.5 The rigid body spider model; (**a**) internal forces on the patella-tibia limb and (**b**) a model with spring and damper elements in the joints for the first phase

- The orientation of the body always remains horizontal during the first jump phase.
- The jump is considered as a plane problem, that is the spider is projected on the moving plane.
- Jumping occurs only with the help of the fourth legs; the mass of the other legs is considered together with the mass of the body.
- Only three lacunae with the radius a_i are in the middle of the leg limbs and the lacuna cross-sectional area remains constant during the movement.

34.3.1 Differential Equations of Motion

For the description of the position and movement of the whole spider in the xy-plane (jump plane) we need three coordinates (Fig. 34.5a). There are angles φ_1, φ_2 and φ_3 between the x axis and the axis of the corresponding leg limbs. The ground, the spider body and leg limbs are numbered from 0 for the body up to 4 for the ground. During the flow of hemolymph in lacunae, the pressure in the system is changed. The pressure loss is calculated using Poiseuille's formula (34.1). Equation (34.2) is used to define the volume flow.

Equations of motion on the basis of the moment of inertia principle describe the rotation of the leg limbs about the centres of mass of corresponding leg limbs with respective masses m_i, with lacuna areas $A_i = 3\pi a_i^2$ and the moments of inertia J_{Si}, $i = 1, \ldots, 3$ (i is the number of equation):

$$J_{Si}\ddot{\varphi}_i = M_{i,i-1} + M_{i,i+1} - M_{Di} - k(\dot{\theta}_i + \dot{\theta}_{i+1}) - M_{Ci} + M_{C,i+1}, \quad i = 1, \ldots, 3. \tag{34.6}$$

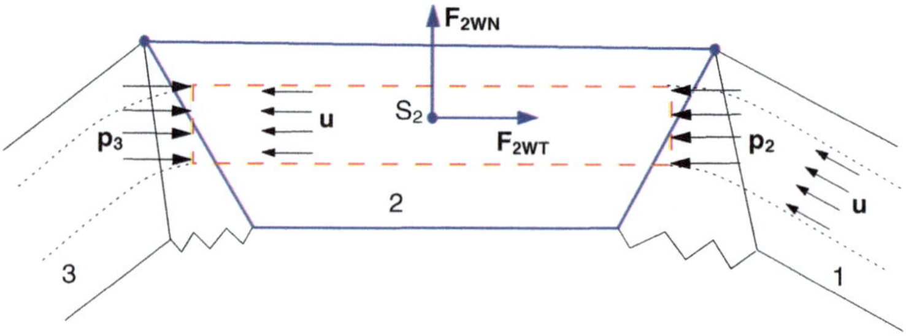

Fig. 34.6 A schematic representation of the acting volume of haemolymph (*discontinuous red line*) for calculating the hydraulic forces and moments; u is the velocity of the haemolymph

The parameters θ_i are the angles between the leg limbs (Fig. 34.5b):

$$\theta_1 = \varphi_1, \quad \theta_i = \pi + \varphi_i - \varphi_{i-1}, \quad i = 2, 3, \quad \theta_4 = 0. \tag{34.7}$$

M_{Wi} are the moments, which are produced by hemolymph pressure at the leg limbs. The equations for these moments are given by the principle of moment of inertia for the acting hemolymph volume. In Fig 34.6, the simplified form of acting volume is shown for lacunae of the patella-tibia limb. Accordingly, hemolymph exerts forces at the leg limb. Particularly, for the patella-tibia limb, the forces from hemolymph can be calculated using the momentum principle according to (34.8):

$$\begin{aligned} F_{2\mathrm{WT}} &= A_2(p_3 - p_2) - \rho\frac{\dot{V}_2^2}{A_2}(1 - \cos(\varphi_2 - \varphi_1)), \\ F_{2\mathrm{WN}} &= \rho\frac{\dot{V}_2^2}{A_2}\sin(\varphi_2 - \varphi_1). \end{aligned} \tag{34.8}$$

The internal forces for the leg limbs are calculated using the momentum principle for each limb and for the spider body. For example, the forces for the patella-tibia limb are the following:

$$\begin{aligned} F_{23y} &= -m_2\ddot{y}_2 - m_2 g + F_{21y} + F_{2\mathrm{WN}}\sin\varphi_2 + F_{2\mathrm{WT}}\cos\varphi_2, \\ F_{23x} &= -m_2\ddot{x}_2 + F_{21x} - F_{2\mathrm{WN}}\cos\varphi_2 + F_{2\mathrm{WT}}\sin\varphi_2. \end{aligned} \tag{34.9}$$

The moments M_{ij} (moment at the limb i from the limb j) can be defined through the internal forces. For the patella-tibia limb, the moments are:

$$\begin{aligned} M_{21} &= -F_{21x}\frac{l_2}{2}\cos\varphi_2 + F_{21y}\frac{l_2}{2}\sin\varphi_2, \\ M_{23} &= -F_{23x}\frac{l_2}{2}\cos\varphi_2 + F_{23y}\frac{l_2}{2}\sin\varphi_2. \end{aligned} \tag{34.10}$$

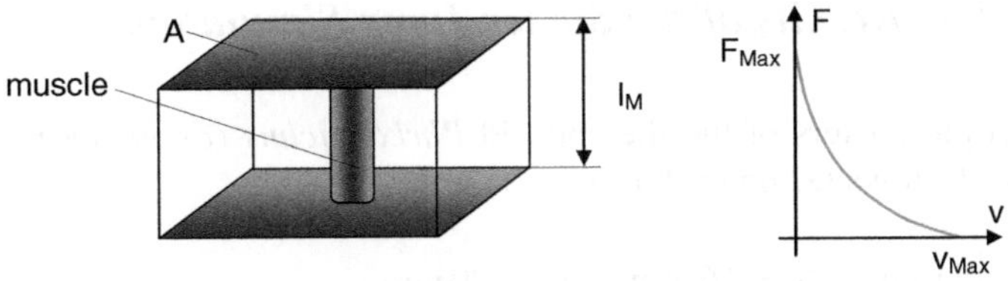

Fig. 34.7 The model of the muscle acting in the prosoma. *Left*: the prosoma as a cube with contraction muscles reduced to one middle muscle. *Right*: the muscle forces dependence on the contraction velocity

The moments of the spring elements between the leg limbs are defined according to (34.4).

34.3.2 Hydraulic Pressure in the Prosoma

The contraction of the muscles in the prosoma causes its volume change. As a consequence, the pressure rises in the prosoma first, and then in the extremities. For the model spider a pressure up to 60 kPa is assumed; separated legs stand this pressure. For the description of the pressure rise in the prosoma the following model is suggested. The form of the prosoma is simplified to a cube. In the middle there is a muscle which contraction produces the volume reduction of the prosoma and consequently the pressure rise (Fig. 34.7, left). Hence, the power of the muscle with the length l_{M} is:

$$p\dot{V} = F_{\mathrm{Mus}} \frac{\mathrm{d}l_{\mathrm{M}}}{\mathrm{d}t} = F_{\mathrm{Mus}} v. \tag{34.11}$$

In this equation, p is the pressure of the haemolymph, F_{Mus} is the force and v is the contraction velocity of the muscle. The muscle force is calculated using Hill's formula:

$$F_{\mathrm{Mus}} = \frac{c}{v+b} - a \tag{34.12}$$

with constants a, b and c. We assume that the muscle area equals $A_{\mathrm{Mus}} = A/5$ and obtain the pressure in the prosoma as follows:

$$p = \frac{5F_{\mathrm{Max}}}{A} \frac{Av_{\mathrm{Max}} - \dot{V}}{Av_{\mathrm{Max}} + 5\dot{V}}, \tag{34.13}$$

where $A = (m_k/\rho_k)^{2/3}$ is the upper or lower area of the prosoma with the mass m_k and the density ρ_k, $k = 1, \ldots, 3$.

34.3.3 The Results of the Spider Jump Simulations

The model parameters of the theraphosid *Phrixotrichus roseus* are measured and assumed as follows (Zentner 2003):

$$\begin{aligned}
&l_1 = 17\,\text{mm}, \quad l_2 = 16\,\text{mm}, \quad l_3 = 20\,\text{mm}, \\
&r_1 = 1.9\,\text{mm}, \quad r_2 = 1.2\,\text{mm}, \quad r_3 = 1\,\text{mm}, \\
&a_1 = 0.41\,\text{mm}, \quad a_2 = 0.33\,\text{mm}, \quad a_3 = 0.37\,\text{mm}, \\
&m_1 = 0.23\,\text{g}, \quad m_2 = 0.095\,\text{g}, \quad m_3 = 0.074\,\text{g}, \quad m_0 = 4\,\text{g}, \\
&c = 0.037\,\text{N mm/rad}, \quad k = 3.5\,\text{kg mm}^2/\text{s}, \quad \rho = 1060\,\text{kg/m}^3. \qquad (34.14)
\end{aligned}$$

Boundary conditions for the differential equations of motion are formulated corresponding to the observations of the leg kinematic:

$$\begin{aligned}
&\varphi_1(0) = 3.1\,\text{rad}, \quad \dot{\varphi}_1(0) = 0, \\
&\varphi_2(0) = 1.3\,\text{rad}, \quad \dot{\varphi}_2(0) = 0, \\
&\varphi_3(0) = 0.3\,\text{rad}, \quad \dot{\varphi}_3(0) = 0. \qquad (34.15)
\end{aligned}$$

The differential equations of motion are integrated as an initial value problem using the software Mathematica®. The modelling results for the angles between the leg limbs $\theta_i(t)$, pressures $p_i(t)$, $i = 1, \ldots, 3$ in the leg limbs and the normal component of the ground force are presented in Fig. 34.8.

For the motion in the ballistic phase, we assume that the spider is a mass point, which moves along the parable curve. The end velocity and the end body position of the first phase are taken as initial conditions for the ballistic phase. As an example, we consider the jumps of two spiders of different sizes. One model spider has the dimensions of a theraphosid and another of a smaller hunting spider (Fig. 34.9). The last spider is eight times smaller as the theraphosid. The simulation shows that the smaller spider jumps ten times further than the big one, which carries out almost no ballistic phase.

34.3.4 Conclusions of the Simulation

The validation of the model was carried out for *Phrixotrichus roseus* (Theraphosidae) and *Cupiennius salei* (Ctenidae). Only the geometrical values and the spiders' weight were adjusted in this model but a good correspondence between the kinematics of the model and the jump of a real spider could be achieved (Karner 1999; Zentner 2003). This agreement has a qualitative character because of numerous assumptions related to the values of the geometric and material parameters.

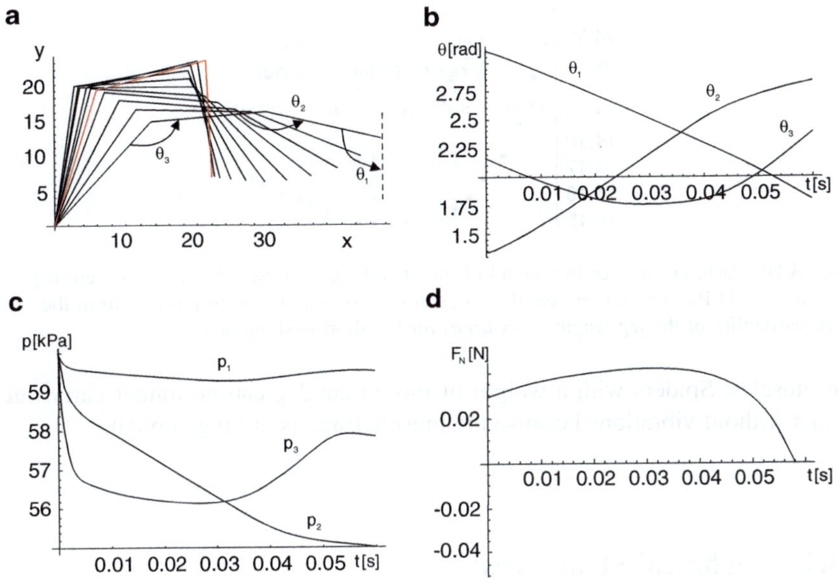

Fig. 34.8 Modelling the jump of a theraphosid spider. (**a**) Simulation of the leg moving during pushing. (**b**) Angles between the leg limbs. (**c**) Pressures in each limb. (**d**) Normal component of the ground force

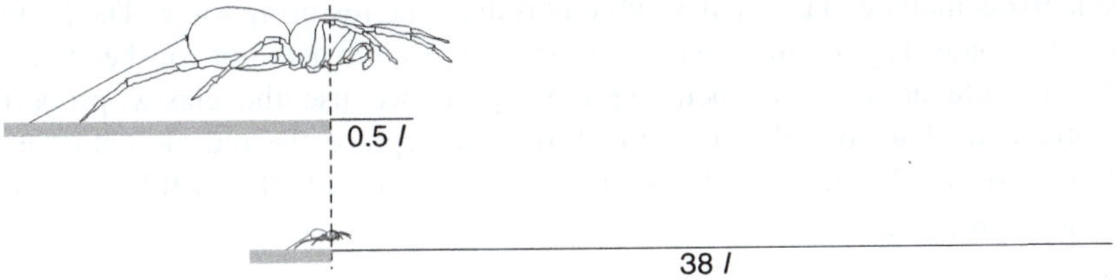

Fig. 34.9 The relation between the jump distance and the leg length for two spiders of different sizes (Theraphosidae, Ctenidae)

In order to answer the question, why the spider legs are so long, the leg lengths were varied and different parameters, for example the speed, acceleration, energy and pushing time etc., were calculated. Only for the pushing time and for bigger pressures from 50 to 70 kPa, we have a minimum and this minimum is in the size range of the real leg length (Fig. 34.10). This means that with the real leg length, the spider has the minimal time to push from the ground. If the legs of the model spider are longer or shorter, then the first phase of the jump is slower. Moreover, the pushing time has a shallow maximum, particularly for *Cupiennius salei*. A spider with a little longer or a little shorter leg has a relatively short pushing phase too.

Model spider *C. salei* with larger weight from 1.5 to 3 g jumps without vibrations only after the increasing of damping in the joints. This damping can be produced by

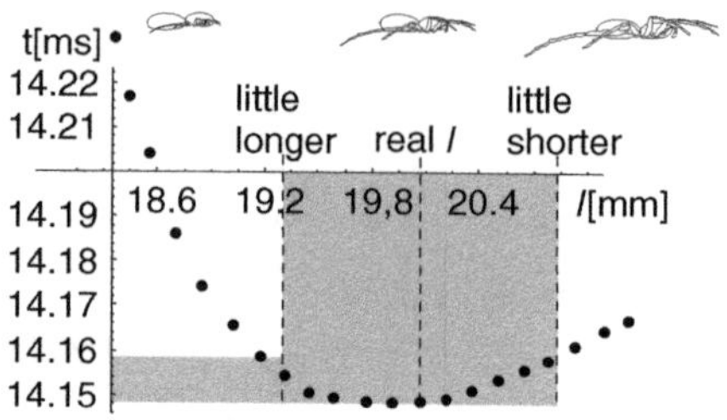

Fig. 34.10 Pushing time for the model of the ctenid *Cupiennius salei* with different legs, inner pressure is 60 kPa; the real leg length guarantees the minimal time to pushing from the ground; large variability of the leg length (*grey area*) for the short pushing phase

the muscles. Spiders with a weight of more than 3 g can no longer carry out the jumps without vibrations because the muscle force is not big enough.

34.4 Technical Structures

To design the technical structures according to the functionality of the spider leg, we use the most important principles of its hydraulic mechanism. As a basis for such technical structures, the compliant mechanisms are applied. The compliant mechanisms include elastic parts, which produce the restoring force. The hydraulic joint of a spider leg contains two actuators: a hydraulic actuator for the stretch-out and a muscle actuator for bending the leg. If we use the elastic property of compliant mechanisms, the restoring force can replace the muscle actuator. The following principles of the hydraulic mechanism of a spider leg will be used for the technical structures:

- Hydraulic mechanism of movement.
- Hollow space of the structure.
- Asymmetrical construction (asymmetry about the length axis).
- Monolithic construction.
- Changing of inner pressure produces the relative movement of structure parts.

Such structures have a tube form with a local joint, which is positioned asymmetrically about the length axis. The joint of a monolithic structure is called a coherent joint. This coherent joint can be performed with the help of geometric or material properties. Under the inner pressure, the local fragment with more elasticity than the rest of the structure is stretched, because of asymmetrically geometric or material properties, and the structure is bent. Because of the elasticity, the structure moves back. Therefore we have a structure containing a coherent joint and only one actuator, which can be either hydraulic or pneumatic. Table 34.1 shows different constructions of structures with coherent joints performed by asymmetrical geometry, by asymmetrical material properties and by their combination.

Table 34.1 Three types of deferent structures with coherent joints performed by asymmetrical geometry, asymmetrical material properties and by their combination

Structures with coherent joints	1st type: asymmetrical geometry	2nd type: asymmetrical material properties	3rd type: combination
Examples	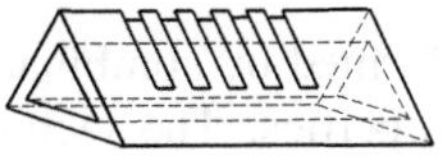 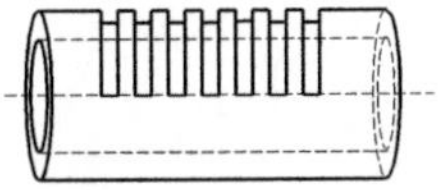	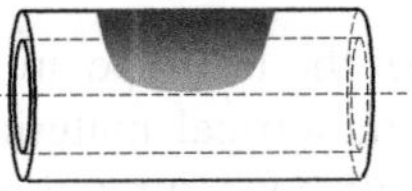	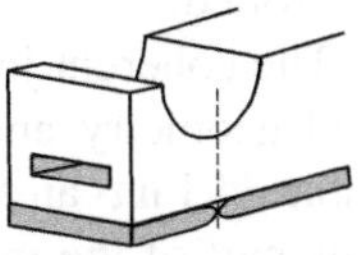 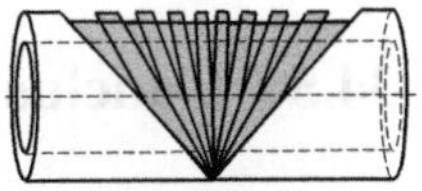
Real structures	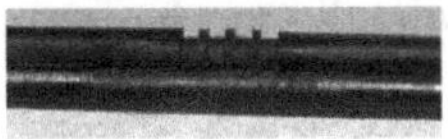 	 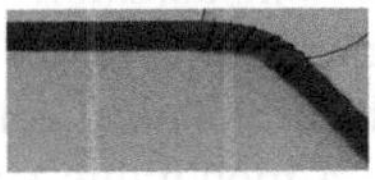	 

The specific geometric asymmetry, concerning the wall thickness or wall shape, which can or cannot be periodical, creates a coherent joint performed with the help of geometric properties (first type). The technical structure manufactured from silicon has a corrugated part, which is stretched under the inner pressure. The bending angle of such structures can grow up to 45°.

The structure of the second type of coherent joints contains a fragment of the material, which is more elastic than the whole structure. The material may possess local high compliance, which can be either constant or variable. In order to reach the second type of coherent joint with constant compliance, materials with different compliance properties are combined with each other. Reaching the joints with variable compliance, the properties of the local structure part must be changed. The technical structure in Table 34.1 is an example for coherent joint with variable compliance. The specific material properties of polymer tubes are changed with the help of the local effect of a heating source realised by a heating wire. Between the wire and one particular side of the tube, the thermal insulation is inserted. The material becomes more compliant at the warmed location in difference to the rest of the structure. The compliant part is extended under the increasing pressure in the tube and the structure is bent. The maximal bending angle is again up to 45°. This bended shape of the tube remains after cooling down. The original shape of the tube can be restored by the renewed warming of the tube without inner pressure. Such joints can be used as monolithic cascade structures; they can also be separately controlled.

The coherent joints of the last type are performed by combination of asymmetrical geometry and asymmetrical material properties. The technical structure in Table 34.1 has an area with concentrated corrugations on both sides of the structure. This part of the material is more elastic than the whole structure. Under an inner pressure, such a structure can be bent up to 90°.

34.5 Conclusions

The model-based simulations of the spider hydraulic mechanism allow reaching a new knowledge about kinematic and dynamic behaviours of the spiders. According to the moving principles of the spider leg, technical compliant structures with coherent joint can be developed and performed. Because of the elasticity of such structures they need only one actuator, which can be either hydraulic, like a spider leg, or pneumatic. The elasticity assumes a role of muscles and moves the structure in the beginning shape. These technical compliant structures have several advantages. For example, it can be a monolithic structure (i.e. consist of only one part) and therefore predestined for miniaturisation. Compliant mechanisms feature less friction and accordingly they require low maintenance and also they can perform a complicated trajectory. Such mechanisms can be used as gripper finger, robot arms, and legs of walking robots and as a sonde in medical technology.

References

Foelix RF (2011) Biology of spiders. Oxford University Press, Oxford

Karner M (1999) Volumenverschiebungen beim Sprung der Jagdspinne *Cupiennius salei*. PhD Thesis, University of Frankfurt am Main

Kropf C (2013) Hydraulic system of locomotion. In: Nentwig W (ed) Spider ecophysiology. Springer, Heidelberg (this volume)

Wirkner C, Huckstorf K (2013) The circulatory system of spiders. In: Nentwig W (ed) Spider ecophysiology. Springer, Heidelberg (this volume)

Zentner L (2003) Untersuchung und Entwicklung nachgiebiger Strukturen basierend auf innendruckbelasteten Röhren mit stoffschlüssigen Gelenken. ISLE, Ilmenau

Chapter 35
Adhesion to Flat Surfaces: From Spiders to Stickers

Tobias Seidl and Renato Vidoni

35.1 Introduction: Spider Ecology and Tarsus Morphology

Similar to other animals, spiders possess structures for handling their surrounding environment. These are mostly claws, located on the pretarsus of all spiders, and—in only a number of taxa—adhesive hairs, the so-called setules. These setules are either distributed over the whole length of the tarsus as in lycosid spiders (Rovner 1978) or located ventrally of the claws as a tuft or the so-called scopula (Hill 1977). While the claws serve for handling threads or rough structured substrate, the scopula is used for attachment to smooth surfaces such as leafs or glass slides. In contrast to insect adhesion, the active secretion of fluids for adhesive purpose is not common and still subject of discussion (see Niederegger 2013).

Spiders follow different strategies to gather food, ranging from web building (and waiting for the prey to get trapped) to freely hunting by approaching and subsequently grabbing their prey. Due to its free-ranging lifestyle, the salticid *Evarcha arcuata* does not have any specialized tool for the handling of spider silk as, for example, the orb weaver *Araneus diadematus*. The hierarchically structured scopula with its single hairs and abundant number of setules is an optimal adaptation toward the ground dwelling, predatory lifestyle of this species. Jumping spiders hunt down their prey in the vegetation moving on slippery plant surfaces, and thereby a firm grip to the substrate is vital, as is the firm attachment to captured prey. Tight attachment to smooth surfaces is ensured by the claw tuft apparatus, the scopula (Fig. 35.1a). The scopula is formed by a number of long hair-like appendages in a parallel configuration—the setae. Each seta is covered by tiny pointed hairs on the dorsal side and by a dense population of hairs on the ventral side, called setules. The ventral

T. Seidl (✉)
Department of Mechanical Engineering, Westfälische Hochschule, Bocholt, Germany
e-mail: tobias.seidl@w-hs.de

R. Vidoni
Faculty of Science and Technology, Free University of Bozen-Bolzano, Bolzano, Italy

W. Nentwig (ed.), *Spider Ecophysiology*,
DOI 10.1007/978-3-642-33989-9_35,

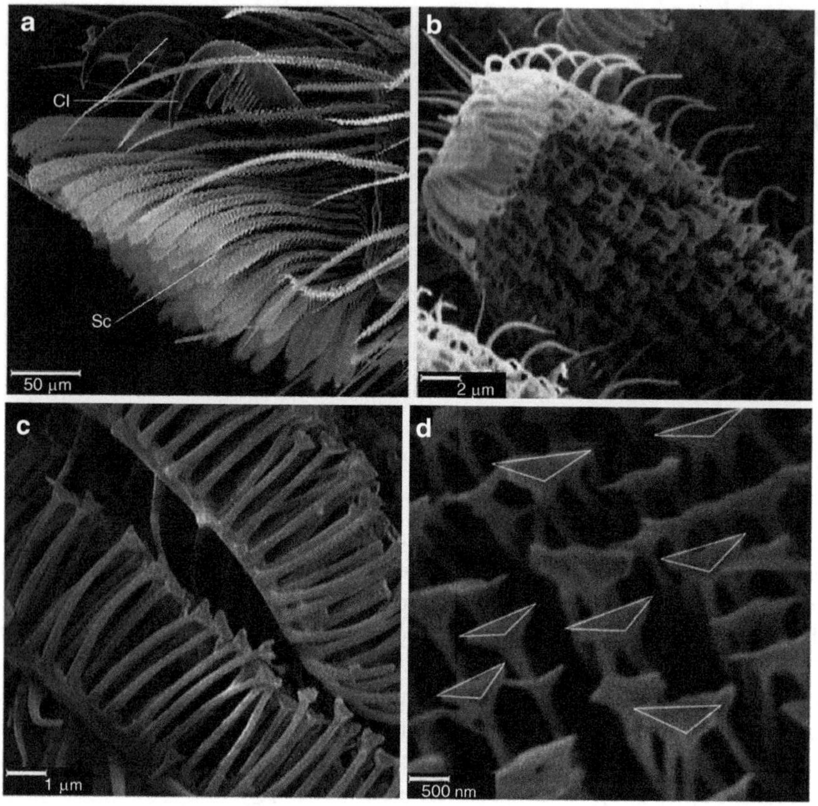

Fig. 35.1 Hierarchical structure of *Evarcha arcuata*'s attachment devices. (**a**) Tarsus with scopula (Sc) and claws (Cl). (**b**) The scopula consists of setae, while each seta is covered by many setules. (**c**) Setules on the ventral side end in a flattened and tilted triangular structures, called spatula. (**d**) Each spatula is about 500 nm wide (SEM micrographs: Andrew Martin, taken from Kesel et al. 2003)

hairs are flattened at the free end, showing an inclined triangular-shaped spatula (Fig. 35.1b–d). These spatulae form the point of contact when adhering to the substrate. *E. arcuata* has an estimate of 624,000 contact elements, a high number in comparison to other arthropods (Martin et al. 2002b).

In contrast to the contacting setules on the ventral side, the function of the tapered processes on the setae's dorsal surface remains unclear. Their size and distribution along the dorsal surface might play a role in the prevention of any adherence of one seta to another. The scopulae on the two frontal leg pairs are considerably smaller in size than those on the hind leg pairs (Martin et al. 2002a). This may be due to the individual role of legs during locomotion and prey capture (Betz and Kölsch 2004). While it appears rather a trivial aspect at first sight, easy detachment from the substrate is crucial for highly motile organisms and indeed a central requisite for the design of any tarsal attachment device be it insect, spider, frog, or gecko (Arzt et al. 2003).

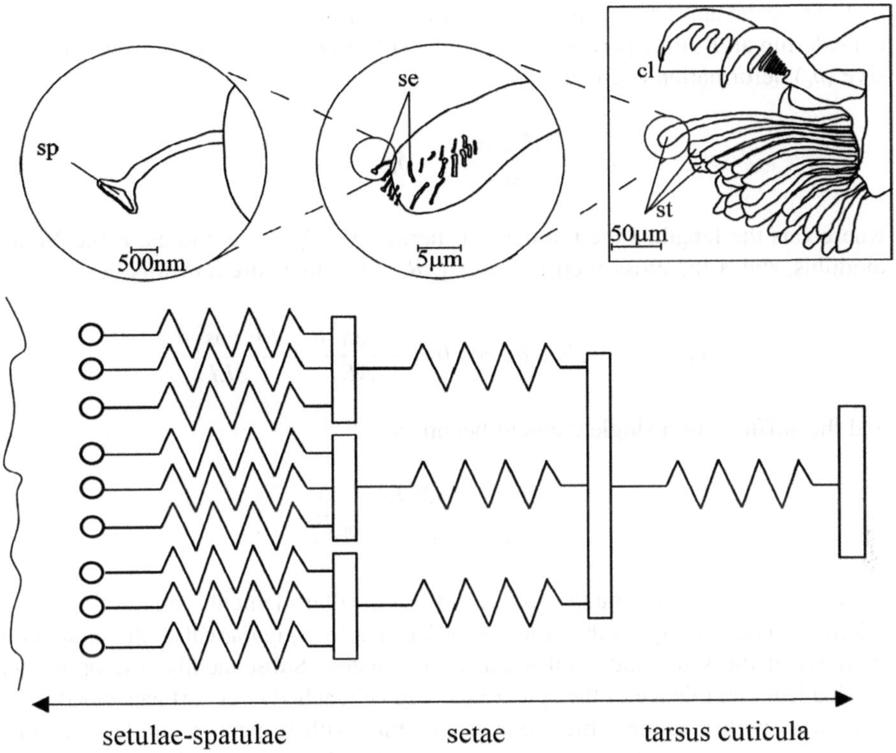

Fig. 35.2 Elasto-mechanical model of the hierarchical adhesive structure of spider claw tufts. sp = spatulae, se = setulae, st = setae, cl = claws. Modified from Gasparetto et al. (2009)

35.2 Hierarchical Design of the Claw Tuft

The use of van der Waals forces for attachment faces two principal challenges: these forces are small and do not range very far. Hence, many contact elements are required and need to be brought close to the substrate, that is, nanometers, since these forces are inversely proportional to the surface distance.

Since real-world substrates are seldom smooth to an industrial standard, it is necessary to employ a mechanism, which allows sufficient compliance of the scopula to bring a sufficient amount of hairs close to potential attachment sites.

The single elements of the scopula exhibit certain mechanical properties: under load, each of the long and thin hairs is subject to bending, storing energy like a cantilever beam, and, hence, can be modeled as a spring in tension or compression (Bhushan et al. 2006; Gasparetto et al. 2009). In consequence, the scopula—with seta, setule, and triangular end—can be described as a three-level spring model made of tarsus (cuticula), setae, and setules (Fig. 35.2).

Let θ be the approaching angle between the fibrillar elements and the surface. By considering an F force perpendicular to the substrate, a bending (δ_b) and compressive (δ_c) deformation is created as

$$\delta_b = \frac{Fl^3 \cos\theta}{3EI}, \quad \delta_c = \frac{Fl \sin\theta}{AE}$$

where L is the length, I the moment of inertia ($\pi R^4/4$), R the radius, E the Young modulus, and A the cross-sectional area (πR^2). The total stress results in

$$\delta_n = \delta_c \sin\theta + \delta_b \cos\theta = \frac{Fl \sin^2\theta}{AE} + \frac{Fl^3 \cos^2\theta}{3EI}$$

and the stiffness of a single element becomes

$$k = \frac{\pi R^2 E}{l \sin^2\theta\left(1 + \frac{4l^2 \cot^2\theta}{3R^2}\right)}$$

By considering the Young modulus of the fibrillar elements (materials such as β-keratin have Young modulus in the order of a few gigapascals), the equivalent stiffness of the setae and setules can be estimated. Subsequently, the optimized mechanical compliance of the spider tarsi can be reached over surfaces of different roughness with a suitable hierarchical structure with respect to single layer dry-adhesive systems simulated and demonstrated by Gasparetto et al. (2009).

35.3 Adhesion with a Nonadhesive Material

Quantifying the adhesive properties of single setule has become possible with the invention of atomic force microscopes and related instruments. Point spectroscopic measurements (Fig. 35.3) with calibrated cantilevers allowed to determine an average attachment force of 38.12 nN per setule in *E. arcuata*, resulting in a theoretically possible total adhesive force of 2.38×10^{-2} N (Kesel et al. 2003). Considering that the mean body mass of *Evarcha arcuata* is 15.1×10^{-3} kg, the maximal adhesion force equals 160 times the body weight and, hence, grants safe attachment even when hunting prey. These scopulae are part of the animal's cuticle exhibiting super-hydrophobic characteristics too.

As prior mentioned, the major acting physical principle is the van der Waals forces, weak and short-ranging (i.e., tens of nanometers) intermolecular forces, possibly supported by capillary forces (Kesel et al. 2003; Autumn and Peattie 2002). Concerning the adhesive force model, if the surface asperities and the adhesive element ends are modeled like spherical tips, the interaction force can

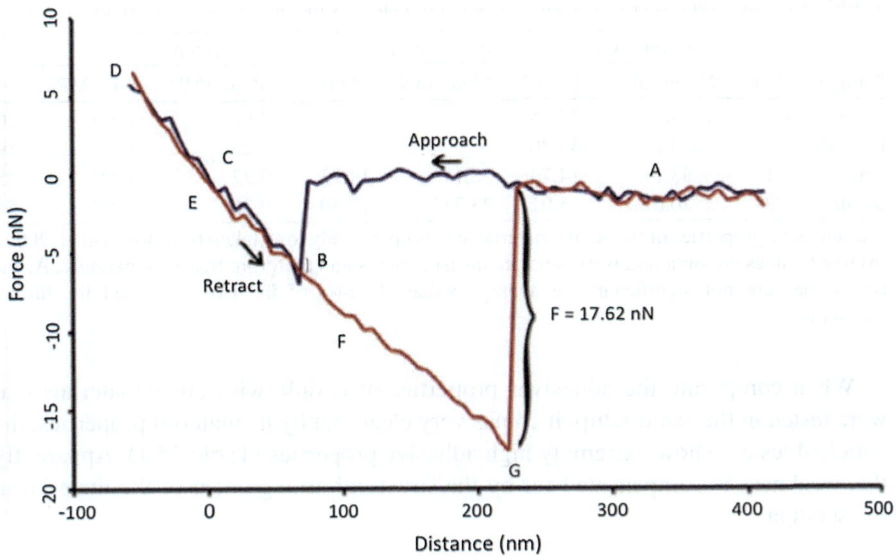

Fig. 35.3 Force-distance curve of an atomic force microscope during point spectroscopy. The *straight vertical line* of the lower curve denotes the lift-off event of the cantilever from the specimen. The force required to break the contact is the attachment force of the specimen to the cantilever. Phases of force spectroscopy: (*A*) cantilever out of contact, (*B*) snap-in due to attractive forces, (*C*) loading of probe onto sample with increasing force, (*D*) begin retraction of cantilever, (*E*) unloading, (*F*) probe remains in contact due to adhesive forces, negative loading, (*G*) force exerted by cantilever spring equals and subsequently exceeds adhesive force of probe-sample contact (from Kesel et al. 2004)

be estimated by considering the interaction of two spheres as formulated by the JKR model (Johnson et al. 1971). Thus, the adhesion force results in

$$F_{\mathrm{ad}} = \frac{3}{2}\pi r_{\mathrm{c}} E_{\mathrm{ad}}$$

where r_{c} is the reduced radius:

$$r_{\mathrm{c}} = \left(\left(\frac{1}{r_1}\right) + \left(\frac{1}{r_2}\right)\right)^{-1}$$

with r_1 the radius of the hemisphere on the tip of the adhesive element and r_2 the radius of the hemisphere that represents the asperity, here assumed as $r_1 = r_2$. Let the energy of adhesion E_{ad} range from 10 to 66 mJ/m^2, that is, the typical range for van der Waals surfaces, and working with the estimated radius of the spider's setules, the adhesion force becomes very close to the experimental ones, that is, 38 nN (Gasparetto et al. 2009).

Table 35.1 Adhesion forces measured on several materials including *Evarcha arcuata* setule

		Absolute force		Force/setule		Force/area		
Sample	Probe	Mean (nN)	SD (nN)	Mean (nN)	SD (nN)	Mean (MPa)	SD (MPa)	*n*
Glass	C1	1315.53	22.85	–	–	3.65	0.06	20
Epoxide	C1	443.72	149.20	–	–	1.23	0.41	19
Setula	C1	38.11	14.73	38.11[a]	14.73	0.22	0.09	45
Setula	C2	20.67	5.91	43.71[a]	12.50	0.26	0.07	50

The adhesive properties of the setule material are comparatively low (taken from Kesel et al. 2004).
[a]Marked values are obtained from separate measurements and therefore treated separately. As the two values are not significant, the average value (41 nN) of the two was used for further calculations.

When comparing the adhesives properties of cuticle with other materials that were tested in the same setup, it comes very clear that by its material properties, the cuticle does not show extremely high adhesive properties (Table 35.1). Apparently this weakness is compensated for by the structural arrangement of the material in the scopula.

35.4 Adhesion and Adhesive Properties

The previously described physical principle is only one element contributing to the design of biomimetic adhesive devices based on the spider attachment performance. In the following we introduce the most important features:

The triangular-shaped distal end on each setule—the so-called spatula—shows anisotropic adhesion to the substrate, meaning that the adhesive force is strongly correlated with the spatula's orientation towards the substrate. This allows the animal to actively control adhesion and, for example, to produce low detachment force when required, as demonstrated nicely by Gao et al. (2005) using finite element modeling techniques. Here, adhesion maximizes at angles around 30°; at smaller angles, sliding occurs and detachment at higher angles. And indeed, the spider's scopula attachment system seems to work completely passively, in contrast to the muscularly mediated digital hyperextension in geckos (Autumn and Peattie 2002). In the latter, an active distal peeling motion allows easy detachment of the adhesive system from the substrate, while *E. arcuata*'s attachment is only controlled via the movement of the spider's legs. To this point there were no active detachment systems observed in living spiders, and since the scopula lies on the tarsus tip, both the pulling of the front and the pushing of the rear legs create a sort of natural and passive rotation motion with a pivot of the touching area. This mechanism allows to gradually detaching the fibrillar elements in contact with the surface without a significant effort.

Two important properties keep these hierarchical structures functional during their lifetime: self-cleaning and anti-bunching. The former prevents loose particles from sticking to the tarsus and consequently blocking adhesion sites and reducing attachment properties (Hansen and Autumn 2005). The latter prevents the densely packed setules from sticking to each other rather than to the substrate. In short, this is achieved by (1) the anisotropic adhesive properties of the spatula and (2) the optimized geometrical conditions and spring constants of the setules preventing their terminal ends to meet in such an alignment that would allow bunching (see, for instance, Sitti and Fearing 2003). Indeed, the anti-bunching effect is a result of a careful choice of geometrical parameters of the fibrillar elements at each level, that is, radius, length, and distance, to create a particle-repellant and bunching-resistant hairy attachment system. These parameters can and need to be taken into account when designing biomimetic attachment devices.

35.5 Technical Application

Spiders are skillful locomotors and, together with *E. arcuata*'s powerful passively controlled attachment and detachment system, great models for biomimetic transfer. The two most appealing approaches will be described in the following.

35.5.1 Bioadhesives

The first attempt is to create bio-inspired dry-adhesive structures for universal application. Such attachment devices would employ the same physical principle and the same observed statistical adhesion with a summed attachment force correlating with the dynamics during attaching. Such "stickers" could function in many environments, even liquid or vacuum, but, contrary to, for example, Velcro, would not need a predefined substrate. They would have no power required to maintain attachment, no dependence upon gravitational forces to realize attachment, and availability on any surface regardless of issues related to surface roughness or friction coefficient. It is easy to imagine that space research is highly interested in creating a bio-inspired artificial adhesive able to mimic the most versatile used repairing and attaching technique in orbit.

Early attempts tried to duplicate the natural dry-adhesive performance by designing and fabricating extremely small hairs (Sitti and Fearing 2003). This resulted in an ineffective adhesion on the macroscale. After that, as reported in Sameoto and Menon (2010) and Zhang et al. (2010), a significant effort has been given to this topic, and effective adhesive systems, even if not yet as powerful as the natural ones, have been realized. In particular, the evolution of synthetic dry-adhesives has taken two main directions: the first focuses on the development of relatively soft elastomer fibers with mushroom-shaped tips, which typically exhibit high normal adhesion strengths, while the second focuses on stiff, very high aspect ratio fibers, which exhibit good friction forces, but either low or zero normal adhesion strength.

Two fabrication methods are preferred for prototyping attachment devices: micro/nanocasting and gas-phase growth or etching (Sameoto and Menon 2010). In the former, the majority of developed adhesives have been realized using materials such as polydimethylsiloxane (PDMS) and polyurethanes (PU), while, in the latter, aligned arrays of carbon nanotubes are usually grown.

Such perfect adhesives can answer to different technical challenges or create new effective and efficient systems for a broad range of applications. Indeed, they could be used for biomedical applications such as safe and sterilized endoscopy and tissue adhesives; micro-electro-mechanical systems (MEMS) and wafer alignment and manipulation; structures to substitute glues, screws, traditional adhesives, reversible adhesive gloves, or rock climbing aids; and robotic systems.

35.5.2 Bio-inspired Spider Robots

Taking the complexity of biomimetic transfer some level higher, the entire subsystem of locomotion and adhesion, that is, the spider's leg, gets into focus. From the locomotion point of view, a spider-inspired legged robot is not a new idea, but the availability of new mechatronic systems and the discovery and study of the spiders' reversible passive adhesive principles have drawn new attention and interest to these arthropods.

Indeed, the idea of using biomimetic, asymmetric adhesives to control attachment to various types of surfaces merely via the kinematics of the legs can be exploited in order to create legged robots that can both overcome obstacles and cope with different kinds of surfaces by means of these synthetic dry-adhesive systems and with a minimal effort.

The European Space Agency has encouraged works in this direction (compare Gasparetto et al. 2008, 2009, 2010; Menon et al. 2008; Li et al. 2012). The first approach deals with modeling and simulating a climbing spider-inspired robot by studying the anatomy, the adhesive and locomotion capabilities of the spider (i.e., an eight-legged system), and the kinematic and dynamic behavior in different operative and slope conditions (Fig. 35.4a), while the second approach deals with the macro-, micro-, and nano-structural design and realization of a miniaturized spider-inspired legged robot capable to negotiate different roughness and slope surfaces and of operating in a space environment (Fig. 35.4b). The realized prototype, called Abigaille II, is a lightweight robot actuated by eighteen miniaturized motors and equipped on the tip of the legs with bio-inspired adhesive patches and is able to climb effectively vertical Plexiglas surfaces.

Fig. 35.4 Bio-inspired spider robots. *Above*: spider robot simulator implementing the cooperation between locomotion and adhesion (Vidoni and Gasparetto 2011). Blue elements represent spiders' leg joints. The tip of the tarsus of each leg is visualized with different colors due to the angle of adhesion and to the condition of the leg: red—in adhesion; green—in detachment or in flight. *Arrows* represent the body reference frame. *Below*: spider robot prototype with dry-adhesive pads developed at Simon Fraser University (Li et al. 2012). The prototype shows six legs in order to reduce the number of actuators and, in the meantime, allow a biomimetic gait

35.6 Conclusions

The attempt to design autonomous devices—commonly known as robots—has led many people to draw inspiration from the broad choice of functional autonomous mobile organisms—commonly known as animals. Among these, hunting spiders have caught the attention of the researchers since they can move fast over a broad range of surfaces and slopes and easily overcome obstacles. Walking on vertical and inverted surfaces is as possible as dynamically hunting down prey. In contrast to insects, spiders employ a dry-adhesive system using van der Waals interactions between their numerous tiny terminal ends of the hairy scopula and hence supply a model for adhesion also in extreme environments like vacuum. Indeed, effective attachment is achieved through elastic and damping properties of both material and hierarchical structure, while detachment is achieved via the angle variation between tarsus and substrate thanks to the macroscopic legs motion (Gasparetto et al. 2010) and hence comes "free of charge" by control of leg posture. Mimicking the adhesive system as well as locomotory capabilities will enable engineers to create effective and innovative technological devices in a broad range of application fields, starting from universally usable reversible adhesives to autonomous space and hazardous environments exploring robots and manipulators.

References

Arzt E, Gorb E, Spolenak R (2003) From micro to nano contacts in biological attachment devices. Proc Natl Acad Sci 100:10603–10606

Autumn K, Peattie AM (2002) Mechanisms of adhesion in geckos. Integr Comp Biol 42: 1081–1090

Betz O, Kölsch G (2004) The role of adhesion in prey capture and predator defence in arthropods. Arthrop Struct Develop 33:3–30

Bhushan B, Peressadko TW, Kim AG (2006) Adhesion analysis of two-level hierarchical morphology in natural attachment systems for "smart adhesion". J Adhes Sci Technol 20: 1475–1491

Gao H, Wang X, Yao H, Gorb S, Arzt E (2005) Mechanics of hierarchical adhesion structures of geckos. Mech Mater 37:275–285

Gasparetto A, Vidoni R, Seidl T (2008) Kinematic study of the spider system in a biomimetic perspective. In: Proceedings of the IEEE/RSJ international conference on intelligent robots and systems, Nice, France

Gasparetto A, Seidl T, Vidoni R (2009) A mechanical model for the adhesion of spiders to nominally flat surfaces. J Bionic Eng 6:135–142

Gasparetto A, Vidoni R, Seidl T (2010) Passive control of attachment in legged space robots. Appl Bionics Biomech 7:69–81

Hansen WR, Autumn K (2005) Evidence for self-cleaning in gecko setae. Proc Natl Acad Sci 102: 385–389

Hill DE (1977) The pretarsus of salticid spiders. Zool J Linn Soc 60:319–338

Johnson KL, Kendall K, Roberts AD (1971) Surface energy and the contact of elastic solids. Proc R Soc London A 324:301–313

Kesel AB, Martin A, Seidl T (2003) Adhesion measurements on the attachment devices of the jumping spider *Evarcha arcuata*. J Exp Biol 206:2733–2738. doi:10.1242/jeb.00478

Kesel AB, Martin A, Seidl T (2004) Getting a grip on spider attachment: an AFM approach to microstructure adhesion in arthropods. Smart Mater Struct 13:512–518. doi:10.1088/0964-1726/13/3/009

Li Y, Ahmed A, Sameoto D, Menon C (2012) Abigaille II: toward the development of a spider-inspired climbing robot. Robotica 30:79–89

Martin A, Dillinger SCG, Kesel AB, Nachtigall W (2002a) The structure of tarsal attachment elements in Araneae. Zoology 105(35 Suppl):95

Martin A, Seidl T, Kesel AB (2002b) Das mikro-mechanische Haftsystem der Spinne *Evarcha arcuata*. In: Wisser A, Nachtigall W (eds) Biona-Report 16:226–230

Menon C, Li Y, Sameoto D, Martens C (2008) Abigaille-I: towards the development of a spider-inspired climbing robot for space use. In: IEEE RAS/EMBS international conference on biomedical robotics and biomechatronics, Scottsdale

Niederegger (2013) Functional aspects of spider scopulae. In: Nentwig W (ed) Spider ecophysiology. Springer, Berlin (this volume)

Rovner JS (1978) Adhesive hairs in spiders: behavioral functions and hydraulically mediated movement. Symp Zool Soc Lond 42:99–108

Sameoto D, Menon C (2010) Recent advances in the fabrication and adhesion testing of biomimetic dry adhesives. Smart Mater Struct 19:103001
Sitti M, Fearing RS (2003) Synthetic gecko foot-hair for micro/nano structures as dry adhesives. J Adhes Sci Technol 18:1055–1074
Vidoni R, Gasparetto A (2011) Efficient force distribution and leg posture for a bio-inspired spider robot. Rob Auton Syst 59:142–150
Zhang H, Guo DJ, Dai ZD (2010) Progress on gecko-inspired micro/nano-adhesion arrays. Chin Sci Bull – Bionic Eng 55:1843–1850. doi:10.1007/s11434-010-3065-z

Chapter 36
Technical and Biomedical Uses of Nature's Strongest Fiber: Spider Silk

Christina Allmeling, Christine Radtke, and Peter M. Vogt

36.1 Introduction

In their unique combination of tensile strength, elasticity, and light weight, spider silk fibers as well as fibers imitating spider silk are considered attractive for a number of technical applications. Low immunogenicity and a high biocompatibility support the notion of an ideal biomaterial which is further encouraged by the fact that spider silk is sterilizable at high temperatures (Gellynck et al. 2008).

Hence humans have been making use of spider silk for thousands of years in both technical and biomedical sense. Several old cultures like the Romans used cobwebs to stop wounds from bleeding (Newman and Newman 1995), and in native societies, like the Aborigines, spider silk was used as fishing lines for small fish, an application still practiced by the people of the Solomon Islands (Heim et al. 2009).

Since then, human use of spider silk has been disregarded, mostly due to the nonindustrial nature of spider silk reeling. Progressive technological developments, however, have prompted the demand of a more sustained production of industrial goods (US Congress 1993) among which the use of biopolymers, like spider silk or derivatives thereof, has a great potential. In contrast to synthetic fibers based on raw materials with limited availability and/or slow and environmental problematic degradation, spider silk in its natural, in processed, or in artificial forms is biodegradable and independent on fossil resources.

In a broader sense spider silk has inspired many adaptive solutions like silk-mimicking polymers or fibrillogenesis. Gosselin and coworkers became inspired to fabricate microfibers from a viscous polymer solution through liquid rope coiling instability which causes viscous solutions like honey to spontaneously form coils on surfaces. The researchers hypothesize that the process induces formation of bonds

C. Allmeling (✉) • C. Radtke • P.M. Vogt
Department of Plastic, Hand and Reconstructive Surgery, Medical School Hannover, Hannover, Germany
e-mail: allmeling.christina@mh-hannover.de

W. Nentwig (ed.), *Spider Ecophysiology*,
DOI 10.1007/978-3-642-33989-9_36, © Springer-Verlag Berlin Heidelberg 2013

which equal the hydrogen bonding observed in spider silk fibers (Gosselin et al. 2012). A recently published study copied the beads-on-a-string morphology of flagelliform silk, which comprises the fibers of the capture spiral, by coating poly (dimethylsiloxane) on nylon fibers. The authors could demonstrate that the coating breaks spontaneously into regularly located droplets. The coated fibers demonstrated higher adhesion to glass plates attributed to the higher fiber–substrate contact area (Sahni et al. 2012).

Materials inspired by spider silk must not necessarily be of fibrillar nature as they might be applied as surface coatings or as thin and tough membranes. Silk protein-based foams and hydrogels with defined pore sizes have applications, e.g., as tissue scaffolds. Dispersed to microspheres, silk-derived peptides may be used for time-controlled drug delivery.

In our chapter we will summarize industrial as well as biomedical applications of spider silk and give a short discourse on the production of recombinant spider silk proteins and their use for spinning fibers. Future inventions and applications are discussed briefly.

36.2 Different Forms of Spider Silk

36.2.1 Natural Spider Silk

Spiders are unique among silk-producing arthropods as they depend on the production of silk lifelong. An outstanding example is the beauty and sophistication of a spider orb web (Fig. 36.1). Not only adapted to capture prey, these weavings practically serve as a sensual elongation of the spider's body helping the animal to orient in their environment; to detect prey, enemies, and mates alike; and their webs have to fulfill these different tasks. As a result spider silk evolved as an outstanding fibrous biomaterial consisting almost entirely of large proteins solely encoded by the spider silk fibroin gene family (Gatesy et al. 2001). Spider silk proteins are of a modular nature consisting of nonrepetitive C- and N-terminal sections among which long repetitive parts consisting mainly of four characteristic amino acid motifs (A_n, GA, GGX, and $GPGX_n$) are found (Gatesy et al. 2001; see also Blackledge 2013). Proteins show remarkable variability depending on allelic variants of which multiples can exist within species and even individual genomes although negative selection pressure keeps protein sizes in a range between 200 and 350 kDa (Chinali et al. 2010). They are expressed in specialized glands located in the opisthosoma of the spider. These glands are connected to the spinnerets by cone-shaped ducts in which the polymerization to fibers takes place. The conformational changes during the spinning process were studied by comparing the Raman spectra of the native silk dope and the dry silk dope probed in the region of the sac of the major ampullate glands of *Nephila clavipes* (Nephilidae) (Lefevre et al. 2011). This includes arrangement of the silk's structural elements to pseudocrystalline regions

Fig. 36.1 *Nephila* sp. in its natural habitat (République des Seychelles)

of antiparallel β-sheets interspersed with elastic amorphous segments. This composition is made responsible for the remarkable mechanical properties of likewise strength, ascribed to the crystalline regions, and extensibility. The assembly of silk fibers has been especially well described for the dragline silk which is formed in the major ampullate glands of Araneidae spiders. Silk proteins secreted by specialized epithelial cells of the glands are stored at high concentrations in the lumen of the gland. On their way out, the proteins undergo a number of structural changes in a pH-regulated self-assembly process (reviewed by Heim et al. 2009). Higher flow rates are advantageous for fiber alignment which argues for a mechanical impact of flow and shear forces. Sponner and coworkers demonstrated that among the two proteins which mainly compose *Nephila* dragline silk, MaSp1 and MaSp2, MaSp2 is predominantly located in core regions of the fibers, whereas MaSp1 appears to be uniformly distributed along the radial axis which might also be important for the spinning process (Sponner et al. 2005).

Technical application of spider silk proteins and biopolymers basically depends on their specific properties resulting from their structure and composition. One of the most striking features of dragline spider silk is its ability to incorporate water into its amorphous parts. This causes entropy-driven recoiling of molecular chains leading to massive but reversible shrinking in contact with water or in a relative humidity greater than 60 % – a phenomenon called supercontraction (Liu et al. 2005). Agnarsson and coworkers exposed single dragline threads from *N. clavipes* to cyclic changes of humidity resulting in cycles of contraction able to produce an

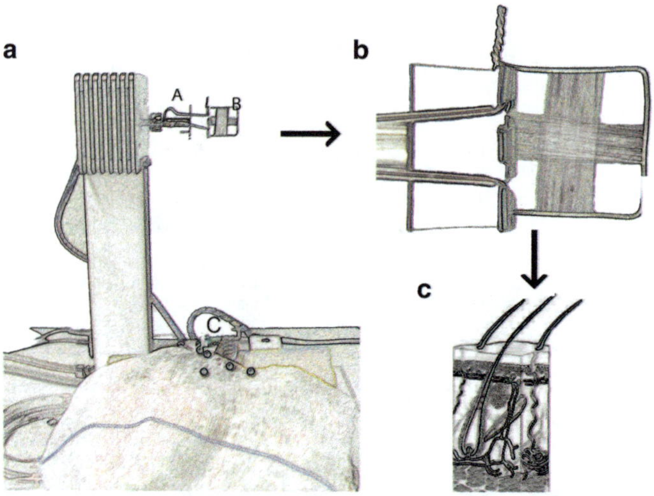

Fig. 36.2 An example for a spider silk reeling machine as developed together by the Leibniz University Hannover and the Medical School Hannover. (**a**) The reeling machine is equipped with a clamping system (*A*) which can hold devices like spools or frames (*B*). The spider (*C*) is fixed by a piece of tissue and pins before the reeling machine. While the device fastened to the reeling machine rotates, dragline silk is pulled out of the spider. (**b**) A steel frame cross-weaved with spider dragline silk. (**c**) Scaffolds of cross-weaved spider silk can be used to study cellular growth and migration of spider silk fibers or for skin tissue engineering

amazing amount of work, e.g., a single 40 mm-long fiber lifted at least 100 mg. The authors propose an energy-efficient and environmentally friendly biomimetic muscle with higher work density and higher stress resistance than most biological muscles (Agnarsson et al. 2009). Assuming a humidity-controlled atmosphere, cyclic spider silk contraction might be used in a variety of applications including sensors and microelectromechanical systems (MEMS) devices and is also interesting in biomedical applications where small steering elements are required (Bai et al. 2006). This is especially interesting as spider silk is the so far only known material which is apt to supercontract at physiological conditions.

Likewise, also a shape–memory effect has been assigned to spider silk. Spider silk fibers return to their original shape independent of external stimuli after a twisting deformation. By comparison, shape–memory nitinol, a nickel–titanium alloy, is needed to be heated to 90 °C under the same experimental conditions before recovery (Emile et al. 2006).

Natural spider silk can be harvested as dragline silk either with special devices, e.g., reeling machines (Fig. 36.2). Other possibilities include harvest of egg sac silk which can be used naturally or as regenerated silk. In the latter case natural silk fibers are solubilized to fibroin extracts which are subsequently processed into artificial fibers or foams. As an example Gellynck and coworkers used 9-M LiBr to dissolve egg sac silk, while hexafluoro-2-propanol and concentrated formic acid are also applicable. Production of scaffolds might include electrospinning procedures and salt-leaching processes (Gellynck et al. 2008; Zhou et al. 2008).

36.2.2 Recombinant Spider Silk

Although cocoon silk from silk moths has been harvested for over 5,000 years, the domestication of spiders for large-scale silk collection is still impractical due to the animals' territorial and cannibalistic nature. Additionally, the amount of silk which can be harvested from one single animal is six times lower in araneids than in the silkworms (*Bombyx mori*). Natural spider silk is subjected to considerable variability, e.g., to spinning forces and humidity which might be the reason for the remarkable differences in tensile strength stated in the published literature (Perez-Rigueiro et al. 2005).

A possible solution for this problem is the development of efficient strategies for synthetic spider silk production. The repetitive sequences encoding for the core structure of a fibroin protein were first identified for major ampullate silk of *Nephila clavipes*. A number of sequences for different types of silks and species followed as summarized, e.g., by Swanson et al. (2009). Transgenic systems used for production of spider silk are manifold and include bacteria, yeast, plants, silkworm, and mice (reviewed by Heim et al. (2009).

A number of studies concentrated on the production of spider silk in bacteria (see Heim et al. 2009 for an overview). The repetitive nature of spider silk caused a challenge to its biotechnological production, first in the construction of the appropriate expression systems as repetitive sequences are impossible to amplify by polymerase chain reaction and corresponding mRNAs are prone to obstructive secondary structures, second in the difficulty to produce sufficiently sized proteins as a consequence of tRNA depletion. In their pioneering work the work group around T. Scheibel developed a cloning strategy of seamlessly joining solid-phase synthesized oligonucleotides which allowed also for the adaption of codon usage (Heim et al. 2009). In their fundamental work, Lazaris and coworkers described the first successful attempt to produce spider silk protein in mammalian cell lines, e.g., baby hamster kidney cells (BHK). Lower molecular weight products predominated possibly due to inefficient transcription allegeable to complex secondary structures or to specific tRNA shortage in these highly repetitive genes. The protein was spun into fibers with a considerable tensile strength of 2.26 gpd (grams per denier: a unit which corresponds to the ultimate tensile strength) at its highest, but the values are still essentially lower than native spider silk which reportedly ranges between 7 and 11 gpd (Lazaris et al. 2002).

Spider silk production in the milk of transgenic goats has been proposed, although no results have been published so far. In spite of that, transgenic mice were generated expressing an artificial dragline analogue in their milk (cited after Heim et al. 2009).

Scheller et al. expressed synthetic genes in transgenic tobacco and potato plants. To this end, fragments adjusted to *N. clavipes* MaSp1 sequence were assembled to large units exploiting the modular nature of silk albeit lacking the nonrepetitive 3′ terminus of natural silk. The expressed proteins accumulated stably in the endoplasmic reticulum of leave and tuber cells and could be purified by chemical extraction followed by a heating step (Scheller et al. 2001). Xia et al. (2010) demonstrated that the strength of recombinantly produced spider silk fibers depends on the monomer proteins.

Recombinant spider silk production has been filed in a number of patents, and industrial production is currently arising. The chemical company DuPont has been interested in the production of artificial spider silk for 20 years. They recently filed a patent together with Toray Industries in which they claim a silk thread produced by transgenic silkworm with an intact fibroin H gene. The hybrid thread might contain spider silk dispersed in the silkworm fibroin or else fused to a polypeptide of the fibroin chain protein or might be inserted between the N-terminal and the C-terminal parts of the fibroin chain protein forming cysteine disulfide bridges fibroin L (Pat. No. US 2008287651 A1). EntoGenetics is also relying on transgenic silkworms raised for ballistic textile production. In their approach insertion of the appropriate spider DNA sequence into the silkworm gene replaces a significant portion of the native fibroin (Pat. No. EP 2244563 A1). Sigma-Aldrich has developed an approach to target specific sequences in the silkworm genome by CompoZr zinc finger nuclease technology (Pat. No. US 2011023154 A1). Together with Kraig Biocraft Laboratories, they announced a project to insert spider silk sequence, for instance, derived from the flagelliform gene to generate hybrid genes. Nexia Biotechnologies Ltd. developed an approach to generate transgenic mammals which they intended to use for production of recombinant spider silk in goats' milk. Their projects were passed to the Advanced Foods and Materials Network since.

For scientific purposes, recombinant spider silk protein generally is intended for spider silk characterization, e.g., for structural investigations. These studies use variations of domain compositions to investigate the influence of protein size or multimer compositions on physicochemical and biomechanical properties or self-assembly into fibers. Other studies exploit the cytocompatible nature of silk for tissue engineering (reviewed by Hsia et al. 2011). The latter application does not necessarily depend on the production of strong fibers but makes use of different formats adapted to the requirements. Chemical and physical parameters, including temperature, pressure, solvents, and composition of the silk proteins, influence the morphology of the protein assemblies. Biomedical interesting forms include capsules and spheres usable as carriers and films and foams for tissue engineering purposes or as biomedical devices like dressings and implants (reviewed by Spiess et al. 2010). As an example, scaffolds of variable porosity were molded by a leaching process using recombinant spider silk proteins, in this case *N. clavipes* spidroin 1 produced in the yeast *Pichia pastoris* (Agapov et al. 2009).

36.2.3 Recombinant Proteins as Chimeric Proteins

An obvious advantage of recombinant spider silk proteins is the possibility to combine spider silk with foreign elements or design variations of the natural sequence to alter its properties. This kind of protein engineering facilitates certain steps of production or allows for adaption to specific applications. Such either the physicochemical properties may be changed or bioactive molecules may be integrated. As an example for the latter, Bini et al. (2006) inserted a CRGD motif into their 15mers based on *Nephila clavipes* MaSp1 to enhance cellular adhesion on their biomaterial. The RGD motif was originally identified on fibronectin and is recognized by cellular adhesion molecules, the integrins. The resulting recombinant protein tended to oligomerize due to the inserted Cys residues and showed enhanced solubility in water. Bone marrow-derived stem cells grown on silk films comprised of the recombinant proteins were able to differentiate into the osteogenic lineage although the measured calcium content stayed behind the respective controls of unmodified 15mers. 15mer repeats fused to recombinant dentin matrix protein (DMP)1 had calcium hydroxyapatite nucleating ability comparable like DMP1 expressed in hard tissues (Huang et al. 2007).

Beside combining spider silk segments with fibroin-derived stretches, Scheller and coworkers also generated a chimeric construct of spider silk and an elastin-based repetitive biopolymer with 100 repeats of the pentapeptide $VPGX_{aa}G$ with X_{aa} standing for either G, V, or A. Human chondrocytes kept on monolayer coating displayed higher cell number in cultures coated with the spider silk-elastin chimera than in any of the included controls. Additionally the cell morphology remained rounded instead of increasing flatness as a sign of dedifferentiation in two-dimensional culture (Scheller et al. 2004).

36.3 Cytocompatibility and Biocompatibility

Safety concerns and the wish to implicate silks in clinical use and as a biomaterial and scaffold for tissue engineering have inspired a number of studies describing cell and body reactions to silk exposure.

Widhe and coworkers used matrices from recombinant spider silk protein consisting of five glycine-rich alternating with four polyalanine stretches connected to a nonrepetitive globular C-terminal domain by a serine- and alanine-rich linker. The recombinant protein produced in *Escherichia coli* self-assembled into fibers which were also used to produce meshes and was cast into films and foams. The primary human fibroblasts used in this study were able to attach and grow on all different matrices almost alike, reaching highest cell counts on combinations of films and meshes (Widhe et al. 2010).

Recombinant protein based on spidroin 1 could be processed into three-dimensional porous scaffolds proposed for use in tissue engineering. 3T3 fibroblasts colonized the 700 μm-thick specimen from the surface, filling the deeper layers through channels between the pores after 14 days (Agapov et al. 2009).

In contrast to a number of studies describing production and cytocompatibility of artificially produced silks, studies reporting of the biocompatibility of native spider silk are limited. In their basic work concerning the local response to spider silk,

Vollrath and coworkers implanted different forms of spider silk in pigs. At the same time spider silks and clinically established wound dressings were used for split-skin wounds over a 15-day period. Implantations were accompanied by lymphoplasma-cellular infiltrations and a high number of giant cells; granulomatous reaction, however, was limited. Immune reaction to epicutaneously used spider silk was also low and comparable to controls. Nevertheless, any advantages by spider silk treatment could not be observed over the duration of the experiment (Vollrath et al. 2002).

In their study about the use of silk scaffolds for chondrocyte cultivation, Gellynck and coworkers processed silkworm cocoon silk, *Araneus diadematus* (Araneidae) dragline and egg sac silk. They produced nonwoven scaffolds from cocoon and egg sac silk by stitching the raw material on yarn grids before processing them to give pure silk fibers. Due to mild processing conditions, fiber structures were kept intact and chondrocytes could be observed on them for up to 6 weeks after seeding. Both types of silk were also completely dissolved and mended into salt-leached porous scaffolds in which the growth patterns of the cells could be influenced by pore sizes after silk particle size (Gellynck et al. 2008).

Human Schwann cells isolated from peripheral nerves adhered sufficiently on native spider silk fibers with high vitality rates after 48 h excluding any toxic or harmful effects on the cells. The silk could be collected directly from the major ampullate gland of adult female *Nephila* spiders and was used without any further treatment. A nerve conduit was developed based on these data consisting of spider silk fibers aligned within a decellularized vein in which the supplemented Schwann cells aligned along the fibers (Allmeling et al. 2006).

36.4 Medical Applications

36.4.1 Spider Silk as a Biomaterial for Surgical Sutures and in Regenerative Medicine

Biomaterials in use for biomedical applications have to meet a number of requirements like biocompatibility and a mild inflammatory response, biodegradability in a reasonable time, and specific structural and mechanical properties. Tensile strength and elasticity of spider silk fibers render them interesting for surgical sutures and tissue engineering purposes (Fig. 36.3), while its biocompatibility is a prerequisite for clinical application.

Despite the fact that silk-based sutures derived from *B. mori* cocoons have been used for over 100 years, questionable biocompatibility has led to displacement by synthetic materials including polydioxanone, polypropylene, and nylon. Nevertheless, silk sutures which provoke a low inflammatory response are still of clinical interest due to their beneficial biodegradability and their mechanical properties. Our group has used spider silk for braiding sutures of about 25 μm corresponding to microsurgical suture material 10–0 after USP definition. The sutures had a tensile

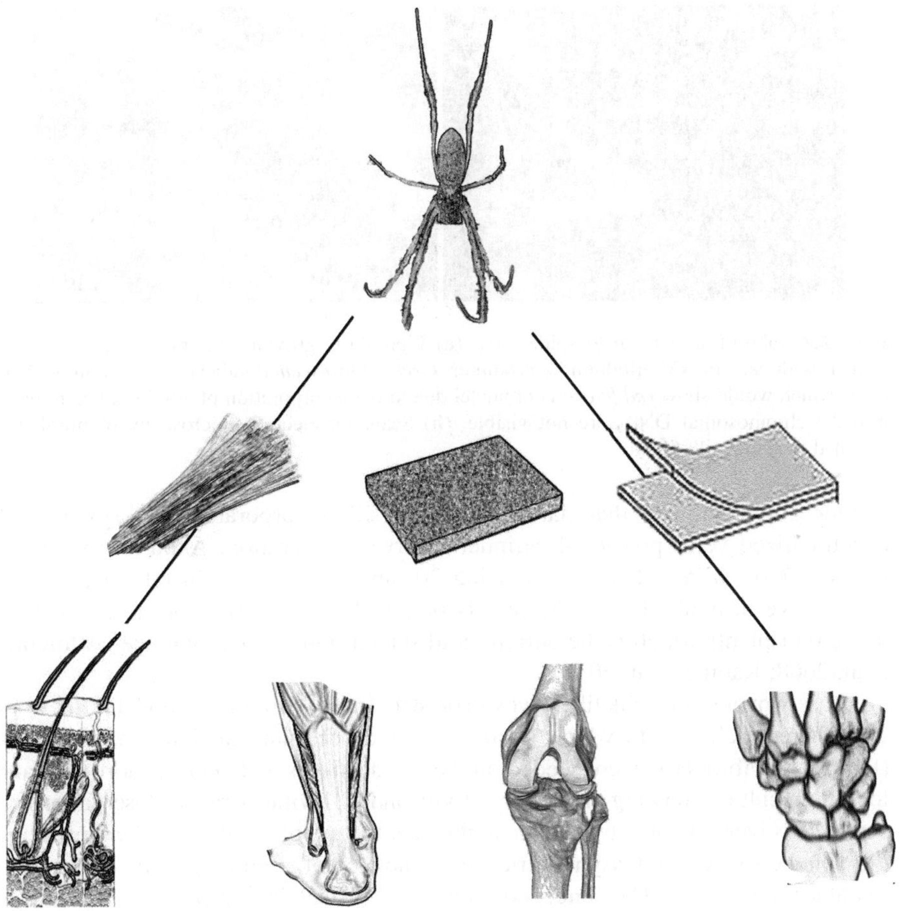

Fig. 36.3 Applications of spider silk in the field of tissue engineering. Spider silk might be used as fibers, porous scaffolds produced from solubilized silk protein, or surface coatings. Its application is interesting for the engineering of a number of tissues among them skin, tendon, cartilage, and bone

strength of 0.7–0.8 MPa depending on the number of strands braided and a stress–strain ratio of 0.13 MPa, while 0.3 MPa tensile strength and 0.08 MPa stress–strain ratio were measured for commercial nylon suture (Kuhbier et al. 2011).

RGD modified 16mers and 32mers recombinantly produced in *E. coli* have been used as a wound dressing in a rat model of second-degree burn wounds. Healing of the burns was compared to equal wounds treated with collagen sponges or left untreated as a negative control. Spider silk proteins significantly promoted wound healing to a regenerated skin which appeared like normal skin on day 21. Differences to the collagen-treated groups, however, were only marginal, constricted mainly to an early reduction of immune cell infiltration (Baoyong et al. 2010).

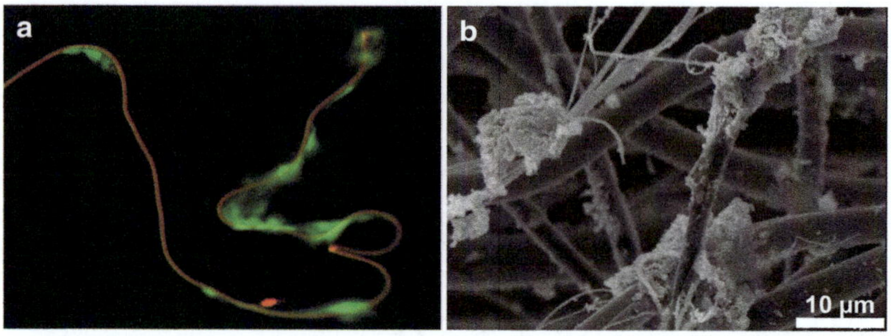

Fig. 36.4 Fibroblasts adhere to spider silk. (**a**) Fibroblasts growing on spider silk have been stained with calcein AM/ethidium heterodimer. *Green fluorescence* indicates cell vitality. Dead cells, which would show *red fluorescent* nuclei due to the incorporation of ethidium heterodimer into the chromosomal DNA, are not visible. (**b**) Scanning electron microscopy of fibroblasts attached to spider silk fibers

Our group showed that natural dragline silk incorporated into isogenic or decellularized veins promoted peripheral nerve regeneration. Axon regrowth and remyelination were achieved, replacing 20 mm sciatic nerve in rats and 60 mm tibial nerve in adult sheep. The results obtained were comparable to autologous nerve transplants which is the current gold standard in clinical practice (Allmeling et al. 2008; Radtke et al. 2011).

Cross-woven spider silk fibers woven on steel frames provided a model framework for adherent cell growth which we used to study cellular reactions on spider silk (Fig. 36.1). Fibroblasts were seeded on the silk scaffolds and kept in short-term and long-term culture showing metabolic activity and high vitality rates. Observations on single cells were enabled by the single thread structure of the devices demonstrating the spindle-shaped and asymmetrical morphology of migrating cells (Fig. 36.4) (Kuhbier et al. 2010). The concept of cross-woven steel frames was later transferred to skin tissue engineering. The approach resulted in a bilayered structure of epidermis and dermis when corresponding cell lines were cultured at the air–liquid interface to induce skin cell differentiation (Fig. 36.1) (Wendt et al. 2011).

The British company Oxford Biomaterials developed an approach to adapt silkworm protein to spider silk properties. Their material was called Spidrex® and is tested in number of biomedical products in form of spinout ventures: Neurotex Ltd. is transferring Spidrex® technology into nerve repair, Suturox Ltd. is focusing on Spidrex®-based sutures, and Orthox Ltd. takes advantage of the material's biomechanical properties for bone and cartilage repair.

36.4.2 Spider Silk as Drug Delivery Vehicles

Drug delivery devices for controlled release of bioactive molecules are of potential interest for pharmaceutical, cosmetic, and food industry (Fig 36.5). Lammel and

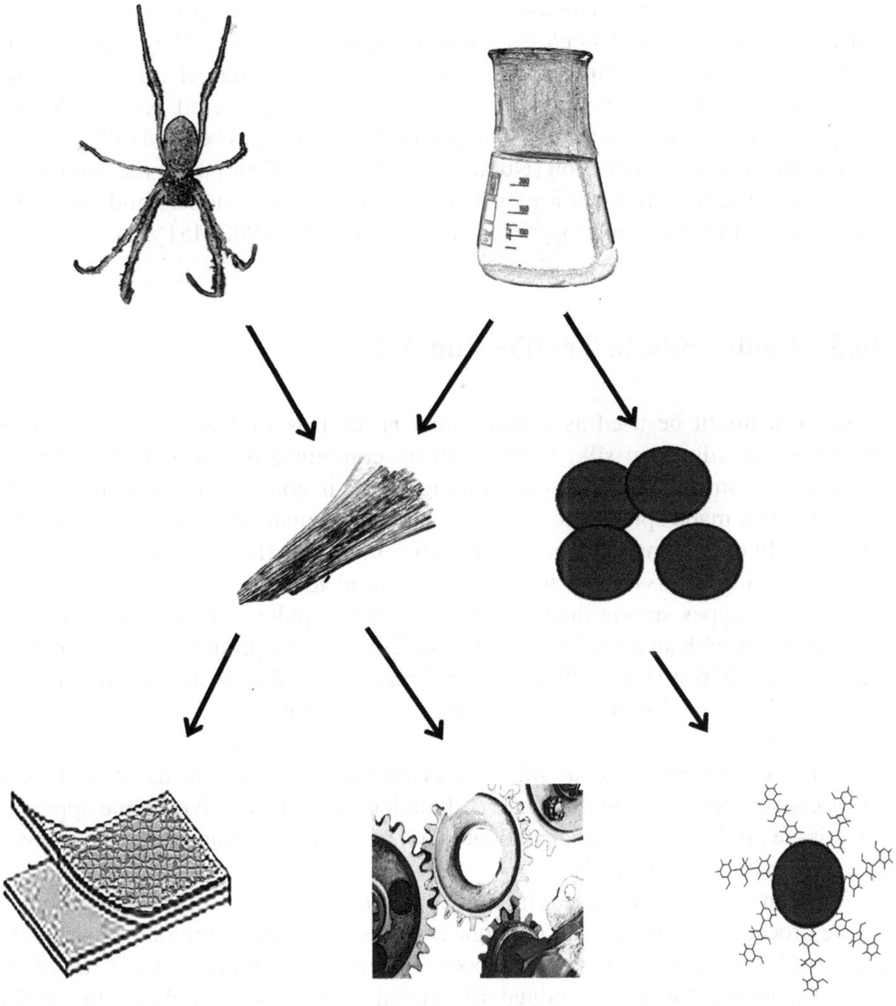

Fig. 36.5 An overview over possible application of spider silk. Technical application of spider silk might involve natural or recombinant silk protein, used as fibers or dispersed. Nanoscaled applications might take use of nanostructured surfaces, microelectromechanical systems, or drug delivery vehicles

coworkers used engineered spider silk particles eADF4(C16) mimicking part of the ADF4 silk protein from *Araneus diadematus*. The generation of such particles can be controlled by potassium phosphate concentration which is an important advantage in biomedical application where often sensitive constituents are involved. At physiological pH these particles were negatively charged such as small molecules with positive net charge can diffuse into the interior. When loading is complete, transfer into aqueous or acidic release media might induce redistribution of the load to the surface accompanied by a constant release (Lammel et al. 2011).

The group of Kaplan made use of a spider silk repeat unit based on a consensus repeat derived from *Nephila clavipes* major ampullate spidroin They expressed the repeats as a 6mer in a fusion with a poly(L-lysine) stretch of 30 residues and different tumor-homing peptides in *E. coli*. The bioengineered proteins formed complexes with reporter gene carrying plasmids which could be used efficiently for in vitro and in vivo transfection (Numata et al. 2011). In their patent file Kaplan and Wang describe the production of silk fibroin microsphere under mild condition, making use of lipids as microsphere templates (US 2010/0028451 A1).

36.5 Spider Silk in Textiles and Art

Spider silk might be used as a tissue fiber at least as well as its more common analogue, the silkworm silk. Lightweight in combination with extreme material strength is promising for creating clothing which is conformable and durable. Its exclusiveness made spider silk tissue a symbol of human talent and inspiration for centuries. In the seventeenth and eighteenth centuries, different approaches tried to use spider silk for tissue production. Bon started to experiment with harvest of different silk types among them egg sacs of native spiders. His efforts resulted in knitted gloves with an astonishing weight of 21 g, while a production with common silk weighed 226 g (Peers 2012). Later, French naturalist Réaumur also tried to make stocking or gloves from spider egg sac silk but gave it up as unpractical (Lewis 1996).

Termeyer harvested egg sac silk but was able to reel spider silk also directly with a self-constructed winding machine. Although these trials received some appreciation, tissue production of spider silk could not be brought to industrial scale and was thus terminated (Peers 2012).

Peers and Godley worked with silk of *Nephila madagascariensis*, weaving a textile brocaded in a traditional Madagascarian way and an embroidered cape using the thread of millions of spiders for both. The resulting shining textiles are of the natural golden color of the undyed silk threads reflecting the light on the skilful patterns created by the artists (Peers 2012).

Spider silk has also inspired music. In their patent file US 2011/0174134 A1 E. und V., Mueller-Zierach describe instrument strings of synthetic spider silk which might especially benefit from the tensile strength of silk fibers. This idea has been realized with native spider silk by Osaki, who used his technical ability to harvest large quantities of spider silk to generate violin strings of distinctive timbre (Osaki 2012).

36.6 Spider Silk in Technical Applications

Less artificial and more profane applications include fabrics whose assignation depends on their elasticity and ability to absorb large quantities of kinetic energy. Examples are the use as protective clothing and body armor, like knife-resistant gloves and bulletproof vests. In their patent file DE 102009013861 A1, Takata (Petri AG) proposed the use of spider silk for gas bags and safety belt for passenger safety in cars. They want to use natural or recombinant spider silk either alone or in combination with silicone, polyurethane, or polyamides.

Ropes and nets from spider silk or spider silk-like fibers promise a great strength and durability as well as sustainability attractive for modern industrial production. Other conceptions depend less on fiber structures but use silk fibroins for production of films and panels or reinforce composite materials, for instance carriages able to absorb the kinetic energy of accidents. Culpepper suggested artificial spider silk stripes as an ankle support integrated into shoes to prevent involuntary lateral movements and ankle injuries (Pat. No. US7587841). Nevertheless, in domains needing high amounts of starting material, cost-benefit calculations render any use of native spider silk impossible which means development of artificial silk production processes which reliably reproduce spider silk properties is a prerequisite for such applications.

Meanwhile, applications making use of spider silk fibers in miniature have already been realized. As an example, spider silk was used as crosshairs in optical targeting devices such as guns and telescopes until the Second World War (Heim et al. 2009; Lewis 1996).

Silk fibroins are also interesting for production of films and coatings or for nanostructuring of surfaces usable, e.g., for nanoscaled steering devices or sensor systems (Fig. 36.5). In this sense, an invention filed recently describes photonic nanoimprinting of fibroin biopolymer films which might be used for different optical devices, like biophotonic sensors or optofluidic devices, or for biomedical applications including drug delivery and tissue engineering (Pat. No. US 2012/0034291 A1). Nanobioconjugates of gold nanoparticles bound on the surface of silk fibers of *Pholcus phalangioides* (Pholcidae) show ohmic electronic conduction. Exposition to the vapors of organic solvents, like methanol, leads to linearly enhanced electronic conduction probably due to fiber contraction. Spider silk prepared in such a manner could thus be used as a methanol sensor system (Singh et al. 2007).

36.7 Conclusion

In conclusion, spider silk is a biopolymer with a huge number of possible textile, technical, and biomedical applications. For several inventions prototype status could be achieved, while most ideas are still of speculative nature. This is especially true for applications which afford quantities of silk which are unrealistic to gain by traditional reeling methods but need industrial large-scale production. This could be realized by production of artificial silk fibers based on recombinant proteins. Recombinant silk production has also the advantage that genetic modifications allow for tailored proteins adapted to specific requirements. These approaches, however, are still thwarted by immense technical challenges due to the size and repetitive nature of silk proteins. Available data so far supports the notion that spider silk is highly cytocompatible and not immunogenic which renders it interesting for biomedical applications. Increasing knowledge about the molecular nature of the fibers and their particular characteristics inspires nanotechnology, e.g., MEMS.

Acknowledgments The authors want to thank Kerstin Reimers and Jörn W. Kuhbier for the help with the manuscript and the images. We also thank Stefanie Michael for the beautiful image of a *Nephila* spider. We are also grateful to Cornelia Kasper and Nicholas Godley for fruitful collaboration and remarkable discussions. We apologize to all authors whose contributions to spider silk research could not be presented due to space limitations.

References

Agapov II, Pustovalova OL, Moisenovich MM, Bogush VG, Sokolova OS, Sevastyanov VI, Debabov VG, Kirpichnikov MP (2009) Three-dimensional scaffold made from recombinant spider silk protein for tissue engineering. Dokl Biochem Biophys 426:127–130

Agnarsson I, Dhinojwala A, Sahni V, Blackledge TA (2009) Spider silk as a novel high performance biomimetic muscle driven by humidity. J Exp Biol 212:1990–1994

Allmeling C, Jokuszies A, Reimers K, Kall S, Choi CY, Brandes G, Kasper C, Scheper T, Guggenheim M, Vogt PM (2008) Spider silk fibres in artificial nerve constructs promote peripheral nerve regeneration. Cell Prolif 41:408–420

Allmeling C, Jokuszies A, Reimers K, Kall S, Vogt PM (2006) Use of spider silk fibres as an innovative material in a biocompatible artificial nerve conduit. J Cell Mol Med 10:770–777

Bai J, Ma T, Chu W, Wang R, Silva L, Michal C, Chiao JC, Chiao M (2006) Regenerated spider silk as a new biomaterial for MEMS. Biomed Microdevices 8:317–323

Baoyong L, Jian Z, Denglong C, Min L (2010) Evaluation of a new type of wound dressing made from recombinant spider silk protein using rat models. Burns 36:891–896

Bini E, Foo CW, Huang J, Karageorgiou V, Kitchel B, Kaplan DL (2006) RGD-functionalized bioengineered spider dragline silk biomaterial. Biomacromolecules 7:3139–3145

Blackledge TA (2013) Spider silk: molecular structure and function in webs. In: Nentwig W (ed) Spider ecophysiology. Springer, Heidelberg (this volume)

Chinali A, Vater W, Rudakoff B, Sponner A, Unger E, Grosse F, Guehrs KH, Weisshart K (2010) Containment of extended length polymorphisms in silk proteins. J Mol Evol 70:325–338

Emile O, Le Floch A, Vollrath F (2006) Biopolymers: shape memory in spider draglines. Nature 440:621

Gatesy J, Hayashi C, Motriuk D, Woods J, Lewis R (2001) Extreme diversity, conservation, and convergence of spider silk fibroin sequences. Science 291:2603–2605

Gellynck K, Verdonk PC, Van Nimmen E, Almqvist KF, Gheysens T, Schoukens G, Van Langenhove L, Kiekens P, Mertens J, Verbruggen G (2008) Silkworm and spider silk scaffolds for chondrocyte support. J Mater Sci Mater Med 19:3399–3409

Gosselin FP, Therriault D, Levesque M (2012) Microfabrication of a spider-silk analogue through the liquid rope coiling instability. Bull Am Physic Soc (APS March Meeting) 57(1):H52

Heim M, Keerl D, Scheibel T (2009) Spider silk: from soluble protein to extraordinary fiber. Angew Chem Int Ed Engl 48:3584–3596

Hsia Y, Gnesa E, Jeffery F, Tang S, Vierra C (2011) Spider silk composites and applications. In: Cuppoletti J (ed) Metal, ceramic and polymeric composites for various uses. InTech, Rijeka

Huang J, Wong C, George A, Kaplan DL (2007) The effect of genetically engineered spider silk-dentin matrix protein 1 chimeric protein on hydroxyapatite nucleation. Biomaterials 28: 2358–2367
Kuhbier JW, Allmeling C, Reimers K, Hillmer A, Kasper C, Menger B, Brandes G, Guggenheim M, Vogt PM (2010) Interactions between spider silk and cells—NIH/3T3 fibroblasts seeded on miniature weaving frames. PLoS One 5:e12032
Kuhbier JW, Reimers K, Kasper C, Allmeling C, Hillmer A, Menger B, Vogt PM, Radtke C (2011) First investigation of spider silk as a braided microsurgical suture. J Biomed Mater Res B Appl Biomater 97:381–387
Lammel A, Schwab M, Hofer M, Winter G, Scheibel T (2011) Recombinant spider silk particles as drug delivery vehicles. Biomaterials 32:2233–2240
Lazaris A, Arcidiacono S, Huang Y, Zhou JF, Duguay F, Chretien N, Welsh EA, Soares JW, Karatzas CN (2002) Spider silk fibers spun from soluble recombinant silk produced in mammalian cells. Science 295:472–476
Lefevre T, Boudreault S, Cloutier C, Pezolet M (2011) Diversity of molecular transformations involved in the formation of spider silks. J Mol Biol 405:238–253
Lewis RV (1996) Unraveling the weave of spider silk. Bioscience 46:636
Liu Y, Shao Z, Vollrath F (2005) Relationships between supercontraction and mechanical properties of spider silk. Nat Mater 4:901–905
Newman J, Newman C (1995) Oh what a tangled web: the medicinal uses of spider silk. Int J Dermatol 34:290–292
Numata K, Reagan MR, Goldstein RH, Rosenblatt M, Kaplan DL (2011) Spider silk-based gene carriers for tumor cell-specific delivery. Bioconjug Chem 22:1605–1610
Osaki S (2012) Spider silk violin strings with a unique packing structure generate a soft and profound timbre. Phys Rev Lett 108:154301
Peers S (2012) Golden spider silk. V&A Publishing, London
Perez-Rigueiro J, Elices M, Plaza G, Real JI, Guinea GV (2005) The effect of spinning forces on spider silk properties. J Exp Biol 208:2633–2639
Radtke C, Allmeling C, Waldmann KH, Reimers K, Thies K, Schenk HC, Hillmer A, Guggenheim M, Brandes G, Vogt PM (2011) Spider silk constructs enhance axonal regeneration and remyelination in long nerve defects in sheep. PLoS One 6:e16990
Sahni V, Labhasetwar DV, Dhinojwala A (2012) Spider silk inspired functional microthreads. Langmuir 28:2206–2210
Scheller J, Guhrs KH, Grosse F, Conrad U (2001) Production of spider silk proteins in tobacco and potato. Nat Biotechnol 19:573–577
Scheller J, Henggeler D, Viviani A, Conrad U (2004) Purification of spider silk-elastin from transgenic plants and application for human chondrocyte proliferation. Transgenic Res 13: 51–57
Singh A, Hede S, Sastry M (2007) Spider silk as an active scaffold in the assembly of gold nanoparticles and application of the gold-silk bioconjugate in vapor sensing. Small 3:466–473
Spiess K, Lammel A, Scheibel T (2010) Recombinant spider silk proteins for applications in biomaterials. Macromol Biosci 10:998–1007
Sponner A, Unger E, Grosse F, Weisshart K (2005) Differential polymerization of the two main protein components of dragline silk during fibre spinning. Nat Mater 4:772–775
Swanson BO, Anderson SP, Digiovine C, Ross RN, Dorsey JP (2009) The evolution of complex biomaterial performance: the case of spider silk. Integr Comp Biol 49:21–31
US Congress (1993) Biopolymers: making materials nature's way. Background paper, Office of Technology Assessment, OTA-BP-E-102. US Government Printing Office, Washington DC
Vollrath F, Barth P, Basedow A, Engstrom W, List H (2002) Local tolerance to spider silks and protein polymers in vivo. In Vivo 16:229–234
Wendt H, Hillmer A, Reimers K, Kuhbier JW, Schafer-Nolte F, Allmeling C, Kasper C, Vogt PM (2011) Artificial skin—culturing of different skin cell lines for generating an artificial skin substitute on cross-weaved spider silk fibres. PLoS One 6:e21833

Widhe M, Bysell H, Nystedt S, Schenning I, Malmsten M, Johansson J, Rising A, Hedhammar M (2010) Recombinant spider silk as matrices for cell culture. Biomaterials 31:9575–9585

Xia XX, Qian ZG, Ki CS, Park YH, Kaplan DL, Lee SY (2010) Native-sized recombinant spider silk protein produced in metabolically engineered Escherichia coli results in a strong fiber. Proc Natl Acad Sci USA 107:14059–14063

Zhou S, Peng H, Yu X, Zheng X, Cui W, Zhang Z, Li X, Wang J, Weng J, Jia W, Li F (2008) Preparation and characterization of a novel electrospun spider silk fibroin/poly(D,L-lactide) composite fiber. J Phys Chem B 112:11209–11216

Chapter 37
Spider Venom Components as Drug Candidates

Vera Oldrati, Estelle Bianchi, and Reto Stöcklin

37.1 Introduction

The biopharmaceutical research community now widely recognizes animal venoms as a rich source of bioactives. Many research groups have focused their studies on animal venoms, resulting in several innovative drug candidates, some of which have been commercialized, with others showing promise in preclinical or clinical development. To date, six venom-derived molecules are already on the market, while more than a dozen are on the right track. Key publications covering the field of venoms to drugs at broad include Lewis and Garcia (2003), Bogin (2005), Fox and Serrano (2007), Saez et al. (2010), Stöcklin and Vorherr (2011), King (2011) and Harvey and Stöcklin (2012). Of these drugs, most are derived from snake venoms, which is most probably due to the ease of access to reasonable amounts of starting material for discovery projects. With some 43,000 spider species (almost all venomous) distributed in 110 distinct families (Platnick 2012) and with new species described regularly, arachnids represent by far the largest group of venomous animal species, making up 20–25 %. Furthermore, as discussed later in this chapter and in other chapters of this book, spider venoms are neither less complex nor less potent than other venoms. On the contrary, they are among the most complex and potent and still have many surprises to reveal. The major limitation in the study of spider venom is accessing sufficient amount of raw material to conduct a discovery project, which, until recently, was huge compared with the amount of venom produced by a single spider. This is indeed clearly reflected in the venom-derived drugs on the market; except for Prialt that was launched in 2004 and originates from a cone snail, all venom-derived drugs or products that made it to market originate from large and easily available species, mostly snakes. In contrast, the drug development pipeline today reflects a clear tendency towards smaller

V. Oldrati • E. Bianchi • R. Stöcklin (✉)
Atheris Laboratories, Case postale 314, 1233 Bernex-Geneva, Switzerland
e-mail: reto.stocklin@atheris.ch

W. Nentwig (ed.), *Spider Ecophysiology*,
DOI 10.1007/978-3-642-33989-9_37,

animals and currently includes nearly a dozen drug candidates undergoing clinical trials in humans that originate from cone snail, scorpion, sea anemone, toad and vampire bat venoms (King 2011; Harvey and Stöcklin 2012).

Using traditional drug discovery strategies, hundreds or thousands of spiders were required to produce enough venom for each study. This meant keeping huge collections of specimens in captivity or organizing large-scale field trips to collect in the wild, which is very expensive, challenging and may raise bioethical issues. This is why, to date, studies have been carried out on at most a few hundred species, including a dozen life-threatening species and the large and easily available theraphosids, which is one among 110 arachnid families with only about 1,000 species total. However, the recent progress in bioanalytical and bioinformatics techniques, and innovative approaches to drug discovery based on venomics (Stöcklin and Favreau 2002; Ménez et al. 2006; Favreau and Stöcklin 2009), now allow the investigation of the composition and biological activity of single constituents present in tiny amount of raw venom (Stöcklin and Favreau 2002; Escoubas and King 2009; Favreau et al. 2010; Jiang et al. 2010; Vetter et al. 2011; Koua et al. 2012; Violette et al. 2012). Successful studies have even been achieved on single living arthropod specimens (Favreau et al. 2006). As a result, during the last decades, numerous bioactive peptides have been discovered in disparate spider venoms, totalling so far around 1,700 bioactive candidates comprising approximately 260 low-molecular-mass compounds (organic acids, nucleosides, nucleotides, amino acids, amines), 200 acylpolyamines (some containing amino acids, some not), 100 linear peptides (mostly cytotoxins and antimicrobial peptides), 900 cysteine-knotted mini-proteins (targeting ion channels and receptors), 220 enzymes (mostly, sphingomyelin phosphodiesterases have been investigated to date, but proteases, peptidases and hyaluronidases were also identified, among others) and 20 neurotoxic proteins (found in theridiid spiders only) (Stöcklin and Cretton 1998; Kuhn-Nentwig et al. 2011; Herzig et al. 2011; Jungo et al. 2012). Bearing in mind that the time from discovery to market ranges between 10 and 20 years for a drug, there is no doubt that the number of drugs originating from smaller or rare species that were previously impossible to study will increase in the future.

This chapter will give an overview of the latest developments and future perspectives in spider venom drug discovery and outline the potential applications of spider venom components in the medical field.

37.2 Potential Applications of Spider Venom

37.2.1 Calcium Channels

Calcium channels are ubiquitous in animals and humans. They are involved in the transduction of the electrical impulse into neurotransmitter release as well as hormone secretion, cell proliferation and gene expression. Calcium channels are implicated in the pain pathway, and it is well established that selective block of these channels results in a long-lasting analgesia. Molecules with this pharmacological activity can thus be employed to treat inflammatory, neuropathic and refractory pain, as an alternative to opioid drugs. Based on their functional characteristics, six types of voltage-sensitive calcium channels (Ca_v) have been characterized (L, N, P, Q, R and T). Each type reacts to different inhibitors and activators and comprises several subtypes, which are named according to their $\alpha 1$ subunit. The L-type Ca_v channel, for example, is the target of the ω-conotoxin-derived analgesic drug Prialt. More than one hundred so-called ω-toxins have been identified in spider venoms. These peptides act selectively as inhibitors of Ca_v, causing a long-lasting blockade. This group of toxins is heterogeneous: molecular masses range from 2,938 to 8,752 Da, and they are found in the venom of spiders belonging to 11 distinct families (Actinopodidae, Hexathelidae, Theraphosidae, Filistatidae, Plectreuridae, Segestriidae, Agelenidae, Ctenidae, Sparassidae, Lycosidae and Oxyopidae). Although further studies are needed, ω-toxins are certainly promising for the development of original analgesic drugs, hopefully without the important side effects observed with existing molecules. Moreover, selectively targeting of Ca_v is also an interesting approach to the development of new diagnostic reagents. At present, the neurological autoimmune disease Lambert–Eaton myasthenic syndrome is characterized by the presence of autoantibodies directed against calcium channels ($Ca_v2.1$ and 2.2). A toxin that selectively targets those channels could be employed to reveal the presence of such antibodies. PnTx3-4, a protein identified in the ctenid *Phoneutria nigriventer* (Brazilian wandering spider) venom, has already been identified as promising for such an application (De Lima et al. 2009; Kuhn-Nentwig et al. 2011).

37.2.2 Potassium Channels

Potassium channels represent the most abundant class of ion channel. They are essentially divided into voltage-sensitive and ligand-sensitive types. Although many subtypes have been identified so far, this class is still far from fully characterized. These channels are found throughout the human body and are modulated by different families of ligands, regulating diverse processes. For example, ATP-sensitive K^+ channels, situated in pancreatic β-cells, regulate insulin secretion. Another class of medically important potassium channels is the voltage-gated potassium channels (K_V). They are constituted of four main subunits and several accessory subunits to form an ion-permeable pore. They are found, for instance, in sensory neurons, where they are involved in nociceptive transmission, and also in heart muscle, where they are implicated in cardiac arrhythmias. Kv1.3 is an interesting target since these channels are overexpressed at the surface of hyperactive effector memory T-cell lymphocytes in some autoimmune diseases (e.g. multiple sclerosis, type 1 diabetes, rheumatoid arthritis and psoriasis). One mini-protein derived from sea anemone venom, ShK-186, is currently undergoing phase I clinical study in the USA by Kineta (Seattle, WA, USA) who acquired the rights from Airmid Corporation

(Redwood City, CA, USA) and the University of California Irvine (Castañeda et al. 1995; Chi et al. 2012). A total of 75 spider toxins belonging to the κ-toxin family that block K_v channels have been identified so far. These toxins are present in the venom of four different spider families (Theraphosidae, Hexathelidae, Sparassidae and Ctenidae), and they have a molecular mass ranging from 3,280 to 6,172 Da. Among them, three heteropodatoxins, isolated from the sparassid *Heteropoda venatoria* (huntsman spider) venom, have been used to study $K_v4.2$ channels in cardiac myocytes. As these channels are selectively blocked by heteropodatoxins, these peptides have been important in defining the physiologic role of $K_v4.2$ channels and investigating their structure (Sanguinetti et al. 1997; Kazic and Gojkovic-Bukarica 1999; Kuhn-Nentwig et al. 2011).

37.2.3 Sodium Channels

Voltage-gated sodium channels are responsible for the rapid depolarization of excitable cells that is required to start an action potential. They are composed of one α and one β subunit; there are 9 α subunits, from $Na_v1.1$ to $Na_v1.9$. Functionally, voltage-gated sodium channels are classified according to their sensitivity to tetrodotoxin. The tetrodotoxin-sensitive $Na_v1.3$, and especially $Na_v1.7$, and the tetrodotoxin-resistant $Na_v1.8$ are involved in pain signalling. Several drugs targeting Na_v channels, such as phenytoin, carbamazepine, lamotrigine and lidocaine, are currently employed to treat epilepsy and neuropathic pain; Na_v channels are hence validated targets for pain treatment. $Na_v1.4$ is selectively located on muscle cells at the neuromuscular junction and offers a promising target for myorelaxant drugs. One cone snail peptide, XEP-018 or CnIIIC from *Conus consors* (Favreau et al. 2012), is currently undergoing preclinical studies for the treatment of dystonia (Atheris Laboratories and the Toxinomics Foundation in Switzerland together with the CNRS in France, in the frame of the CONCO project www.conco.eu). Three toxin types that target voltage-gated sodium channels have been identified in spider venom: the β-toxins shift the voltage dependence of channel activation, the δ-toxins delay channel inactivation and the μ-toxins inhibit Na_v channels. Among these toxins, those that block human $Na_v1.7$ channels are particularly interesting for the development of innovative pain-killing drugs. To date, three peptides isolated from Theraphosidae spider venom have been shown to produce such an effect. Huwentoxin-IV (HWTX-IV) from the Chinese bird spider *Haplopelma schmidti* (sub *Selenocosmia huwena*, Liang 2004) and Protoxin-2 (ProTx-II) from the Peruvian green velvet tarantula *Thrixopelma pruriens* (Middleton et al. 2002) have attracted the interest of several major pharmaceutical companies for their ability to modulate $Na_v1.7$ channels. In addition, a toxin isolated from *Phoneutria nigriventer* venom, δ-ctenitoxin-Pn2a, has been shown to modulate Na_v channel activity. The pharmacology of this toxin has not been fully elucidated, but its action on Na_v channels results in an increased calcium influx, which leads to increased synthesis of nitric oxide. Nitric oxide is a vasodilator, and its injection into the corpus cavernosum at an extremely low concentration results in erection. Since neither toxic nor systemic effects are observed, δ-ctenitoxin-Pn2a appears promising for the treatment of erectile dysfunction, particularly compared with current drugs that have several, not insubstantial, side effects (Lewis and Garcia 2003; Nunes et al. 2008; Saez et al. 2010; Kuhn-Nentwig et al. 2011).

37.2.4 Acid-Sensing Ion Channels

Acid-sensing ion channels (ASICs) belong to the epithelial sodium channel/degenerin superfamily. They are activated by a high extracellular proton concentration and are mainly distributed in neuronal cells. Seven subunits have been identified to date: ASIC1a, 1b, 1b2, 2a, 2b, 3 and 4, with the ASIC1a subunit the most abundant in the central nervous system. Although further investigation of this channel family is needed, it has nevertheless been established that ASICs are involved in nociception, perception of taste and mechanical stimuli. As an important pH decrease is observed in several pathological conditions, such as ischemia, arthritis, osseous fractures and tumours, these channels emerge as promising therapeutic targets. So far, only one potent and selective inhibitor of ASIC1a has been validated: Psalmotoxin-1a (Pi-theraphotoxin-Pc1a, Pi-TRTX-Pc1a or PcTx1) (Fig. 37.1). This toxin, constituted of 40 amino acids, was isolated from the venom of the Trinidad chevron tarantula *Psalmopoeus cambridgei* (Theraphosidae) and inhibits ASIC1a by increasing the apparent channel affinity for H^+ when the toxin is bound. Recent studies show that this toxin is an effective analgesic if injected intrathecally. Moreover, its peripheral administration following a stroke results in a neuroprotective effect. In addition to contributing to the functional and structural understanding of ASIC1a channels, this toxin, under the code THA901, is currently in preclinical development by Theralpha in France as an innovative analgesic and neuroprotective drug (Chen et al. 2005; Poirot et al. 2006; Saez et al. 2010).

37.2.5 Transient Receptor Potential Vanilloid Cation Channel

Transient receptor potential vanilloid cation channels (TRPVs) are known to be involved in sensory signalling, including thermosensation and nociception. In particular TRPV1, which is activated by capsaicin, as well as by several inflammatory agents and noxious heat, may constitute an interesting and original target for analgesic drugs. In order to better investigate the function of TRPVs, powerful and selective biochemical tools are needed. Several so-called vanillotoxins have been characterized from the venom of *Haplopelma schmidti* (sub *Ornithoctonus huwena*) and *Psalmopoeus cambridgei* (Theraphosidae). Owing to their selective activation of certain TRPV

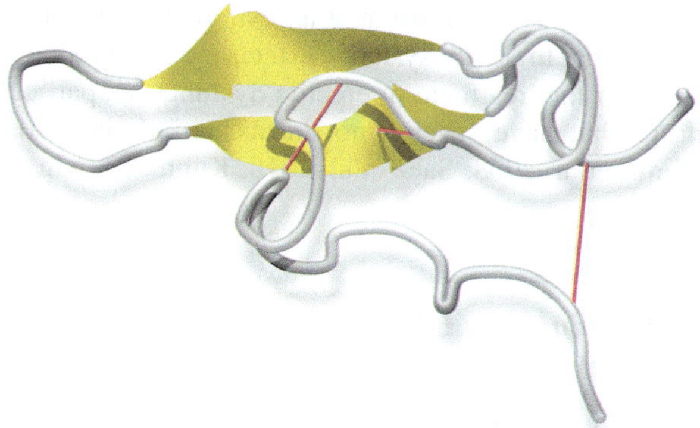

Fig. 37.1 Three-dimensional structure of psalmotoxin-1, a 40 amino acid mini-protein from the venom of the Trinidad chevron tarantula (*Psalmopoeus cambridgei*) containing three disulfide bridges. As far as we know, this lead candidate is the only spider venom compound presently undergoing preclinical studies, apparently with its native sequence

channel subtypes, vanillotoxins represent a useful tool for the structural characterization and pharmacological study of these channels (Bohlen et al. 2010; Saez et al. 2010).

37.2.6 Mechanosensitive Ion Channels

Mechanosensitive ion channels are ubiquitous and mediate various physiological functions. In sensory neurons, they are involved in mechanotransduction of external stimuli, such as pressure and touch, while in the heart, they have been implicated in pathological conditions, such as atrial fibrillation. A selective and potent blocker of mechanosensitive ion channels has been found in the venom of the Chilean rose tarantula *Grammostola rosea* (sub *Grammostola spatulata*). This 34-residue peptide, GsMTx4, has been shown to block those channels without any cytotoxic effect. This toxin is currently being used to study the structure and pharmacology of mechanosensitive ion channels. Considering the implication of these channels in pain sensation, GsMTx4 could be a novel drug candidate to alleviate mechanical hyperalgesia and was suggested as a drug candidate to reduce mechanical and neuropathic pain (Gottlieb et al. 2004; Park et al. 2008). Furthermore, as these channels are involved in arrhythmia, the use of GsMTx4 for the treatment of heart disease can also be considered. An application of this toxin in treating muscular dystrophy has recently been proposed. This hereditary disorder is characterized by defects in muscle proteins that impair movement and lead to respiratory insufficiency. Selective blockade of mechanosensitive ion channels appears to improve muscle performance and palliate this pathological condition. GsMTx4 was recently found to block the Piezo1 cation selective mechanosensitive channel that is also associated in hereditary xerocytosis (Saez et al. 2010; Bae et al. 2011; Zarychanski et al. 2012).

37.2.7 Acetylcholine Receptors

Acetylcholine is a neurotransmitter that binds to two receptor superfamilies: the nicotinic (ionotropic) and muscarinic (metabotropic) acetylcholine receptors. These receptors are ubiquitous in the human body and are involved in numerous physiological and physiopathological processes. To date, there is no characterized spider toxin that has been shown to be active at acetylcholine receptors. However, in collaboration with Daniel Bertrand at HiQScreen SARL (Geneva, Switzerland), we have identified a fraction of the venom of the Sydney funnel-web spider *Atrax robustus* (Hexathelidae) that selectively inhibits the α7 nicotinic acetylcholine receptor subtype (Favreau et al. 2010). The mechanism and the specific venom component responsible for this antagonism have not yet been elucidated. As the α7 nicotinic acetylcholine receptor is involved in cognitive functions, neuronal plasticity and mental disorders such as schizophrenia, further investigation of this spider venom could lead to innovative bioactives for unmet medical needs.

37.2.8 N-Methyl-D-Aspartate Receptors

The *N*-methyl-D-aspartate (NMDA) receptor is an ionotropic glutamate receptor that is present in central nervous system. Binding of glutamate to the NMDA receptor causes opening of calcium channels, which leads to entry of calcium and sensitization and hyperexcitability of spinal cord neurons. NMDA receptors are involved in pain transmission and have been implicated in several diseases of the central nervous system. Increased intracellular calcium levels caused by dysregulation of NMDA receptors is thought to play a role in neurodegeneration in Parkinson's and Alzheimer's disease. NMDA receptors are also a validated target for epilepsy treatment. Considering the central role of this receptor in the functioning of the central nervous system, many research groups have focused their efforts on identifying ligands that can modulate these receptors. In particular, a small molecule NMDA antagonist (NPS-1506 or delucemine: 3,3-bis(3-fluorophenyl)-*N*-methylpropan-1-amine) (Fig. 37.2) was designed based on argiotoxin 636, an NMDA antagonist isolated from the venom of the araneid *Argiope aurantia*, an orb-weaver spider (Adams et al. 1987; Guth et al. 1990). NPS-1506 was designed and first developed by NPS Pharmaceuticals in Salt Lake City, Utah (USA), as a neuroprotective drug for ischemic stroke, haemorrhagic stroke and head trauma (Mueller et al. 1991; Moe et al. 1998,) to prevent excessive calcium influx during the consequent ischemia. The project was abandoned, and the molecule was further developed under the name delucemine for the treatment of major depressive disorder. Delucemine successfully passed clinical phase I study in humans, but on 6 May 2005, NPS confirmed that it has suspended development of delucemine and was evaluating alternative indications and strategies for the future development of it. In 2006, NPS entered into an agreement with Johnson & Johnson Pharmaceuticals regarding intellectual property related to delucemine, who apparently kept the project on hold (Berressem 1999; Bogin 2005).

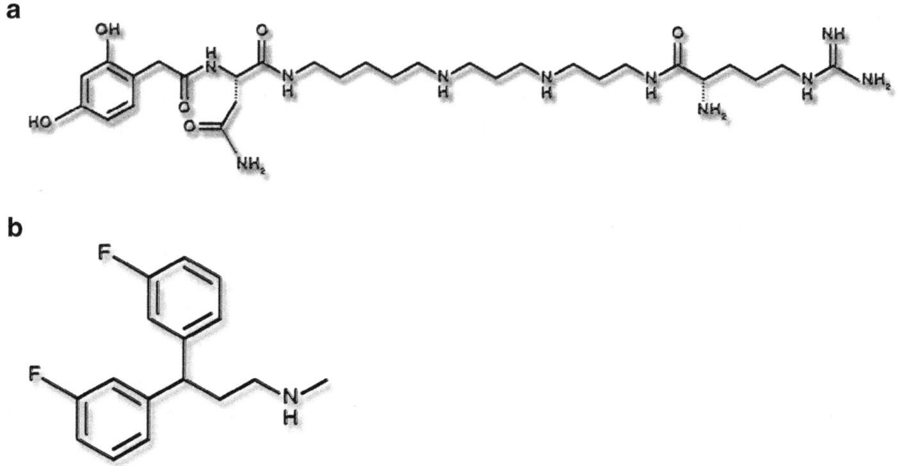

Fig. 37.2 NPS-1506 or delucemine (**b**) is the only spider venom-derived compound that entered into human clinical trials. Its design was based on argiotoxin 636 (**a**). Argiotoxins contain a hydrophilic basic domain of arginine, a polyamine, and asparagine is connected to an aromatic moiety contributed either by 4-hydroxyindole-3-acetic acid or 2,4-dihydroxyphenylacetic acid

37.2.9 Latrophilin Receptor

The latrophilin-1 receptor is a G-protein-coupled receptor that is mainly expressed at the presynaptic level in brain neurons and in endocrine cells of the kidneys and pancreas. This receptor was recently discovered as a result of several studies on a toxin extracted from the venom from the Mediterranean black widow *Latrodectus tredecimguttatus* (Theridiidae), named α-latrotoxin. This toxin of 131,500 Da was shown to have a potent stimulant effect on neuronal and endocrine cells; however, the underlying mechanism was completely unknown. Today its receptor has been structurally characterized and has been demonstrated to be involved in a number of physiological processes. Primarily, as it is found in pancreatic β-cells, it was suggested that the latrophilin-1 receptor could be involved in insulin secretion and therefore in insulin-related diseases, such as diabetes or obesity. However, there is also evidence to suggest that this receptor may be implicated in mental disorders, such as schizophrenia and bipolar disorder. It has been shown that mice lacking the latrophilin-1 receptor have a behaviour that corresponds to schizophrenia phenotypes. In addition, further studies on latrophilin genes led to the characterization of two homologous receptors: latrophilin-2 and 3. These receptors have been studied in much less detail, but it has been shown that latrophilin-2 is ubiquitous in mammalian tissues, and its expression may be correlated with incidence of breast cancer. These findings are examples of usefulness of spider toxins in the medical research not only as lead candidates for the design of new drugs but also in receptor deorphanization and the elucidation of physiological pathways (Lang et al. 1998; Ushkaryov et al. 2008; Silva et al. 2009; Silva and Ushkaryov 2010).

37.2.10 Purinergic P2X Receptors

P2X purinergic receptors are ATP-gated ion channels that are permeable to the cations Na^+, K^+ and Ca^{2+}. Seven receptor subtypes, $P2X_1$ to $P2X_7$, have been identified, and among these, subtypes $P2X_3$, $P2X_4$ and $P2X_7$ are involved in pain signalling. $P2X_3$ receptors are found on ascending nociceptive sensory neurons, where they are involved in acute, inflammatory, neuropathic, visceral and cancer-related pain, as well as in migraine. This subunit is hence a suitable target for novel polyvalent analgesic drugs. Purotoxin-1, a 35 amino acid toxin isolated from the venom of the lycosid *Geolycosa* sp., has been found to be a potent and extremely selective modulator of $P2X_3$ receptors and causes receptor desensitization. Preliminary results with this peptide are promising and may lead to a novel analgesic drug targeting this receptor (Savchenko et al. 2009; Grishin et al. 2010; Saez et al. 2010).

37.2.11 Antibacterial and Antifungal Action

Despite the 17 classes of antibiotic compounds existing to date, research into antimicrobial agents must perpetually evolve. The emergence of nosocomial infections due to antibiotic resistance of pathogens such as *Staphylococcus aureus* or *Streptococcus pneumonia* has made the search for new antibiotics a priority. Daptomycin, a natural lipopeptide but not from animal venom that has recently been commercialized for the treatment of Gram-positive infections, highlights the relevance of antimicrobial peptides as a valid alternative to smaller organic molecules. So far, a multitude of membrane-acting antimicrobial peptides have been isolated from the venom of araneomorph and mygalomorph spiders, together with members of other families such as defensins (Gao et al. 2005; see also Kuhn-Nentwig and Nentwig 2013). Most of these peptides are characterized by an amphipathic α-helical structure, which enables them to interact with bacterial cell membranes, perturbing their function. This interaction is nonspecific and may cause a general cytolytic effect; however, experiments have demonstrated that this adverse effect can be overcome by truncation of the peptide, without significant loss of antimicrobial activity. An example of an antimicrobial peptide with a different structure is GsMTx-4, which is isolated from *Grammostola rosea* (sub *Grammostola spatulata*) venom and is described above for its activity on mechanosensitive ion channels. This peptide is constituted of a triple-stranded antiparallel β-sheet, stabilized by three disulfide bridges. The antimicrobial effect of GsMTx-4 is greater on Gram-positive bacteria, such as *Staphylococcus*, versus Gram-negative bacteria. This effect may be due to an interaction with the lipid packing of the bacterial membrane; however, another hypothesis for the antimicrobial effect of GsMTx-4 is its activity at mechanosensitive ion channels in bacteria. Developing antimicrobial peptides for pharmaceutical use, however, remains challenging and is typically limited by production costs. Since most toxins are active in the micromolar range and multiple daily injections over 2 more weeks would be required for a course of treatment, large amounts of peptide would be required, making the cost of treatment extremely high (Bulet et al. 2004; Bulet and Stöcklin 2005; De Lima et al. 2009; Saez et al. 2010).

37.2.12 Antimalarial Action

Malaria is caused by *Plasmodium* sp. infection, spread by *Anopheles* mosquitoes, with *Plasmodium falciparum* being the most virulent species. Chloroquine is currently considered as the reference treatment, but resistance of some *Plasmodium* strains limits its efficacy. Two peptides, psalmopeotoxin I (U1-theraphotoxin-Pc1a, TRTX-Pc1a, 33 amino acids) and II (U2-theraphotoxin-Pc1a, U2-TRTX-Pc1a, 28 amino acids), have emerged as possible alternatives to existing antimalarial drugs. These peptides have been isolated from the venom of *Psalmopoeus cambridgei*, and they bind selectively to *Plasmodium falciparum*-parasitized erythrocytes. While the molecular target of psalmopeotoxins remains unknown, it has been established that they are extremely selective and do not cause haemolytic or cytotoxic side effects. These peptides may represent a great opportunity to treat this major disease that is responsible for more than 650,000 deaths every year (Choi et al. 2004; De Lima et al. 2009; Saez et al. 2010).

37.3 Conclusions and Perspectives

In conclusion, it has been established that animal venoms are a significant and important source of bioactive compounds, and spider venom is no exception. Venom-derived peptide libraries are thus increasingly employed in drug discovery. A multitude of compounds that are active at many different targets have been identified in venom of all classes of venomous animals, including spiders. Recent advances in spider venom research have been made possible by significant improvements in venomics drug discovery strategies, in particular in analytical techniques and bioinformatics, as well as the emergence of next-generation sequencing technology giving access to genomic and venom gland transcriptomic studies. In view of the huge phylogenetic and molecular diversity they offer, there is no doubt that spiders still hold many surprises. Undoubtedly, new developments in pharmacological research technologies will enable us to fully exploit this rich source, in a manner that is ecological, effective and sustainable.

Acknowledgments We are grateful to the Swiss Commission for Technology and Innovation (CTI) and to the Swiss Initiative in Systems Biology (SystemsX.ch) for financial support. Part of the tools and strategies presented herein were developed and validated in the frame of CONCO, the cone snail genome project for health (www.conco.eu) within the 6th Framework Program

(LIFESCIHEALTH-6 Integrated Project LSHB-CT-2007, contract number 037592), with Dr. Torbjörn Ingemansson as scientific officer. We would like to express our deepest gratitude to Xavier Sprungli (The Toxinomics Foundation) for his ongoing help and creativity; he prepared the figures of this article. We are grateful to Nicolas Hulo (Atheris Laboratories), Roman Mylonas (Atheris Laboratories and Swiss Institute of Bioinformatics), Frédérique Lisacek (Swiss Institute of Bioinformatics), Lucia Kuhn-Nentwig and Wolfgang Nentwig (University of Bern) for fruitful collaborations and ongoing help. We would like to thank Ron Hogg of OmniScience SA for editorial support. This article is a tribute to our friend Vincent Deryck who left us after a long and courageous fight against cancer. Vincent was instrumental in the development of venom production at Alphabiotoxine in Belgium, and he will be missed a lot.

References

Adams ME, Carney RL, Enderlin FE, Fu ET, Jarema MA, Li JP, Miller CA, Schooley DA, Shapiro MJ, Venema VJ (1987) Structures and biological activities of three synaptic antagonists from orb weaver spider venom. Biochem Biophys Res Commun 148:678–683

Bae C, Sachs F, Gottlieb PA (2011) The mechanosensitive ion channel Piezo1 is inhibited by the peptide GsMTx4. Biochemistry 50:6295–6300

Berressem P (1999) From bites and stings to medicines. Chem Br 35(4):40–42

Bogin O (2005) Venom peptides and their mimetics as potential drugs. Modulator 19:14–20

Bohlen CJ, Kalbacher H, Gründer S (2010) A bivalent tarantula toxin activate the capsaicin receptor, TRPV1, by targeting the outer pore domain. Cell 141:834–845

Bulet P, Stöcklin R (2005) Insect antimicrobial peptides: structure, properties and gene regulation. Protein Pept Lett 12:3–11

Bulet P, Stöcklin R, Menin L (2004) Antimicrobial peptides: from invertebrates to vertebrates. Immunol Rev 198:169–184

Castañeda O, Sotolongo V, Amor AM, Stöcklin R, Anderson AJ, Harvey AL, Engström A, Wernstedt C, Karlsson E (1995) Characterization of a potassium channel toxin from the Caribbean sea anemone *Stichodactyla helianthus*. Toxicon 33:603–613

Chen X, Kalbacher H, Gründer S (2005) The tarantula toxin Psalmotoxin 1 inhibits acid-sensing ion channel (ASIC) 1a by increasing its apparent H^+ affinity. J Gen Physiol 126:71–79

Chi V, Pennington MW, Norton RS, Tarcha EJ, Londono LM, Sims-Fahey B, Upadhyay SK, Lakey JT, Iadonato S, Wulff H, Beeton C, Chandy KG (2012) Development of a sea anemone toxin as an immunomodulator for therapy of autoimmune diseases. Toxicon 59:529–546

Choi SJ, Parent R, Guillaume C, Deregnaucourt C, Delarbre C, Ojcius DM, Montagne JJ, Célérier ML, Phelipot A, Amiche M, Molgo J, Camadro JM, Guette C (2004) Isolation and characterization of psalmopeotoxin I and II: two novel antimalarial peptides from the venom of the tarantula *Psalmopoeus cambridgei*. FEBS Lett 572:109–117

De Lima ME, Monteiro de Castro Pimenta A, Martin-Eauclaire MF, Zingali RB, Rochat H (eds) (2009) Animal toxins: state of the art, perspectives in health and biotechnology. Editoria ufmg, Brazil

Escoubas P, King GF (2009) Venomics as a drug discovery platform. Expert Rev 6:221–224

Favreau P, Menin L, Michalet S, Perret F, Cheneval O, Stöcklin M, Bulet P, Stöcklin R (2006) Mass spectrometry strategies for venom mapping and peptide sequencing from crude venoms: case applications with single arthropod specimen. Toxicon 47:676–687

Favreau P, Stöcklin R (2009) Marine snail venoms: use and trends in receptor and channel neuropharmacology. Curr Opin Pharmacol 9:594–601

Favreau P, Neveu E, Maver M, Stöcklin R, Bertrand D (2010) New tools to discover active molecules in venoms: from theory to practice. In: Barbier J, Benoit E, Marchot P, Mattéi C, Servent D (eds) Advances and new technologies in toxinology, SFET edn. Gif sur Yvette, France. http://www.sfet.asso.fr

Favreau P, Benoit E, Hocking HG, Carlier L, D' Hoedt D, Leipold E, Markgraf R, Schlumberger S, Córdova MA, Gaertner H, Paolini-Bertrand M, Hartley O, Tytgat J, Heinemann SH, Bertrand D, Boelens R, Stöcklin R, Molgó J (2012) A novel μ-conopeptide, CnIIIC, exerts potent and preferential inhibition of Na(V) 1.2/1.4 channels and blocks neuronal nicotinic acetylcholine receptors. Br J Pharmacol 166:1654–1668

Fox JW, Serrano SM (2007) Approaching the golden age of natural product pharmaceuticals from venom libraries: an overview of toxins and toxin-derivatives currently involved in therapeutic or diagnostic applications. Curr Pharm Des 13:2927–2934

Gao L, Zhang J, Feng W, Bao N, Song D, Zhu BC (2005) Pharmacological characterisation of spider antimicrobial peptides. Protein Pept Lett 12:507–511

Gottlieb PA, Suchyna TM, Sachs F (2004) Mechanosensitive ion channels as drug targets. Curr Drug Targets CNS Neurol Disord 3:287–295

Grishin EV, Savchenko GA, Vassilevski AA, Korolkova YV, Boychuk YA, Viatchenko-Karpinski VY, Nadezhdin KD, Arseniev AS, Pluzhnikov KA, Kulyk VB, Voitenko NV, Krishtal OO (2010) Novel peptide from spider venom inhibits P2X3 receptors and inflammatory pain. Ann Neurol 67:680–683

Guth SL, Scapini DA, Drescher MJ, Drescher DG (1990) Argiotoxin-636 blocks effects of N-methyl-D-aspartate on lateral line of *Xenopus laevis* at concentrations which do not alter spontaneous or evoked neural activity. Life Sci 47:1437–1445

Harvey AL, Stöcklin R (2012) From venoms to drugs: introduction. Toxicon 59:433

Herzig V, Wood DL, Newell F, Chaumeil PA, Kaas Q, Binford GJ, Nicholson GM, Gorse D, King GF (2011) ArachnoServer 2.0, an updated online resource for spider toxin sequences and structures. Nucleic Acids Res 39(Database issue):D653–D657, Accessed 20 June 2012

Jiang L, Zhang D, Zhang Y, Peng L, Chen J, Liang S (2010) Venomics of the spider *Ornithoctonus huwena* based on transcriptomic *versus* proteomic analysis. Comp Biochem Physiol 5:81–88

Jungo F, Bougueleret L, Xenarios I, Poux S (2012) The UniProtKB/Swiss-Prot Tox-Prot program: a central hub of integrated venom protein data. Toxicon 60:551–557, Accessed 20 June 2012

Kazic T, Gojkovic-Bukarica L (1999) Ion channels and drug development, focus on potassium channels and their modulators. Facta Universitatis, Med Biol 6:23–30

King GF (2011) Venom as a platform for human drugs: translating toxins into therapeutics. Expert Opin Biol Ther 11:1469–1484

Koua D, Brauer A, Laht S, Kaplinski L, Favreau P, Remm M, Lisacek F, Stöcklin R (2012) ConoDictor: a tool for prediction of conopeptide superfamilies. Nucleic Acids Res 40(W1): W238–W241

Kuhn-Nentwig L, Nentwig W (2013) Main components of spider venoms. In: Nentwig W (ed) Spider ecophysiology. Springer, Heidelberg (this volume)

Kuhn-Nentwig L, Stöcklin R, Nentwig W (2011) Venom composition and strategies in spiders: is everything possible? Adv In Insect Phys 40:1–86

Lang J, Ushkaryov Y, Grasso A, Wollheim CB (1998) Ca^{2+}-independent insulin exocytosis induced by α-latrotoxin requires latrophilin, a G protein-coupled receptor. EMBO J 17: 648–657

Lewis RJ, Garcia ML (2003) Therapeutic potential of venom peptides. Nat Rev 2:790–802

Liang S (2004) An overview of peptide toxins from the venom of the Chinese bird spider *Selenocosmia huwena* Wang [=*Ornithoctonus huwena* (Wang)]. Toxicon 43:575–585

Ménez A, Stöcklin R, Mebs D (2006) 'Venomics' or: the venomous systems genome project. Toxicon 47:255–259

Middleton RE, Warren VA, Kraus RL, Hwang JC, Liu CJ, Dai G, Brochu RM, Kohler MG, Gao YD, Garsky VM, Bogusky MJ, Mehl JT, Cohen CJ, Smith MM (2002) Two tarantula peptides inhibit activation of multiple sodium channels. Biochemistry 41:14734–14747

Moe ST, Smith DL, Chien Y, Raszkiewicz JL, Artman LD, Mueller AL (1998) Design, synthesis, and biological evaluation of spider toxin (argiotoxin-636) analogs as NMDA receptor antagonists. Pharm Res 15(1):31–38

Mueller AL, Albensi BC, Ganong AH, Reynolds LS, Jackson H (1991) Arylamine spider toxins antagonize NMDA receptor-mediated synaptic transmission in rat hippocampal slices. Synapse 9:244–250

Nunes KP, Costa-Gonçalves A, Lanza LF, Cortes SF, Cordeiro MN, Richardson M, Pimenta AMC, Webb RC, Leite R, De Lima ME (2008) Tx2-6 toxin of the *Phoneutria nigriventer* spider potentiates rat erectile function. Toxicon 51:1197–1206

Park SP, Kim BM, Koo JY, Cho H, Lee CH, Kim M, Na HS, Oh U (2008) A tarantula spider toxin, GsMTx4, reduces mechanical and neuropathic pain. Pain 137:208–217

Platnick NI (2012) The world spider catalog. Version 12.5. http://research.amnh.org/iz/spiders/catalog/. Accessed 20 June 2012

Poirot O, Berta T, Decosterd I, Kellenberger S (2006) Distinct ASIC currents are expressed in rat putative nociceptors and are modulated by nerve injury. J Physiol 576:215–234

Saez NJ, Senff S, Jensen JE, Er SY, Herzig V, Rash LD, King GF (2010) Spider-venom peptides as therapeutics. Toxins 2:2851–2871

Sanguinetti M, Johnson J, Hammerland L, Kelbaugh P, Volkmann R, Saccomano N, Mueller A (1997) Heteropodatoxins: peptides isolated from spider venom that block Kv4.2 potassium channels. Mol Pharmacol 51:491–498

Savchenko HA, Vassilevski AA, Pluzhnykov KA, Korolkova YV, Mamenko MV, Volkova TM, Maksymiuk OP, Boychuk YA, Grishin EV, Kryshtal OO (2009) Peptide components of *Geolycosa* spider venom modulate P2X receptor activity of rat sensory neurons. Fiziol Zh 55(2):11–16

Silva JP, Suckling J, Ushkaryov Y (2009) Penelope's web: using α-latrotoxin to untangle the mysteries of exocytosis. J Neurochem 111:275–290

Silva JP, Ushkaryov YA (2010) The latrophilins, "split-personality" receptors. Adv Exp Med Biol 706:59–75

Stöcklin R, Cretton G (1998) VENOMS—the ultimate database on venomous animals. Professional edition, Version 1.0, CD-ROM 1 & 2, Atheris Laboratories, Geneva, Switzerland

Stöcklin R, Favreau P (2002) Proteomics of venom peptides. In: Ménez A (ed) Perspectives in molecular toxinology. Wiley, Chichester

Stöcklin R, Vorherr T (2011) Future perspectives of venoms for drug discovery. PharManufacturing: The International Peptide Review 22–25

Ushkaryov Y, Rohou A, Sugita S (2008) α-Latrotoxin and its receptors. Handb Exp Pharmacol 184:171–206

Vetter I, Davis JL, Rash LD, Raveendra A, Mobli M, Alewood PF, Lewis RJ, King GF (2011) Venomics: a new paradigm for natural product-based drug discovery. Amino Acid 40:15–28

Violette A, Biass D, Dutertre S, Koua D, Piquemal D, Pierrat F, Stöcklin R, Favreau P (2012) Large-scale discovery of conopeptides and conoproteins in the injectable venom of a fish-hunting cone snail using a combined proteomic and transcriptomic approach. J Proteomics 75(17):5215–5225

Zarychanski R, Schulz VP, Houston BL, Maksimova Y, Houston DS, Smith B, Rinehart J, Gallagher PG (2012) Mutations in the mechanotransduction protein PIEZO1 are associated with hereditary xerocytosis. Blood 120(9):1908–1915

Appendix. Spider Phylogeny

Since many aspects mentioned in this book are better understandable within a phylogenetic context, a phylogenetic cladogram of spiders is given here. However, currently no generally accepted system is available and therefore the approach which is presented here reflects the best compromise of current knowledge and opinions. This system bases on the cladograms provided by Coddington (2005) and Jocqué and Dippenaar-Schoeman (2006) and includes key results from important further publications such as Silva (2003), Hedin and Bond (2006), Benjamin et al. (2008), Rix et al. (2008), Alvarez-Padilla et al. (2009), Arnedo et al. (2009), Blackledge et al. (2009), Miller et al. (2009, 2010), Dimitrov et al. (2011), Lopardo et al. (2011), Bond et al. (2012), Griswold et al. (2012), and Labarque and Ramirez (2012).

The here presented system includes 110 families. Sinopimoidae are not included; they comprise only one species and may show up to belong to Linyphiidae. Cycloctenidae are also not included; they belong to the RTA clade but their further position is unclear. The best approach to include Miturgidae results in different positions for two parts: the subfamily Eutichurinae (including *Cheiracanthium*) close to Corinnidae and the subfamily Miturginae close to Zoridae (Silva 2003). If confirmed, this may point to an upcoming splitting of the family. Micropholcommatidae (Micropholc.) may become a subfamily of Anapidae (Lopardo et al. 2011).

Thanks for discussing this phylogenetic system go to Miquel Arnedo, Jonathan Coddington, Gustavo Hormiga, Rudy Jocqué, Christian Kropf, Jeremy Miller, and Nikolaj Scharff.

W. Nentwig (ed.), *Spider Ecophysiology*,
DOI 10.1007/978-3-642-33989-9, © Springer-Verlag Berlin Heidelberg 2013

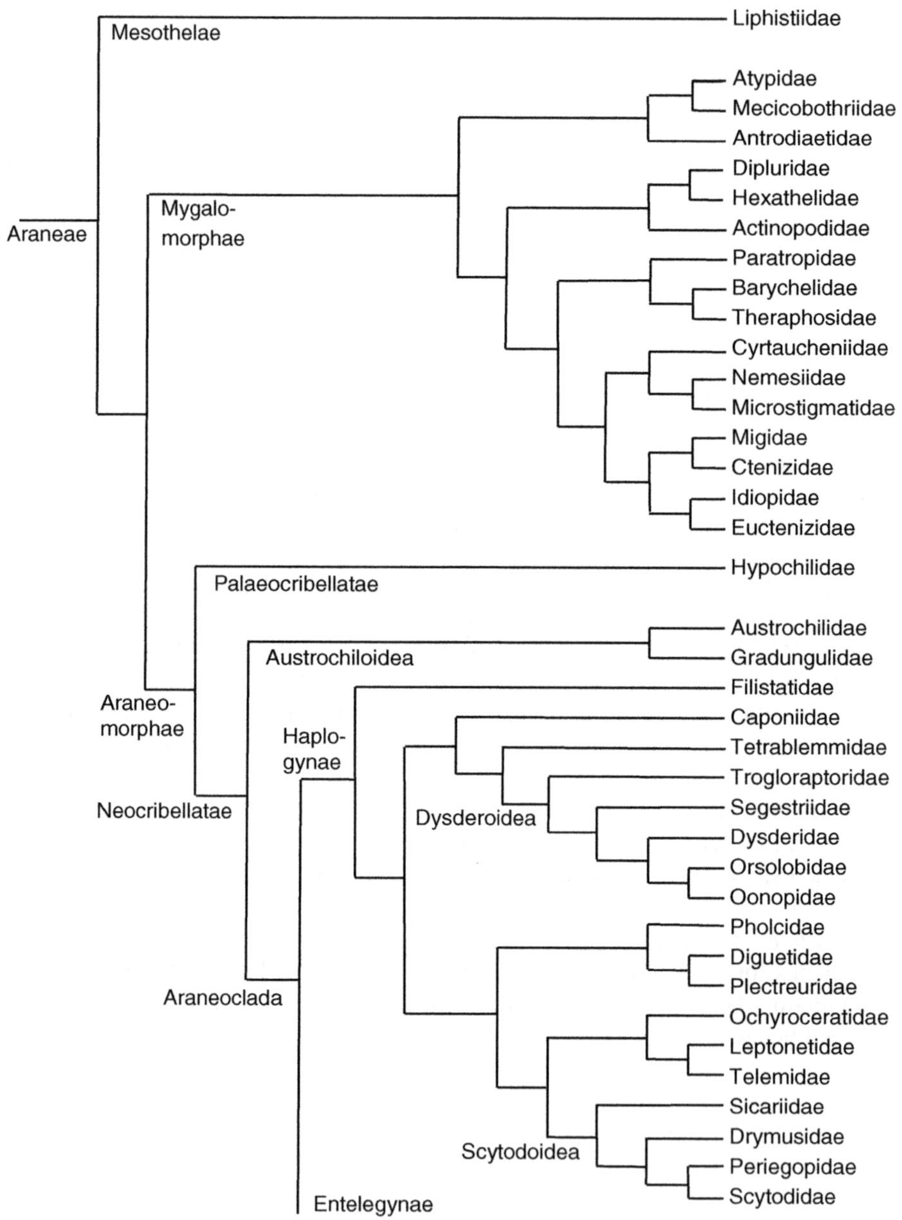
Liphistiidae
Mesothelae
Atypidae
Mecicobothriidae
Antrodiaetidae
Dipluridae
Hexathelidae
Actinopodidae
Paratropidae
Barychelidae
Theraphosidae
Cyrtaucheniidae
Nemesiidae
Microstigmatidae
Migidae
Ctenizidae
Idiopidae
Euctenizidae
Araneae
Mygalo-
morphae
Hypochilidae
Palaeocribellatae
Austrochilidae
Gradungulidae
Austrochiloidea
Filistatidae
Caponiidae
Tetrablemmidae
Trogloraptoridae
Segestriidae
Dysderidae
Orsolobidae
Oonopidae
Araneo-
morphae
Haplo-
gynae
Neocribellatae
Dysderoidea
Pholcidae
Diguetidae
Plectreuridae
Ochyroceratidae
Leptonetidae
Telemidae
Sicariidae
Drymusidae
Periegopidae
Scytodidae
Araneoclada
Scytodoidea
Entelegynae

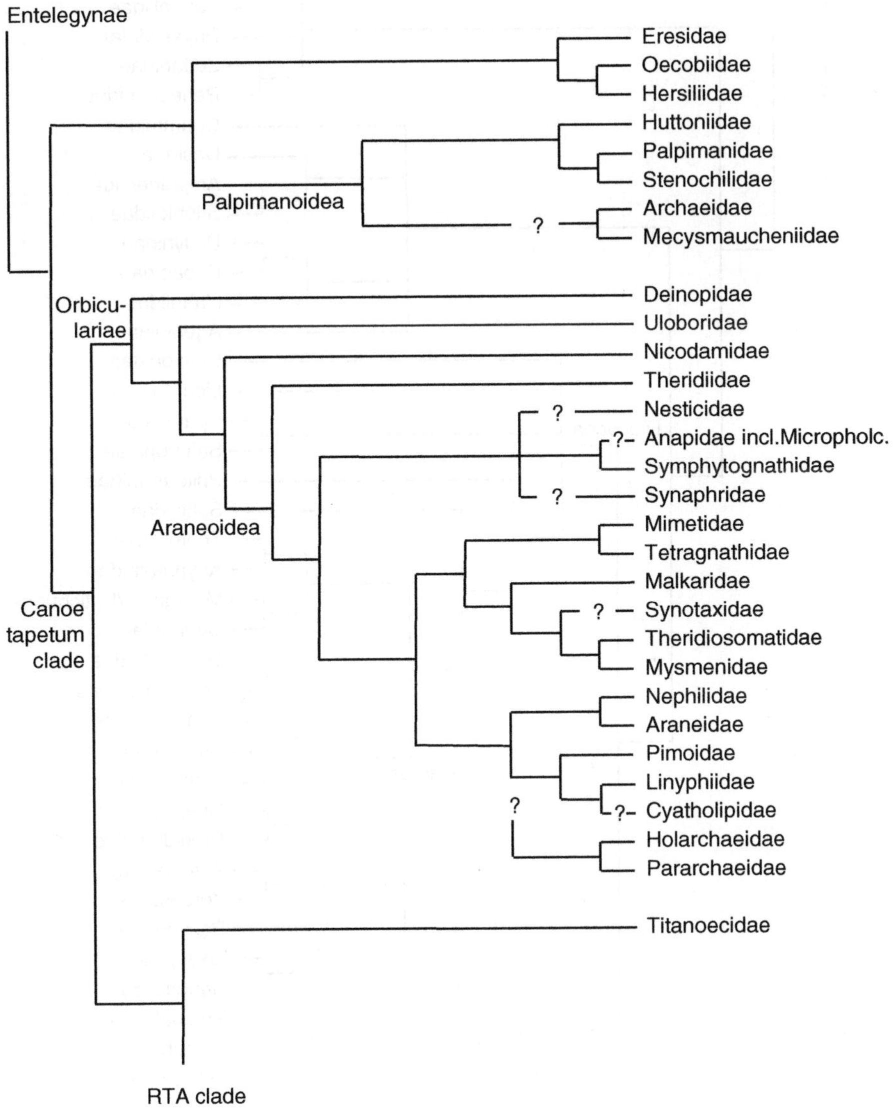
Entelegynae
Eresidae
Oecobiidae
Hersiliidae
Huttoniidae
Palpimanidae
Stenochilidae
Archaeidae
Mecysmaucheniidae
Palpimanoidea
?
Orbicu-
lariae
Deinopidae
Uloboridae
Nicodamidae
Theridiidae
Nesticidae
Anapidae incl.Micropholc.
Symphytognathidae
Synaphridae
Araneoidea
Mimetidae
Tetragnathidae
Malkaridae
Synotaxidae
Theridiosomatidae
Mysmenidae
Canoe
tapetum
clade
Nephilidae
Araneidae
Pimoidae
Linyphiidae
Cyatholipidae
Holarchaeidae
Pararchaeidae
Titanoecidae
RTA clade

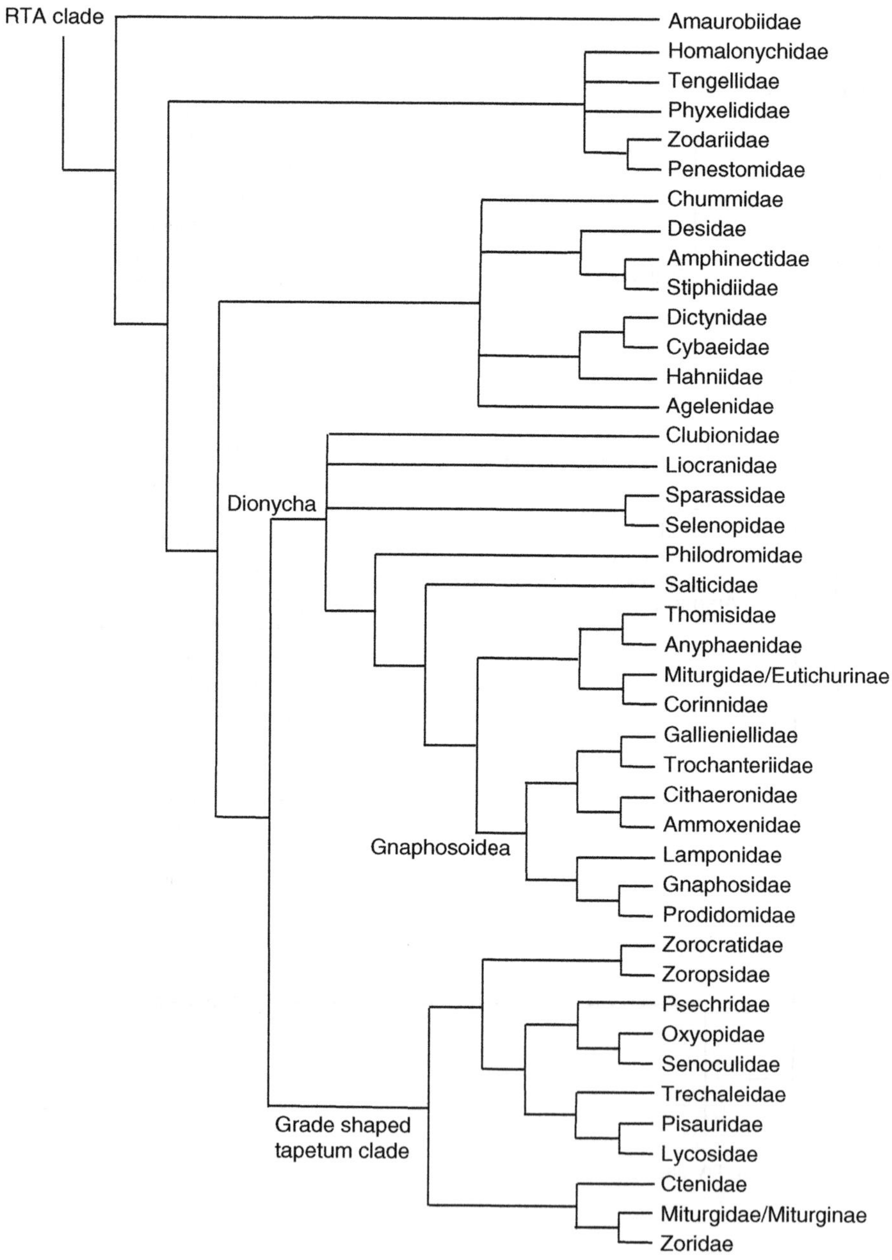
RTA clade
Amaurobiidae
Homalonychidae
Tengellidae
Phyxelididae
Zodariidae
Penestomidae
Chummidae
Desidae
Amphinectidae
Stiphidiidae
Dictynidae
Cybaeidae
Hahniidae
Agelenidae
Clubionidae
Liocranidae
Dionycha
Sparassidae
Selenopidae
Philodromidae
Salticidae
Thomisidae
Anyphaenidae
Miturgidae/Eutichurinae
Corinnidae
Gallieniellidae
Trochanteriidae
Cithaeronidae
Ammoxenidae
Gnaphosoidea
Lamponidae
Gnaphosidae
Prodidomidae
Zorocratidae
Zoropsidae
Psechridae
Oxyopidae
Senoculidae
Trechaleidae
Grade shaped tapetum clade
Pisauridae
Lycosidae
Ctenidae
Miturgidae/Miturginae
Zoridae

References

Alvarez-Padilla F, Dimitrov D, Giribet G, Hormiga G (2009) Phylogenetic relationship of the spider family Tetragnathidae (Araneae, Araneoidea) based on morphological and DNA sequence data. Cladistics 25:109–146

Arnedo MA, Hormiga G, Scharff N (2009) Higher-level phylogenies of linyphiid spiders (Araneae, Linyphiidae) based on morphological and molecular evidence. Cladistics 25:231–262

Benjamin SP, Dimitrov D, Gillespie RG, Hormiga G (2008) Family ties: molecular phylogeny of crab spiders (Araneae: Thomisidae). Cladistics 24:708–722

Blackledge TA, Scharff N, Coddington JA, Szüts T, Wenzel JW, Hayashi CY, Agnarsson I (2009) Reconstructing web evolution and spider diversification in the molecular era. Proc Natl Acad Sci USA 106:5229–5234

Bond JE, Hendrixson BE, Hamilton CA, Hedin M (2012) A reconsideration of the classification of the spider infraorder Mygalomorphae (Arachnida: Araneae) based on three nuclear genes and morphology. PLoS One 7:e38753

Coddington JA (2005) Phylogeny and classification of spiders. In: Ubick D, Paquin P, Cushing PE, Roth V (eds) Spiders of North America: an identification manual. American Arachnological Society, Worcester

Dimitrov D, Lopardo L, Giribert G, Arnedo MA, Alvarez-Padilla F, Hormiga G (2011) Tangled in a sparse spider web: single origin of orb weavers and their spinning work unravelled by denser taxonomic sampling. Proc R Soc B 279:1341–1350. doi:10.1098/rspb.2011.2011

Griswold CE, Audisio T, Ledford JM (2012) An extraordinary new family of spiders from caves in the Pacific Northwest (Araneae, Trogloraptoridae, new family). ZooKeys 215:77–102

Hedin M, Bond JE (2006) Molecular phylogenetics of the spider infraorder Mygalomorphae using nuclear rRNA genes (18S and 28S): conflict and agreement with the current system of classification. Mol Phylogenet Evol 41:454–471

Jocqué R, Dippenaar-Schoeman AS (2006) Spider families of the world. Royal Museum for Central Africa, Tervuren

Labarque FM, Ramirez MJ (2012) The placement of the spider genus *Periegops* and the phylogeny of Scytodoidea (Araneae: Araneomorphae). Zootaxa 3312:1–44

Lopardo L, Giribet G, Hormiga G (2011) Morphology of the rescue: molecular data and the signal of morphological characters in combined phylogenetic analyses—a case study from the mysmenid spiders (Araneae, Mysmenidae), with comments on the evolution of web architecture. Cladistics 27:278–330

Miller JA, Griswold CE, Yin CM (2009) The symphytognathoid spiders of the Gaoligongshan, Yunnan, China (Araneae, Araneoidea): systematics and diversity of micro-weavers. ZooKeys 11:9–195

Miller JA, Carmichael A, Ramirez MJ, Spagna JC, Haddad CR, Rezac M, Johannesen J, Kral J, Wang X-P, Griswold CE (2010) Phylogeny of entelegyne spiders: affinities of the family Penestomidae (NEW RANK), generic phylogeny of Eresidae, and asymmetric rates of change in spinning organ evolution (Araneae, Araneoidea, Entelegynae). Mol Phylogenet Evol 55:786–804

Rix MG, Harvey MS, Roberest D (2008) Molecular phylogenetics of the spider family Micropholcommatidae (Arachnida: Araneae) using nuclear tRNA genes (18S and 28S). Mol Phylogenet Evol 46:1031–1048

Silva D (2003) Higher-level relationships of the spider family Ctenidae (Araneae: Ctenoidea). Bull Am Mus Nat Hist 274:1–86

Subject Index

A
Abdominal sac, 44, 46, 53f, 368
Acanthoscurria
  A. gomesiana, 6, 84, 86
  A. natalensis, 88
Acanthoscurrin, 82, 87
Acari, 5, 9, 112
Acaricide, 415ff, 442
Acceleration, 459
Accessory gland, 174, 181f
Acentric, 161
Acetamiprid, 419
Acetylcholine, 193, 206, 261, 286, 420, 497
Acetylcholinesterase, 420
Acetylputrescine, 287f
Acetyltaurine, 287
Achaearanea, 198
Achiasmatic, 167
Achromatic vision, 321, 326–330
Acidification, 271, 411
Acid-sensing ion channel, 208f, 211, 495
Acinifom gland/silk, 273f, 283
Acremonium, 117
Acrocentric, 160–164
Acrosomal, 176–180
Actinopodidae, 493, 506
Aculepeira ceropegia, 396
Acylpolyamine, 88, 192–195, 200f, 206, 213, 492
Adaptive, 44, 71, 76, 81, 103, 143, 145, 150, 323, 379, 411, 475
Adaptive immune system, 81
Adenylate energy charge index, 408
Adhesive, 61ff, 271ff, 276–278, 295–297, 308ff, 463ff
ADP, 206, 408
Adrenaline, 193
Aerial dispersal, 68ff
Aerobic metabolic, 36
Aerobic–anaerobic partition, 35
Aerodynamic, 68, 70, 276
Agelena, 262
  A. consociata, 130
  A. labyrinthica, 18, 404–410
  A. orientalis, 218
Agelenidae, 7, 18, 31f, 48, 86, 96, 101, 127ff, 143f, 164ff, 192ff, 218, 262, 387, 404, 412, 493, 508
Agelenopsis, 96, 262
  A. aperta, 143f, 194f, 218
Agglutinin, 85
Agglutionation system, 84f
Aggregate gland/silk, 243f, 271, 283ff
Aggressive mimicry, 150f
Agonistic, 132ff, 147
Agriculture, 72, 393–398, 430, 441
Agrobiont, 71f
Akanthomyces, 110, 114f
Alaninamide, 286f
Alanine, 268, 270, 275, 286f, 481
Albumin, 11
Alcohol, 127, 129, 133, 441
Aldehyde, 127, 129
Aliphatic compound, 284, 286
Aliphatic hydrocarbon, 127
Alkane, 133
Alkyl methyl ester, 288
Allelochemical, 126, 142
Allocosa
  A. alticeps, 147
  A. brasiliensis, 147
Allomone, 126, 142
Allosteric effector, 12
Allosteric modifier, 210

W. Nentwig (ed.), *Spider Ecophysiology*,
DOI 10.1007/978-3-642-33989-9, © Springer-Verlag Berlin Heidelberg 2013

Allostery, 10
Alopecosa pulverulenta, 98
Alpaida veniliae, 418, 422–424
α-helix, 195f, 217ff, 270, 307, 499
α-latrocrustotoxin, 198
α-latrotoxin, 198, 230, 498
Alzheimer's disease, 497
Amaurobiidae, 51, 127, 194, 197, 508
Amaurobius ferox, 51
Amblypygi, 5f, 174, 179
Amidation, 218, 220
Amine, 192f, 206, 492
Amino acid, 4, 7, 9, 85–89, 128, 142, 192–201, 206, 217ff, 244, 256, 267–275, 287–292, 376, 387f, 397, 476, 492, 495–500
Aminoproprionamide, 287
Ammoxenidae, 58, 191, 508
Amorphous, 267ff, 364, 477
AMP, 206, 408
Amphinectidae, 509
Amphipathic, 86, 499
Ampullate gland/silk, 68, 269–281, 283, 291, 476–486
Anaeorobiosis, 36
Anaerobic, 51
- capacity, 35
- metabolism, 36, 408

Analgesic, 213, 261, 493ff
Anamorph, 110, 112
Anapidae, 507
Anclystaceae, 112
Anelosimus eximius, 130f
Anoteropsis hilaris, 417–420
Antibiotic, 100, 109, 229, 499
Antifeedant, 415, 422, 441
Antimicrobial, 11, 81ff, 195, 290, 293, 492
Antioxidative, 408f
Antivenom, 256–261
Antrodiaetidae, 506
Anyphaenidae, 31, 58, 68, 388, 394, 435, 443, 508
Aorta, 15f, 18ff, 50
Aphantochilinae, 191, 350
Aphid, 93, 374–382, 389, 432, 444
Aphonopelma
- *A. hentzi*, 4ff, 10, 17, 22ff, 32, 36, 47f, 50f, 85, 88,
- *A. marxi*, 23, 82f
- *A. seemanni*, 59ff

Apis mellifera, 320, 355
Apodem, 30
Apoptosis, 408
Application, 268, 415ff, 436, 442ff, 451ff, 469ff, 475ff, 492ff
Apposition eye, 325
Appressorium, 109
Arachnidicolae, 108, 113
Arachnophobia, 253
Araneidae, 7, 34f, 86, 96, 101, 143ff, 160ff, 191ff, 245, 271ff, 283ff, 319ff, 343, 362ff, 380, 385ff, 394, 407, 416ff, 433f, 442, 477ff, 507
Araneism, 258
Araneoclada, 16, 506
Araneoidea, 194, 243, 273ff, 283ff, 507
Araneomorphae, 7, 11, 29ff, 45, 52, 68, 86ff, 159ff, 180f, 191, 203, 225, 253, 271, 274, 303, 313, 344, 499, 506
Araneus, 194, 325–330
- *A. aurantia*, 146, 288, 290, 295, 497
- *A. cavaticus*, 285f
- *A. diadematus*, 18, 96, 101, 286, 365–367, 385f, 405–410, 418, 422, 434f, 463, 482, 485
- *A. trifolium*, 276
- *A. ventricosus*, 96, 114, 160

Archaeidae, 507
Arginine, 195, 223, 498
Argiope, 86, 161, 104, 363, 442
- *A. aemula*, 322f
- *A. auratia*, 288, 295
- *A. bruennichi*, 143ff
- *A. flavipalpis*, 362f
- *A. keyserlingi*, 376
- *A. lobata*, 194, 288
- *A. reinwardti*, 363
- *A. trifasciata*, 271, 288

Argiotoxin, 497
Argyrodes, 293
Argyroneta aquatica, 17–19
Ariadna lateralis, 159
Arolia, 57
Aromatic, 194, 206, 284. 288, 498
Arrhythmia, 261, 493, 496
Arsenophonus, 95, 99
Artery, 47
Artificial diet, 393ff
Asci, 113
Ascomycota, 109, 12f
Ascospore, 110, 113
Ascostroma, 113
Asexual stage, 112
Aspergillus, 116
Associative learning, 149
Astacin, 236

Ataxia, 417, 421
ATP, 35f, 206, 260, 408, 493, 499
Atracotoxin, 256
Atrax, 6, 254–256
Attachment system, 57ff
Atterocopus, 267
Atypidae, 159–165, 174, 506
Atypus, 6
  A. affinis, 159, 165
Austrochilidae, 17, 160, 506
Austrochiloidea, 16, 29, 506
Austrochilus, 17, 159
Autoimmune disease, 493
Autosome, 160–169
Auxiliary spiral, 273
Avicularia
  A. metallica, 200
  A. minatrix, 161
  A. versicolor, 334
Axoneme, 178–180
Azadirachta indica, 441
Azadirachtin, 419, 441ff

B
Bacillus thuringiensis, 429ff, 445
Bacteria, 74, 83ff, 93ff, 152, 193, 223f, 234, 289, 429, 479, 499
Bacteroidetes, 93
Badumna
  B. insignis, 309
  B. longinqua, 304, 309f
Bagheera kiplingi, 388
Ballistic, 454, 458, 480
Ballooning, 67ff, 100, 148, 245, 272
Barychelidae, 58, 506
Beauveria bassiana, 113
Benzoylurea, 422
β-caryophyllene, 148
β-1,3 glucans, 84
β-hairpin, 87, 196
Betaine, 286–292
β-sheet, 207, 270, 477, 499
β-toxin, 197, 494
Bioaccumulation, 405–412
Bioactive, 481, 484, 491ff
Bioadhesive, 469
Biocompatible, 475, 481f
Bioconcentration, 405
Bioconcentration factor, 406
Biogenic amine, 193, 206
Bioinsecticide, 204, 208
Biological control, 393f, 398, 430
Biomaterial, 475ff
Biomedical, 470, 475ff
Biomimetic, 278, 468–471, 478
Biopolymer, 475ff
Biotin, 396
Bite, 125, 135, 204, 231, 233, 237, 243, 254–263
Blood cells, 82
Body condition, 74ff
Body condition index, 380
Body size, 8, 24f, 32, 34, 37, 101, 137, 262, 277, 380, 424f, 432
Bola spider, 144, 150f, 191
Bombyx mori, 479
Book lung, 9, 16, 21, 29–35, 50f, 303
Brachypelma
  B. albopilosum, 127, 129, 135–137
  B. vagans, 135
Bradykinin, 288
Bryobia, 101
Bt crop, 386, 429ff
Bt israelensis, 429
Bt kurstaki, 429ff, 445
Bt protein, 431ff
Bt tenebrionis, 429, 435
Buprofezin, 419
Butyric acid, 143

C
Cabal, 212f
Cadaverine, 193, 218
Cadmium, 403ff
Calamistrum, 305ff
Calcium, 87, 286, 396, 403, 407, 481, 497
Calcium channel, 87, 197f, 208ff, 220, 492ff
Calmodulin, 224
Calocera, 114
Camouflage, 110, 320–323, 349, 361, 365
Candida albicans, 87
Cannibalism, 69, 133–137, 146, 176, 375–382, 398
Canoe tapetum clade, 507
Capillarization, 19
Capillary, 36
Capillary force, 64, 173, 290, 296, 308–313, 466
Caponiidae, 30f, 58, 506
Capture spiral, 273–276, 476
Carbamate, 415–417
Carbohydrate, 396
Carbon dioxide, 21
Carbosulfan, 444
Carboxylesterase, 409f

Carcinoscorpius rotundicauda, 5
Cardiac artery, 17ff
Cardiac cycle, 22ff
Cardiac disturbance, 260f, 493, 496
Cardinium, 95ff
Cardiomyocyte, 16f
Cardioregulatory nerve, 23
Casein, 397
Catalase, 408–410
Catecholamine, 261
Cav channel, 87, 197f, 208ff, 220, 261, 492ff
cDNA, 4f, 244, 270, 274
cDNA library, 84, 86, 218, 225,
Cellular defence, 109
Central nervous system, 19, 22, 206, 495, 497
Centriole, 176, 178f
Centromere, 160f, 163, 169
Cephalosporium, 117
Cheiracanthium, 262, 442
  C. inclusum, 262, 394f, 434
  C. mildei, 262, 442f
  C. punctorium, 221, 262
Chelicerae, 18, 144f, 152, 191, 203, 242–246, 253–256, 260–262, 303, 334–336, 342–345, 350
Chemical communication, 125ff
Chemoreceptor, 65, 125, 128
Chemosensitive, 128, 144, 153, 197
Chiasmata, 163, 165, 167
Chileotaxus, 177f
Chilobrachys guangxiensis, 212
Chloride, 286
Chlorpyrifos, 418f, 444
Cholesterol, 127, 136, 376, 396f
Choline, 193, 233, 287, 289, 292
Cholinesterase, 417, 420
Chondroitin sulphate, 224f
Chromatic vision, 321–330
Chromatin, 164, 166f, 179
Chromosome, 159ff
Chrysomelidae, 429ff
Chrysso scintillans, 363f
Chummidae, 508
Circadian, 421
Circulatory organ/system, 2, 15ff
Cithaeronidae, 508
Citric acid, 143, 145, 193, 206
Clathroconium, 115
Clavicipitaceae, 112
Clavicipitales, 112
Cleistospermia, 181f
Clomazone, 418f
Clotting, 81ff
  cascade, 81ff
  enzyme, 11, 85
Clubiona, 444
C. phragmitis, 404
Clubionidae, 7, 58, 68, 96, 191, 387, 394, 404, 444, 508
Coagulation, 81
Coagulation cascade, 83
Coagulin, 84
Coagulogen, 84
Cobweb, 258, 274, 278, 283, 291–293, 434, 475
Cocoon, 128, 130, 350, 411, 479, 482
Coelotes terrestris, 133, 136
Coenospermia, 181f
Coevolution, 109, 267, 277, 303, 355,
Coiling process, 180
Coleosoma octomaculatum, 99
Collagenase, 199
Collembola, 180, 374ff, 393, 432
Coloration, 319ff, 335ff, 353ff, 362–366
Color-blind, 325
Colour change, 351, 353f, 361ff
Competition, 69, 103
Complement system, 233
Compound eye, 320, 325
Cone snail, 212f, 491–494
Conidiobolus, 12
Conidium, 110
Conspicuous, 245, 261, 297, 319ff, 334, 349f, 356, 364f
Contact pheromone, 126, 136, 143ff
Contamination, 403ff
Copper, 403ff
Copper protein, 3, 11
Cordyceps, 108ff
  C. cylindrica, 110, 113
  C. singeri, 110, 113
  C. thaxteri, 118
Corinnidae, 394, 505, 508
Corynebacterium, 234
Cosmophasis, 341
Courtship, 142ff, 344f, 375, 423
C-reactive protein, 85
Cribellate, 271, 278, 284, 296, 303ff
Crustacea, 3, 18
cry gene, 430ff
cry protein, 429ff
Crypsis, 231, 351f, 378
Cryptochiasmatic, 167
Cryptocyanin, 3
Crystalline, 83, 243, 267–277, 334, 343, 349, 353–356, 364, 429ff, 476f
CsTx, 217–222

Ctenidae, 7f, 17, 47f, 58, 61f, 82ff, 112, 144, 193ff, 217ff, 259ff, 458ff, 508
Ctenidin, 82, 87
Ctenitoxin, 494
Ctenizidae, 58, 110, 113, 118, 194, 506
Cupiennin, 195, 222–224
Cupiennius salei, 7ff, 17ff, 36, 47f, 61ff, 82ff, 112, 144, 148, 153, 193–196, 217ff, 397, 458–460,
Cuticle, 3, 11, 21, 32ff, 49ff, 85, 125ff, 243f, 247, 290f, 308, 311, 335ff, 353ff, 365, 369, 420, 466, 468
Cuticular compounds, 125ff
Cuticular lamella, 335ff
Cuticular lipid, 127, 132, 135, 137
Cyanocytes, 82f
Cyatholipidae, 507
Cybaeidae, 17, 32, 96, 508
Cycloctenidae, 505
Cyclodiene, 415, 420
Cyclosa conica, 96
Cyhalothrin, 418
Cylindrical gland, 283
Cypermethrin, 418f, 445
Cyriopagopus, 334
Cyrtarachne, 286
Cyrtaucheniidae, 506
Cyrtophora cicatrosa, 362–364
Cysteine, 86, 89, 192ff, 218ff, 229, 237, 407, 480, 492
Cystine, 207f, 213, 256
Cyto-insectotoxin, 195
Cytolytic, 193–200, 217ff, 499
Cytoplasmic incompatibility, 95, 99f-101

D

Damper element, 453, 455
Damping, 275f, 453, 459, 471
Dasumia carpathica, 160
Deacidification, 411
Decapoda, 3, 10, 22
Decapsulation, 183
Defensins, 82, 86
Deferent duct, 174f, 181
Degranulation, 82, 84
Deinopidae, 284, 306, 507
Delena cancerides, 162, 165
δ-atracotoxin, 256
δ-ctenitoxin, 494
Deltamethrin, 418f
δ-toxin, 197, 212, 494
Depolarization, 193, 206, 209–212, 261, 494
Dermonecrosis, 230–237, 243, 258, 262
Desiccation, 127
Desidae, 96, 304, 309f, 508
Detoxification, 409f, 420
Deuteromycota, 115
Dextrose, 397
Diacylhydrazine, 416
Diaea
D. circumlita, 99
D. evanida, 352, 355
D. lactea, 352
Diakinesis, 164, 167f
Diasterometric ratio, 143
Diastole, 23f
Diazinon, 418
Dicentric, 161, 163
Dicofol, 418
Dictynidae, 31, 162, 174, 307f, 417, 508
Diet, 127, 150–152, 231, 271, 277, 288, 373ff, 387–390, 393ff, 434
Diflubenzuron, 419
Diguetidae, 31, 165, 506
Dimethoate, 418
Dimethyl citrate, 144
Dimethylalkane, 127
Dimethylhentriacontane, 134
Dionycha, 508
Diplocephalus
D. latifrons, 96
D. picinus, 96
Diploid number, 159ff
Dipluridae, 112, 163, 167, 506
Diptera, 57, 374, 429–431
Dispersal, 67ff, 100–103, 113, 133f, 350, 421
Disulphide bridge, 87, 196, 218, 220, 222
Diuresis, 420
Diurnal, 141, 319ff, 344f
Diversiculum, 364–369
Diversity, 102, 118, 145, 159, 166, 174ff, 192, 194f, 197, 200, 213, 225, 229–233, 271, 303, 307, 393, 410, 412, 444, 500
D-lactate, 36
DNA, 95, 112, 162, 290, 480, 484
Domain, 88, 212, 221, 234, 269, 290, 480f, 487f
Dopamine, 193
Dragline, 68, 126, 271, 273, 288, 312, 350, 477ff
Drug, 204, 229, 260, 476, 484–487, 491ff
Drymusa, 232
Drymusidae, 165, 231, 506
Dummy, 324, 329f
Dysdera, 8f, 11, 32, 175
D. crocata, 167

Dysdera (cont.)
D. erythrina, 34
D. ninnii, 168
Dysderidae, 31f, 45, 58, 96, 160–168, 174, 506
Dysderoidea, 30, 160f, 169, 506
Dyspnoea, 256

E

Ebrechtella tricuspidata, 387f
Ecdysone, 11, 397, 441
Ecdysteroid, 136
Ecological niche, 406, 412
Ecological web, 380
Ecribellate, 283f, 286, 312
Ectomone, 126f
Egg sac, 144f, 274, 292f, 380, 423f, 478, 482, 486
8-methyl-2-aonanone, 143
Ejaculation, 173
Ejaculatory duct, 174
Elaphomyces, 113
Elasticity, 43, 243, 247, 289, 295–298, 460–462, 475ff
Electrophysiological, 129, 153, 197
Electrostatic, 109, 195, 223, 226
11-methylpentatriacontane, 133
Enantiomer, 143
Encapsulation, 81, 83
Endeis spinosa, 5
Endoplasmic reticulum, 9, 175, 479
Endosternite, 44–46
Endosulfan, 418
Endosymbiont, 74, 76, 93 ff
Engyodontium, 17
Enhance, 10, 51, 64, 74, 134, 167, 193, 198, 217–222, 290, 311, 336, 374ff, 394, 408, 417, 481, 487
Enhancer peptide, 221f
Enoplognatha ovata, 99
Ensifera, 57
Entapophyse, 30
Entelecara flavipes, 96
Entelegyne, 7, 30f, 58, 95, 142, 145, 159ff, 173ff, 221, 506
Enterococcus faecalis, 88
Entomogenae, 108
Entomophthorales, 112
Envenomation, 204, 226, 231, 242, 257, 260f
Enzyme, 3, 11, 84–86, 109, 192–194, 198–201, 204, 207, 213f, 217, 224f, 229ff, 243, 258, 263, 378, 408–412, 417, 420, 492
Ephebopus cyanognathus, 334–337
Epicuticle, 109, 127
Epidermis, 32f, 46, 152, 484
Epilepsy, 494, 497
Epithelium, 173, 175, 181, 293
Eresidae, 130, 164, 443, 507
Erigone, 71f
E. arctica, 74, 96
E. atra, 68, 73f, 95f, 100–103, 418, 422, 424
E. dentipalpis, 68, 74f, 96
E. longipalpis, 97
E. promiscua, 97
Eriovixia cavaleriei, 96
Eschar formation, 257
Escherichia coli, 87f
Euathlus, 334
Euchelicerata, 5
Euctenizidae, 506
Eukaryotic, 162, 195, 220–224
Euplantulae, 57
Eurypelma californicum. See Aphonopelma hentzi
Euryphagous, 374
Eurypterid-like, 33
Evaporation, 34
Evarcha, 164
E. arcuata, 61ff, 463–468
E. culicivora, 144, 147, 149
E. falcata, 98
Evolution, 3ff, 25, 29, 33f, 67–70, 85, 94, 102, 112–115, 130, 133, 137, 142, 150, 162ff, 174, 176ff, 196, 199f, 207, 219, 229ff, 244f, 253f, 267ff, 310ff, 330f, 353, 356f, 374, 376, 411, 469
Excretion, 364, 404, 406, 420
Exocuticle, 109
Exocytosis, 84
Exoskeleton, 21, 85, 109
Extensor muscle, 43–47, 53, 246, 451
Extracellular, 180f, 198f, 207, 209, 226, 236, 495
Extrafloral nectary, 389
Exuvia, 407

F

Factor C, 84
Farnesyl acetate, 145
Fasciculation, 256f
Fatality, 204, 254ff
Fat body, 33
Fatty acid, 127f, 130, 133, 136f, 288, 290, 376
Fecundity, 76, 102, 397, 410f, 416, 424f, 435
Feminization, 95, 99

Fenithrothion, 419
Fenvalerate, 418
Fertility, 21, 424
Fertilization, 136, 145, 183
Fiber, 206, 212, 268ff, 283ff, 306ff, 469, 475ff
Fibril, 178, 243, 269, 271, 302ff
Fibroin, 244, 267–271, 289f, 476ff
Filistata
    F. hibernalis, 17
    F. insidiatrix, 8
Filistatidae, 17, 31, 48, 160, 165, 178, 181, 304–309, 493, 506
Fitness, 69–74, 100, 376–381, 389, 403, 409, 411
5-hydroxytryptamine, 193
5-methylhentriacontane, 132
Flagelliform gland/silk, 245, 273f, 283, 306, 476, 480
Flavobacterium, 99
Flexor muscle, 44–54, 246, 421, 451
Floronia bucculenta, 362–369
Folate, 396
Folsomia candida, 374
Foraging, 37, 67, 134–136, 148–151, 320–330, 350, 373–379, 404, 415
Formulation, 394–398, 415ff, 431ff, 443
Frontinella communis, 386, 418–421, 434
Fructose, 387
Functional diversity, 200, 412
Functional response, 381, 416, 422–425
Fungal pathogen, 107ff
Fungi, 84, 87, 88, 108ff, 152, 386
Fungicidal, 88
Fungicide, 415ff. 442
Fungivory, 386
Funnel web, 256, 309, 312

G

GABA, 193, 286f, 292
GABamide, 286–292
Gallieniellidae, 508
Gammaproteobacteria, 93
Gas exchange, 33, 36
Gasteracantha, 363f
GDPD, 234
Gea heptagon, 363f
Gecko, 57, 62, 64, 464, 468
Gelanor, 177f
Gene, 4–8, 71, 94, 101, 206–209, 224, 230ff, 244, 267ff, 290, 292
Gene flow, 75, 100, 103,
Generalist predator, 149, 203, 213, 373f, 381f, 385
Genetic drift, 103
Genetically engineered plant, 429ff
Genome, 94, 234, 476, 480
Geolycosa, 218, 499
Ghelna canadensis, 98
Gibellula, 111, 114, 116f
Gladicosa gulosa, 394
Gland. *See* Assessory gland; Midgut gland; Silk gland; Venom, gland; Wax gland
Glucose, 387, 397
Glucuronic acid, 224
Glutamate, 193f, 213, 497
Glutamate receptor, 194, 213, 497
Glutamic acid, 206
Glutathione, 408–410
    oxidase, 408–410, 420
    s-transferase, 408, 420
Glyceride, 127
Glycine, 84, 86f, 217, 268, 270–275, 286–288, 292, 481
Glycogen, 180, 286
Glycolipid, 199
Glycolysation, 9, 290
Glycoprotein, 221, 243, 269, 290–297
Glyphosate, 152, 418
Gnaphosidae, 7, 31, 58, 96, 191, 508
Gnaphosoidea, 508
Golgi apparatus, 9, 175f
Gomesin, 82, 87
Gonad, 17, 174, 406
Gonatium rubens, 97
Gonopore, 173f
Grade shaped tapetum clade, 508
Gradungulidae, 506
Grammostola rosea, 208, 496, 499
Gram-negative bacteria, 84ff, 499
Gram-positive bacteria, 86ff, 499
Granules, 81, 407
Granulocytes, 82f, 86, 224
Granulomanus, 110f, 114, 116
Gregarious, 129f
Guanine, 343, 354–356, 361ff
Guanocyte, 343, 353f, 361ff
Guild, 403, 412
Gumfoot threat, 274, 278, 291ff

H

Habrocestum pulex, 98, 149
Habronattus, 164
    H. hallani, 339–344
Hadronyche, 254–256
Haemocyanin. *See* Hemocyanin

Haemolymph. *See* Hemolymph
Hahniidae, 31, 508
Hamiltonella defensa, 93
Haplogyne, 8f, 30, 58, 95, 145, 159ff, 178ff, 197, 231, 234, 506
Haplopelma
 H. haianum, 200
 H. lividum, 335
 H. minax, 205
 H. schmidti, 208, 211, 494f
Harpactea, 45
Headache, 262
Head-to-tail, 195
Heart, 15ff, 43, 47, 50
 beat, 23, 52
 rate, 24f, 37, 204, 417
 wall, 81ff
Heat shock protein, 408–410
Heavy metal, 403ff
Helix-hinge-helix structure, 223
Hemerythrin, 3
Hemmingsen's equation, 35
Hemocyanin, 3ff, 33–37, 82–85
Hemoglobin, 3, 10
Hemolymph, 10f, 15ff, 21f, 32ff, 43ff, 81ff, 127ff, 193, 206, 217, 226, 420, 451ff
 channel, 49
 pressure, 21–25, 35f, 43ff, 173, 245–247, 451ff
Hemolysis, 257
Heparan sulphate, 224f
Hepatopancreas, 86, 285
Heptathela kimurai, 159, 176
Herbicide, 152. 415ff, 442
Herbivory, 385f
Hersiliidae, 507
Heterochromatin, 160–167
Heterogametic, 167
Heteropoda, 177, 262
 H. venatoria, 494
Heteropodatoxin, 494
Heteropycnotic, 164f, 168
Heterothele gabonensis, 335
Hexadecane, 136
Hexadecyl acetate, 145
Hexamer, 3ff
Hexamerin, 3, 11
Hexathelidae, 194, 197, 254, 261, 493–497, 506
Hibana
 H. futilis, 388
 H. velox, 394f, 435, 443
Hibernation, 72, 381
Hippasa partita, 225
Hirsutella, 116
Histamine, 193, 206, 213, 218, 236, 261
Histamine-releasing factor, 236
Hogna, 175, 262
 H. carolinensis, 18
 H. helluo, 144, 149, 151, 416
Holarchaeidae, 191, 253, 507
Holocentric, 161
Holocnemus pluchei, 98, 102
Holokinetic, 160–169
Hololena, 262
 H. curta, 195
Holoplatys sp., 98
Holothele, 335
Homalonychidae, 58, 508
Homeostatis, 15, 408
Homogametic, 166
Honeybee, 290f, 320–323, 350–355
Hormesis, 416, 424f
Host, 74, 81, 93–95, 99–103, 107–118, 148, 389, 445
Hsp70, 409f
Human population, 255
Humidity, 63, 65, 289, 296, 308–313, 420, 477–479
Humoral defence, 109
Humoral response, 81
Hunger, 152, 373ff
Huttoniidae, 507
Huwentoxin, 208, 494
Hyaluronic acid, 199, 224
Hyaluronidase, 199, 207, 213, 224–226, 236, 492
Hydraulic, 15, 20–25, 35, 43ff, 60, 246, 451ff
Hydrocarbon, 130, 133
Hydrophobic, 196
Hydrostatic, 46, 51, 64, 245f
Hygroscopic, 289
Hylaphantes graminicola, 97, 417f, 422–425
Hyluronan, 199, 224f
Hymenoptera, 57, 320–328, 353
Hymenostilbe, 116
Hyperpolarization, 210
Hypertension, 256, 260
Hypha, 109, 115f
Hyphomycetes, 112, 115
Hypochilidae, 165, 178, 303, 305, 506
Hypochilus, 308
 H. pococki, 305
Hypocreales, 112f
Hypodermis, 353–355, 361, 367–369
Hypolimnas, 102

Hypomma
H. bituberculatum, 97
H. cornutum, 97
Hyptiotes, 307, 312
H. cavatus, 305, 309
H. gertschi, 309
Hysteresis, 275
Hysterocrates gigas, 210

I
ICK motiv, 196f, 207f, 213, 219–222, 237, 256, 492
Idiopidae, 262, 506
Imidacloprid, 418, 444f
Immigration, 70
Immune defense, 15
Immune response, 3, 11, 233, 482, 484
Immune system, 81ff
Imperfect stage, 112
Implantation fossa, 178–180
Inbreeding, 69, 74, 76, 103
Indicoblemma lannaianum, 53f
Indolic, 194
Infection, 81ff, 93ff, 108ff, 258, 263, 499f
Inflammation, 257
Inhibition, 23, 87f, 134, 137, 193, 197f, 208–213, 221, 224f, 237, 261, 403, 408, 417, 420–423, 441, 493–497
Inhibitor cysteine knot motif, 196f, 207f, 213, 219–222, 237, 256, 492
Injury, 82
Innate immune system, 81
Insect growth regulator, 415f, 425, 441
Insecticidal, 195, 204, 208, 210, 217–222, 236f, 416, 418, 423, 429ff, 441ff
Insecticide, 210, 415ff, 429, 436f, 441ff
Integument, 127, 342, 365
Interference, 333ff
Interphase, 165f
Ion channel, 192f, 197f, 208f, 211, 213, 217, 221, 225, 236f, 261, 263, 431, 492ff
Ion concentration, 193, 217, 286
Ionotropic receptor, 193, 206, 213, 497
Iridescence, 334ff
Iridopelma, 335
Iron, 396, 403ff
Isaria, 16
I. arachnophila, 108
Ischaemia, 257, 260
Ischnothele guianensis, 112
Isethionic acid, 287–292

J
Jacksonoides queenslandicus, 144, 149

K
Kaira alba, 151, 288
Kairomone, 126, 142, 148–151, 422, 424
κ-toxin, 198, 212, 494
Karyotype, 159ff
Kevlar, 268
Kinase, 224
Kinetic(s), 9f, 267, 275f, 486f
Kinetochore, 161, 168
Kishinouyeus, 114
Knottin, 207
Kukulcania, 307
K. hibernalis, 48, 51, 160, 304–310
Kunitz-type fold, 207f, 212
Kv channel, 193, 197f, 208–213, 261, 493f
Kynurenine, 353f

L
Labidognatha, 11
Lachesana tarabaevi, 195f, 223
Lactic acid, 193, 342
Lacunae, 15, 21, 47, 54, 451ff
Lamponidae, 508
Large protein, 192, 198, 229, 476
Larinia
L. argiopiformis, 96
L. tabida, 96
Larinioides
L. patagiatus, 162
L. sclopetarius, 416–418, 422
Laser Doppler vibrometry, 248
Laser trans-illumination technique, 24
Lasiodora 88
Latarcin, 195, 223
Latouchia, 114
Latrocrustotoxin, 198
Latrodectism, 259f
Latrodectus, 198, 229f, 258ff, 287, 292f
L. curacaviensis, 259
L. geometricus, 259
L. hasselti, 142, 176, 198, 259
L. hesperus, 148, 200, 259, 292f, 295
L. indistinctus, 259
L. mactans, 198, 259, 292f
L. tredecimguttatus, 193, 198, 259, 498
Latroinsectotoxin, 198
Latrophilin receptor, 498

Latrotoxin, 198, 230, 498
LD50, 200, 219–222, 416, 425
Lead, 403ff
Leberidocytes, 83
Lecanicillium, 114
Lectin, 85, 208f, 212, 294
Leg autotomy, 43
Leishmania amazonensis, 87
Lepidoptera, 150, 328f, 429ff
Lepthyphantes minutus, 97
Leptoneta infuscata, 165
Leptonetidae, 165, 506
Lethal, 193, 204, 209, 253ff, 416f
Leucauge, 325, 362
    L. blanda, 363
    L. celebesiana, 363
    L. dromedaria, 98
    L. magnifica, 322f, 329
    L. subgemmea, 362f
Leukemic cell, 224
Life cycle, 4, 35, 112, 364, 374–382
Life history traits, 67, 75, 403f, 417, 425
Lightning-strike cabal, 212
Limulus polyphemus, 5, 10
Linear peptide, 192, 195, 200, 492
Linoleic acid, 136
Linyphia, 177f, 293f, 365, 442
    L. triangularis, 96, 142, 367, 405–410
Linyphiidae, 7, 31f, 68, 73ff, 96ff, 142, 178f, 194, 283ff, 362ff, 386f, 407ff, 417ff, 433ff, 442, 507
Liocranidae, 58, 98, 508
Liphistiidae, 17, 177, 274, 506
Liphistius, 128, 175, 177
    L. desulator, 25
    L. malayanus, 17f
Lipopolysaccharide, 84, 87, 195
Lipoprotein, 109
Locomotion, 21f, 32, 36, 43ff, 61, 67, 81, 197, 420f, 451, 464, 470f
Longevity, 387f, 397, 407
Low molecular mass compound, 141, 192, 195, 199f, 204, 206, 217, 221, 224, 233, 286f, 292, 492
Loxosceles, 175, 193, 199, 230ff, 257ff, 263
    L. deserta, 200
    L. gaucho, 257
    L. intermedia, 174, 180, 237, 257
    L. laeta, 230, 237, 257
    L. parrami, 257
    L. rufescens, 164, 230, 257
Loxoscelism, 257f
Lung, 9, 16, 21, 29–35, 50f, 303
Lycosa, 262, 436
    L. singoriensis, 218
Lycosidae, 7, 31f, 58, 71, 98, 112, 144, 160ff, 178, 184, 195ff, 218, 225, 262, 387, 393f, 404ff, 416ff, 435f, 442, 463, 493, 499, 508
Lysine, 195, 222f, 486
Lysoglycerophospholipid, 199
Lysophosphatic acid, 199
Lysophospholipid, 233

M

Macrarugs rufus, 97
Maevia inclemens, 98
Magnesium, 396f, 403
Magnesium sulphate, 397
Major ampullate gland/silk, 269ff, 283, 476ff
Malaria, 212, 224, 500
Malathion, 418f, 444
Male-killing, 95, 100–103
Malkaridae, 507
Mallos, 307
Malthonica
    M. ferruginea, 165, 168
    M. silvestris, 164
Mandibulata, 9
Maratus
    M. splendens, 339–344
    M. volans, 339–341
Marpissa lineata, 98
Mass rearing, 394ff
Mastigoproctus giganteus, 45
Mastophora
    M. cornigera, 150
    M. hutchinsoni, 150f
    M. stowei, 144
Matachia livor, 309, 312
Maternal effect, 74–76
Mating, 72f, 100f, 127, 130, 142–147, 176, 261, 375, 395, 398, 415f, 423
Matriphagy, 376
Mechanics, 62, 245, 278, 451ff
Mechanosensitive receptor, 197, 208–210, 496–499
Mechatronic, 470
Mecicobothriidae, 506
Mecysmaucheniidae, 507
Medical importance, 199, 228, 230–233, 254ff, 462, 470, 475ff, 492f, 497f
Megalepthyphantes collinus, 97
Meioneta rurestris, 97
Meiosis, 159ff

Melanin, 81–85
Melanisation, 85
Membrane, 23f, 43, 45, 48f, 54, 85f, 167f, 180, 192–200, 207, 209, 217, 220–226, 233, 247, 253, 258, 260, 263, 367, 417, 431, 452, 476, 499
Membranolytic, 195, 220–223
Mercury, 403ff
Mesocuticle, 109
Mesothelae, 6, 16, 29, 34, 159ff, 173ff, 191, 199, 253, 304, 506
Meta
  M. menardi, 86
  M. mengei, 99
Metabolic rate, 25, 34f
Metacentric, 160ff
Metacordyceps, 113
Metalloprotease, 193, 199
Metallothionein, 407–411
Metaphase, 160–168
Metaphidippus, 389
Metapopulation, 75
Metarhizium, 113
Metellina
  M. merianae, 118
  M. segmenta, 99, 405, 409f
Methamidophos, 418
Methionine, 376
Methoxyalkane, 288. 292, 294
Methoxyfenozide, 419
Methyl palmitate, 136
Methyl stearate, 136
Methylester, 130
Methyl-malonyl-coa, 127
Methyloctadecadienoate, 132
Methyloctadecanoate, 133f
Methylpentacosane, 133
Methyltetradecanoate, 133
Methyltriacontane, 133
Mexitilia, 307
  M. trivittata, 307
Miagrammopes, 307, 309f, 312
  M. animotus, 304f, 309f
Micrargus herbigadus, 97
Micrococcus luteus, 87f
Microelectromechanical system, 478, 485
Microlinyphia, 294
  M. imprigra, 97
  M. pusilla, 97
Microneta viaria, 97
Micropholcommatidae, 505
Microstigmatidae, 506
Microtrichia, 57
Microtubules, 168, 176–179
Microvilli, 175
Midgut gland, 17, 343, 361ff, 405–409
Migidae, 506
Migraine, 499
Mimetidae, 178, 191, 507
Mimicry, 150f, 396
Mini-protein, 192–201, 206, 217, 492–495
Minor ampullate gland/silk, 273, 283
Misionella, 307
Missulena, 262
Misumena vatia, 148, 350, 352, 365f, 406
Misumenoides formosipes, 387
Mitochondrium, 36, 176, 408
Miturgidae, 221, 434, 442, 505, 508
Modisimus, 112
Mollicutes, 93
Molting, 11, 43, 82f, 146, 441
Monocentric, 160f
Monocrotophos, 443
Monogamy, 176
Monomethyalkane, 127
Monophagous, 191
Mortality, 70, 108, 117f, 374, 415ff, 434f, 442–445
Motor cabal, 212f
Moulting, 11, 43, 82f, 146, 441
Mucopolysacharide, 199
Mucor hiemalis, 12
Muscle, 16, 18, 24f, 31ff, 43ff, 86, 173, 197, 204, 206, 209f, 224f, 246f, 256, 261, 267, 269, 271, 367–369, 421, 451ff, 478, 493f, 496
Muscular dystrophy, 496
Musculi laterales, 44–46, 53
μ-toxin, 197, 494
Mutualism, 94, 148, 388–390
Myastenic syndrome, 493
Mycelium, 109f
Mygalin, 88
Mygalomorphae, 6, 16ff, 29, 35f, 58, 68, 86–88, 110, 113, 159ff, 173ff, 191ff, 203, 225, 253ff, 274, 303, 499, 506
Myocardium, 22–24, 81
Myogenic, 22
Mysmenidae, 507

N

N-acetylgalactosamine, 290, 293f
NADPH oxidase, 224
Nanocrystal, 267–270
Nanospring, 270, 274

Nausea, 204, 256, 260f
Nav channel, 193, 197, 208–213, 256, 261, 494f
Necrotic, 199, 230, 233, 243, 257f, 262f, 408
Nectar, 330, 385ff, 395
Neem, 441ff
N-eicosane, 132
Nemesia, 6
Nemesiidae, 58, 262, 506
Neocribellatae, 506
Neolana pallida, 307, 309, 313
Neon neli, 98
Neonicotinoid, 422f
Neoscona, 325–330
  N. pratensis, 418, 422–424
  N. punctigera, 328–330
  N. scylla, 162
Neosparassus, 262
Neoxyphinus, 175
Nephila, 175, 194, 284ff, 325, 477, 482
  N. clavata, 98
  N. clavipes, 286ff, 476–481, 468
  N. inaurata, 7
  N. madagascariensis, 486
  N. pilipes, 288, 322, 324, 328
  N. plumipes, 98
Nephilengys, 175, 194
Nephilidae, 7, 98, 175, 192ff, 284, 286, 319ff, 387, 476, 507
Nereine
  N. clathrata, 96f
  N. montana, 97
  N. peltata, 97
Nerolidol, 148
Nerve, 19, 22f, 50, 86f, 128, 153, 197f, 225f, 261, 453, 482, 484
Nesticidae, 291, 507
Nesticodes rufipes, 115
Neuroactive substance, 30
Neurogenic, 22
Neurotoxin, 192f, 197, 199f, 203, 208, 217ff, 229f, 236–238, 256, 415, 419f, 423, 492
Neurotransmitter, 192f, 197–200, 209, 244, 257, 260, 286, 417, 420, 422, 492, 497
N-heneicosane, 133
N-hentriacontane, 133
N-heptacosane, 133
N-heptadecane, 133
N-heptatriacontane, 134
Niacin, 396
Nicodamidae, 507
Nigma, 175
  N. flavescens, 174
Nirvana cabal, 213
Nitric oxide synthase, 224, 494
Nitrogen, 381
NMDA receptor, 213, 497
N-nonacosane, 132
Nociception, 211, 493ff
N-octacosane, 132
N-octadecane, 132f
Nocturnal, 141, 151, 261, 319, 324ff, 334
Nodulation, 81, 83
Nomuraea atypicola, 113, 117
Nonacosane, 136
Non-newtonian liquid, 245f
Nontarget, 415
Noradrenaline, 193
N-pentacosane, 133
N-tricosane, 133
N-tritriacontane, 132
Nucleolus organiser region,, 162
Nucleoplasm, 179
Nucleoside, 192f, 492
Nucleotide, 192, 206, 479, 492
Nucleus, 82f, 164–169, 176–180, 367, 369
Nurscia albofasciata, 99
Nutrient, 21, 137, 373ff, 387, 395ff
Nutrition, 21, 81, 134–137, 271, 289, 373ff, 387, 394ff, 420, 434

O

O_2
  binding, 8ff
  debt, 36
  transport, 35
Ochyroceratidae, 506
Octadecadienoic acid, 132
Octadecenoid acid, 132f
Octonoba, 277
  O. sinensis, 309
Octopamine, 18f, 193, 206
Odour, 129, 147ff
Oecobiidae, 7, 180, 507
Oedothorax
  O. apicatus, 97, 418–424
  O. fuscus, 97
  O. gibbosus, 97, 100
  O. retusus, 97
Oenocytes, 83
Oleic acid, 133, 136
Olfactory, 71, 126–128, 141ff, 388, 397, 422
Oligo-hexamer, 3, 11
ω-toxin, 197, 493
Ommatidium, 325f

Ommochrome, 338, 353, 356
Oogenesis, 423
Oonopidae, 31, 174, 506
1-docosanol, 133
1-eicosanol, 133
1-octadecanol, 133
Open circulatoy system, 15, 36, 81
Ophiocordyceps, 113
  O. caloceroides, 110
  O. engleriana, 110
Ophiocordycipitaceae, 113
Opiliones, 5, 9, 34, 112
Opisthosoma, 15ff, 29ff, 43ff, 68, 101, 109–111, 127, 173, 243ff, 267, 285, 294, 304, 322, 327, 330, 334, 341f, 353f, 361ff, 396, 407, 476
Optical density, 333
Optimal foraging, 373, 376
Orbiculariae, 175, 270ff, 507
Orb web, 245, 273ff, 283ff, 308ff, 319ff, 385ff, 422f, 476f
Organic acid, 192f
Organochlorine, 415, 422
Organophosphate, 415, 417, 420–425
Orsolobidae, 58, 506
Oscillation, 50, 243, 247–249, 276
Osmotic pressure, 10
Ostia, 17f, 22
Oviposition, 395, 416, 424
Oxidative phosphorylation, 408
Oxygen transport, 9f
Oxyopes, 436, 444
  O. salticus, 149
  O. sertatus, 98
Oxyopidae, 7, 98, 162, 178, 195ff, 387, 444, 493, 508
Oxyopinin, 223

P

Pachygnatha
  P. degeeri, 99
  P. listeri, 99
Pachytene, 164, 167
Paecilomyces, 116
Pain, 193, 204, 206, 211, 213, 230, 254ff, 324, 492–499
Pain theory, 193
Palaeocribellatae, 16, 506
Palmitic acid, 133f, 136
Palpal organ, 43, 173f
Palpigradi, 5
Palpimanidae, 507
Palpimanoidea, 507
Pamphobeteus, 335
Pandinus imperator, 6, 10
Panthotenic acid, 396
Pantopoda, 5
Paracribellar, 306
Paralysis, 150, 193f, 197–199, 206, 210, 212f, 253, 417, 421
Paraphysa scrofa, 211
Pararchaeidae, 507
Parasexuality, 112
Parasocial, 134f
Parasporal crystal, 429, 431
Parasteatoda tepidariorum, 271
Paratropidae, 506
Pardosa
  P. agrestis, 98
  P. agricola, 98
  P. amentata, 112, 376, 380, 416–422, 425
  P. astigera, 406, 411
  P. bifasciata, 98
  P. hyperborea, 98
  P. laura, 412
  P. lugubris, 32, 98, 404–410
  P. milvina, 144, 149, 151f, 416–421, 424, 436
  P. monticola, 98
  P. palustris, 405, 408–410, 416–421
  P. prativaga, 376–380, 418–420
  P. pseudoannulata, 418–422, 425, 435, 442f
  P. pullata, 98
  P. purbeckensis, 71
  P. riparia, 98
  P. saltans, 405, 407, 412
  P. sphagnicola, 98
  P. wagleri, 160
Parkinson's disease, 497
Parthenogenesis, 95, 99f
Patent, 480ff
Pathogen, 11, 81ff, 107ff, 224, 258, 499
Pax islamita, 164
Peckham's mimicry, 150
Pedicel, 18, 50f
Pedipalpi, 180
Pelegrina proterva, 98
Pellenes tripunctatus, 164
Penestomidae, 508
Pentadecane, 136
Peponcranium ludicrum, 97
Perania nasuta, 52f
Perfect stage, 112
Perforatorium, 178
Pericardial membrane, 23
Pericardial sinus, 16, 22ff

Pericentric, 162, 165
Perichomosome, 164
Periegopidae, 231, 506
Peristaltic heart contraction, 22
Perithecium, 110–113
Peroxidase, 408–410, 420
Pesticide, 393f, 415ff, 429f, 441ff
Petiolus, 29, 31
pH dependence, 10
Phagocytosis, 81ff
Pharmacological, 23, 204, 207, 229, 493–496
Phenolic, 194
Phenoloxidase, 3, 11, 82, 85
Pheromone, 126ff, 141ff
Phidippus, 342, 344
  P. audax, 61, 98, 342
  P. pulcherrimus, 345
  P. regius, 342
  P. rimator, 61
Philodromidae, 7, 31, 58, 65, 174, 350, 363ff, 416, 419, 442, 508
Philodromus, 442
  P. cespitum, 416, 419, 422, 443
  P. dispar, 65, 174
  P. spinitarsis, 363f
Philoponella arizonica, 309
Pholcidae, 7, 31, 47, 98, 102, 112, 143–145, 165, 177ff, 250, 406, 487, 506
Pholcus, 177
  P. beijingensis, 143ff
  P. crypticolens, 98
  P. phalangioides, 18, 47, 98, 406, 487
Phoneutria, 259–261
  P. fera, 261
  P. keyserlingi, 261
  P. nigriventer, 195, 218, 261, 493f
  P. reydyi, 86, 261
Phonognatha graeffei, 363f
Phormictopus, 335
Phoroncidia studo, 151
Phosphamidon, 444
Phosphate, 195, 284, 286, 297, 397, 485
Phosphatidylcholine, 199, 223
Phosphodiesterase, 199, 492
Phosphoglycoprotein, 293
Phospholipase, 193, 199–201, 225f, 233f, 258
Phospholipid, 195, 198f, 217, 221, 223, 233, 235f, 258
Phosphoric acid, 193
Phosphorus, 396
Photo-damage, 355f
Photoreceptor, 320–327, 349f
Phototransduction, 321
Phrixotrichus roseus, 451, 453
Phylloneta impressa, 432–436
Phylogeny, 5f, 68f, 93f, 102, 113–115, 176, 183, 230–234, 272, 303, 323, 339, 343, 350, 356, 505ff
Phyxelidae, 508
Pigment, 333ff, 353ff, 361, 364
Pikelinia, 307
Piloerection, 256
Pimoidae, 179, 507
Pirata
  P. piraticus, 404–406, 410
  P. piratoides, 416–419, 422–425
  P. subpiraticus, 435
Piriform gland/silk, 273f, 283
Pirimicarb, 418
Pisaura mirabilis, 148
Pisauridae, 7, 148, 178, 194, 387, 508
Pityohyphantes phrygianus, 97, 100–103
Plasmatocytes, 82f
Plasmodium, 224, 500
Plectreuridae, 31, 493, 506
Plesiomorphic, 17, 25, 178, 199, 304, 313
Pleura, 44, 52
Pocadicnemus juncea, 97
Poecilopachys australasiae, 96
Poecilotheria, 204
  P. formosa, 159
  P. metallica, 335–340, 344
  P. regalis, 205
Polenecia, 307
Pollen 310, 385ff, 431ff
Pollinator, 148, 323, 328, 350–357, 430
Pollution, 403ff
Polyalanine, 268, 270, 274f, 481
Polyamine, 192–194, 206, 218, 498
Polyandrous, 145
Polybetes pythagoricus, 86, 417, 419
Polydimethylsiloxane, 470, 476
Polygamy, 176
Polyphagous, 149, 191
Polyploidisation, 163
Polyurethane, 470, 487
Population, 71ff, 100–103, 118, 152, 161–165, 288, 293, 379–382, 394f, 403ff, 415, 430, 435f, 443–445
Portia
  P. fimbriata, 144, 149
  P. labiata, 152
Postsynaptic, 194
Posttranslational, 218, 222f
Potassium, 193, 197, 217, 286, 396f, 485
Potassium channel, 193, 197f, 208–213, 493f

Precursor, 82–86, 223, 243, 353, 397
Predator–prey, 136, 149, 319
Preoral digestion, 199, 207
Presynaptic, 197, 498
Previtellogenesis, 136
Prey specialist, 191
Pritha, 307
Prochloraz, 418
Prodidomidae, 508
Prohemocytes, 81ff
Prokaryotic, 195, 220, 222
Proline, 199, 270, 273–275, 287, 290, 292
Prometaphase, 164, 167
Propeptid, 219
Prophenoloxidase, 9, 85
Propionyl-CoA, 127
Propylester, 130
Prosoma, 15ff, 43ff, 127, 191, 203, 241ff, 253, 322, 340, 378, 343, 451, 454ff
Protease, 84f, 207–213, 236, 291, 420, 425, 431, 492
Protease inhibitor, 208f, 211, 213
Proto-silk, 245f, 249f
Psalmopeotoxin, 500
Psalmopoeus cambridgei, 205, 211f, 495–500
Psalmotoxin, 495f
Psechridae, 508
Pseudocrystalline, 476
Pseudoflagelliform, 296, 306
Pseudohemocyanin, 3, 11
Pseudoscorpiones, 5, 9, 57, 180
Psychodidae, 151
Pterinochilus chordatus, 205
Pulmonary oedema, 256, 261
Pulmopericardial sinus, 22ff
Pulvilli, 57
Purine, 364
Purotoxin, 499
Putrescine, 193, 218, 287f
Pycnogonida, 4f
Pyrethroid, 415ff
Pyridoxine hydrochloride, 396

Q

Quantum, 321, 326

R

Rabidosa rabida, 168, 419–423
Receptor signal, 320ff
Recombinant, 233, 270, 476ff
Recycling of silk, 286, 288f, 311f, 385f, 431
Reflectance, 320–327, 336–340, 349–357
Refractive index, 333ff
Renal failure, 257, 260
Reoxygenation, 21
Repetitive sequence, 268, 270, 274, 290, 476ff
Reproduction, 67–75, 93ff, 125f, 134, 142ff, 373, 381f, 394–397, 403ff, 423
Reproductive parasite, 93ff
Reproductive stage/state, 112, 137, 142ff, 364
Reproductive system, 20, 173ff
Resorption, 173, 271, 364, 366
Respiration, 2, 29ff, 81
Respiratory organ, 10, 29ff
Reverse Bohr effect, 10
Riboflavin, 396
Ricinulei, 5, 46, 180
Rickettsia, 94ff
Rickettsiales, 93
Robot, 278, 462, 470f
Rondonin, 82, 88
RTA clade, 8, 11, 178, 180, 272, 507f
Runcinia, 352

S

Salticidae, 7, 31ff, 47, 58ff, 68, 98, 111, 144ff, 164, 191, 334ff, 387f, 419ff, 463, 508
Salticus, 95
S. scenicus, 32f, 98, 419–421
Saturated fat, 396
Scale hair, 334ff
Schizocosa, 177f, 382
  S. duplex, 242
  S. humilis, 443
  S. malitiosa, 164, 174, 182, 184
  S. ocreata, 419–421
Sclerite, 45, 53f
Sclerotisation, 3, 85
Scopula, 57ff, 463ff
Scorpiones, 5f, 33, 255
Scytodes, 147, 152, 231f, 242ff, 291
  S. thoracica, 242ff
  S. velutina, 243
Scytodidae, 37, 147, 165, 174, 191, 231, 241ff, 253, 506
Scytodoidea, 165, 231, 506
Secondary copulatory organ, 173
Segestria florentina, 166
Segestriidae, 31f, 159ff, 174, 493, 506
Selene, 420
Selenopidae, 58, 508
Semicrystalline, 267
Semi-hydraulic machine, 43

Seminal fluid, 173ff
Semiochemical, 126f, 141f, 152
Senoculidae, 508
Sepsis, 257
Sericophily, 135
Serine, 84f, 209–213, 290, 481
Serine protease, 84f, 209, 211–213
Serotonin, 18f, 193, 206, 213
Seta, 57ff, 305ff, 338, 463ff
Setae, 57ff, 305ff, 338ff, 463ff
Setal comb, 291, 306
Setula, 463ff
Sex
  chromosome, 159ff
  pheromone, 137, 142f, 149, 151
  ratio, 70, 76, 100, 103
Sexual stage, 112
Shear force, 246, 269–171, 477
Sheet web, 32, 283, 293f, 312f, 386, 434
Sialic acid 224
SIBD-1, 88
Sicariidae, 31, 37, 164f, 174–180, 192f, 199f, 229ff, 257, 263, 506
Sicaritoxin, 237
Sicarius,, 230ff, 263
SicTox, 230, 233–237
Sidymella, 352
  S. longipes, 351
Sieve tracheae, 30
Signal, 34, 125ff, 141ff, 198, 209, 320ff, 344, 355, 367, 494ff
Signal peptide, 9, 86, 219
Silk, 60ff, 67ff, 126ff, 141ff, 241ff, 256f, 267ff, 283ff, 311ff, 350, 463, 475
Silk gland, 17, 65, 68, 127, 241ff, 267ff, 283ff, 476ff
Silkworm, 479ff
Silometopus elegans, 97
Sinopimoidae, 505
Sinus, 15ff
Siratoba referena, 309
Sit-and-wait, 35f, 148, 325, 331, 350, 388
Sitticus pubescen, 47
Sleep sickness, 224
Small cationic peptide, 195, 217, 223
SMase D, 230–237
Snake, 234, 255, 491
Social, 70, 95, 125ff
Sodium, 193, 197, 286, 396
Sodium channel, 193, 197, 208–213, 256, 494f
Solifugae, 5, 9, 57
Sparassidae, 7, 58, 86, 162ff, 177, 193, 197, 262, 417, 419, 493f, 508
Spasm, 256
Spatula, 57ff, 339, 464ff
Spectral, 320f, 337
Sperm, 143, 145f, 173ff
  conjugate, 180–183
  storage, 145, 174, 181f
  web, 173
Spermatheca, 146
Spermatid, 175–179
Spermatogenesis, 175f, 423
Spermatogenic, 174
Spermatogonial, 160, 164f, 175f
Spermatophore, 174, 181f
Spermatozoa, 173ff
Spermidine, 88, 193
Spermine, 193
Spermophor, 173
Sphingomyelin, 199, 233, 235, 258, 492
Sphingomyelinase, 199, 230, 237f, 243, 258, 263
Spider-insect interaction, 319ff
Spidroin, 268ff, 480ff
Spigot, 65, 68, 245, 271, 274, 284, 291, 304ff
Spinneret, 29, 53, 65, 68, 152, 245, 266, 274, 283, 285f, 304ff, 322, 476
Spinosad, 418f, 445
Spinosyn, 422–424
Spiracle, 30, 37
Spiroplasma, 94ff
Spitting spider, 37, 147, 152, 241ff
Spore, 109ff, 386, 429, 431
Sporotrichum 117
Spreading factor, 199, 225f
Spring element, 453, 457f, 465
Stabilimentum, 273f
Stable isotop analysis, 388
Staphylococcus aureus, 87, 258, 499
Starch, 397
Starvation, 35, 37, 71–76, 373, 380, 387, 424
Stearic acid, 133
Steatoda, 175, 198, 293
  S. bipunctata, 45
Stegodyphus, 95, 130
  S. lineatus, 133, 164, 376, 378
  S. sarasinorum, 443
Stemonyphantes lineatus, 97
Stenochilidae, 507
Stenophagous, 374
Stephanopsis scabra, 357

Sterol, 397
Stickiness, 243, 286ff, 303ff
Sticky spiral, 245, 258, 284–297
Sticky substance, 150, 243, 245, 283ff, 303
Stiphiidae, 508
Stomach, 18
Streptococcus, 258, 499
Stromata, 110
Stromatopelma calceatum, 209
Stromatoxin 209
Structural color, 333ff
Structural diversity, 178–181, 194, 200
Subesophageal ganglion, 18f
Subiculum, 110f
Sublethal, 410, 415ff, 425, 435
Submetacentric, 160f
Subsocial, 95, 133
Subtelocentric, 160f
Sucking stomach, 18
Sucrose, 397
Supercontraction, 477
Superoxide dismutase, 408f
Superposition eye, 325, 329f
Supraesophageal ganglion, 18
Survival, 69, 75, 93, 134, 146, 254, 257, 324, 374ff, 387f, 393ff, 403, 409f, 415f, 434f, 443
Sweating, 256, 260f
Symbiosis, 94
Symphytognatidae, 31, 191, 253, 507
Symptomatic treatment, 260f
Synanthropic, 257, 262
Synaphridae, 507
Synaptic, 19, 198, 260
Synaptonemal, 164, 166
Synema globosum, 352
Synergistic, 192f, 212f, 217–221, 225, 291, 411, 420
Synnemata, 110f
Synotaxidae, 178, 507
Synspermia, 180–182
Synthetic pesticide, 415ff, 444f
Systemic reaction, 257
Systole, 17, 23f

T

Tachycardia, 256
Tachykinin, 261
Tachylectin, 85
Tachypleus tridentatus, 11, 84, 86
Tachystatin, 85
Taenidia, 32
Tangaroa beattyi, 305
Tarsal organ, 128f
Taurine, 193, 218, 286f
TCTP, 236
Tegenaria, 32
  T. agrestis, 262
  T. atrica, 48, 86, 128, 130ff, 193
  T. domestica, 18, 136, 166
  T. duellica, 96, 101
Teichoic acid, 195
Telema tenella, 175
Telemidae, 30, 175, 181, 506
Teleomorph, 110, 112
Telomeric, 160, 162, 166
Temperature, 10, 25, 71–76, 102, 207, 211, 289, 420, 475, 480
Temporary spiral, 273
Tengella radiata, 312
Tengellidae, 508
Tennesseellum formicum, 386, 434
Tenuiphantes
  T. tenuis, 97, 418, 421f
  T. zimmermanni, 97
Termite, 191
Testis, 174–176, 181
Tetrablemmidae, 31, 52–54, 506
Tetracosane, 136
Tetradecane, 136
Tetragnatha, 177–179, 362f
  T. eurychasma, 363
  T. filicifilia, 363
  T. mandibulata, 445
  T. maxillosa 445
  T. montana 99
  T. paludicola 363
Tetragnathidae, 7, 86, 98, 118, 177, 194, 283, 294, 319ff, 343, 362ff, 387, 408, 445, 507
Tetranychus urticae, 100
Thanatosis, 364f
Thelytokous parthenogenesis, 95
Theraphosa blondi, 203
Theraphosidae, 6, 17, 32, 47ff, 58ff, 82ff, 127ff, 159, 161, 193ff, 203ff, 262, 334ff, 451ff, 492ff, 506
Theraphotoxin, 205–212, 495, 500
Theridiidae, 7, 35, 45, 99, 115, 130f, 142, 151, 174ff, 191ff, 258ff, 271, 283ff, 343, 363–365, 387, 417, 432, 492, 498, 507
Theridiosomatidae, 294, 507
Thermosensitive receptor, 197

Thiamin, 396
Thomisidae, 7, 31ff, 58, 60, 99, 148, 191, 349ff, 365, 386f, 406, 508
Thomisus, 353, 355
T. onustus, 352, 386f
T. spectabilis, 148, 351f, 354f
3-methylheptacosane, 133
Thrixopelma pruriens, 210, 494
Tidarren argo, 175
Tiptoe behaviour, 68
Tiso vagans, 97
Titanoecidae, 99, 507
Tomography, 17
Torrubiella, 110f, 114f, 117
Toxicity, 87, 206, 212, 217, 220f, 225, 291, 374, 405, 415, 435, 442ff
Toxin, 109, 192ff, 203ff, 217ff, 229ff, 256–260, 284, 288, 373–375, 382, 420, 429, 431, 492ff
Trachea, 8f, 21, 29ff
Trachelas volutus, 394f
Tracheoles, 33
Trade-off, 277, 403, 410f
Transcriptome, 196, 212, 236, 500
Transgenic, 430, 435, 479f
Transient receptor potential channel, 208f, 211, 495
Transmission, 93f, 101f, 127, 198, 212, 277, 326, 493, 497
Trechaleidae, 508
Triacontane, 136
Triadimenol, 418
Trigonotarbid-like ancestor, 33
Trimethyl methylcitrate, 143
Triterpenoid, 441
Trochanteridae, 508
Trochosa parthenus, 149
Troglorapteridae, 506
Troxochrus scabriculus, 97
Trypanosome, 224
Tubuliform gland/silk, 273, 283
Tumour cell, 87, 224, 495
20-hydroxyecdysone, 136
2-methylhexacosane, 132
Tylorida striata, 363
Tyramine, 193

U
Uloboridae, 31–34, 99, 160, 191, 253, 277, 304ff, 507
Uloborus
U. danolius, 160
U. globosis, 309
Ultraviolet/UV, 320ff, 336ff, 349ff, 431
Ummeliata insecticeps, 435
Uroctea durandi, 180
Uropygi, 5f, 174

V
Valve, 22f, 36, 50–54, 245, 249, 269, 271
van der Waals forces, 62–64, 308, 313, 465–471
Vascular constriction, 257
Vascularization, 18
Vasodilation, 261, 494
Velcro, 308, 469
Venom, 18, 25, 35, 148, 191ff, 203ff, 217ff, 229ff, 241ff, 253ff, 491ff
duct, 241, 244f, 250, 253
gland, 18, 148, 191, 193, 199, 203, 212, 218, 225, 235, 241–244, 249f, 253, 291, 500
Venomics, 492, 500
Verticillium, 117
Viscerocutaneous, 257
Viscid, 271, 277
Viscous, 243, 275, 295, 306, 310–313, 475
Visual signal, 141, 320–325, 344, 355
Vitamin, 375f, 388, 396f
Vitellogenesis, 136
Volatile, 72, 126, 129f, 141f, 441
Voltage activated channel, 197f, 208–210, 220, 493f
Vomiting, 204, 256f, 260–262

W
Waitkera waitakerensis, 304, 309
Walckenaeria
W. antica, 97
W. atrotibialis, 98
W. obtusa, 98
Wavelength, 320, 326, 333ff, 349, 353
Wax, 109
Wax gland, 127
Web, 29, 32, 35, 58, 127, 129–133, 137, 142–148, 231, 267ff, 283ff, 303ff, 405, 417, 422, 432
Wind speed, 71
Wolbachia, 94ff
W. pipientis, 93
Wound healing, 85

X
X chromosome, 160–169
Xenesthis immanis, 335
Xenobiotic, 403, 409

Xerolycosa nemoralis, 409f
Xiphosura, 5, 82, 84
Xysticus, 352

Y
Y chromosome, 160–165
Yeast, 87
Yolk, 82

Z
Zinc, 396, 403ff
Zodariidae, 149, 164, 191ff, 223, 508
Zodarion, 191
Zoridae, 58, 505, 508
Zorocratidae, 508
Zoropsidae, 508
Zosis geniculatus, 99, 304
Zygiella
  Z. atrica, *286*
  Z. x-notata, 96, 380
Zygometis lactea, 351f
Zygomycota, 12
Zygotene, 166
Zymogen, 84